POLYURETHANE

Edited by **Fahmina Zafar** and **Eram Sharmin**

Polyurethane
http://dx.doi.org/10.5772/2416
Edited by Fahmina Zafar and Eram Sharmin

Contributors
Eram Sharmin, Fahmina Zafar, Ahmed Issam Mohammed, Mohamed Rashidah Hamidi, Ruslan Davletbaev, Ilsiya Davletbaeva, Olesya Gumerova, Nataly Kozak, Eugenia Lobko, Suzana M. Cakić, Ivan S. Ristić, Olivera Z. Ristić, Ahmadreza Gharehbagh, Zahed Ahmadi, Valentina Cauda, Furio Cauda, Yerkesh Batyrbekov, Rinat Iskakov, V. Shim, J. Boheme, C. Josten, I. Anderson, Maria Butnaru, Ovidiu Bredetean, Doina Macocinschi, Cristina Daniela Dimitriu, Laura Knieling, Valeria Harabagiu, Abhay K. Mahanta, Devendra D. Pathak, Rafael Vasconcelos Oliveira, Valfredo Azevedo Lemos, M.S. El-Shahawi, A.A. Al-Sibaai, H.M. Al-Saidi, E.A. Assirey, Jan Bodi, Zoltan Bodi, Jiri Scucka, Petr Martinec, Mariana Paulino, Filipe Teixeira-Dias, Ahmed Tawfik, Žiga Voršič, Eram Sharmin, Fahmina Zafar, Sharif Ahmad, Nataly Kozak, Anastasyia Hubina, Khairiah Haji Badri

Published by InTech
Janeza Trdine 9, 51000 Rijeka, Croatia

Publishing Process Manager Vedran Greblo
Technical Editor InTech DTP team
Cover InTech Design team

Additional hard copies can be obtained from orders@intechopen.com

Polyurethane, Edited by Fahmina Zafar and Eram Sharmin
p. cm.
ISBN 978-953-51-0726-2

For EU product safety concerns:
IN TECH d.o.o., Prolaz Marije Krucifikse Kozulić 3, 51000 Rijeka, Croatia,
info@intechopen.com or visit our website at intechopen.com.

Meet the editors

Dr Fahmina Zafar is Senior Research Associate/Scientist Pool [Under Scientists' Pool Scheme, Council of Scientific and Industrial Research (C.S.I.R.), India] at the Department of Chemistry, Jamia Millia Islamia (J.M.I.) (A Central University), New Delhi, India. Dr Zafar has received her Ph.D in Chemistry from the same University in 2006. She was a Research Associate and Senior Research Fellow (C.S.I.R., India) at J.M.I. She holds M.Sc. in Organic Chemistry from J.M.I. in 1999 and B.Sc. in Chemistry and Textile Clothing from Mahila Vidyalaya Degree College, Lucknow University, Uttar Pradesh in 1995. She has more than 80 publications in conference Proceedings, refereed journals and Books. Her pioneer work is in the development of bio-based Polymers, metallopolymers, organic-inorganic hybrid and nanocomposites for Green Environment in different fields of applications. Dr Zafar is expert in synthesis and characterization of seed oil derived polymers such as polyurethanes, polyesteramides, polyurethaneamides for antimicrobial and corrosion protective applications.

Dr Eram Sharmin, Senior Research Associate [Under Scientists' Pool Scheme, Council of Scientific and Industrial Research (CSIR), India] at the Department of Chemistry, Jamia Millia Islamia (JMI) (A Central University), New Delhi, India, has persued her PhD (JMI, 2007), MSc (Aligarh Muslim University-AMU, Aligarh, UP, India 2000) and BSc (AMU, 1998) degree in Chemistry. Having previously worked as Senior Research Fellow and Research Associate (CSIR, India) at JMI, she has about 65 publications in conference proceedings, refereed journals and books. Her research interests include green organic-inorganic hybrid and composite materials from bio-based epoxies, polyols, polyurethanes as antimicrobial and corrosion resistant coatings.

Contents

Preface

Polyurethane [PU] is a class of polymers built up of carabamate linkages, which provide special characteristics to the material. It is the outcome of pioneering research work of Otto Bayer and his coworkers in 1937 at I.G. Farben laboratories, in Leverkusen, Germany. Initial PU consisted of foams and fibres, while today, innovations and researches in the past decades have brought colossal changes in the world of PU. Attention is focussed on green PU such as from non-isocyanate technologies, from biobased polyols and isocyanates, PU hybrids, PU composites and so on. PU find versatile applications as foams, adhesive, surface coatings, and sealants, to name a few. The enchanting and worthy world of PU beckoned us to bring forth the book titled "Polyurethane". The book is divided into three sections: structures, properties and characterization of PU, applications of PU and a separate section on Biobased PU, covering the research and development in these areas. Each contributed chapter handles new and interesting topics introducing the reader to the wider known and unknown applications of PU such as PU for medical, urological stenting practice, carriers of antituberculosis drugs, orthopaedic, fuel binder, extraction of metals, grouting technologies, crashworthiness, isolation for covered conductor, treatment of industry wastewater, cast elastomers, alkanolamide PU coatings and foams, and others. The book aims to cater a larger audience comprising of readers from polymer chemistry, materials chemistry, and industrial chemistry.

It is an immense pleasure to see the book "Polyurethane" in its final shape; the credit goes together to the authors, contributors, and the technical staff of InTech Open Access Publisher, particularly, to Mrs. Marija Radja and Mr. Vedran Greblo, Editor Relations Consultants, the funding agency Council of Scientific and Industrial Research, New Delhi, India for Senior Research Associateship (Under Scientists' Pool Scheme, CSIR). Without their help and dedication, it would be impossible to have this book published so efficiently. Time has come to thank them all.

Eram Sharmin and Fahmina Zafar, Ph.D.
Senior Research Associates, Department of Chemistry,
Jamia Millia Islamia, A Central University, New Delhi,
India

Section 1

Introduction

Chapter 1

Polyurethane: An Introduction

Eram Sharmin and Fahmina Zafar

Additional information is available at the end of the chapter

http://dx.doi.org/10.5772/51663

1. Introduction

1.1. History of polyurethane

The discovery of polyurethane [PU] dates back to the year 1937 by Otto Bayer and his coworkers at the laboratories of I.G. Farben in Leverkusen, Germany. The initial works focussed on PU products obtained from aliphatic diisocyanate and diamine forming polyurea, till the interesting properties of PU obtained from an aliphatic diisocyanate and glycol, were realized. Polyisocyanates became commercially available in the year 1952, soon after the commercial scale production of PU was witnessed (after World War II) from toluene diisocyanate (TDI) and polyester polyols. In the years that followed (1952-1954), different polyester-polyisocyanate systems were developed by Bayer.

Polyester polyols were gradually replaced by polyether polyols owing to their several advantages such as low cost, ease of handling, and improved hydrolytic stability over the former. Poly(tetramethylene ether) glycol (PTMG), was introduced by DuPont in 1956 by polymerizing tetrahydrofuran, as the first commercially available polyether polyol. Later, in 1957, BASF and Dow Chemical produced polyalkylene glycols. Based on PTMG and 4,4′-diphenylmethane diisocyanate (MDI), and ethylene diamine, a Spandex fibre called Lycra was produced by Dupont. With the decades, PU graduated from flexible PU foams (1960) to rigid PU foams (polyisocyanurate foams-1967) as several blowing agents, polyether polyols, and polymeric isocyanate such as poly methylene diphenyl diisocyanate (PMDI) became available. These PMDI based PU foams showed good thermal resistance and flame retardance.

In 1969, PU Reaction Injection Moulding [PU RIM] technology was introduced which further advanced into Reinforced Reaction Injection Moulding [RRIM] producing high performance PU material that in 1983 yielded the first plastic-body automobile in the United States. In 1990s, due to the rising awareness towards the hazards of using chloroalkanes as blowing agents (Montreal protocol, 1987), several other blowing agents outpoured in the market (e.g., carbon dioxide, pentane, 1,1,1,2-tetrafluoroethane, 1,1,1,3,3-pentafluoropropane). At the same time, two-pack PU, PU- polyurea spray coating technology came into foreplay, which bore significant advantages of being moisture insensitive with fast reactivity. Then blossomed the strategy of the utilization of vegetable oil based polyols for the development of PU. Today, the world of PU has come a long way from PU hybrids, PU composites, non-isocyanate PU, with versatile applications in several diverse fields. Interests in PU arose due to their simple synthesis and application protocol, simple (few) basic reactants and superior properties of the final product. The proceeding sections provide a brief description of raw materials required in PU synthesis as well as the general chemistry involved in the production of PU.

2. Raw materials

PU are formed by chemical reaction between a di/poly isocyanate and a diol or polyol, forming repeating urethane groups, generally, in presence of a chain extender, catalyst, and/or other additives. Often, ester, ether, urea and aromatic rings are also present along with urethane linkages in PU backbone.

2.1. Isocyanates

Isocyanates are essential components required for PU synthesis. These are di-or polyfunctional isocyanates containg two or more than two –NCO groups per molecule. These can be aliphatic, cycloaliphatic, polycyclic or aromatic in nature such as TDI, MDI, xylene diisocyanate (XDI), meta-tetramethylxylylene diisocyanate (TMXDI), hydrogenated xylene diisocyanate (HXDI), naphthalene 1,5-diisocyanate (NDI), p-phenylene diisocyanate (PPDI), 3,3′-dimethyldiphenyl-4, 4′-diisocyanate (DDDI), 1,6 hexamethylene diisocyanate (HDI), 2,2,4-trimethylhexamethylene diisocyanate (TMDI), isophorone diisocyanate (IPDI), 4,4′-dicyclohexylmethane diisocyanate (H_{12}MDI), norbornane diisocyanate (NDI), 4,4′-dibenzyl diisocyanate (DBDI). Figure 1 shows examples of some common isocyanates.

The isocyanate group bears cumulated double bond sequence as R-N=C=O, wherein the reactivity of isocyanate is governed by the positive character of the carbon atom (Scheme 1), which is susceptible to attack by nucleophiles, and oxygen and nitrogen by electrophiles.

If R is an aromatic group, the negative charge gets delocalized into R (Scheme 2), thus, the aromatic isocyanates are more reactive than aliphatic or cycloaliphatic isocyanates. In case of aromatic isocyanates, the nature of the substituent also determines the reactivity, i.e., electron attracting substituents in ortho or para position increase the reactivity and electron donating substituents lower the reactivity of isocyanate group. In diisocyanates, the presence of the electron attracting second isocyanate increases the reactivity of the first isocyanate; para substituted aromatic diisocyanates are more reactive that their ortho analogs primarily attributed to the steric hindrance conferred by the second –NCO functionality. The reactivities of the two-NCO groups in isocyanates also differ with respect to each other, based on the position of –NCO groups. For example, the two-NCO groups in IPDI differ in their reactivity due to the difference in the point of location of –NCO groups. TMXDI serves as an aliphatic isocyanate since the two isocyanate groups are not in conjugation with the aromatic ring. Another isocyanate of increasing interests is vinyl terminated isocyanate since along with the –NCO group, the extra vinyl group provides sites for crosslinking (Figure 2).

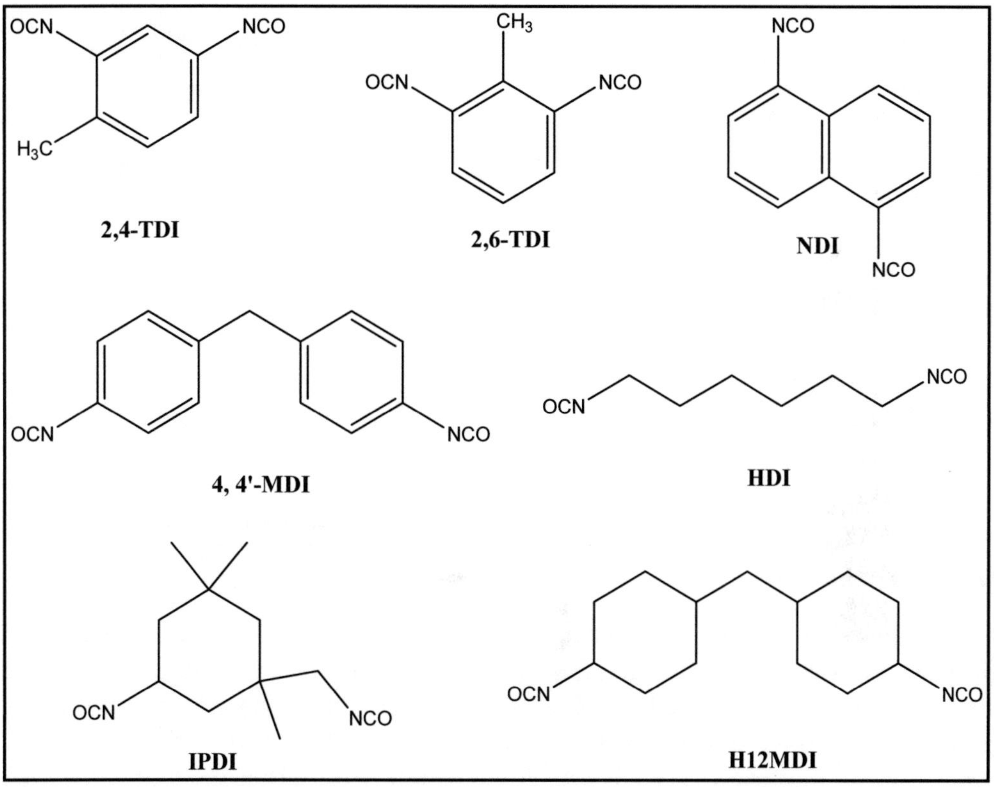

Figure 1. Common isocyanates

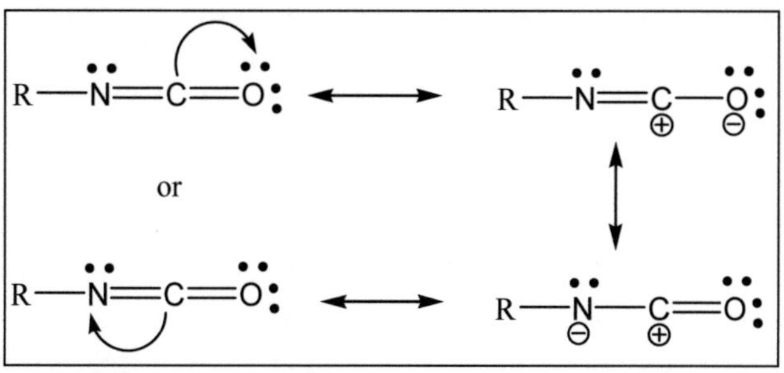

Scheme 1. Resonance in isocyanate

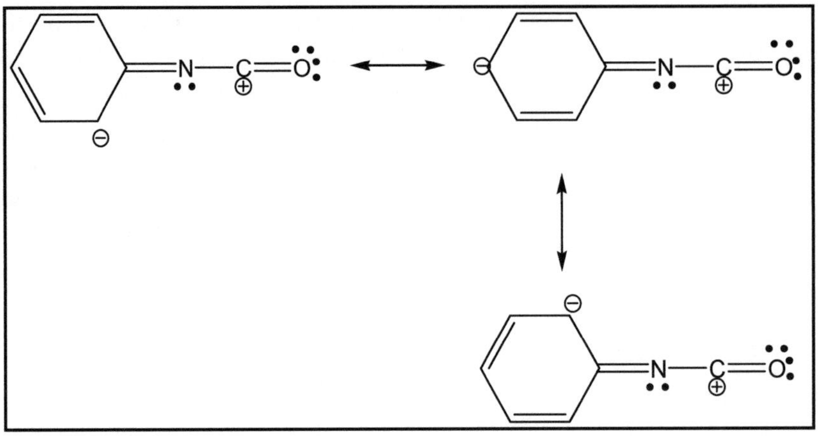

Scheme 2. Resonance in aromatic isocyanate

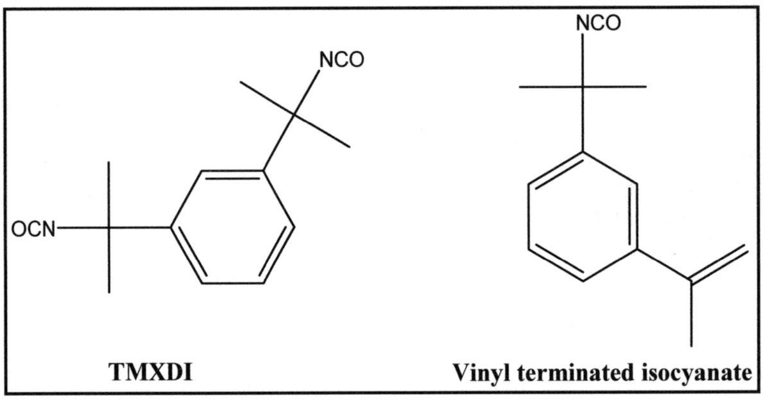

Figure 2. Other isocyanates

Polyisocyanates such as triisocyanates derived as TDI, HDI, IPDI adducts with trimethylolpropane (TMP), dimerized isocyanates termed as uretdiones, polymeric MDI, blocked isocyanates (where alcohols, phenols, oximes, lactams, hydroxylamines are blocking agents) are also used in PU production. Lately, fatty acid derived isocyanates are also prepared via Curtius rearrangement with view to produce entirely biobased PU. The choice of the isocyanate for PU production is governed by the properties required for end-use applications. To prepare rigid PU, aromatic isocyanates are chosen, however, PU derived from these isocyanates show lower oxidative and ultraviolet stabilities.

2.2. Polyols

Substances bearing plurality of hydroxyl groups are termed as spolyols. They may also contain ester, ether, amide, acrylic, metal, metalloid and other functionalities, along with hydroxyl groups. Polyester polyols (PEP) consist of ester and hydroxylic groups in one backbone. They are generally prepared by the condensation reaction between glycols, i.e., ethylene glycol, 1,4-butane diol, 1,6-hexane diol and a dicarboxylic acid/anhydride (aliphatic or aromatic). The properties of PU also depend upon the degree of cross-linking as well as molecular weight of the starting PEP. While highly branched PEP result in rigid PU with good heat and chemical resistance, less branched PEP give PU with good flexibility (at low temperature) and low chemical resistance. Similarly, low molecular weight polyols produce rigid PU while high molecular weight long chain polyols yield flexible PU. An excellent example of naturally occurring PEP is Castor oil. Other vegetable oils (VO) by chemical transformations also result in PEP. PEP are susceptible to hydrolysis due to the presence of ester groups, and this also leads to the deterioration of their mechanical properties. This problem can be overcome by the addition of little amount of carbodiimides. Polyether polyols (PETP) are less expensive than PEP. They are produced by addition reaction of ethylene or propylene oxide with alcohol or amine starters or initiators in presence of an acid or base catalyst. PU developed from PETP show high moisture permeability and low Tg, which limits their extensive use in coatings and paints. Another example of polyols is acrylated polyol (ACP) made by free radical polymerization of hydroxyl ethyl acrylate/methacrylate with other acrylics. ACP produce PU with improved thermal stability and also impart typical characteristics of acrylics to resultant PU. These PU find applications as coating materials. Polyols are further modified with metal salts (e.g., metal acetates, carboxylates, chlorides) forming metal containing polyols or hybrid polyols (MHP). PU obtained from MHP show good thermal stability, gloss and anti-microbial behavior. Literature reports several examples of VO based PEP, PETP, ACP, MHP used as PU coating materials. Another example is VO derived fatty amide diols and polyols (described in detail in chapter 20 Seed oil based polyurethanes: an insight), which have served as excellent starting materials for the development of PU. These PU have shown good thermal stability and hydrolytic resistance due to the presence of amide group in the diol or polyol backbone.

2.3. Additives

Along with a polyol and an isocyanate, some additives may also be required during PU production, primarily to control the reaction, modify the reaction conditions, and also to finish or modify the final product. These include catalysts, chain extenders, crosslinkers, fillers, moisture scavengers, colourants and others. In PU production, catalysts are added to promote the reaction to occur at enhanced reaction rates, at lower temperatures, for deblocking the blocked isocyanates, for decreasing the deblocking and curing temperatures and times. A number of aliphatic and aromatic amines (e.g., diaminobicyclooctane-DABCO), organometallic compounds (e.g., dibutyltin dilaurate, dibutyltin diacetate), alkali metal salts of carboxylic acids and phenols (calcium, magnesium, strontium, barium, salts of hexanoic, octanoic, naphthenic, linolenic acid) are used as catalysts. In case of tertiary amines, their catalytic activity is determined by their structure as well as their basicity; catalytic activity increases with increased basicity and decreases with the steric hindrance on the nitrogen atom of amine. They promote their catalytic action by complex formation between amine and isocyanate, by donating the electrons on nitrogen atom of tertiary amine to the positively charged carbon atom of the isocyanate. Metal catalysts bear superiority over tertiary amines because they are comparatively less volatile and less toxic. Metals catalyse the isocyanate-hydroxyl reaction by complex formation with both isocyanate and hydroxyl groups. The positive metal centre interacts with electron rich oxygen atom of both the isocyanate and hydroxyl groups forming an intermediate complex, which by further rearrangement results in the formation of urethane bonds. Difunctional low molecual weight diols (ethylene glycol, 1,4-butanediol, 1,6-hexanediol), cyclohexane dimethanol, diamines, hydroxyl-amines (diethanolamine and triethanolamine) are used as chain extenders in PU synthesis while those with functionality 3 or > 3 are used as crosslinkers. Since isocyanates are too sensitive to moisture or water even in traces, moisture scavengers, which react more readily with water than an isocyanate, are incorporated to cut off/eliminate the involvement of water during PU synthesis, e.g., oxazolidine derivatives, zeolite type molecular sieves. Blowing agents are used to produce PU foams with cellular structures by foaming process (e.g., hydrocarbons, CO_2, hydrazine).

3. Chemistry of PU

PU are carbonic acid derivatives. The older term for them is an ester of a substituted carbamic acid, polycarbamate, from carbamic acid. PU are formed by (i) the condensation polymerization reaction of bischloroformates with diamine (Scheme 3) and (ii) addition polymerization reaction of diisocyanates with di or polyfunctional hydroxy compounds, or other compounds having a plurality of active hydrogen atom (Scheme 4). The latter method is more important from the industrial point of view since in this method no by-product is formed.

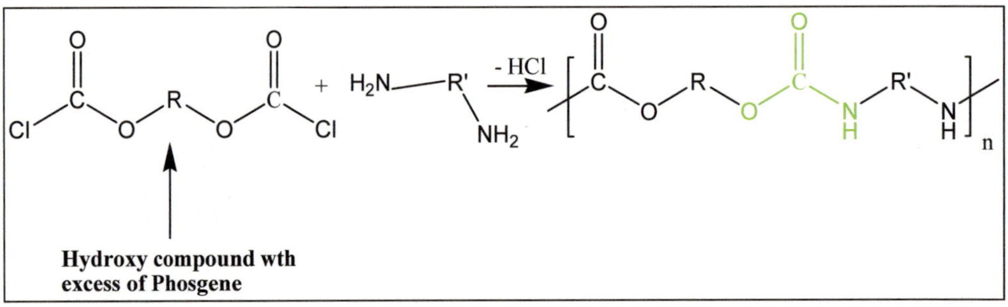

Scheme 3. Reaction of bischloroformate with diamine

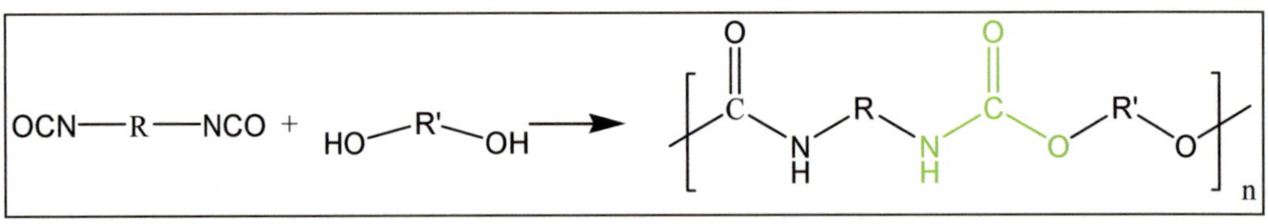

Scheme 4. Reaction of diisocyanate with di or poly hydroxy compound

The isocyanate reaction offers the possibility of producing tailor-made polymeric product ranging from fibres to rubber. Generally, the isocyanate reactions are divided into two classes, (a) addition (primary and secondary) reaction with compound containing active hydrogen (Schemes 5 and 6), (b) self-addition reaction (Scheme 7). In some of the reactions, CO_2 is released which assists in the formation of PU foams.

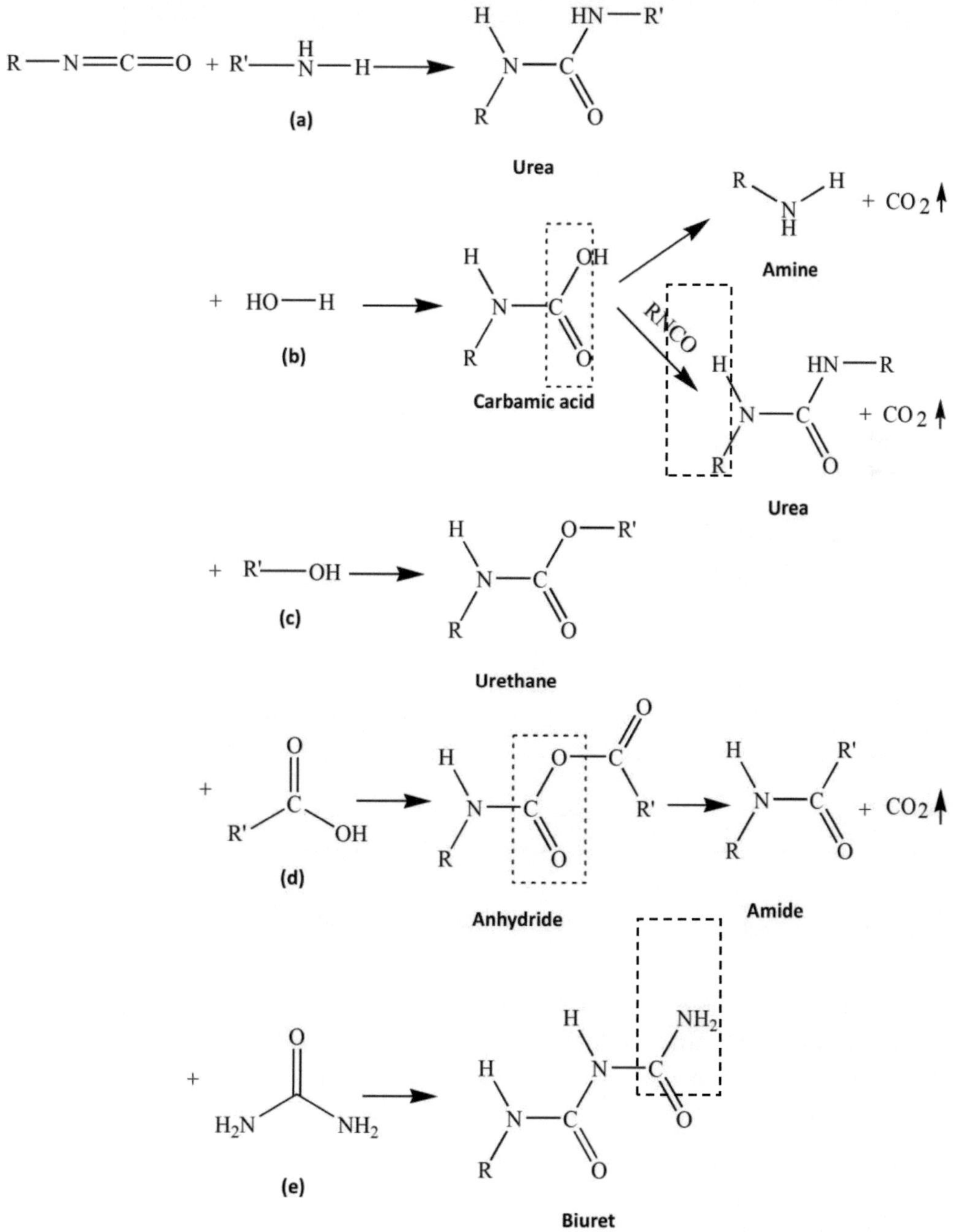

Scheme 5. Primary addition reactions of isocyanate with (a) amine, (b) water, (c) alcohol, (d) carboxylic acid, (e) urea.

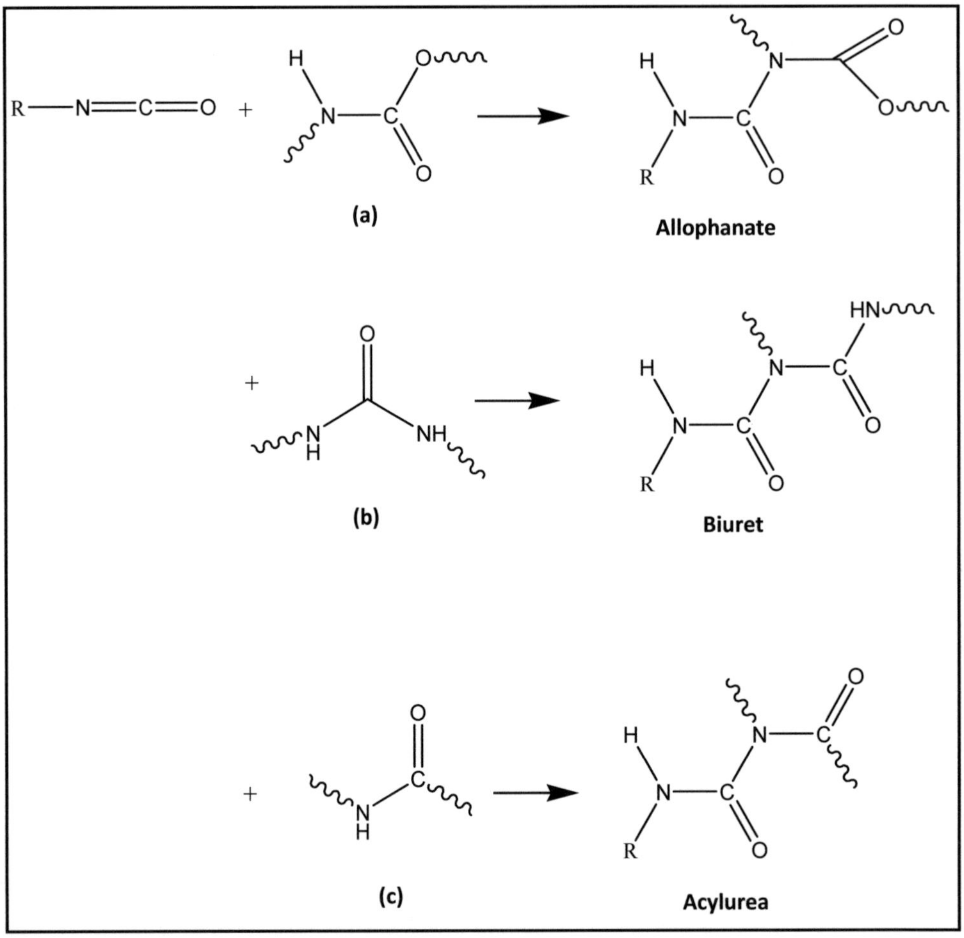

Scheme 6. Secondary addition reactions of isocyanate with (a) polyurethane, (b) polyurea and (c) polyamide

Wurts in 1848 discovered the basic reaction of isocyanate (Scheme 4). He found that isocyanates having the structure R-N=C=O, where R= alkyl or aryl group, react rapidly at room temperature with compounds containing active hydrogen atoms, like amine, water, alcohol, carboxylic acid, urethanes and ureas (Scheme 8).

It is observed that a linear PU is formed when a diisocyanate react with diol whilst branched or cross-linked PU results with the reaction of polyhydric compound (polyol). The branched or cross- linked PU are also formed when a compound containing three or more isocyanate groups reacts with a diol; however, this approach is of limited commercial importance.

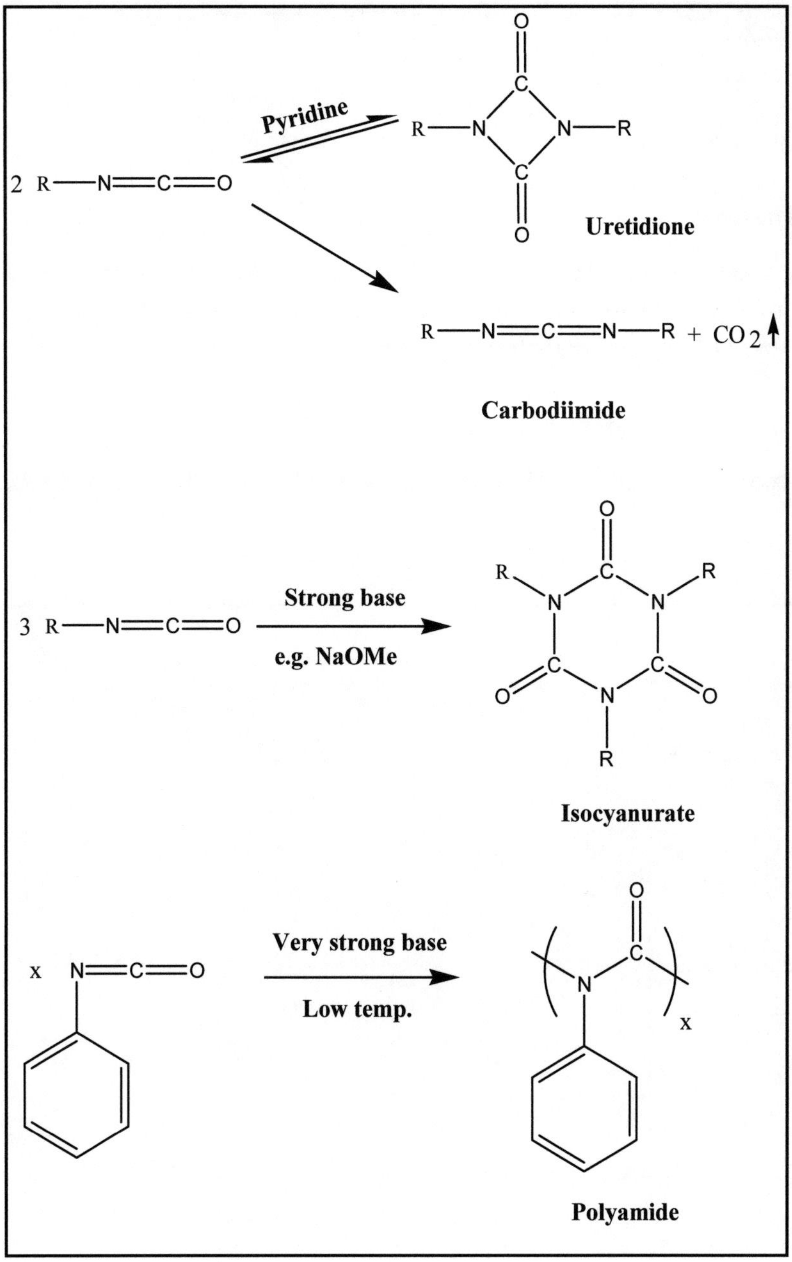

Scheme 7. Self -addition reactions of isocyanate

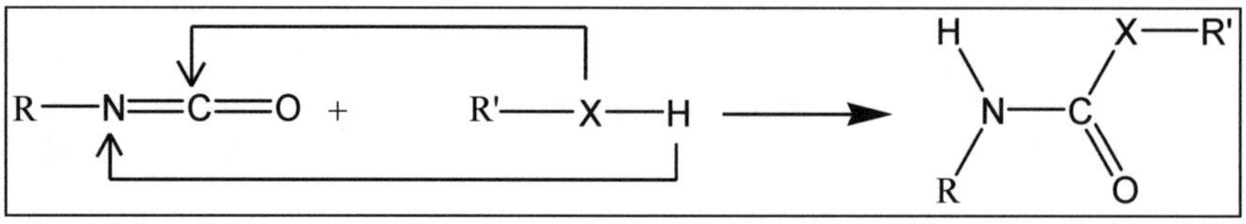

Scheme 8. Reaction of isocyanate with active hydrogen compound

4. Mechanism

The reaction of an isocyanate with active hydrogen compounds is carried out with or without a catalyst. The self-addition reactions of isocyanates do not usually proceed as readily as reactions with active hydrogen compounds.

4.1. Reaction in the absence of a catalyst

The active compound itself acts catalytically in the reaction as follows (Scheme 9).

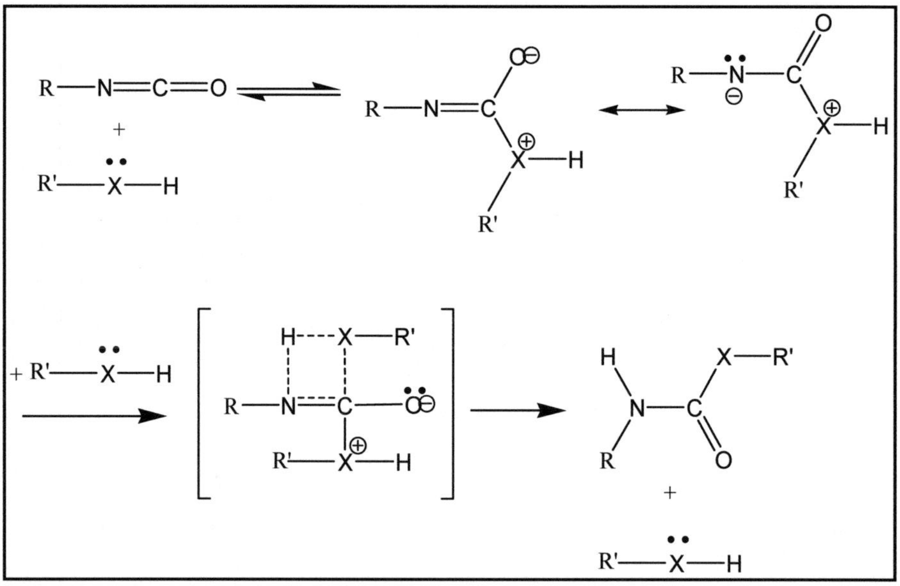

Scheme 9. Isocyanate reaction in the absence of a catalyst

As given in Scheme 9, in the reactions proceeding in the absence of a catalyst, the electrophilic carbon of the isocyanate is attacked by the nucleophilic centre of the active hydrogen compound; hydrogen is added to –NCO group. The reactivity of the –NCO groups is increased due to the presence of the electron withdrawing groups, and decreases by the electron donating groups. While the aromatic isocyanates are more reactive than the aliphatic isocyanates, steric hindrance at –NCO or HXR′ groups reduce the reactivity.

The order of reactivity of active hydrogen compounds with isocyanates in uncatalyzed systems is as follows:

Aliphatic amines> aromatic amines> primary alcohols> water>secondary alcohol> tertiary alcohol> phenol> carboxylic acid> ureas> amides>urethanes.

4.2. Reaction in the presence of a catalyst

The isocyanate reactions of class (a) are also extremely susceptible to catalysis. The various isocyanate reactions are influenced to different extents by different catalysts. Many commercial applications of isocyanates utilize catalysed reactions. Tertiary amines, metal compounds like tin compounds (as mentioned earlier in the chapter) are most widely used catalysts for the reaction (Schemes 10 and 11). The mechanisms are similar to that of the uncatalyzed reaction (Scheme 9).

The tertiary amines and metal salts catalyse the reaction as follows:

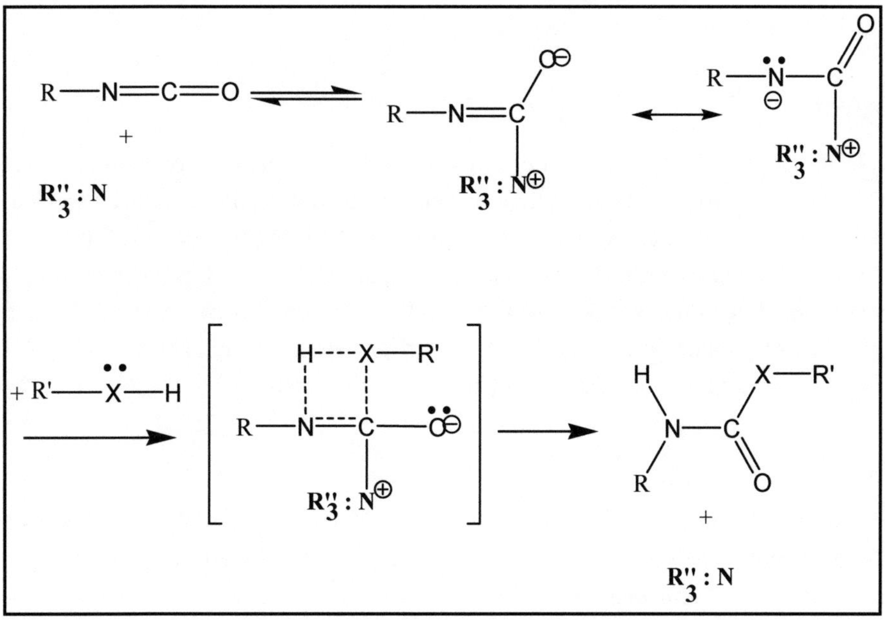

Scheme 10. Tertiary amine catalysed reaction

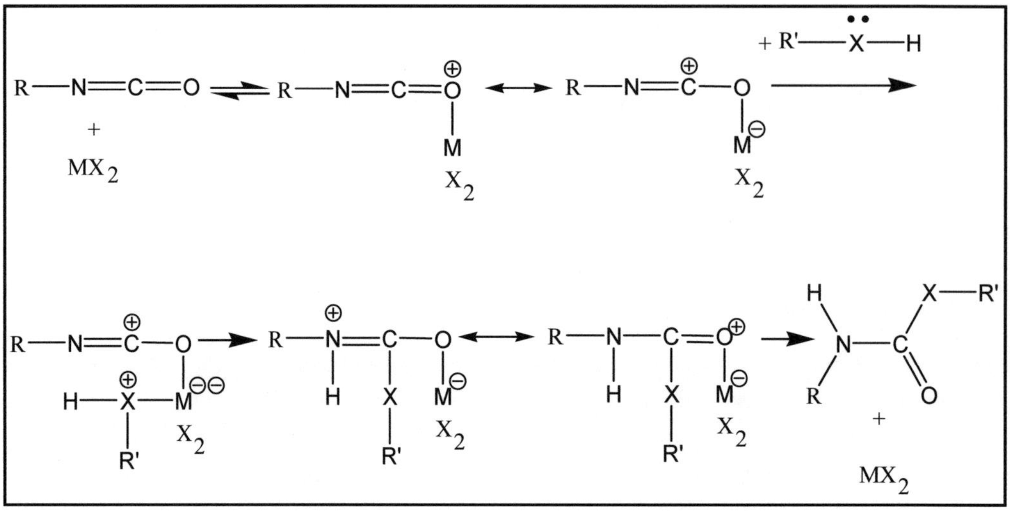

Scheme 11. Metal salts catalysed reaction

The catalytic activity of amines closely parallels to the base strength of the amines except when steric hindrance becomes pronounced. This catalyst is also effective for self-addition reactions while metal salt compounds generally have less influence; tin compounds are particularly poor catalysts in these reactions.

5. Hazards

Although PU are chemically inert in their fully reacted form , the risks of asthmatic symptoms arise on human exposure even in smaller concentrations due to the volatility associated with isocyanates arise the risk of asthmatic symptoms on 12 human exposure, even in smaller concentrations. On exposure to flames, hazards of ignition are feared. Isocyanates may also be sensitive on our skin. Some isocyanates may also be anticipated as carcinogens. Thus, persons working with isocyanates must be equipped with proper protection devices such as gloves, masks, respirators, goggles, and others, as precautionary measures.

6. Conclusion

PU are thermoplastic and thermoset in nature. The type, position, and structure of both the isocyanate and polyol determine the progress of PU forming reactions as well as their properties and end-use applications. Hydrogen bonding also plays a key role in determining the properties of final PU product. Due to the associated health hazards, complete precautions are necessary while working with isocyanates. PU are available as one-pack or two-pack PU. PU dispersions, waterborne PU, PU Interpenetrating

Networks PU, hybrids and composites are used in various applications such as paints and coatings, adhesives, sealants, foams, absorbents, flame retardants, fuel binders, in automobiles, in biomedical applications (urological stenting practices, carriers of antituberculosis drugs, orthopaedics), extraction of metals, grouting technologies, crashworthiness, treatment of industry wastewater, cast elastomers, and others as also discussed in proceeding chapters.

Author details

Eram Sharmin and Fahmina Zafar*
Materials Research Laboratory, Department of Chemistry,
Jamia Millia Islamia (A Central University), New Delhi, India

Acknowledgement

Dr Fahmina Zafar (Pool Officer) and Dr Eram Sharmin (Pool Officer) acknowledge Council of Scientific and Industrial Research, New Delhi, India for Senior Research Associateships against grant nos. 13(8385-A)/2010-POOL and 13(8464-A)/2011-10 POOL, respectively. They are also thankful to the Head, Department of Chemistry, Jamia Millia Islamia (A Central University), for providing support to carry out the work.

7. References

Chattopadhyay D.K., Raju K.V.S.N. Structural Engineering of Polyurethane Coatings for High Performance Applications. Progress in Polymer Science 2007; 32: 352–418.

Desroches M., Escouvois M., Auvergne R., Caillol S., Boutevin B. From Vegetable Oils to Polyurethanes: Synthetic Routes to Polyols and Main Industrial Products. Polymer Reviews 2012; 52 (1): 38–79.

Lligadas G., Ronda J.C., Galia`M., Cadiz V. Plant Oils as Platform Chemicals for Polyurethane Synthesis:Current State–of–the–Art. Biomacromolecules 2010; 11: 2825–2835.

Malcolm P S. Polymer Chemistry An Introduction. 3rd Edn. New York: Oxford University Press, Oxford; 1999.

Nylen P., Sunderland E. Modern Surface Coatings. London: John Wiley & Sons; 1965.

Paul, S. Surface Coating–Science and Technology. New York: John Wiley & Sons; 1985.

Petrović Z. S. Polyurethanes from Vegetable Oils. Polymer Reviews 2008; 48:109–155.

* Corresponding Author

Pfister D.P., Xia Y., Larock R.C. Recent Advances in Vegetable Oil–based Polyurethanes. Chem Sus Chem 2011; 4(6):703–17.

Saunders K J. Organic Polymer Chemistry. 2nd Edn. New York: Chapman & Hall; 1981.

http://en.wikipedia.org/wiki/Polyurethane (accessed on 11th July 2012)

Section 2

Synthesis and Properties

Chapter 2

New Liquid Crystalline Polyurethane Elastomers Containing Thiazolo [5,4d] Thiazole Moiety: Synthesis and Properties

Issam Ahmed Mohammed and Mohamed Rashidah Hamidi

Additional information is available at the end of the chapter

http://dx.doi.org/10.5772/36784

1. Introduction

Originally there are three states of matter; solid, liquid and gas. The emergence of an exotic and extraordinary form of matter, which is known as liquid crystal has been considered as one of the major breakthrough in polymer science. Liquid crystal can be defined as an intermediate of solid (crystal) and liquid (Knight & Vollrath, 2002) where the molecules have the capabilities to flow like a liquid (mobility) as well as possessing the common property associated to solid, which is the degree of order (Doldeny & Alder, 1998). In addition, liquid crystal materials are self assembling by nature and can offer a very elegant and effective way of controlling and tuning the physical properties that ultimately define the self-organizing and self assembly process (Zhang et.al., 2008). One of the exciting developments involving this unique material is the introduction of liquid crystalline behavior in polyurethane elastomers (PUE) where the first of this kind was synthesized by Iimura in 1981 (Lin et.al., 2001).

Polyurethane [PU] is one of the most versatile class of polymeric materials known today. Wide variety of structural changes can be produced with the different hydroxyl compounds and isocyanates leading to a wide spectrum of properties and applications (Yeganeh et.al., 2007). It contains a high concentration of polar groups, in particular the urethane group, resulting from isocyanate-hydroxyl reactions. The interactions between these polar entities are of great importance in determining the properties of PU of all types (Lee et.al., 1999) besides the composition and characteristic of the polyol, diisocyanates and the additives utilized during the synthesis (Pachecho et.al., 2009).

High toughness, excellent wear and tear properties and good oil resistance are among the advantages displayed by PUE (Wright & Cumming, 1969). Moreover, not only they have good mechanical and physical properties, PUE are also benefited with biocompatibility characteristics for biomedical applications (Barikani et.al., 2009). Despite all the great aforementioned properties, modifications and improvements are done to conventional PUE in order to meet the qualities in more advanced applications.

Diisocyanates, polyol and low molecular weight diamine or diol (chain extender) are the basic building blocks of conventional PUE (Yeganeh & Mehdizadeh, 2009). In order to synthesize liquid crystal polyurethane elastomers (LCPUE), the low molecular weight diamine or diol used in conventional PUE was substituted with the mesogenic unit. Incorporation of geometrically anisotropic moieties (mesogenic unit) within polymer architecture can drive the formation of liquid crystalline phase from strictly steric repulsion considerations (Abe & Ballauf, 1991; Rowan & Mather, 2007). Furthermore, the insertion of mesogenic unit in the backbone of PUE will impart unique physical properties to the polymer and also improve its mechanical, optical and electrical characteristics (Jia et.al., 1996).

Various mesomorphic behaviors are exhibited with different types of mesogenic units in preparation of LCPUE. In this research work, mesogens consumed were thiazolo [5.4d] thiazoles based and it is known as an important class of biycyclic aromatic molecule comprising two fused thiazole rings (Knighton et.al., 2010). Thiazolothiazole rigid fused ring can enhance the rigidity of the polymer and the conjugation (Osaka et.al, 2007) which makes it a best candidate to be part of the hard segment in the LCPUE network. The hard segments consisted of either 2,5-bis(4-hydroxyphenyl) thiazolo-[5,4d] thiazole or 2,5-bis(4-hydroxy-3-methoxyphenyl) thiazolo [5,4d] thiazole and 4,4'- methylene diphenyl diisocyanate (MDI). As for the soft segments, polyethylene glycol (PEG) 1000, 2000 and 3000 were involved.

The ultimate aim of this work is to synthesize new LCPUE with the presence of thiazolo-[5,4d] thiazole as a chain extender. Study and analysis were carried out to determine the effects and consequences of the introduction of thiazolo-[5,4d] thiazole moiety and the influence of various lengths of polyols on the properties of LCPUE.

2. Experimental

2.1. Materials

Vanillin and 4, 4'-methylene diphenyl diisocyanate (MDI) were purchased from Aldrich Co. (United States). Rubeanic Acid (dithiooxiamide) and 4-hydroxybenzaldehyde were obtained from MERCK Co. (Germany). Polyethylene glycol with molecular weight of 3000, 2000 and 1000 (PEG: Mn= 3000, 2000 and 1000) were purchased from Fluka Chemica (Switzerland). All the chemicals were utilized as received without any further purification. N,N-Dimethylformamide purchased from Aldrich (United States) was distilled over Calcium Hydride (CaH_2) through vacuum distillation before being used.

2.2. Synthesis of monomers and polymers

2.2.1. Synthesis of 2,5-bis(4-hydroxyphenyl)thiazolo-[5,4d] thiazole (I)

Briefly, 3 g (25 mmol) of dithiooxamide (Rubeanic acid) and 15 g (123 mmol) of 4-hydroxybenzaldehyde with the presence of 9 g (97 mmol) of phenol were charged all at once in a 500 ml round bottom flask fitted with condenser and left to be refluxed for 2h. Precipitates were obtained by pouring the hot mixtures to the cold water. Subsequently, the yield was filtered off and washed with ethanol followed by ether. The product obtained was dried at 70°C in a vacuum oven for 24 hours. Recrystallization from cyclohexanone was performed giving an orange-yellowish powder. Yield: 35% with melting point 364°C. Fourier transform infrared (FTIR; KBr, cm^{-1}): 3492 (-OH), 1606 (C=N), 1596 (C=C), 855 (*p*-substituted benzene). ^{1}H-NMR (400 MHz, DMSO-d_6 ppm): δ_H 7.12 (m, aromatic protons), 9.8 (s, -OH). Elemental analysis: Found: C, 59.16; H, 3.28; N, 8.84, $C_{16}H_{10}N_2O_2S_2$ Calc.: C, 58.89; H, 3.09, N, 8.59.

2.2.2. *Synthesis of 2,5-bis(4-hydroxy-3-methoxyphenyl) thiazolo [5,4d] thiazole (II)*

The same procedure was applied to the synthesis of 2,5-bis(4-hydroxy-3-methoxyphenyl) thiazolo [5,4d] thiazole except that 4-hydroxybenzaldehyde was substituted with vanillin. Orange-yellowish powder was obtained as the end product. Yield: 26% with the melting point of 259°C. Fourier transform infrared (FT-IR; KBr disc): 3534 cm^{-1} (OH), 1608 cm^{-1}(C=N), 1510 cm^{-1} (C=C), 842 cm^{-1} (-CH out of plane). ^{1}H-NMR (400 MHz, DMSO-d_6 ppm): δ_H 7.09 (m, aromatic protons), 9.5 (s, -OH) and 3.87 (s, OCH_3). Elemental analysis: Found: C, 55.60; H, 4.03; N, 6.89, $C_{18}H_{14}N_2O_4S_2$ Calc.: C, 55.95; H, 3.62, N, 7.25.

2.2.3. *Synthesis of liquid crystalline polyurethane elastomers (LCPUE)*

Preparation of LCPUE was achieved by two steps solution polymerization reaction, where isocyanate terminated pre-polymer was synthesized initially in the first stage. To produce pre-polymer, 0.01 mol of PEG (Mw = 1000, 2000, and 3000) and 0.02 mol of MDI were mixed in 500ml of reactor flask equipped with condenser, thermometer, nitrogen inlet and mechanical stirrer. The mixture was allowed to be stirred and heated for 4h at 70°C in the presence of 15 ml of DMF as solvent and nitrogen gas was kept flowing to provide inert atmosphere. The reaction was followed by chain extension process, using either compound (I) or (II), where the chain extender was added dropwise within 1h to complete the formation of LCPUE. Subsequently, the temperature was increased to 100°C and the reaction continued for another 9 hr. The hot viscous solution was then poured into 200ml of cold water for precipitation, before subjected to filtration. Later, the filtered product was washed with ethanol several times and finally with ether, before being dried overnight in a vacuum oven at 60° C.

2.4. Measurements

100mg mixture of samples and KBr (grounded) were pressed into translucent disc before being subjected to Nicolar Avatar Model 360 Fourier Transform infrared spectrometer devices to obtain FT-IR spectra. Data was collected in the range of 4000-400cm^{-1}. ^{1}H-NMR and ^{13}C-NMR spectra were obtained using Bruker 400 MHz NMR spectrometer consuming DMSO-d_6 as solvent and TMS as internal standard. Thermal stability of LCPUE was determined by thermogravimetric analyzer (Perkin Elmer Pyris series 6) under nitrogen purge and with 10°C/min of heating rate and the heating was done up to 800°C. Liquid crystalline behavior was verified by means of differential scanning calorimetry (DSC) to observe the behavior of polymers such as glass transition point (T_g), melting point (T_m) and isotropic temperature (T_i). It was conducted utilizing Perkin Elmer Pyris Series 7 thermal analyzer under Nitrogen flux at 10⁰C/min rate of heating. Textures of mesomorphic phases were displayed by Nikon Eclipse E600 polarized microscope equipped with MS600 Linkam Hot stage and SONY CCD-IRIS Color Video Camera. The heating rate was 5°C/min and 10°C/min for the cooling rate. Sample was placed between two thin round glasses and it was then transferred onto microscope fitted with the hot stage to be analyzed. Siemens X-ray Diffractometer model D5000 equipped with a CuKα target at 40KV and 40mA was used in obtaining X-ray scattering curve. Tensile strain properties of LCPUE films were measured by Instron Testing instrument at a constant speed of 500mm/min (speed) where the measurements were performed at room temperature. Brookfield viscometer was used to measure the fluid viscosity where suitable spindle and speed were chosen and it was also performed at room temperature.

3. Results and discussion

3.1. Preparation of chain extender

The preparation of 2,5-bis(4-hydroxyphenyl)thiazolo-[5,4d] thiazole and 2,5-bis(4-hydroxy-3-methoxyphenyl) thiazolo [5,4d] thiazole were conducted according to the reaction shown in Scheme 1. The starting reagent involved for the synthesis of both the compounds were rubeanic acid and either 4-hydroxybenzaldehyde or vanillin with the presence of phenol. Subsequently, both chain extenders prepared were being used in the preparation of LCPUE. Identification of the chemical structures of the aforementioned products was monitored primarily with FT-IR spectroscopy and further confirmation was carried out by ¹H-NMR spectrophotometer.

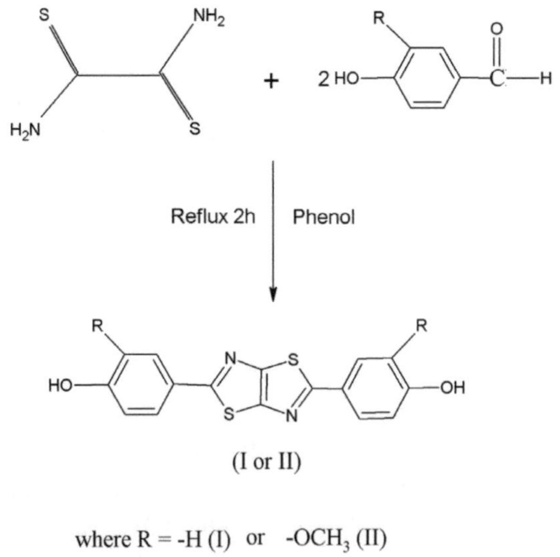

Scheme 1. Preparation of Compound I and II

3.2. Polymer synthesis

LCPUE based on thiazolo [5,4d] thiazoles moiety were synthesized from long chain of diol (PEG 3000, 2000 and 1000) with an excess of diisocyanate (MDI) via addition reaction to give the terminal reactive group which results in the formation of 'extended diisocyanate' or isocyanate pre-polymer. Then, 2,5-bis(4-hydroxyphenyl) thiazolo-[5,4d] thiazole [I] and 2,5-bis(4-hydroxy-3- methoxyphenyl) thiazolo [5,4d] thiazole [II] were added acting as a chain extender in order to convert the pre-polymer into long chain LCPUE. The general route for the preparation of LCPUE was outlined in Scheme 2, yield and viscosity of LCPUE were listed in Table 1 and the data showed that the range of the viscosities and yields obtained were 10,744 to 40 692 cP and 76-87 %, respectively. Range of the viscosities obtained also provides the information of the molecular weight of each polymer synthesized where high value of viscosity indicates high molecular weight of the polymer produced and vice versa (Bagheri & Pourmoazzen, 2008). In this case, all LCPUE samples displayed fairly high molecular weight in accordance with the results demonstrated.

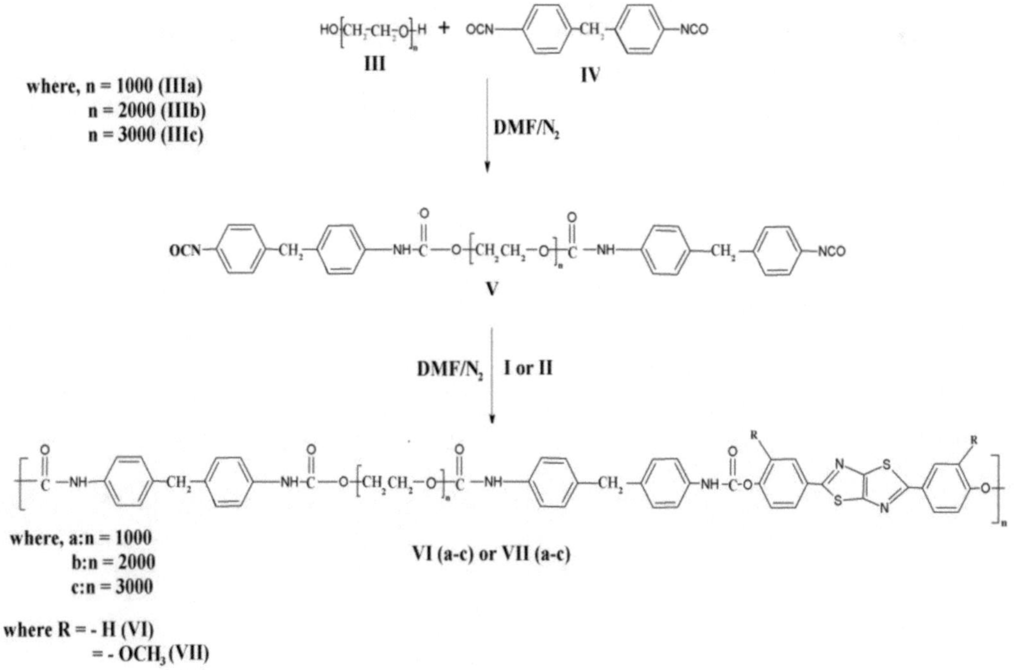

Scheme 2. General route for the preparation of LCPUE VI (a-c) and VII (a-c)

3.3. Structural elucidation

FT-IR was employed to verify functional groups of the pre-polymer, compound I and II, and LCPUE. Prior to the formation of LCPUE which is referring to the pre-polymer state, in the region of 2270 cm^{-1} a peak was observed which was assigned to –N=C=O-

(diisocyanate) whereas according to the IR spectra of compound I and II, a peak was found at 3492 cm^{-1} and 3334 cm^{-1} which corresponds to –OH functional group in the chemical structure. The disappearance of both the bands of -N=C=O- in pre-polymer and –OH of compound I and II, indicates the completion of the reaction of preparation of LCPUE and this fact was also supported with the appearance of new absorption bands at 3356.84cm^{-1} (N-H- stretching) and 1782.5cm^{-1} (carbonyl group) which were attributed to the urethane linkage, –NHCOO- (Zhang et al., 2008; Issam, 2007). Furthermore, the peak at 2884.89 cm^{-1} was ascribed to –CH stretching, whereas the band representing C=C aromatic can be found at 1598.59 cm^{-1}. Figure 1 displayed the FTI-R spectrum of LCPUE VIIa and based on the results obtained, the characteristic absorption bands of FT-IR spectra for the other LCPUE were almost identical to one another. The fact that differentiates LCPUE VI and LCPUE VII was the presence of the methoxy group and it was proven in the FTIR spectrum of LCPUE VIIa, where a peak displayed at the region of 1024.27 cm^{-1} corresponded to the methoxy group.

Further confirmation of chemical composition of LCPUE produced was carried out by means of Nuclear Magnetic Resonance spectroscopy (NMR). ^{1}H-NMR spectrum of LCPUE VIIa was illustrated in Figure.2. A singlet peak centered at 8.76 ppm was assigned to –NHCOO- and this proved the formation of urethane linkage. The appearance of multiplet peaks at 7.53-6.99 ppm and singlet peak at 3.87 ppm was attributed to the aromatic protons and the protons in methoxy group, respectively. Aliphatic chain of polyol (PEG 1000) was detected in the region of 1.23-1.64 ppm.

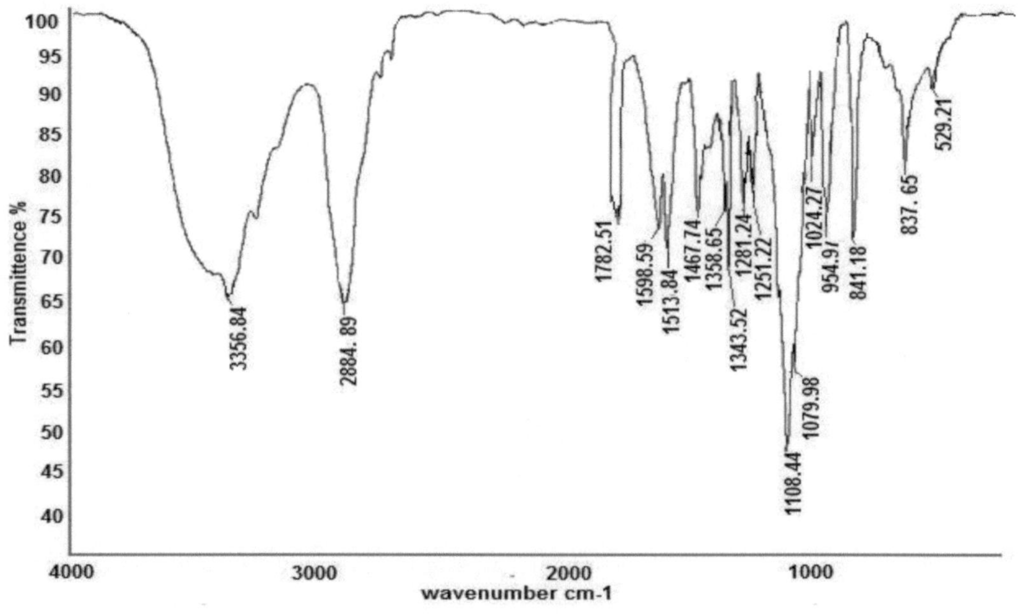

Figure 1. FTIR spectrum of LCPUE VIIa

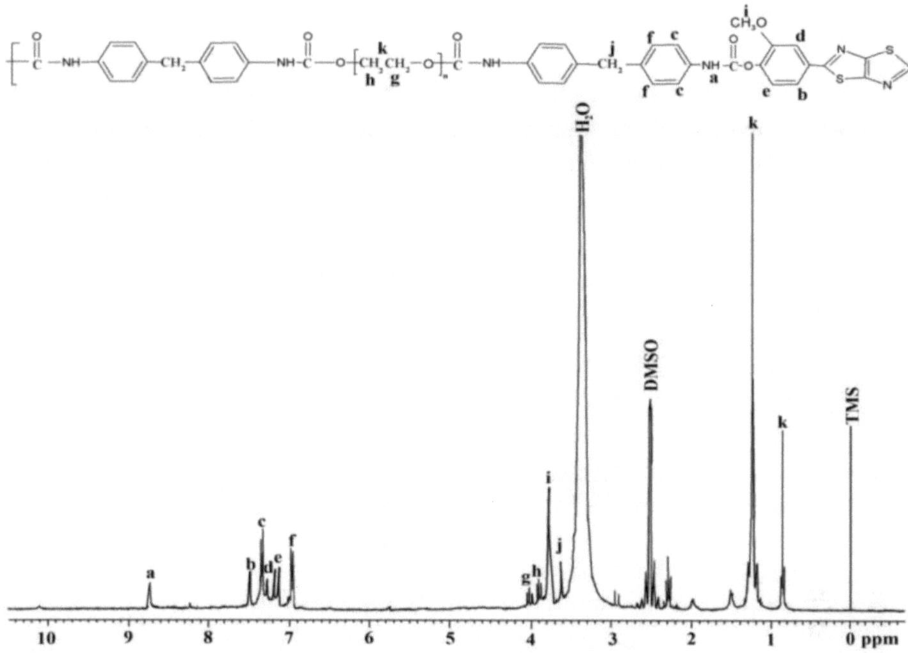

Figure 2. ^{1}H-NMR spectrum of LCPUE VIIa

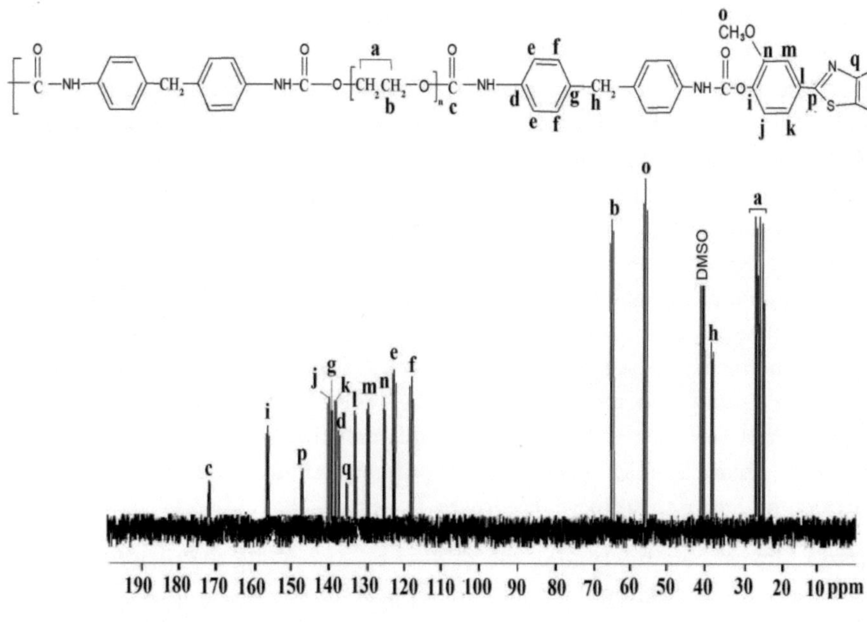

Figure 3. ^{13}C-NMR spectra of LCPUE VIIa

Other than FT-IR and ^{1}H-NMR analysis, ^{13}C-NMR was performed in order to clarify the structure of LCPUE prepared. ^{13}C-NMR spectra portrayed in Figure.3 which represents LCPUE VIIa shows that the formation of urethane linkage (NHCOO) was determined by the appearance of the peak at 173.4 ppm. The methylene group presence in the soft segment of PEG can be seen as a sharp and intense peak at 25-29 ppm. More peaks can be observed at 117.8 to 158.7 ppm and 56.2 ppm where they were assigned to the aromatic carbons and the carbon in methoxy group respectively. Significant peaks in all characterization analysis (FT-IR, ^{1}H-NMR and ^{13}C-NMR) were consistent and adequately provide the evidences to support the fact that the reaction of all materials took place and LCPUE was successfully prepared.

3.4. Thermal and liquid crystalline behavior of polymers

The DSC analysis was conducted at a heating rate of 10°C to understand phase separation behavior of all synthesized LPCUE where the transition occurs, observed under polarizing optical microscope (POM) equipped with heating stage and the results obtained from both measurements were listed in Table 1. Based on the DSC thermograms, upon heating, one step transition and two endothermic peaks were detected where each of them indicates glass transition (T_g), melting endotherm, (T_m) and isotropic endotherm (T_i) respectively, which is also the evidence of the existence of mesophase. LCPUE derived from 2,5-bis(4-hydroxy-3-methoxyphenyl) thiazolo [5,4d] thiazole have transition temperatures lower than those derived from 2,5-bis(4-hydroxyphenyl)thiazolo-[5,4d]thiazole. Methoxy group, which acts as a substituent attached to the phenyl ring has the capability to lower the melting and isotropization temperature and caused thermal suppression of the molecule to occur (Al-Dujaili et.al., 2001). The fact was supported by the results illustrated in Fig.4 where it depicts the DSC thermograms of LCPUE. LCPUE VIIa displayed melting point (T_m) at 164°C and isotropization temperature (T_i) at 187°C whereas for LCPUE VIa, T_m was detected at 176°C and T_i at 205°C. The substituent could also act to reduce the coplanarity of adjacent mesogenic groups and increase the diameter or decrease the axial ratio of the mesogens [Li and Chang, 1991]. Due to the higher range between T_m and T_i of LCPUE VIa, the thermal properties of this polymer are higher and more stable compared to LCPUE VIIa. The types of diisocyanates also contribute to the thermal behavior of LCPUE, where MDI based PU was known for having better order of the rigid chain that approaches the decomposition temperature, giving high melting point to the polymer produced (Jieh & Chou, 1996). As for glass transition, it involves mobility of the chain segments and the Tg will be affected by the mobility restriction on the chain segments, (Suresh et.al., 2008) it therefore explains the varying pattern of the T_g values displayed in Table 1. The decreasing values of T_g can be observed as the length of soft segments increases, indicating that the long chain of polyol gave great flexibility characteristics towards the polymer chains where less mobility restrictions occurred and hence resulting in the lower T_g values.

POM was utilized to investigate the type of mesophase by displaying the phase transition that occurred, subsequently providing the polarizing optical microphotographs of the target compounds. The morphology observed on heating and transition temperatures obtained were given in Figures 5 and 6 and the results were summarized in Table 1. It was revealed that all LCPUE showed mesophases upon melting temperature where the thread texture of the nematic phases can be seen. From the photographs taken by POM, the crystal to mesophase transition occurred at temperature ranging from 129 to 181°C. The samples were further heated after the crystal-nematic transition temperature, and resulted in the disappearing of the texture when reaching the isotropization stage. There were no traces of mesophase transition during the cooling process from POM indicating all samples possessed thermotropic type of liquid crystal. Phase transition temperatures observed through POM were found to be consistent with the corresponding DSC thermograms.

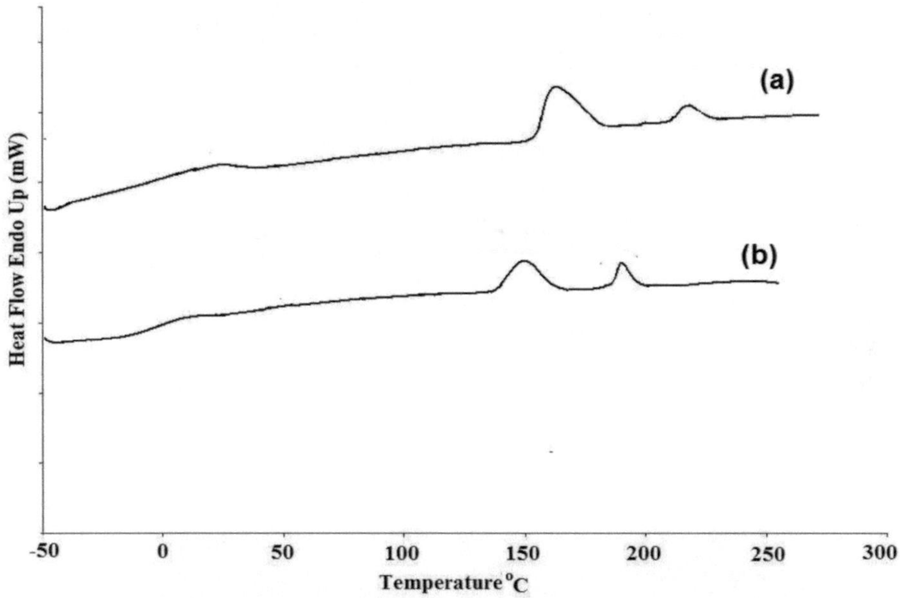

Figure 4. DSC traces of (a) LCPUE VIa (b) LCPUE VIIa

SAMPLE	PEG MOLECULAR WEIGHT	Yield (%)	Viscosity cP	DSC			POM	
				T_g (°C)	T_m (°C)	T_i (°C)	T_m (°C)	T_i (°C)
LCPUE VIa	1000	85	11 108	25.1	176	205	181	200
LCPUE VIb	2000	83	26 456	22.5	153	174	162	180
LCPUE VIc	3000	77	40 692	19.1	139	156	133	161
LCPUE VIIa	1000	76	10 744	15.2	164	187	170	193
LCPUE VIIb	2000	80	22 453	11.8	143	163	148	170
LCPUE VIIc	3000	87	39 981	10.4	125	142	129	149

Table 1. Thermal properties of LCPUE VI (a-c) and LCPUE VII (a-c) by DSC and POM

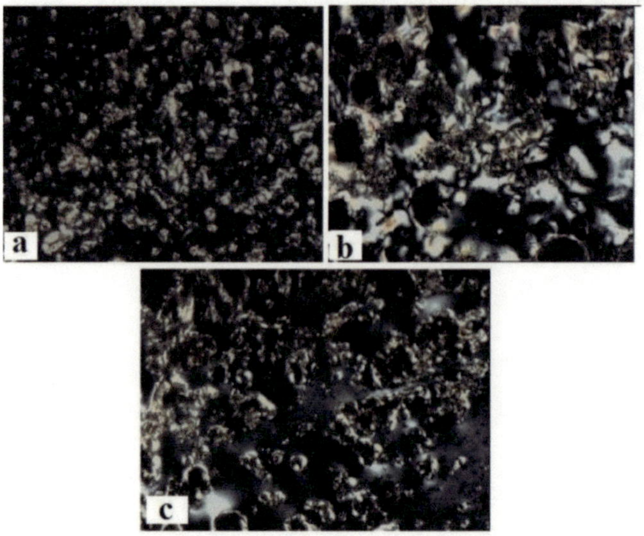

Figure 5. Polarized optical images of (a) LCPUE VIa (181°C), (b) LCPUE VIb (162 °C) and (c) LCPUE VIc (133 °C)

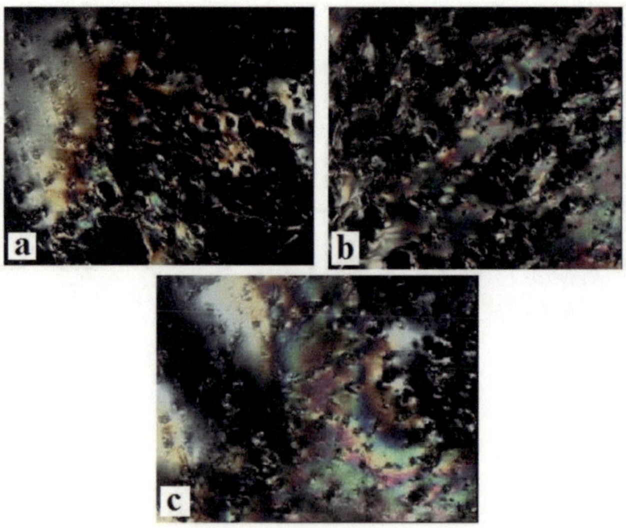

Figure 6. Polarized optical images of (a) LCPUE VIIa (170 °C), (b) LCPUE VIIb (148 °C) and (c) LCPUE VIIc (129 °C)

X-ray diffraction analysis of LCPUE was conducted at room temperature to obtain information on both the mesophase structure and crystallinity of LCPUE. The measurements exhibited several peaks in the range of 2θ= 15 – 25° as observed in Figure 7 and this indicated semi crystalline character possessed by LCPUE. The results obtained in above range also provide details related to the d-spacing of 3.56 and 4.92 Å, thus supporting the characteristic of nematic liquid crystalline phase (Jeh & The, 1994) as displayed through POM.

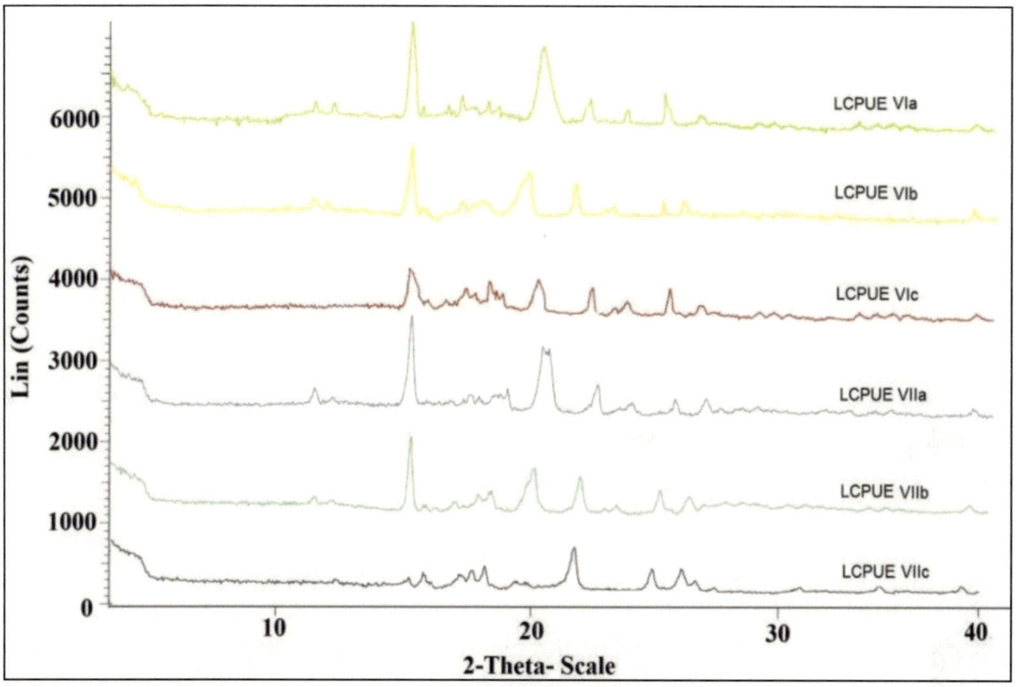

Figure 7. X-ray diffraction scales of LCPUE VI (a-c) and LCPUE VII (a-c)

Thermal stability of prepared LCPUE was investigated by thermogravimetric analysis (TGA). Incorporation of liquid crystalline properties into the polymer structure would enhance the thermal properties (Jahromi et.al., 1994) and the theory has proved to be applicable from the results obtained. This may be partly due to favorable interactions between hard domain interface and the liquid crystalline phase. All synthesized LCPUE possessed good thermal stabilities, however, PU elastomers eventually undergo thermal degradation when exposed to high temperatures. Degradation process occurred in two step pattern where the initial degradation occurs in the hard segment involving the urethane linkages, while the second stage indicated the degradation of soft segments. TGA curves in Figure 8 demonstrated the thermal degradation of all LCPUE prepared where 10% weight loss of LCPUE occurred at about 315-341°C and the maximum degradation temperature was in the range of 430-470°C, signifying a high thermal stability property. Furthermore, it can be observed that LCPUE VIIc demonstrated the lowest degradation temperature among the others and this proved that the length of polyethylene glycol (soft segment) influenced the thermal stability of LCPUE where the order of LCPUE due to their thermal stability can be arranged as LCPUE VIa>VIIa>VIb>VIIb>VIc>VIIc.

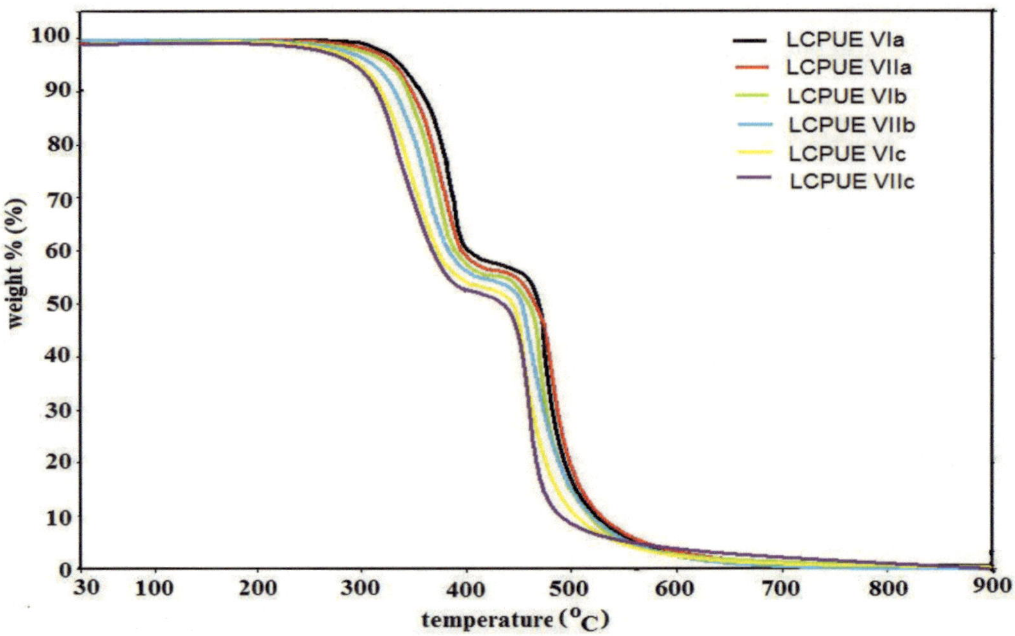

Figure 8. TGA curve of LCPUE VI (a-c) and LCPUE VII (a-c)

3.5. Tensile properties

Table 2 demonstrates tensile properties of the synthesized LCPUE. As seen, all of the polymers possessed good elastic properties with high elongation at break. Due to the data listed, the higher the molecular weight of the soft segments, the greater the elongation at break, but decrease of tensile strength and tensile modulus can be observed. When the molecular weight of polyol increased, the number of urethane groups in the polyol chain was reduced at the same time, and hence the number of rigid segments is lower, consequently, the possible number of intermolecular hydrogen bonds goes down in which –NH and C=O groups are active (Kro & Pitera, 2008). However, the presence of enhanced rigid and high aspect ratio mesogenic unit as part of hard segment in the synthesized LCPUE, is able to give both high strength and good elastic properties to LCPUE even with long soft segments, which

Sample	Tensile modulus (Mpa)	Tensile strength (Mpa)	Elongation at break (%)
LCPUE VIa	17.1	28.2	290
LCPUE VIb	13.4	24.1	450
LCPUE VIc	11.5	19.9	570
LCPUE VIIa	17.5	28.3	330
LCPUE VIIb	13.7	24.3	460
LCPUE VIIc	11.2	19.8	560

Table 2. Mechanical properties of LCPUE VI (a-c) and LCPUE VII (a-c)

is unusual in conventional PUE (Jeong et.al., 2000). Better phase separation will lead to good mechanical properties; hence the introduction of the mesogens unit as chain extender into LCPUE can be said to easily induce the matter (phase separation) to occur.

Author details

Mohammed Ahmed Issam and Hamidi Mohamed Rashidah
University Sains Malaysia, Malaysia

Acknowledgement

The author would like to thank University Sains Malaysia for short term grant no.304.PTEKIND.6311031 and the fellowship scheme for funding the research.

4. References

Abe, A. & Ballauf, M. (1991). Liquid crystallinity in Polymers. John Wiley & Sons Inc, New York, USA

Al-Dujaili, A.H.; Atto, A.T. & Al-Kurde, A.M. (2001). Synthesis and Liquid Crystalline Properties of Models and Polymers containing Thiazolo[5,4-d]thiazole and Siloxane Flexible Spacers. European Polymer Journal Vol.37, pp. 927-932

Bagheri, M. & Pourmoazzen, Z. (2008). Synthesis and Properties of New Liquid Crystalline Polyurethanes containing Mesogenic Side Chain Reactive & functional. Polymers, Vol.68, pp. 507–518

Barikani, M.; Honarkar, H. & Barikani, M. (2009). Synthesis and Characterization of Polyurethane Elastomers based on Chitosan and Poly(e-caprolactone). Journal of Applied Polymer Science, Vol.112, pp. 3157–3165

Doldeny, J.D. & Alder, P.T. (1998). The Mesogenic Index: An Empirical Method for Predicting Polymeric Liquid Crystallinity. High Performance Polymers, Vol.10, pp. 249–272

Issam, A.M. (2007). Synthesis of Novel Y-Type Polyurethane containing Azomethine Moiety, as Non-linear Optical Chromophore and Their Properties. European Polymer Journal, Vol.43, pp. 214-219.

Jahromi, S.; Lub, J. & Mol, G.N. (1994). Synthesis and Photoinitiated Polymerization of Liquid Crystalline Diepoxides. Polymer, Vol. 35, No.3, pp. 622-629

Jeh, C.T. & Teh, C.C. (1994). Study on Thermotropic Liquid Crystalline Polymers -I. Synthesis and Properties of Poly(azomethine-urethane)s. European Polymer Journal, Vol.30, pp. 1059-1064

Jeong, H.M.; Kim, B.K. & Choi, Y.J. (2000). Synthesis and Properties of Thermotropic Liquid Crystalline Polyurethane Elastomers. Polymer,Vol.41, pp. 1849-185

Jia, X.; He, X.D. & Yu, X.H. (1996). Synthesis and Properties of Main-Chain liquid Crystalline Polyurethane Elastomers with Azoxybenzene. Journal of Applied Polymer Science, Vol.62, pp. 465-47

Jieh, S.S. & Chou, C.T. (1996). Studies on Thermotropic Liquid Crystalline Polyurethanes.III.Synthesis and properties of polyurethane elastomers by using various mesogenic units as chain extender. Journal of polymer science part A: Polymer chemistry, Vol.34, pp. 771-779

Knight, D.P. & Vollrath, F. (2002). Biological Liquid Crystal elastomers. Philosophical Transactional Royal Society London. B, Vol.357, pp. 155–163

Knighton, R.C.; Hallett, A.J.; Kariuki, B.M. & Pope, S.J.A. (2010). A One-step Synthesis towards New Ligands based on Aryl-functionalized Thiazolo[5,4-d]thiazole Chromophores. Tetrahedron Letters, Vol.51, pp. 5419–5422

Kro, P. & Pitera, B.P. (2008). Mechanical Properties of Crosslinked Polyurethane Elastomers Based on Well-Defined Prepolymers. Journal of Applied Polymer Science, Vol.107, No.3, pp. 1439–1448

Lee, D.J.; You, S.H. & Kim, H.D. (1999). Synthesis and properties of thermotropic liquid crystalline polyurethane elastomers (II): Effect of Structure of Chain Extender Containing Imide Unit. Korea Polymer Journal, Vol.7, No.6, pp. 356-363

Lin, C.K.; Kuo, J.F. & Chen, C.Y. (2001). Synthesis and Properties of Novel Polyurethanes containing the Mesogenic Moiety of a-Methylstilbene Derivatives. European Polymer Journal, Vol.37, pp. 303-313

Li, C.H.; & Chang, T.C. (1991). Thermotropic Liquid Crystalline Polymer:III: Synthesis and Properties of Poly(amide-azomethine-ester). Journal of Polymer Science Part A: Polymer Chemistry, Vol.29(3), pp. 361-367

Osaka, I.; Sauvé, G.; Zhang, R.; Kowalewski, T. & McCullough, R.D. (2007). Novel Thiophene-Thiazolothiazole Copolymers for Organic Field-Effect Transistors. Advance Material, Vol.19, pp. 4160–4165

Pacheco, M.F.M.; Bianchi, O.; Fiorio, R.; Zattera, A.J.; Giovanel, M.Z.M. & Crespo, J.S. (2009). Thermal, Chemical, and Morphological Characterization of Microcellular Polyurethane Elastomers. Journal of Elastomers and Plastics, Vol.41, pp. 323

Rowan, S.J. & Mather, P.T. (2008). Supramolecular Interactions in the Formation of Thermotropic Liquid Crystalline Polymers. Structure and bonding, Vol.128, pp. 119-149

Suresh, K.I.; Tamboli, J.R.; Rao, B.S.; Verma, S. & Unnikrishnan G. (2008). Effect of Core Group Substituents on the Monomer Mesophase, Photocuring, and Film Viscoelastic Properties of Mesogenic Diacrylates. Polymers for Advanced Technologies, Vol.19, pp.1323-1333

Wright, P. & Cumming, A.P.C. (1969). Solid Polyurethane elastomers. Mclaren and Sons, London

Yeganeh, H. & Mehdizadeh, M.R. (2004). Synthesis and Properties of Isocyanate Curable Millable Polyurethane Elastomers Based on Castor Oil as a Renewable Resource Polyol. European Polymer Journal, Vol.40, pp. 1233–1238

Yeganeh, H.; Talemi, P.H. & Jamshidi, S. (2007). Novel Method for Preparation of Polyurethane Elastomers with Improved Thermal Stability and Electrical Insulating Properties. Journal of Applied Polymer Science, Vol.103, pp. 1776–1785

Zhang, C.; He, Z .; Wang, J.; Wang, Y. & Ye, S. (2008). Study of Mesogenic Properties and Molecular Conformation from a Heterogeneous Tetramer with a Triphenylene Centre Core and Three Cyanobiphenyl Tails. Journal of Molecular Liquids, Vol. 138, pp. 93–99

Zhang, H.; Chen, Y.; Zhang, Y.; Sun, X.; Ye, H. & Li, W. (2008). Synthesis and Characterization of Polyurethane Elastomers. Journal of Elastomers and Plastics, Vol.40, No.2, pp. 161-177

The Modification of Polyurethanes by Highly Ordered Coordination Compounds of Transition Metals

Ruslan Davletbaev, Ilsiya Davletbaeva and Olesya Gumerova

Additional information is available at the end of the chapter

http://dx.doi.org/10.5772/47990

1. Introduction

One of the ways to influence the chemical structure of polyurethanes is to use metal complex systems based on transition metal chlorides for their synthesis. The significance of this trend is conditioned by the ability of metal complexes to order the macromolecular chains, as well as affect the electrical properties of polyurethanes (Davletbaeva et al., 1996, 2001).

The synthesis of metal coordination polymers is a way of affecting the processes of crosslinking of macrochains; interchain and intraionic interactions; and, thereby, preparing polymer materials with special properties (Dirk et al., 1986; Kingsborough & Swager, 1999; Reynolds et al., 1985; Thuchide & Nishide, 1977; Wang & Reynolds, 1988). From the standpoint of designing materials with electric and magnetic properties, it is promising to form in a polymer matrix chains of transition metal ions bound by exchange interaction.

Conventional methods for the creation of interactions of this type in a polymer are primarily based on the presence of certain units in a macromolecule, e.g., those including the phthalocyanine and azomethine moieties. For example, metal atoms in metal phthalocyanine liquid crystalline complexes are bound to one another by chloride bridges and play the role of a spacer between phthalocyanine units, thus promoting overlap of electronic orbitals of parallel molecules (Shirai et al., 1977, 1979).

As a result, the electric conductivity of metal-coordination polymers obtained on the basis of these complexes is increased by a few orders of magnitude relative to undoped systems. The coordination bonding of comb-like liquid crystalline polymers can also give rise to stacked structures. Interaction between metal ions is revealed in such polymers, which is realized owing to the association of metal ions in an indirect manner, through ester oxygen bridges.

However, this approach is seriously limited and cannot be used for the creation of stacked metal-coordinated fragments in a disordered polymer matrix (Brostow, 1990; Carrher, 1981; Serrano & Oriol, 1995).

Metal complex structuring is promising in terms of the influence on properties of polyurethanes. The significance of this trend is conditioned by the ability of metal complexes to order the macromolecular chains, as well as affect the electrical properties of polyurethanes.

2. Coordination compounds based on the aromatic isocyanates and copper (II) chloride for the synthesis of polyurethanes

In the early studies (Davletbaeva et al., 1998) devoted to imparting special properties to polyurethanes by their coordination bonding, it was shown that the interaction of aromatic isocyanates with copper chloride ($CuCl_2$) in the acetone medium in the presence of trace amounts of water proceeded as a sequence of chemical transformations including simple addition and decomposition reactions, redox processes, and subsequent complexation (Fig. 1).

The ultimate products are polynuclear complexes of azoaromatic compounds, in which copper ions occurring in two variable oxidation states are connected by chloride bridges (Fig. 1). As it is seen from the structural formulas, some copper ions are stabilized at the initial degree of oxidation due to the formation of heterovalent pairs connected by chloride bridges. Free isocyanate groups present in these compounds are able to react with oligodiols of different nature. Polyurethanes obtained in such manner form polymer network by coordination bonding of urethane groups and azogroups which are the part of a macrochain.

It is shown that the chloride bridges are replaced by heteroatoms that are present in polymer chains while metal ions are coordinatively bound to macromolecules to crosslink them and, occurring in two interacting variable oxidation states, to form local centers of exchange interactions. As a result of electron transfer from one local coordinated unit to another, which is mediated by electron-donating groups, such as an ester group, the conductivity of polyurethane increases by several orders of magnitude.

3. Reactions of aromatic isocyanates and urethane prepolymers with coordination compounds of iron (III)

Further studies established that similar structural units could be formed even in the polyurethane matrix itself as a result of its modification with metal complexes synthesized for this purpose. A characteristic feature of these coordination compounds is the presence of chloride-bridged metal ions in their structure. One of such crosslinking metal complex system was prepared by the reaction of iron chloride ($FeCl_3$) with ethanolamine (EA). It was found that reactions involving aromatic isocyanates and EA in the coordination sphere of the iron ion (III) led to the formation of azoaromatic compounds shown in figure 2.

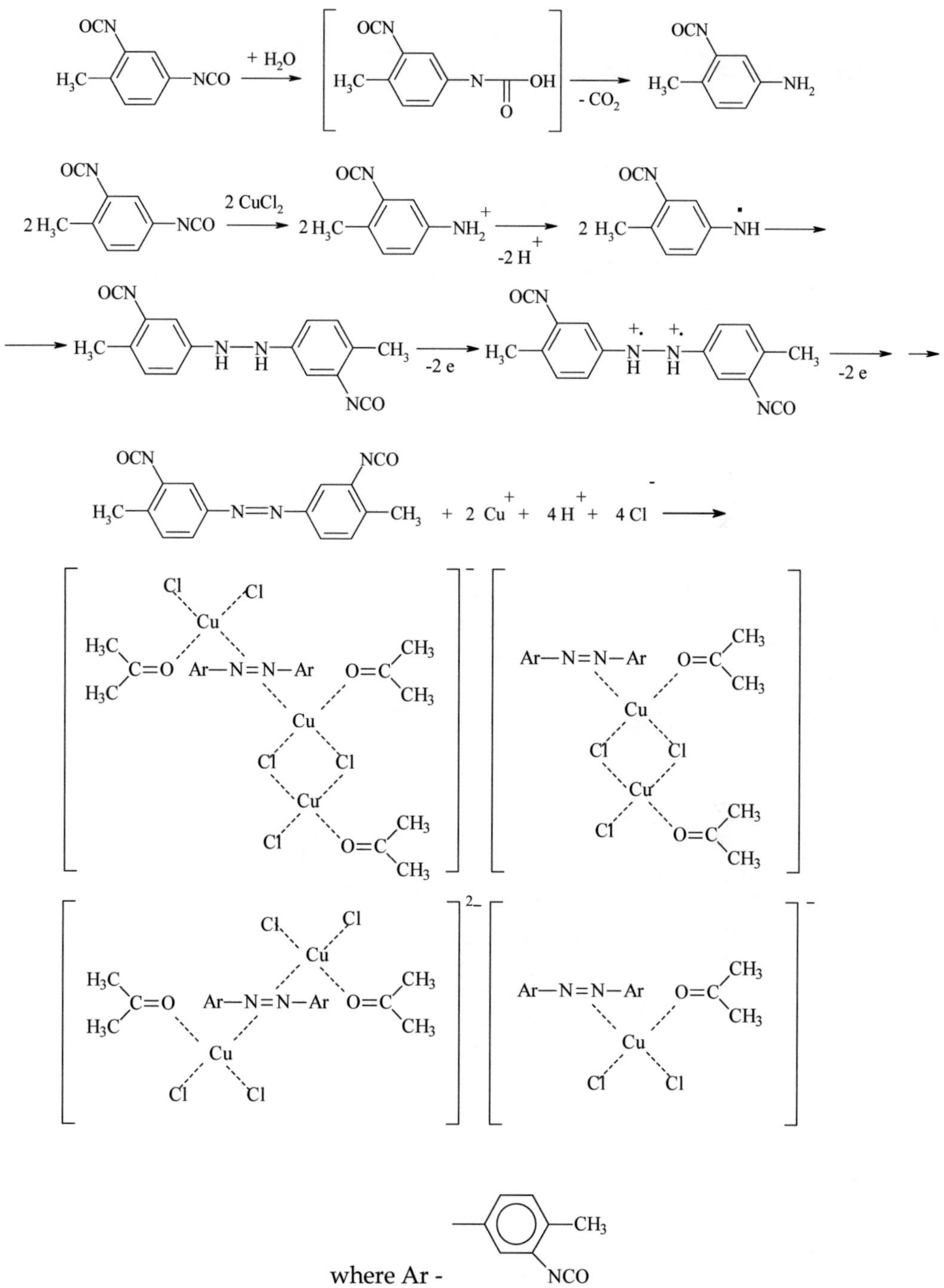

Figure 1. Formation of polynuclear complexes of azoaromatic compounds.

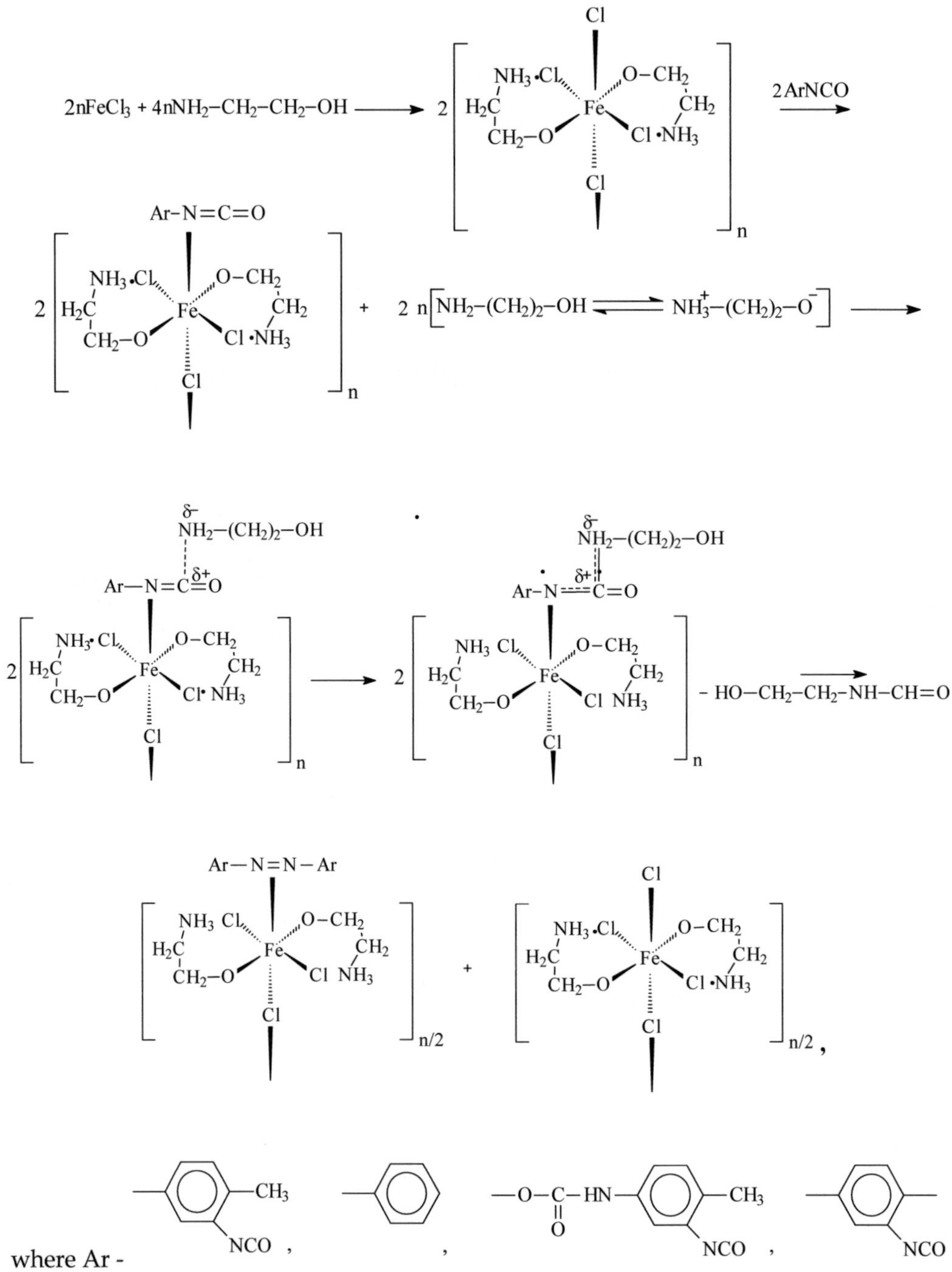

Figure 2. Scheme of formation of azoaromatic compounds.

The continuation of this reaction is the formation of stack coordination compounds in which metal ions form coordination bonds with azogroups (Fig. 3).

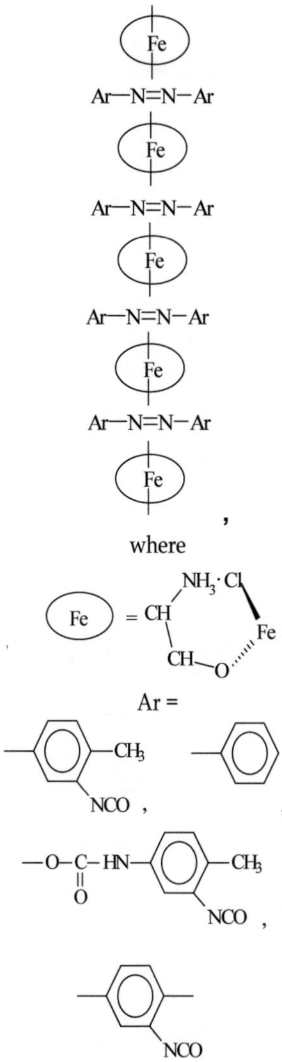

Figure 3. Formation of stack coordination compounds.

Mössbauer studies of iron coordinated compounds reveal that the magnetic ordering is observed at relatively high temperature (80 K). Mössbauer spectrum obtained in transmission geometry at the temperature of the sample being equal to 80 K consists of two components, namely, a doublet in the middle of the spectrum corresponding to the residues of the initial $FeCl_3$, dissolved in the matrix (less than 10% of the total area under the spectrum) and the magnetically ordered component with a hyperfine field average value of about 430 kE and the isomer shift, indicating a high-spin state of the Fe (III) ion. The absence of partial component in the middle of the spectrum is caused by the long average size of supramolecular structures chains (Fig. 4).

The Mössbauer study confirms the columnar structure of the obtained metal complexes, the possibility of their fixation in a flexible-chain polymer matrix containing electron-donating groups, and the existence of magnetic ordering at temperatures below 70 K. Sizes of very thin magnetic fields correspond to high-spin state of iron(III) ions (S=5/2).

It is known that to achieve the effect of magnetic ordering it is necessary that the chain of interacting ions of iron (III) should be long enough and combine up to a few thousand ions. The criterion for judging the length of the chain of magnetic ordered iron ions is relatively high blocking temperature of supermagnetism and the barrier value of effective anisotropy field. In our view, due to significant anisotropy in the structure of the complex the most likely assumption is the increasing of the potential barrier with the increase of chain length. The longer the chain, the higher its strength and the ability to build columns in a polymer matrix, and the specific properties of the structured polymer matrix are more pronounced as well.

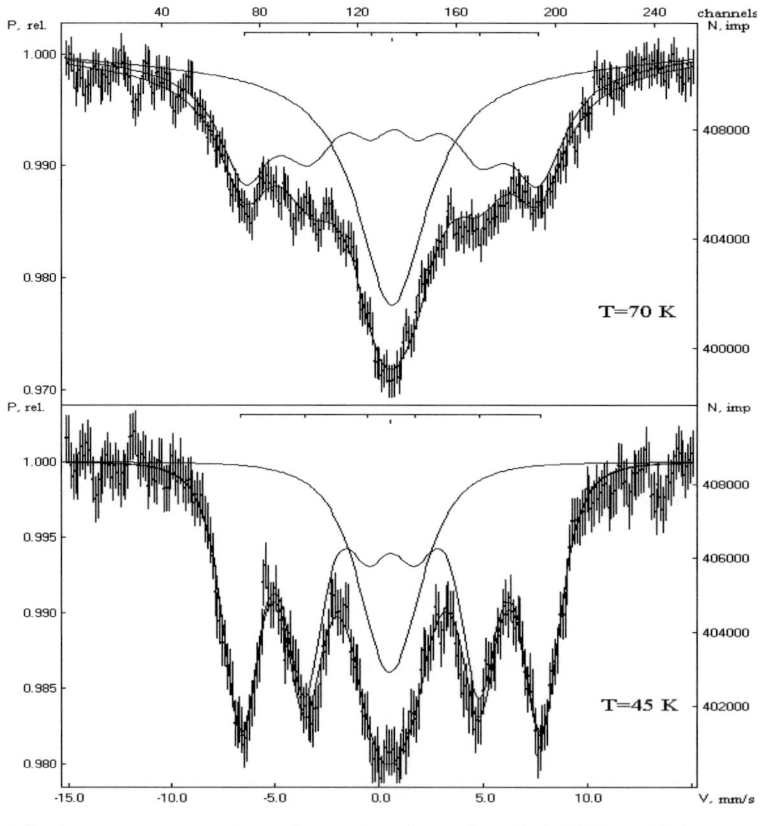

Figure 4. The Mössbauer spectrum of metal complex system based on $FeCl_3$ and EA.

The metal complex system is used for structuring urethane prepolymer containing terminal isocyanate groups. Considering the high flexibility of the urethane prepolymer chain, it can be assumed that the urethane groups will be coordinately bound with iron ions. The result of this interaction should be the formation of columnar structures directly in the polymer matrix. Mössbauer studies of urethane prepolymer confirm these assumptions. The resulting spectra are also of superparamagnetic nature at temperatures below 47K (Davletbaeva et al., 2006).

When polyurethanes are modified by coordination compounds synthesized on the basis of $FeCl_3$ and EA, the minimum values of specific volume electrical resistance (about 10^8 Ohm•sm) are recorded in the concentration area of 0,1% in terms of iron chloride. It should be noted that the ions of iron (III) in the above reactions do not change the oxidation level.

4. Modification of urethane prepolymer by coordination compounds of copper (I, II)

In one of the worked out highly ordered coordination compounds of transition metals for modification of polyurethanes N,N'-Diethylhydroxylamine (DEHA) is used as a ligand exhibiting the properties of a reducing agent. The most appropriate transition metal compound is copper (II) chloride. Some of the Cu (II) ions interacting with DEHA reduce the oxidation level and turn into Cu (I) (see Fig. 5).

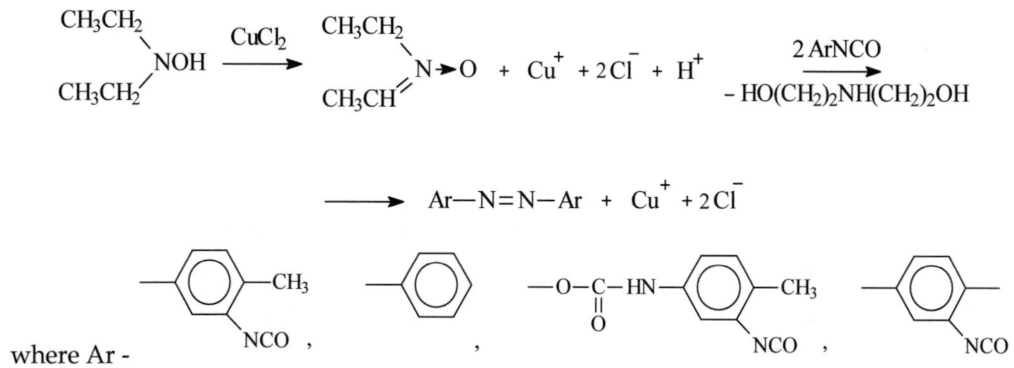

Figure 5. The mechanism of interaction of copper (II) chloride with DEHA.

It was ascertained that the metal complex system showed the ability to interact with aromatic isocyanates to form azoaromatic compounds. The result was the formation of columnar coordination compounds (Fig. 6).

It was established, that metal-complex modification of polyurethanes results in the change of their physicomechanical properties and spasmodic reduction (by 3-4 orders) of volumetric and superficial electric resistance. The reduction of the specific bulk electrical resistance by 3-4 orders is the most significant effect accompanying the metal-complex binding of polyurethanes. In this case the electrical resistance falls spasmodically depending on the nature of flexible chains of polyurethanes, the range of transition -metal ions concentrations. If the content of metal-complex modifying agent is increased further, the electrical resistance increases to some extent. The main role in the mechanism of charge transmission in metal-coordinated polyurethanes is assigned to electron-donating groups which are included in the structure of flexible chains of polyurethane matrix and the presence of transition metal ions having two degrees of oxidation in it.

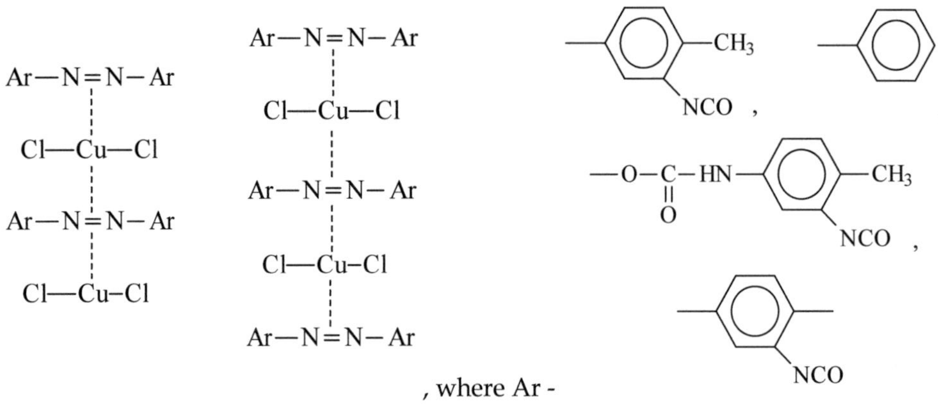

Figure 6. Formation of columnar coordination compounds.

5. Modification of polyurethanes by heteronuclear complexes based on molybdenum (V) and copper (II) chlorides

To study the effect of the coordination compounds structure on the electrophysical properties of the modified polyurethanes the heteronuclear metal complexes based on transition metals of IV and V periods were synthesized. $CuCl_2$ was used as the chloride of 3d-metal, and $MoCl_5$ was used as the chloride of the 4d-metal. DEHA was used as a ligand. Cu (II) has 3d-orbitals that are involved in coordinating binding, in the case of Mo (V) this role is performed by 4d-orbitals. Therefore, it is assumed that when heteronuclear complexes of the columnar structure are formed, where the $CuCl_2$ is in excess, the ions of molybdenum could cause the interruption of chains of exchange interactions between 3d-ions (Fig. 7). It is found that the metal complex system, obtained on the basis of $[CuCl_2]$:[DEHA] = 1:0,7 has ρ_v = 2600 Ohm·sm at room temperature, while the system based on $[MoCl_5]$:[DEHA] = 1:0,7 has ρ_v = 1300 Ohm·sm. The heteronuclear complex obtained at the ratio of $[CuCl_2]$:$[MoCl_5]$:[DEHA] = 0,9:0,1:0,7 at room temperature exhibits ρ_v = 5250 Ohm·sm, while the complex obtained at ratio of $[CuCl_2]$:$[MoCl_5]$:[DEHA] = 0,8:0,2:0,7 has already ρ_v = 73800 Ohm·sm.

Homonuclear and heteronuclear metal complexes were used for polyurethane modification. It was found that both modifying systems were able to react with the urethane prepolymer. When the metal complex structuring of polyurethanes by heteronuclear complexes based on 3d-and 4d-ions took place ϱv values remained similar to the unmodified sample. With the increase of the concentration of heteronuclear complex the specific volume electrical resistance of polyurethanes actually increased slightly (Fig. 8). Another situation is observed in the case of polyurethanes modification by homonuclear coordination compounds. In this case the introduction of metal complexes based on copper led to the decrease of ϱv by three orders. However, when polyurethanes were modified by homonuclear metal complex compounds based on Mo (V) ϱv decreased by 5 orders.

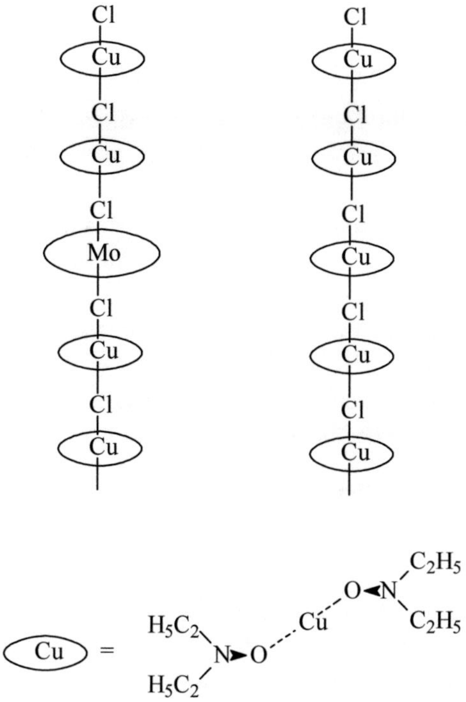

Figure 7. Scheme of the formation of columnar homonuclear and heteronuclear coordination compounds of transition metal.

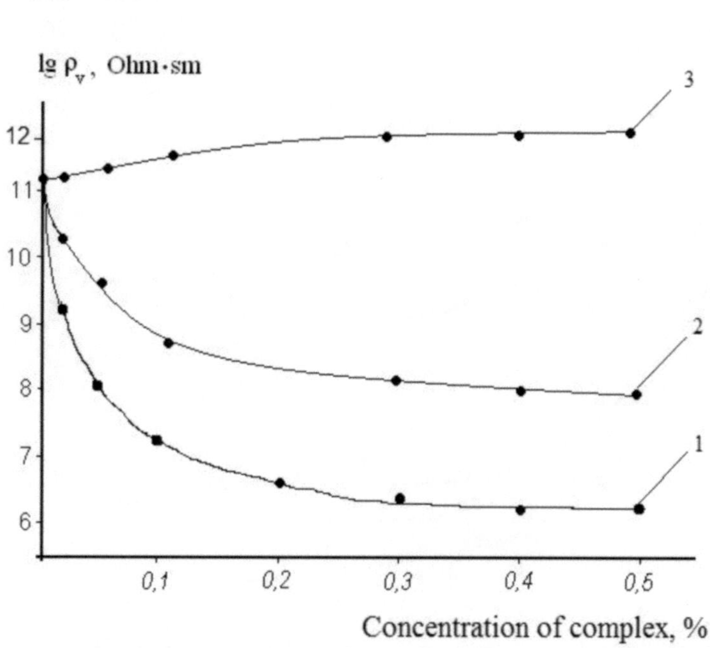

Figure 8. Dependence of ρ_v (Ohm·sm) of polyurethanes on the concentration (%) of metal complexes based on: 1-$MoCl_5$-DEHA; 2-$CuCl_2$-DEHA; 3-$CuCl_2$-$MoCl_5$-DEHA.

6. Catalytic properties of coordination compounds of copper in the reaction with isocyanate and urethane groups

The next step was to change the ligand composition of metal complex modifying system based on CuCl2 and DEHA. The aminopropyltriethoxysilane (AGM) was used as a part of modifying system. The use of the AGM as an additional component to the DEHA was caused by some reasons. The first reason is that the AGM is able to take part in the reactions of sol-gel synthesis, resulting in the hydrolysis of ethoxy-component and subsequent polycondensation of the forming silanol groups. The second reason is the presence of electron-donor amine groups in the AGM which are able to form complexes. Besides, amines can lead to reduction of copper (II) to copper (I). Thereby this substance is interesting in terms of the influence on the reactivity of isocyanate groups and the supramolecular structure of polyurethane, which has domain nature.

Titrimetric determination of concentration of isocyanate groups in UPTI during its interaction with metal complex system based on $CuCl_2$, DEHA and AGM at 100ºC, was carried out. UPTI is industrial prepolymer synthesized on the base of 1 mol polyoxitetramethyleneglicol and 2 mols 2,4-toluene diisocyanate.

It was found that at relatively low concentrations of metal complex (0.01 and 0.05% in terms of $CuCl_2$) in the first ten minutes from the start of the reaction process, the concentration of isocyanate groups started to rise, and only after that it fell. When the content of metal complex was 0.1, 0.5 and 0.75% in terms of $CuCl_2$ the concentration of isocyanate groups began to fall significantly (see fig. 9).

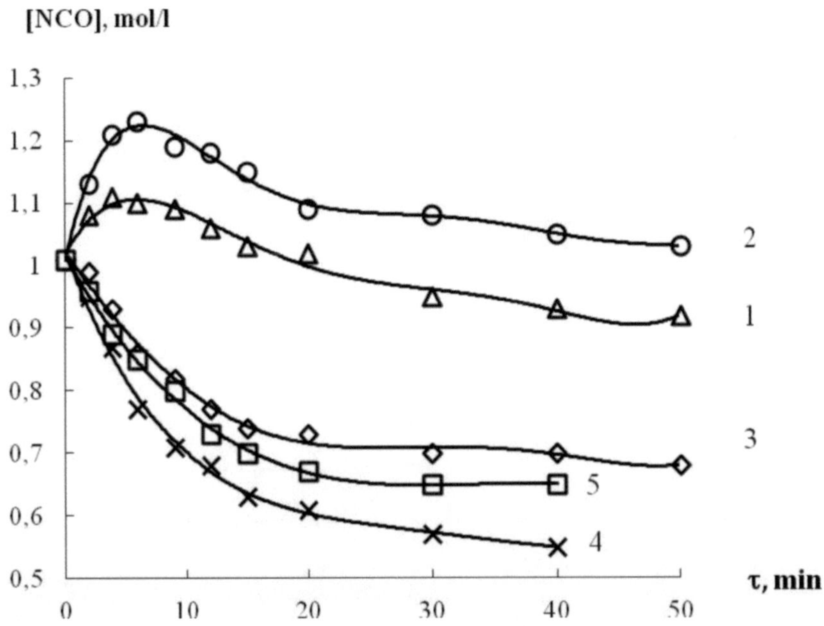

Figure 9. Isocyanate groups consumption curves in system UPTI – $CuCl_2$-DEHA-AGM, T=100°C, at a content of $CuCl_2$: 1 - 0,05%; 2 - 0,1%; 3 - 0,25%; 4 - 0,5%; 5 - 0,75% (wt.).

The titration data confirm the results of IR-spectroscopic studies. It is established that the absorption band at 1731 sm-1 due to the stretching vibrations of carbonyl component of urethane group decreases at low concentrations of metal complex in the first ten minutes from the start of the reaction process. At the same time the intensity of the absorption band at 2273 sm^{-1} due to the stretching vibrations of isocyanate group grows. Later the growth of the intensity of the absorption band at 1731 sm^{-1} and the decrease at 2273 sm^{-1} are observed. Besides in the first ten minutes from the start of the reaction the IR-spectroscopy shows the reduction of the intensity of the absorption band at 3293 sm^{-1} due to the stretching vibrations of N-H-bond that is a part of urethane group. At relatively high concentrations of metal complex (≥0.25%) the interaction is accompanied by the growth of the intensity of the absorption band at 2120 sm-1 due to the formation of carbodiimide group.

The research suggested that at relatively low concentration of metal complex the urethane group dissociates to isocyanate and hydroxyl groups, while at high concentrations of metal complex the isocyanate groups consume to the formation of carbodiimide groups. The part of isocyanate groups is hypothetically consumed on the formation of the azoaromatic groups. It is known that it is impossible to analyze azoaromatic groups using infrared spectroscopy. In this regard, studies were carried out using electron spectroscopy.

Electronic spectrum (Fig. 10) showed absorption at 350 nm, typical for trans-azoaromatic compounds. The absorption in the area 480 nm characterizes the coordination compounds of copper (II).

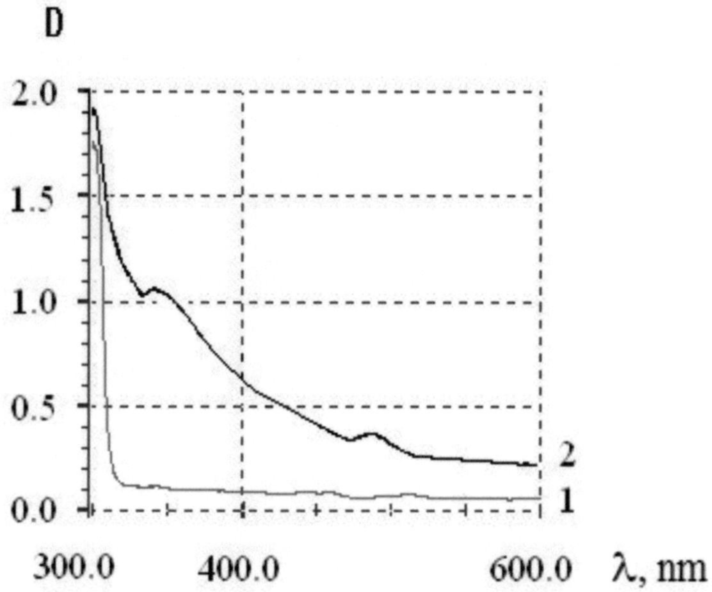

Figure 10. Electron spectrum of urethane prepolymer (1) and prepolymer (2) modified by 0.5% (wt.) metal complex system based on $CuCl_2$ - DEHA – AGM.

Metal complex system, derived on the basis of $CuCl_2$, DEHA and AGM was further used to modify polyurethanes. We measured the dependence of the volume resistivity (ρ_v) of polyurethanes on the concentration of metal complex modifier (Fig. 11).

It turned out that the use of the worked out metal complex system caused the leap of ρ_v by 4 orders (10 000 times) observed at low concentrations of metal complex 0.01%. Here we should note that while using metal complex system based on $CuCl_2$ and DEHA (no AGM) a stepwise drop of ρ_v was observed at much higher concentration of the complex - 0.1%.

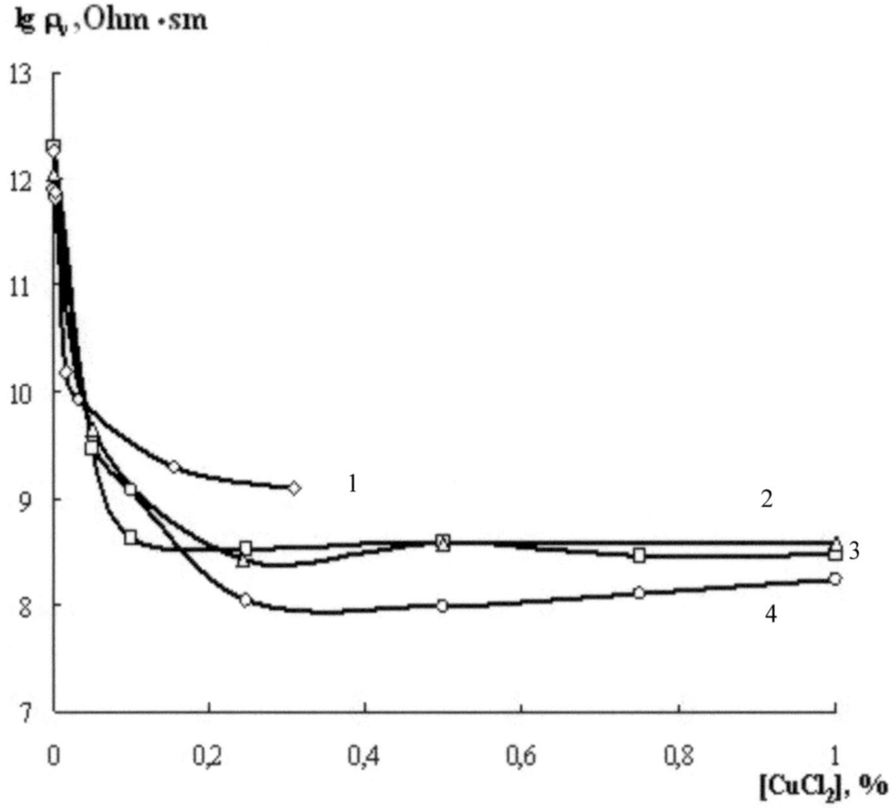

Figure 11. Volume resistivity-concentration diagram of modified polyurethanes under molar ratio of [UPTI]: [Diamed-X] = 1:Y:

1. $[CuCl_2]$:[AGM]=1:4 (Y=0.9);
2. $[CuCl_2]$:[DEHA]:[AGM]=1:1,48:0,25 (Y = 0.9);
3. $[CuCl_2]$:[DEHA]:[AGM]=1:1,48:0,25 (Y = 0.7);
4. $[CuCl_2]$:[DEHA]:[AGM]=1:1,48:0,25 (Y = 0.5).

It was found that the use of metal complex systems based on $CuCl_2$, DEHA and AGM could significantly reduce the dosage of curing agent 4,4-methylene-bis-o-chloroaniline (Diamed-X) for urethane forming system based on UPTI.

7. The use of highly ordered coordination compounds of copper for receiving the rigid polyurethane foam

Metal complexes derived from $CuCl_2$, DEHA and AGM were also tested as modifiers of the polyol component used in the manufacturing of rigid polyurethane foam. It was found that metal complex system based on $CuCl_2$, DEHA and AGM had a significant impact on the rise and curing time of foam, reducing it (Fig. 12).

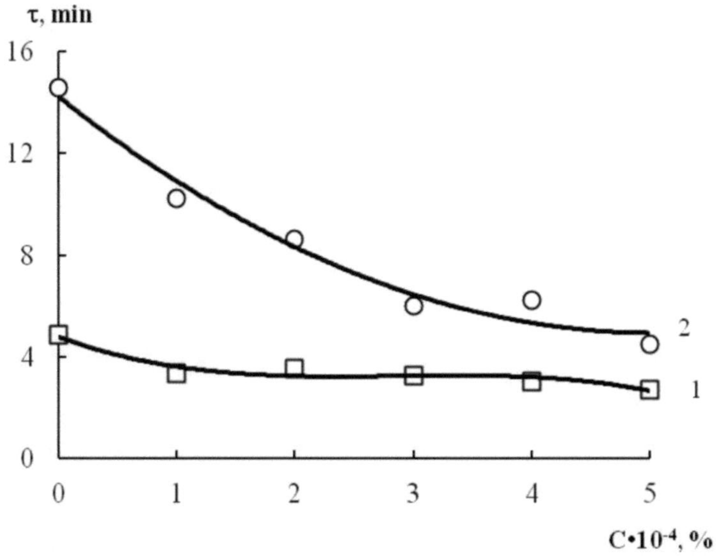

Figure 12. Rise (1) and curing (2) time of foam as a function of metal complex concentration based on $[CuCl_2]$:[DEHA]:[AGM]=1:1,48:0,25 in terms of $CuCl_2$ (%)

It was also established that the increasing of the molar ratio of the AGM in metal complex system led to even greater decrease in rise and curing time of foam (Fig. 13).

In order to establish the role of the AGM in the foaming process it was loaded alone in polyol component (Figure 14). It was found that the AGM also reduces the rise and curing time of the foam. However, these parameters were more than two times higher than the parameters that caused the addition of metal complex system.

We also used metal complex system based on $CuCl_2$ and DEHA as the control modifying system. In this case, in the wide range of concentrations of modifier the foam "collapsed". That is, the foam rose and the subsequently settled out. It should be also mentioned that the density of foam produced using metal complex system $CuCl_2$ - DEHA - AGM did not change in comparison with polyurethane foam obtained by the unmodified polyol component.

Thus, these studies show a significant catalytic effect of the metal complex modifier on foaming. In this connection it should be noted that the polyol component is a complicated balanced system that contains catalysts of amine nature and organotin compounds already. Our results suggest that the metal complex systems act as a cocatalyst.

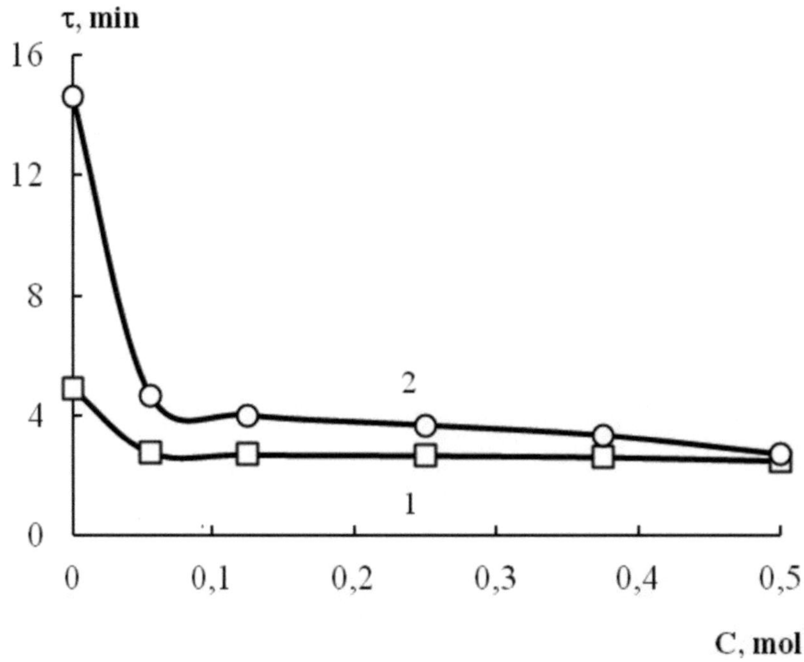

Figure 13. Rise (1) and curing (2) time of foam as a function of modifier concentration based on [$CuCl_2$]:[DEHA]:[AGM]=1:1,48:X, where X is a mole fraction of AGM in the metal complex overall concentration (mol)

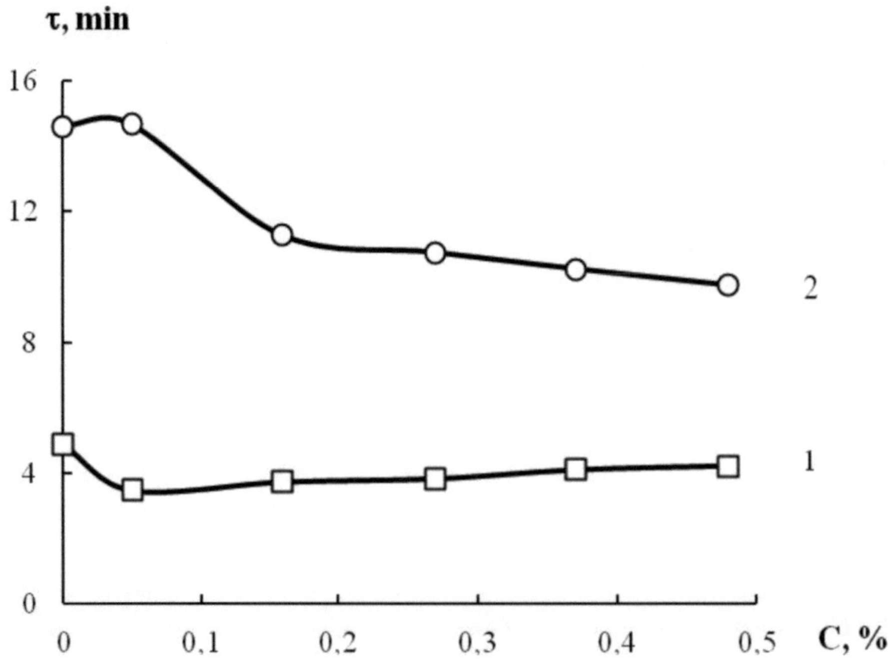

Figure 14. Rise (1) and curing (2) time of foam as a function of AGM concentration.

The next step was the research of such polyurethane foam key indicators as moisture absorption (Fig. 15) and water absorption (Fig. 16-17). It was found that the modified foam had enhanced characteristics as compared with unmodified polyurethane foam.

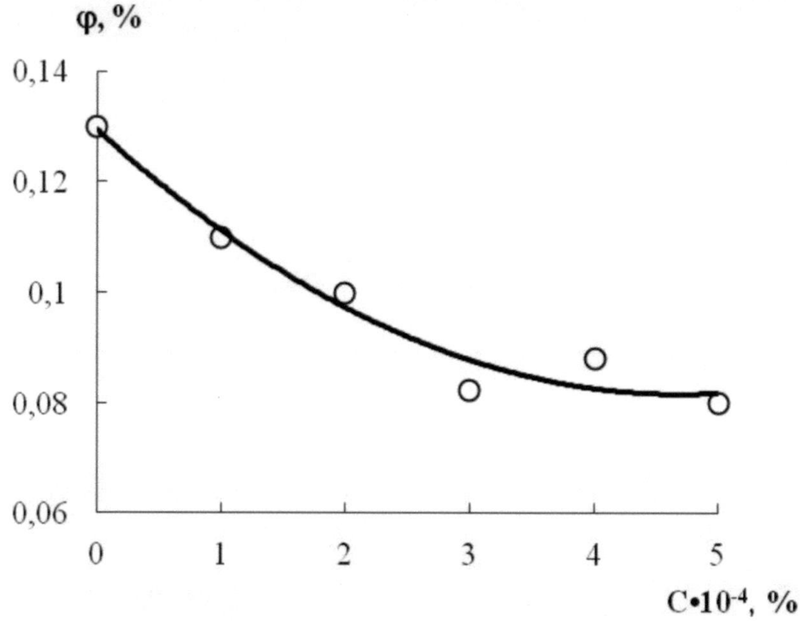

Figure 15. Moisture absorption of rigid foam as a function of metal complex concentration based on $[CuCl_2]$:[DEHA]:[AGM]=1:1,48:0,25 in terms of $CuCl_2$ (%)

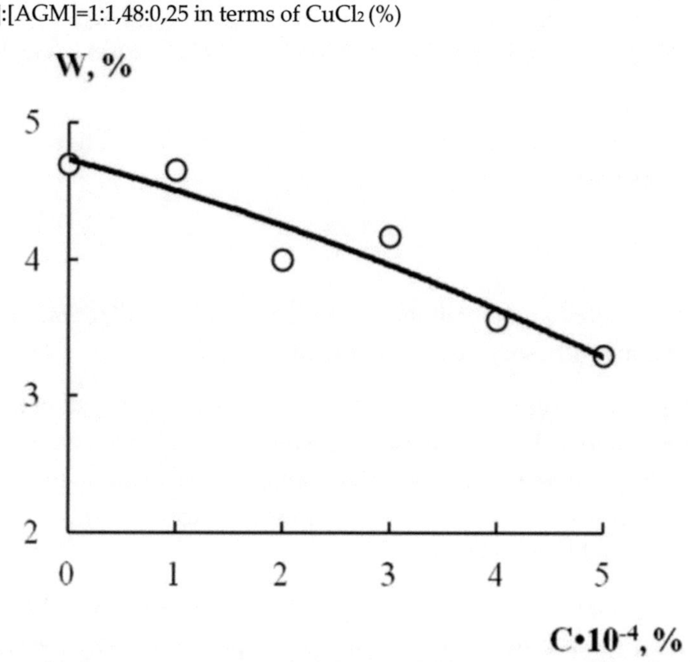

Figure 16. Water absorption of rigid foam as a function of metal complex concentration based on $[CuCl_2]$:[DEHA]:[AGM]=1:1,48:0,25, where in terms of $CuCl_2$ (%)

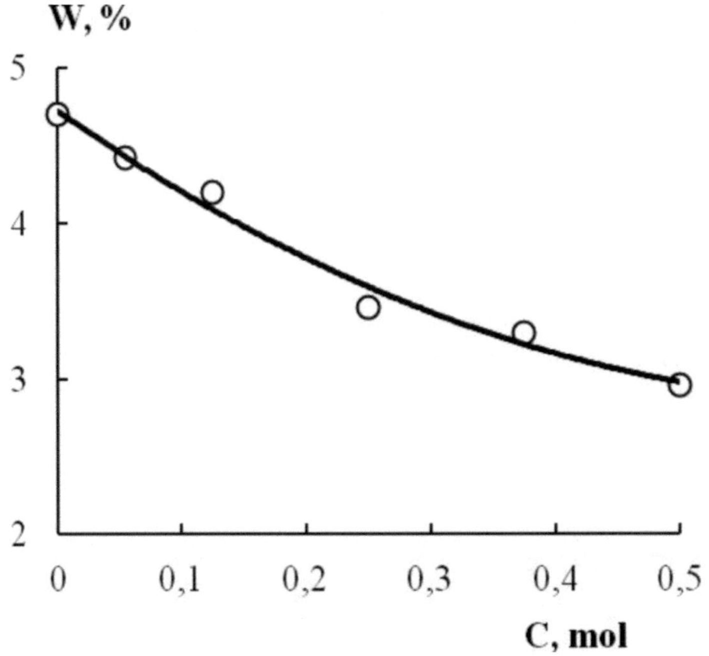

Figure 17. Water absorption of rigid foam as a function of modifier concentration based on [$CuCl_2$]:[DEHA]:[AGM]=1:1,48:X, where X is a mole fraction of AGM in the metal complex overall concentration (mol)

In conclusion, it should be noted that in order to achieve a positive result very small amounts of modifiers are required.

8. Conclusion

We considered the methods of obtaining transition metal coordination compounds that were active in reactions with isocyanate and urethane groups.

The feature of these metal complexes is that the metal ions are arranged in a chain of atoms linked together by chloride bridges. It is established that the chain of exchange-coupled transition metal ions remains in the polyurethanes structured by metal complex compounds. This circumstance is the cause of stepwise decrease in the specific volume resistivity of the modified polyurethanes.

It seems to be interesting for further research in this field to study the effect of metal complex binding on the physical and mechanical properties of polyurethanes. The most promising materials in terms of improving strength properties and heat resistance are thermoplastic urethanes.

Author details

Ruslan Davletbaev, Ilsiya Davletbaeva and Olesya Gumerova
Kazan National Research Technological University, Russia

9. References

Brostow W. (1990). Reaches of the liquid crystalline systems. *Polymer,* Vol.31, pp. (979-1023)

Carrher C.E. (1981). The structure of LC polymeric systems. *J. Chem. Ect.,* Vol.58, pp.(921-929)

Davletbaeva I.M., Kirpichnikov P.A. & Rakhmatullina A.P. (1996). Synthesis and investigation of liquid polyurethane metal complexes. *Macromolecular Symposia,* Vol.106, pp.(87-90)

Davletbaeva I.M, Shkodich V.F., Ismagilova A.I. & Parfenov V.V. (2001). Electro-physical properties of mesogenic metal-coordinated polyuretanes. *Russian polymer news,* Vol.6, No.4, pp.(36-38)

Dirk C.W., Bousseau M., Barret P.H., Moraes F., Wudl F. & Heeger A.J. (1986). Metal Poly(benzodithiolenes). *J. Macromolecules,* Vol.19, pp.(266-268)

Kingsborough R.P, Swager T.M. (1999). Polythiophene Hybrids of Transition-Metal Bis(salicylideninine)s: Correlation Between Structure and Electronic Properties. *J. Am. Chem. Soc.,* Vol.121, pp.(8825-8834)

Reynolds R., Karasz F.E., Lillya C.P. & Chien J.C.W. (1985). Electrically Conducting Transition Metal Complexes of Tetrathiooxalates. *J. Chem. Soc., Chem. Commun, pp.(268-269)*

Serrano J.-L., Oriol L. (1995). Metallomesogenic polymers. *J. Adv. Mater.,* Vol.7, No.4, pp.(348-369)

Shirai H., Vagi S., Suzuli A. (1977). Functional metal-porhyrazine derivatives and their polymers. 1. Synthesis of metal-phtalocyanine derivatives. *J. Macromol.Chem.,* Vol.178, No.7, pp.(1889-1895)

Shirai H., Kobayashi K. & Takemae V. (1979). Organometallic polymers. *J. Mac-romol.Chem.,* Vol.180, pp.(2073-2084)

Thuchide E.,Nishide H. (1977). Polymer-metal complexes and their catalytic activity. *Advances in Polymer Science,* Vol.24, pp.(2-87)

Wang F., Reynolds J.R. (1988). Soluble and electroactive nickel bis(dithiolene) complex polymers. *Macromolecules,* Vol.21, No.9, pp.(2887-2889)

Davletbaeva I.M., Ismagilova A.I, Tyut'ko K.A., Burmakina G.V. & Kuzaev A.I. (1998) Reactions of isocyanates with the system based on $CuCl_2$ - N,N'-Diethylhydroxylamine. *Russian Journal of General Chemistry,* Vol.68, No.6, pp.(1021-1027)

Davletbaeva I. M., Pyataev A. V., Kalachev K. E., Sadykov E. K. & Manapov R. A. (2006) Mössbauer study of structurally ordered iron coordination compounds and polyurethanes crosslinked by them. *Polymer Science,* Ser. A, 2006, Vol. 48, No. 6, pp.(612–617)

Chapter 4

Bottom-Up Nanostructured Segmented Polyurethanes with Immobilized *in situ* Transition and Rare-Earth Metal Chelate Compounds – Polymer Topology – Structure and Properties Relationship

Nataly Kozak and Eugenia Lobko

Additional information is available at the end of the chapter

http://dx.doi.org/10.5772/48002

1. Introduction

The formation of the polyurethanes (PU) with immobilized *in situ* co-ordinating metal compounds allows obtain structurally homogeneous systems with uniform dispersed nanosize metal containing sites. Aggregation of these metal chelate compounds is prevented due to complexing with polar groups of the polymer matrix.

At the same time due to complex formation between the metal compound and polymer functional groups, the structuring of the forming matrix occurs on a nanoscale level. As a result, in the presence of small amounts of metal chelate compound (0,5-5%wt) both change of the polyurethane structure and properties can be observed. To understand the nature of above phenomena the influence of the weak interactions «macromolecule - metal» were analyzed on the metal-containing PUs structure, molecular dynamics and properties.

The present study investigates the formation of nanostructured linear and cross-linked polyurethanes (LPUs and CPUs, respectively) with immobilized *in situ* mono- and poly-heteronuclear chelate compounds of rare-earth and transition metals. Influence of PU topology on self organization processes in polymer matrix and its properties is also subject of analysis.

1.1. Materials, methods and instrumentations

Polypropylene glycol (PPG, MW 1000) was dried under vacuum at 120 °C for 2 h. Tolylene diisocyanate (mixture 80/20 of 2,4- and 2,6- isomers) (TDI) was distilled under vacuum.

Diethylene glycol (DEG) was distilled under vacuum at 105 °C. Trimethylol propane (98%) (TMP) was dried under vacuum at 40-45 °C for 2-4 h. Dichloromethane (CH_2Cl_2), 1,4-dioxane and N,N′- dimethylformamide (DMF) were distilled at 40 °C, 101 °C, 153 °C, respectively. The following chelate compounds of transition and rare-earth metals as PU modifier were used:

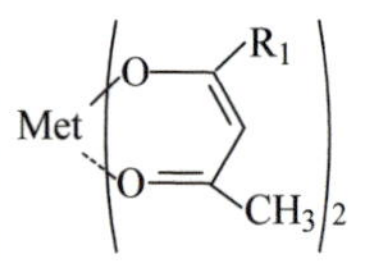	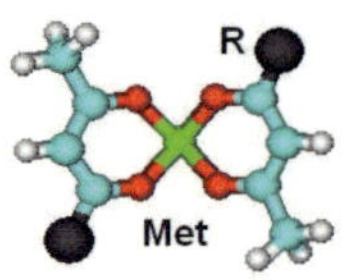 	*(R=-CH_3)* Cu(acac)$_2$ - Copper(2+) acetylacetonate Ni(acac)$_2$ – Nickel(2+) acetylacetonate *(R = -OC_2H_5)* Cu(eacac)$_2$ - Copper (2+) ethyl acetoacetate *(R = --CF_3).* Cu(tfacac)$_2$ - Copper (2+)trifluoro acetylacetonate
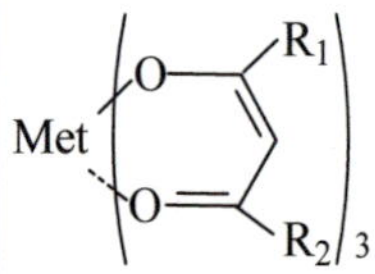	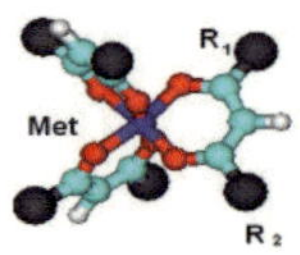 	*(R_1=R_2=-CH_3)* Co(acac)$_3$ – Cobalt (3+) acetylacetonate Cr(acac)$_3$ – Chromium (3+) acetylacetonate Gd(acac)$_3$ – Gadolinium (3+) acetylacetonate Nd(acac)$_3$ – Neodymium (3+) acetylacetonate Er(acac)$_3$ – Erbium (3+) acetylacetonate *(R_1 =-C(CH_3)$_3$, R_2 =-(CF_2)$_2$-CF_3)* Eu(fod)$_3$ – Europium (3+) tris(6,6,7,7,8,8,8-heptafluoro-2,2-dimethyl-3,5-octanedionate) *(R_1 =- thiophene , R_2 =- CF_3)* Eu(TTA)$_3$ – Europium (3+) thenoyltrifluoroacetonate
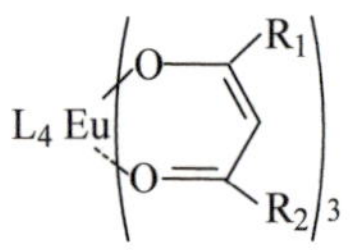	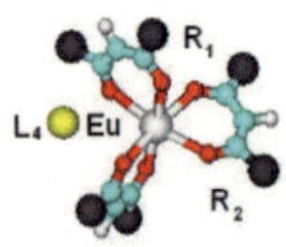 	*(R_1 =- thiophene , R_2 =- CF_3; L_4= phen)* Eu(TTA)$_3$ phen – Europium (3+) tris(thenoyltrifluoroacetonate) phenantroline *(R_1 =- thiophene , R_2 =- CF_3; L_4= triphenylphosphine oxide)* Eu(TTA)$_3$ TPPO –Europium (3+) tris (thenoyltrifluoroacetonate) (triphenylphosphine oxide)
$\left[Met_k^1 Met_m^2 Met_n^3 R_p^1 R_q^2 R_r^3\right]\cdot Sol_t$ $p,q,k,m=1,2,3,4;$ $n,r,t=0,1;$ where Me_2Ea = deprotonated residue of dimethyl aminoethanol Dea = doubly deprotonated residue of diethanolamine	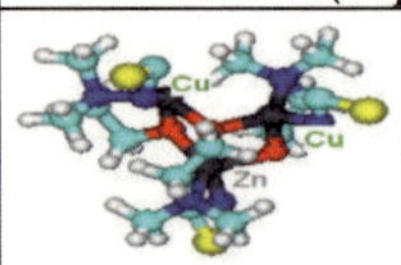 	*k*=2, *m*=1, n=0, *p*=3, *q*=3, r=0, *t*=1, R^1=NCS, R^2=Me_2Ea, Sol= CH_3CN [$Cu_2Zn(NCS)_3(Me_2Ea)_3$].CH_3CN
	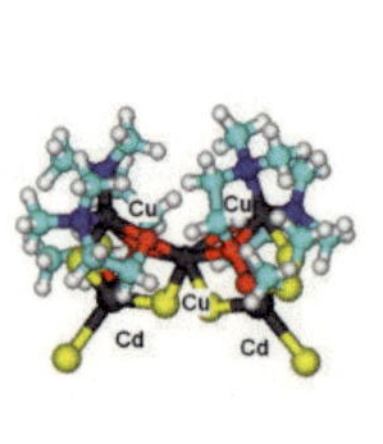 	*k*=2, *m*=3,n=0, p=6, *q*=4, r=0, *t*=2, R^1=Br, R^2=Me_2Ea, Sol= dmso [$Cd_2Cu_3Br_6(Me_2Ea)_4(dmso)_2$]
		k=1, *m*=2, *n*=2, *p*=3, *q*=4, r=4, t=0, R^1= H_2Dea, R^2= NCS, R^3= Dea [Ni(H^2Dea)^2][CoCu(Dea)^2(H^2Dea)(NCS)]^2(NCS)^2

Table 1. The PU modified chelate compounds of transition and rare-earth metals.

In metal chelate compounds used as PU modifier metal ions are already surrounded with organic ligands. This facilitates solvation of modifier in polymer. The listed above transition and rare-earth metal chelate compounds are commercial products (Aldrich). The heteroligand rare-earth metal compounds were synthesized by Professor Svetlana B. Meshkova's group (A. V. Bogatsky Physic-Chemical Institute of National Academy of Sciences of Ukraine, Odessa). Polyheteronuclear metal complexes of Cu (2+), Cd (2+), Zn (2+), Ni (2+) and Co (3+), described in (Skopenko et al., 1997; Vinogradova et al., 2002), were provided by Prof. V. Kokozay's goup (Kiev Taras Shevchenko University). Polyheteronuclear metal chelate compounds can realize unexpected coordination states of transition metal ions. That, in turn, can give new properties to a polymer formed in their presence.

PUs were synthesized in two stages according to standard procedure described in detail elsewhere (Saunders&Frish,1968; Wirspza,1993) using PPG-1000 and TDI based prepolymer. DEG was used as chain extender to obtain LPU (Scheme 2). TMP was used as cross-linking agent to obtain CPU (Scheme 3). Metal chelate compounds were added into reaction mixture as solution in CH_2Cl_2, 1,4-dioxane or DMF to obtain the metal containing PUs with homogeneous distribution of modifier (from 0,5 to 5 %wt.) in polymer matrix. High ability of metal chelate compound to complex formation leads to enrichment of PU matrix with heteroligand macro complexes of 3d- and 4f-metal with prevalence of outer-sphere coordination of macro chains. Such macro complexes act like coordination linkages between polymer chains and form "coordination nodes" in PU (Scheme 1).

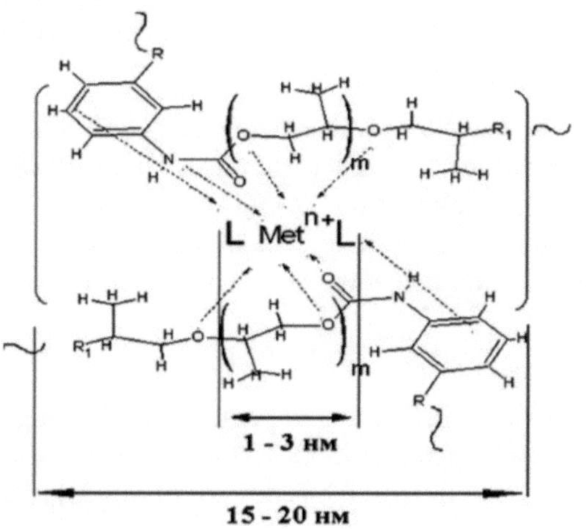

Scheme 1. The coordination junction of PUs networks.

Thus, in the LPU (Scheme 2) in the presence of chelate metal compounds the "coordination nodes" can form.

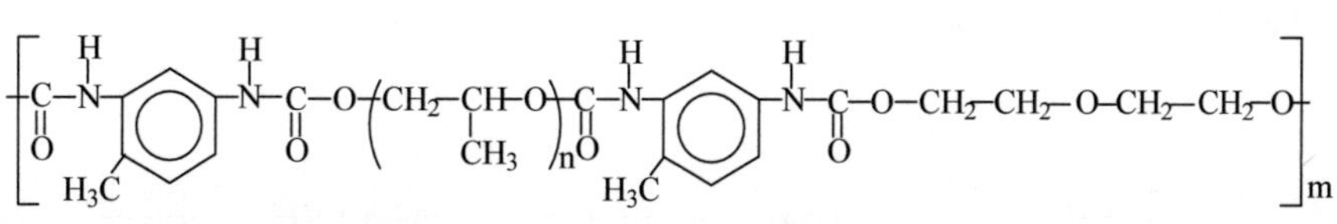

Scheme 2. The general formula of LPU.

In the metal containing CPU both the chemical linkages (Scheme 3) and the "coordination nodes" can form (Scheme 1).

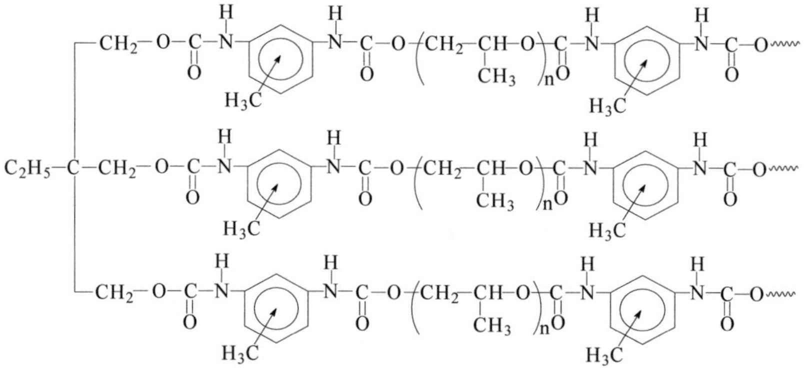

Scheme 3. The fragment of PU network with cross-linkage.

Wide-angle X-ray scattering (WAXS) profiles of studied samples were recorded on a Dron-4-07 diffractometer with Ni-filtered Cu-K$_\alpha$ radiation and Debay-Sherer optical schema. Distance between PU atomic layers (*d*) was estimated using the Bragg equation:

$$\lambda = 2d\sin\theta \tag{1}$$

where λ – the X-ray wave length (λ = 0,154 nm); θ - the diffraction maximum angular position, degrees.

Small-angle X-ray scattering (SAXS) profiles were recorded using KPM-1 X-ray camera (Kratky et al., 1966). The Schmidt's method (Schmidt & Hight, 1960) was used to smooth out the SAXS-profiles to point collimation. X-ray measurements are carried out using monochromatic Ni-filter of Cu-K$_\alpha$ radiation at temperature 22±2 °C. The Bragg's period of uniform electronic density scattering elements was estimated through the equation:

$$D = 2\pi / q \tag{2}$$

The X-band EPR-spectra were recorded at temperature 20°C using radio spectrometer PE-1306 equipped with frequency meter ChZ-54. The magnetic field was calibrated using 2, 2-diphenil-1-pycrilhydrazyl (DPPH) (g=2,0036) and ions of Mn(2+) in MgO matrix (g=2,0015).

Stable nitroxide radical 2,2,6,6-tetramethylpiperidinyl-1-oxy (TEMPO) was used as paramagnetic spin probe (SP). Nitroxide SP was introduced into PU films via diffusion of its saturated vapor at 30°C for 2 hours with subsequent keeping at 20°C for 24 hours.

Correlation time (τ) of SP rotational diffusion in the range of its fast motion ($10^{-11} < \tau < 10^{-9}$s) was calculated according (Vasserman & Kovarskii, 1986) as follows:

$$\tau = 6,65\Delta H_{(+1)}(\sqrt{(I_{+1} / I_{-1})} - 1) \times 10^{-10} c, \tag{3}$$

where $\Delta H_{(+1)}$ – is width of the low-field- component of TEMPO EPR-spectrum, I_{+1} and I_{-1} - are intensities of low-field and high-field components of the spectrum, respectively.

The differential scanning calorimetry in temperature interval from 223 to 750 K was performed using Perkin Elmer DSC 2 instrument with the IFA GmbH's software. The heating rate was 0,05-2 grad/min.

Micro images in light transmission were obtained using an optical microscope XY-B2 (NS Instr. Co.) equipped with digital video ocular ICM 532 and AMCAM/VIDCAP (Microsoft) image processing system.

The surface tension of PUs (γ_{sg}) was determined according to Elton's equation (Tavana et al., 2004) using measurement of contact wetting angle with ethyleneglycol (EG) as wetting liquid at 20°C:

$$\gamma_{sg} = 0{,}5\gamma_{lg}(1 + \cos\theta) \tag{4}$$

where γ_{sg} and γ_{lg} are the surface tension on solid-gas and liquid-gas boundaries, respectively; θ is the boundary wetting angle; solid is PU; liquid is EG.

The mean value of γ_{sg} was calculated as average of 5 different measurements and error of measurements did not exceed the value of 0,5 mN/m.

The spectra of luminescence were obtained using the luminescent spectrometer SDL-1 (LOMO) in an excitation by the mercury lamp. The emission of the most intensive line with the maximum on 365 nm was selected with light filter UFS-2.

Two-electrode method measurements of conductivity at a direct current (dc) were conducted using a Hiresta UP high resistivity meter (Mitsubishi Chemicals, Japan). A dc voltage of 10 V was applied across the sample thickness. The samples were dried over night in an oven at 40°C under vacuum and then kept in dried environment, for the elimination of any moisture effects.

Dielectric relaxation analysis was performed using dielectric spectrometer on the base on alternating current bridge R5083. Complex dielectric permittivity, $\varepsilon^* = \varepsilon' - i\varepsilon''$, of disc-like specimens (diameter: 20 mm) sandwiched between gold-coated brass electrodes was measured over the frequency window from 102 to 105 Hz in the temperature interval from --40 to 120 °C. They have been analyzed from the traditional point of view (Pathmanatham & Johari, 1990; Pissis & Kanapitsas, 1996). Additional formalisms such as: complex admittance σ*, electrical modules M* and impedances Z', Z" were used according to formulas.

$$\varepsilon' = C_1 / C_o, \mathrm{tg}\delta = \omega R C_1 \text{and } \varepsilon'' = \varepsilon' \cdot \mathrm{tg}\delta \tag{5}$$

$$\sigma^* = \sigma' + \sigma'', \ \sigma' = \omega\varepsilon'' \ , \ \sigma'' = \omega\varepsilon', \tag{6}$$

$$M^* = M' + M'', M' = \varepsilon''/\left(\varepsilon'^2 + \varepsilon''^2\right), M'' = \varepsilon'/\left(\varepsilon'^2 + \varepsilon''^2\right) \tag{7}$$

$$Z' = M''/(\omega C_o), \ Z'' = M'/(\omega C_o) \tag{8}$$

C_o and C_1 – are instrument and standard capacitor capacities, ω – cyclic frequency.

The electron spectra of the copper (2+) containing PU films and of copper (2+) chelate compounds solutions in dichloromethane (c = 10^{-2}M) in the ultra-violet and visible region were recorded using the spectrometer Specord UV-VIS.

The quasi-elastic neutron scattering (QENS) was recorded using the multi detector spectrometer "NURMEN" on the atomic reactor BBP-M (The institute of the nuclear research of the NAS of Ukraine). The self-diffusion of chloform used as low molecular probe liquid in swelled PU films was analyzed.

2. Heterogeneity of metal containing polyurethanes

2.1. Structural heterogeneity of PU according to X-ray data

Formation of a polymer matrix in the presence of metal chelate compounds favours creation of a new hierarchy in structural organization of the polymer as compared with metal free system. This effect is caused by complex formation between metal chelate compound and functional groups of the forming polymer (Ying, 2002; Kozak et al., 2000).

Figure 1 represents the WAXS and Figure 2 presents SAXS intensity profiles of metal-free PU and PU modified with metal β-diketonate. The asymmetric diffuse diffraction maxima (Figure 1) point on the amorphous structure of the metal-free and metal containing CPU and LPU. For the LPUs the short-range order parameter *d* (equation 1) is equal to 0.44 nm and don't depend on the metal chelate compound amount (table 1). For the CPU the Bragg's period (*d*) changes from 0.44 to 0.46 nm with increasing of the modifier amount from 0,5 to 5% wt.

The PU's SAXS profiles are characterized by the presence of one amorphous maximum with q_m positions varying from 1,7 to 2,0 nm^{-1} (Figure 2). Such maximum points on the existence of changeover period of uniform electron density scattering elements and areas of uniform distribution of hard and flexible blocks in PU. The Bragg's period (*D*) falls from 3,7 to 3,1 nm with increasing of the modifiers amount from 0,5 to 5% wt. (table 1).

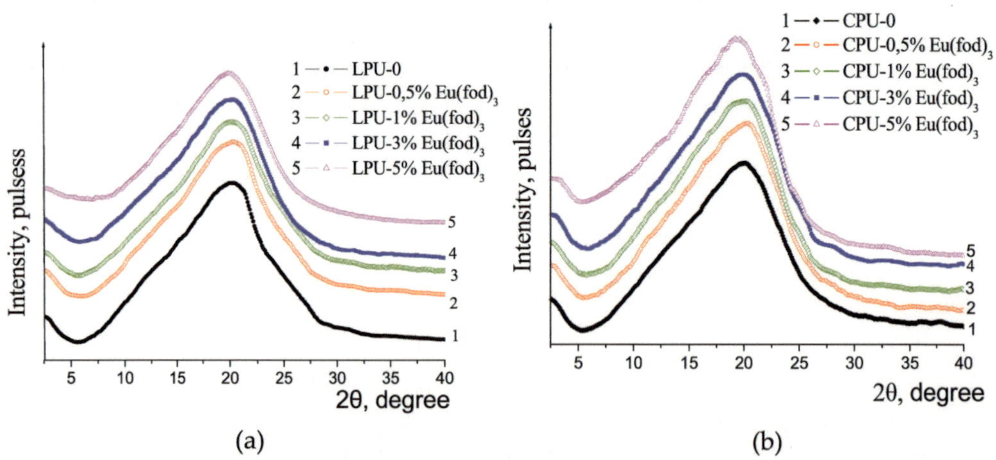

Figure 1. The WAXS intensity profiles of CPU (a) and LPU (b): metal-free (1), modified with 0,5% (2), 1% (3), 3% (4) и 5% (5) Eu(fod)$_3$.

System	2θ, degree	*d*, nm	q_m, nm^{-1}	D, nm
CPU-0	20	0.44	1.7	3.7
CPU-0,5% Eu(fod)$_3$	20	0.44	1.7	3.7
CPU-1% Eu(fod)$_3$	19.9	0.45	1.76	3.6
CPU-3% Eu(fod)$_3$	19.9	0.45	1.76	3.6
CPU-5% Eu(fod)$_3$	19.4	0.46	2.0	3.1
LPU-0	20	0.44	1.7	3.7
LPU-0,5% EEu(fod)$_3$	20	0.44	1.7	3.7
LPU-1% Eu(fod)$_3$	20	0.44	1.9	3.3
LPU-3% Eu(fod)$_3$	20	0.44	1.8	3.5
LPU-5% Eu(fod)$_3$	20	0.44	1.9	3.4

2 θ - the diffraction maximum angular position, degrees;
d – distance between PU atomic layers from WAXS, nm;
q_m - value at maximum intensity of *I*(*q*) relationship, nm^{-1};
D - changeover period of uniform electronic density scattering elements from SAXS, nm.

Table 2. X-ray structural characteristic of LPU and CPU

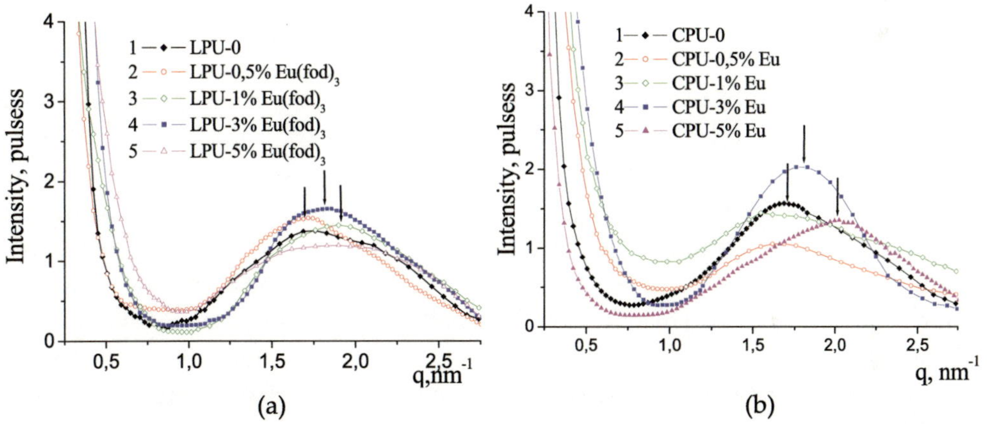

Figure 2. The SAXS intensity profiles of CPU (a) and LPU (b): metal-free (1), modified with 0,5% (2), 1% (3), 3% (4) and 5% (5) Eu(fod)$_3$.

Analysis according (Porod, 1982) of heterogeneity range (l_p) and average diameter (l_1, l_2) of different scattering elements in CPU-0, CPU-Cr and CPU-Co indicate existence of two types of nanosize heterogeneities in the bulk of PU. The first one (with $l_1 < D$) is inherent to segmented PU. The second one (with $l_2 > D$) is generated in the presence of transition metal chelate compound. We can define the latter structures as "metal chelate compound – polyurethane" complexes with polymer chains as macro ligands (Kozak et al., 2006; Nizelskii & Kozak, 2006) (Scheme 1).

Thus, the immobilization *in situ* of metal chelate compounds in polyurethane is accompanied with enrichment of polymer matrix with the nanosize heteroligand macro complexes of metal formed simultaneously with organic nanosize structures typical for metal-free polymer.

2.2. Dynamic heterogeneity of PU according to EPR data

The structural heterogeneity of PU influences the local segmental mobility of macro chains, resulting in "dynamic heterogeneity" of the systems. The analysis of mobility of SP introduced into the polymer gives information concerning such heterogeneity.

Calculated values of τ are listed in the table 2. They characterize the hindered rotation of SP in PUs of different topology. The greater value of τ is, the harder rotation of the probe occurs in polymer matrix.

System	$\tau \cdot 10^{-10}$, c	System	$\tau \cdot 10^{-10}$, c
CPU-0	45	LPU-0	48
CPU-1%Cu(eacac)$_2$	43	LPU-1%Cu(eacac)$_2$	69
CPU-1%Ni(acac)$_2$	42	CPU-1%CuCd	45
CPU-1%Cr(acac)$_3$	50	CPU-1%CuZn	32
CPU-1%Co(acac)$_3$	49	CPU-1%CuNiCo	51

Table 3. The correlation time of TEMPO in CPUs and LPUs, modified with 1% of metal chelate compounds.

As it can be seen from the table 2, in CPU modified with 1%wt. Co(3+) and Cr(3+) chelate compounds the values of τ increase indicating reduction of SP mobility as compared with metal-free CPU. In the contrary, for CPUs modified with 1%wt. Cu(2+), Ni(2+) values of τ decrease as compared with metal-free CPU. This means that Co(3+) and Cr(3+) containing CPUs have more dense macro chain packing as compared with metal free CPU. Where as Cu(2+) and Ni(2+) containing CPUs possess looser macro chain packing. Similarly to (Lipatov et al., 2000), the effect we can relate to difference in metal chelate compounds electron configuration and symmetry. In addition, the influence of metal chelate compound on PU dynamic depends also on the polymer topology. For example, it can be seen the opposite influence of Cu(2+) chelate compounds on the macro chain mobility in LPU and CPU (Table2).

The analysis of SP EPR-spectrum shape and hyperfine splitting (HFS) gives additional information concerned probed medium. In PU matrices that contain metal chelate compounds the EPR spectra of SP have asymmetric shape (Figure 3). In all of the spectra occur essential increasing of central component and broadening of all components as compared with TEMPO spectrum in homogeneous glycerol matrix. In many spectra there is noticeable splitting of low-field and/or high-field components of SP spectrum.

The peculiarities observed are most likely the result of signal superposition of "fast" and "slow" probes located in polymer regions with different mobility. In conformity with above supposition the temperature increasing brings on enhancement of the SP EPR spectra isotropy (Figure 3). Initially asymmetric ESR spectrum becomes more isotropic while heating the sample. The spectrum components narrow and the intensity of central component diminishes.

As a result of heating the equalizing of polymer segments mobility and „unfreezing" of „slow" SP rotation diffusion occurs. The correlation time decreases with the rise of temperature due to increasing of molecular mobility and "softening" of PU matrix. Figure 4 represents the relationship $\tau(T)$ for CPU.

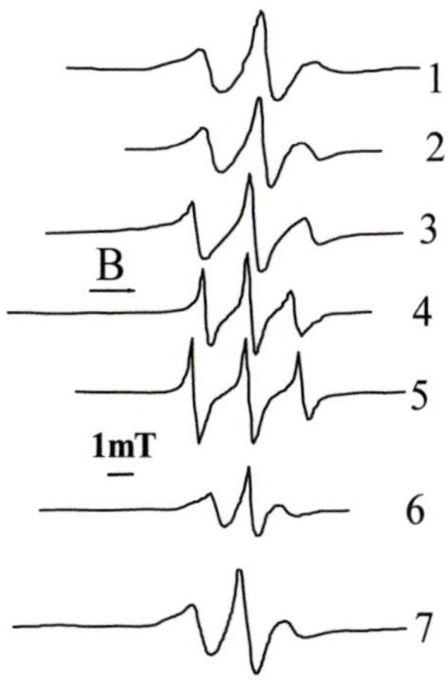

Figure 3. The spectra of the TEMPO introduced in CPU+%Er(acac)3 at the various temperatures: 18 °C (1); 26 °C (2); 44 °C (3) ; 90 °C (4); 114 °C (5); 21 °C (30 min after thermal heating) (6); 18 °C (2 days after thermal heating) (7).

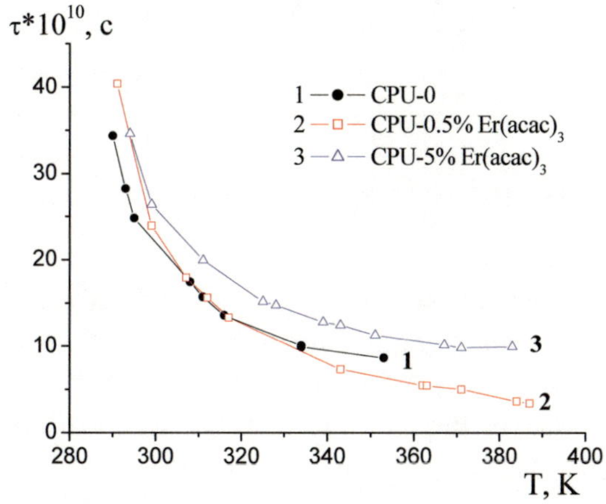

Figure 4. The thermal dependence of correlation time of the TEMPO in CPU-0 (1), CPU, modified with 0,5%wt. (2) and 5%wt. (3) of Er(acac)3.

2.3. Thermodynamic heterogeneity of PU according to DSC data

The PU's thermodynamic heterogeneity is closely associated with above discussed types of heterogeneities. The influence of the metal chelate compounds on the thermodynamic heterogeneity and thermo-physic properties of PUs was analyzed by DSC. Figure 5 illustrates the temperature dependences of specific heat capacity of CPUs modified with 0,5; 1; 3; 5%wt. of $Cu(acac)_2$. The thermo-physic characteristics of copper-containing CPUs are given in Table 3.

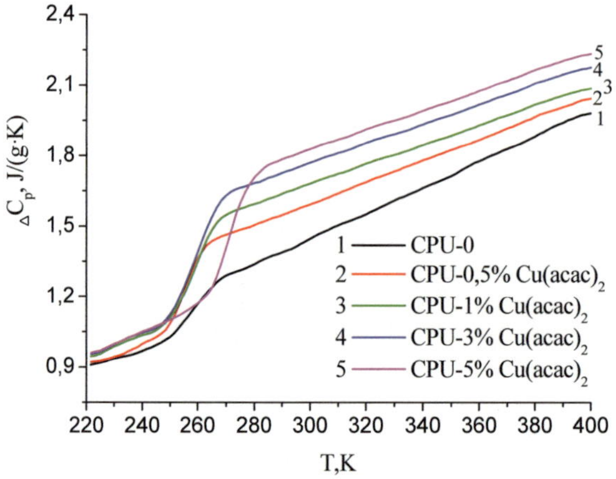

Figure 5. Temperature dependence of specific heat capacity for copper-containing CPU.

System	T_g, K	ΔT, K	ΔC_p , J/ (g·K)	$\frac{\Delta C_{p(CPU-Cu)}}{\Delta C_{p(CPU-0)}}$
CPU-0	258	18	0,25	1
CPU-0,5%$Cu(acac)_2$	256	16	0,38	1,52
CPU-1%$Cu(acac)_2$	258	18	0,43	1,72
CPU-3%$Cu(acac)_2$	260	21	0,50	2,00
CPU-5%$Cu(acac)_2$	271	20	0,55	2,20

Table 4. The thermo-physical properties of copper-containing CPU.

It is evident from fig. 5 and table 3 that for the CPUs the specific heat capacity (ΔCp) grows with increasing of Cu (2+) chelate content from 0,5 to 5% wt. comparing with CPU-0. In addition, the high temperature shifting of glass temperature (T_g) and the broadening of the temperature interval of glassing (ΔT) for CPU- 3%Cu and CPU-5%Cu are observed. The similar effect was discussed in (Lipatov et al., 1999) for CPUs, modified with 1%wt of various transition metals chelate compounds. That effect we can relate to formation of coordination bonds between functional groups of CPU and copper (2+) chelate compound.

Thus, growth of T_g and ΔC_p values with increasing of $Cu(acac)_2$ amount corresponds to rise of polymer segments with decreased mobility due to complexing.

The ratio of $\Delta C_{p(CPU-Cu)}$ to $\Delta C_{p(CPU-0)}$ allows estimate the degree of PU's thermodynamic heterogeneity (Bershtein. & Yegorov, 1990) and analyze the influence of metal chelate modifier content on this type of heterogeneity (table 3). As it can be seen the thermodynamic heterogeneity degree of CPU correlates with modifier amount in the system. This result agrees with X-ray data (section 2.1).

2.4. The formation of ordered micro regions in metal containing PUs

The segregation of metal containing micro crystals in CPU-5% Co and CPU-5%Cr was revealed in (Kozak et al., 2006):. Such unexpected segregation seemed unlikely due to homogeneous dispersion of metal chelate compound solution in reaction mixture (see 2.1) and coordination immobilization of metal chelate compounds in PU matrix. Nevertheless, the further X-ray study of CPU-5%Cu, LPU-5%Cu (fig. 6) and microscopy data (see 2.5) confirm partial segregation of metal-containing sites in PU matrices. This effect can be explained by different complex ability of segmented PU soft and hard components towards metal chelate compound as well as by higher mobility of PU's soft component.

The Scherer's equation (Stompel & Kercha, 2008) for the average diameter (L) of crystallite in amorphous media allows estimate dimensions of the particles in metal-containing PU.

$$L = k\lambda \,/\, (\beta cos\theta_m) \tag{9}$$

Here X-ray wavelength λ = 1,54 Å, k is the shape factor assigned to 0,9, L is the average diameter of the crystals in angstroms, θ_m is the Bragg's angle in degrees, and β is the half-height of diffraction angle in radians. The value of L is equal to 3 nm for CPU-1%Co, it is equal to 4 nm for CPU-1%Cr and it is equal to 10 nm in LPU-1%Eu. The evaluated dimensions of the aggregates in copper containing PUs are ranged from 8 to 12 nm.

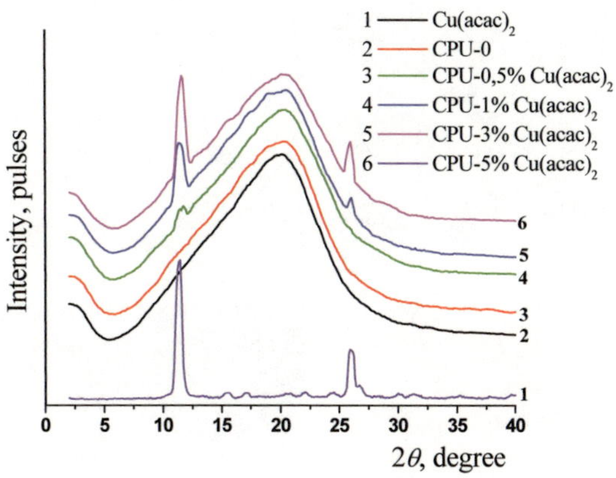

Figure 6. The WAXS diffractograms of the CPU-%Cu(acac)$_2$ films.

Segregation of the micro crystals detected via WAXS study has been also fixed by optical light transmission microscopy and by the scanning electron microscopy (SEM) (Figure 7). The micro crystals detected by optical microscopy are coloured like metal chelate compounds used as PU modifier. Such colouring indicates enrichment of the crystals with corresponding metal ions. The crystalline regions can be formed by the modifier itself and/or by complexes of modifier with PU chains as macro ligand. The last conclusion agrees with the X-ray data that register several discrete peaks in Cu(2+), Cr(3+) and Co(3+) containing PUs (Figure 6).

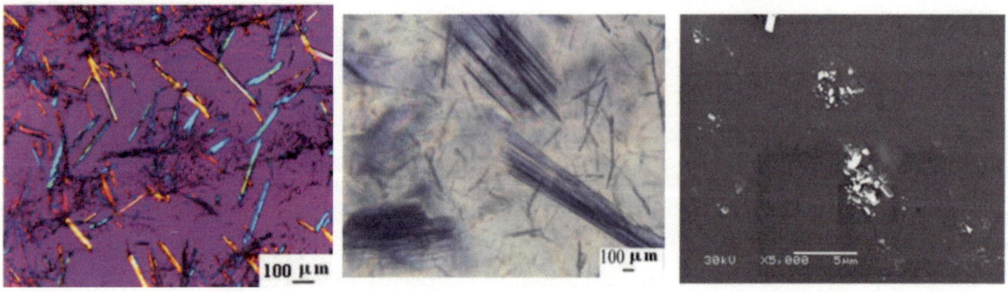

(a) (b) (c)

Figure 7. The optical microscopy (a, b) and SEM microscopy micro images of the LPU-0,5%Cu(acac)$_2$ (in polarized light) (a), LPU-5%Cu(acac)$_2$ (b) and CPU-5%Cr(acac)$_3$.

Optical microscopy allows obtain information concerning two surfaces of one PU film. One of them formed on the boundary "polymer-support" (the PU's surface formed on the Teflon support) and another formed on the boundary "polymer-air" (the PU's surface formed on the air).

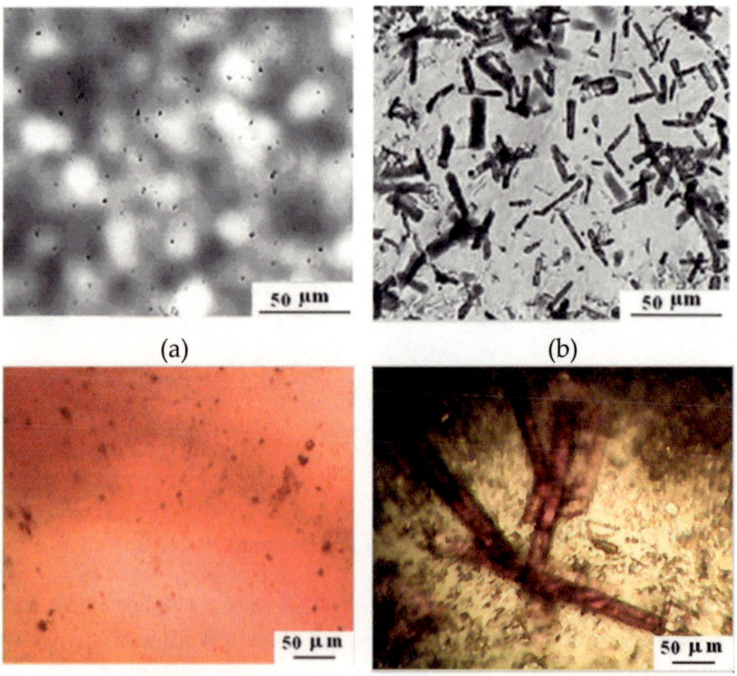

(c) (d)

Figure 8. Micro images of LPU-1%Eu(fod)$_3$ (a,b) and CPU-1%Cr(acac)$_3$ (c, d) surfaces formed at the "polymer-air" boundary (a, c) and the "polymer- support" boundary (b, d).

Figure 8 illustrates the typical differences in surfaces of PU films. As it can be seen, at surface formed at the boundary "polymer-support" (fig. 8, *a*) the size and quantity of crystals are larger. Where as, at surface formed at the "polymer-air" boundary (fig. 8, *b*) the size and quantity of crystals are significantly smaller. For example, the mean size of crystals in LPU-1%Eu changes from the one surface to another from 20 μm to 0,5 μm.

Detailed analysis of PUs surface properties depending on the boundary nature can give additional information.

3. Influence of metal chelate modifiers on surface properties of polyurethanes

The presence of metal chelate compounds in reaction mixture can influence the surface tension of the formed polyurethane. In (Lipatov, 1997) the surface properties were studied of PU with metal ions introduced through in four different ways. There are filling, metal ion cross-linking, metal ion chain-extending and diffusion of metal chelate compound from its solution to polymer being formed earlier. It has been shown that the surface properties of metal containing PU depend on metal quantity much less than on the way of metal chelate compound introduction in polymer. For example, the γ_{sg} of PU filled with $Cr(acac)_3$ (0.18% wt.) changes up to 8 mN/m. On the contrary the γ_{sg} of Pb (15% wt) cross-linked PU changes up to 0.3 mN/m as compared with metal free PU.

Obviously, the PU's surface structure depends on the boundary "polymer-support" or "polymer-air". Data of ESCA and IR-spectroscopy by (Lipatova et al., 1987; Lipatova & Alexeeva, 1988) point on possibility of the chemical unequivalence of the polymer surfaces formed at the different boundaries. In addition in (Kozak et al., 2010) it was observed substantial difference in luminescence intensity at different surfaces of the PU films modified with europium (3+) chelate compounds. Therefore, the surface properties of europium containing LPU and CPU were compared for surfaces formed at the "polymer-air" and "polymer-support" boundary using measurement of contact wetting angle. The data obtained are listed in the table 4.

The values of surface tension of metal containing PU obtained using Wilgelmy method(with water as wetting liquid) (Lipatov et al., 1997) are consistent with values of the surface tension calculated using measurement of contact wetting angle (Table 5) of standard liquid.

The wetting angles at the „polymer-air" boundary for all of CPU and LPU are from 5.5 to 15.5 degrees less than the wetting angles at the „polymer-support" boundary (table 4). The difference between relative values of surface tension (γ_1-γ_2) takes values from 2.18 to 5.59 mN/m. As it is known, the higher compound polarity is the greater surface energy and surface tension it possesses. Obtained results allow conclude that PU surface formed at the „polymer-air" boundary is enriched with more polar groups (e.g. urethane) and PU surface formed at the „polymer-support" boundary is enriched with less polar groups (e.g. glycol segments).

Concentration of PU less polar groups that form the weak complexes with metal chelate compound at the „polymer-support" boundary can facilitate the partial segregation of metal containing centres at this boundary. That conclusion is consistent with microscopic data and photoluminescence measurements.

System	θ, degree		γ_{EG-PU}, mN/m		$\Delta\theta = \theta_2 - \theta_1$, degree	$\Delta\gamma = \gamma_1 - \gamma_2$, mN/m
	θ_1 (the „polymer-air" boundary)	θ_2 (the „polymer-support" boundary)	γ_1 (the „polymer-air" boundary)	γ_2 (the „polymer-support" boundary)		
CPU-0	55	65	38,06	34,50	10	3,56
CPU-1% Eu	53	64	38,74	34,67	11	4,07
CPU-3% Eu	51	66	39,39	33,95	15	5,44
CPU-5% Eu	56	64	37,70	34,77	8	2,93
LPU-0	58	70	37,00	32,45	12	4,55
LPU-1% Eu	61	67,5	35,95	33,43	6,5	2,52
LPU-3% Eu	67,5	73	33,43	31,25	5,5	2,18
LPU-5% Eu	59	74,5	36,63*	30,64	15,5	5,99

θ_1, θ_2 – the wetting angles at the „polymer-air" and the „polymer-support" boundaries, respectively, degree;
γ_1, γ_2 – the surface at the „polymer-air" and the „polymer-support" boundaries, respectively, mN/m;
* the unbalanced wetting angles

Table 5. The contact wetting angle (θ) and surface tension (γ) of PU films. The standard liquid is ethylene glycol (EG) $\gamma_{EG\text{-}air}$ = 48,36 mN/m.

Varying of the metal containing modifier amount (from 0,5 to 5% wt.) in CPU practically does not affect surface tension. In the contrary, change of metal chelate compound content in LPU from 0,5 to 3%wt. lead to decreasing of both γ_1 and γ_2.

The difference of the tendency in changing of surface tension in LPU and CPU clearly depend on polymer topology. Different PU topology results in different segmental mobility of the polymer, that agrees with DRS data. This effect described detailed in Section 5 and Section 2.4. At that time we can't formulate the certain reason for non monotonous influence of the modifier's amount on the surface tension.

4. The influence of polymer topology and modifier content on the luminescent properties of segmented polyurethanes

According to (Lobko et al., 2010) the PU matrix can intensify the photoluminescence of europium chelate compounds introduced into polymer *in situ*. Taking into account that immobilization *in situ* of metal chelate compounds in polymer matrix can influence both structure and properties of the hybrid system (Nizelskii & Kozak, 2006; Nizelskii et al., 2005) the investigation of rare-earth metal compounds in polymeric environment is a way for creation of new optically active materials.

LPUs and CPUs modified with Eu(3+) chelates when exposed in 365 nm UV-light demonstrate the intensive photoluminescence in red region. Figure 9 represents the luminescent spectra of LPU and CPU, modified with various amount of $Eu(fod)_3$.

The luminescent spectra of europium containing PU are diffuse, while luminescence spectrum of $Eu(fod)_3$ is enough well-resolved. According to (Poluectov et al., 1989) the luminescence spectra of europium β-diketonate solutions contain bands corresponding to the $^5D_0$-^7F_i -transitions (where i = 0,1,2,3,4). The spectra of Eu β-diketonate in PU matrices demonstrate the intensive wide band of photoluminescence in the region of λ=610-635 nm ($^5D_0$-$^7F_2$-transition), narrow band λ=660 nm ($^5D_0$-$^7F_3$ –transition) and bands of $^5D_0$-$^7F_{0,\ 1,\ 4}$-transitions (580, 600,700 nm, accordingly) of low-intensity. It is possible to explain the diffuse spectrum of luminescence of europium containing PU in the region of λ =610-635 nm by distortion of the Eu (3+) chelate geometry in PU due to complex "polymer-metal chelate compound" formation and due macroligand steric hindrances.

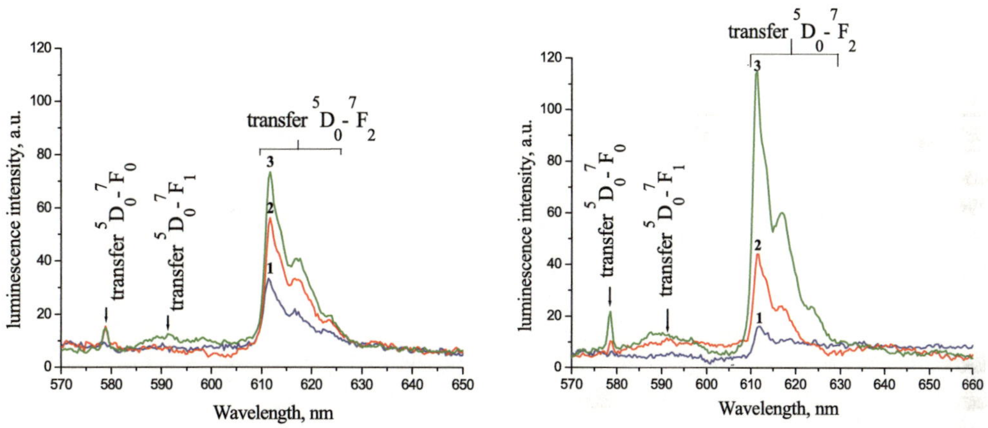

(a) (b)

Figure 9. The spectra of luminescence of LPU *(a)* and CPU *(b)*, modified with europium chelate (λ_{UV} = 365 nm): (1) 0.5%; (2) 1%; (3) 5%.

The intensity of PU-Eu luminescence depends both on the europium chelate content and polymer topology. The luminescence intensity increases with increasing of europium chelate compound content. The luminescence intensities of $^5D_0 \rightarrow {}^7F_2$ transition (λ=612nm) for LPU with 05%, 1% and 5%wt. of $Eu(fod)_3$, correspond as 1:1,8:2,4. The relationship of luminescence intensity *vs.* modifier percentage in CPUs is linear (1:3,3:9,2). The CPU-Eu with low modifier content has the lower luminescent intensity as compared with LPU-Eu. Where as CPU-5%Eu luminescence intensity is 1,5 higher, than LPU-5% Eu luminescence intensity. Taking into account data of Sections 2, 3, 6 we can suppose that due to difference in PU topology this effect is associated with higher concentration of polymer photo transmitting sites near the modifier in CPU as compared with LPU.

The tetra coordinated Eu (3+) chelate compounds with different additional ligands in an external coordination sphere were used to analyse the influence of additional coordination of europium chelate compounds on the intensity of their luminescence. The fig. 10, a illustrates the luminescent spectra of isolated Eu (3+) chelate compounds. The fig. 10, b represents the spectra of luminescence of CPU films, modified with 1%wt. of Eu (3+) compounds.

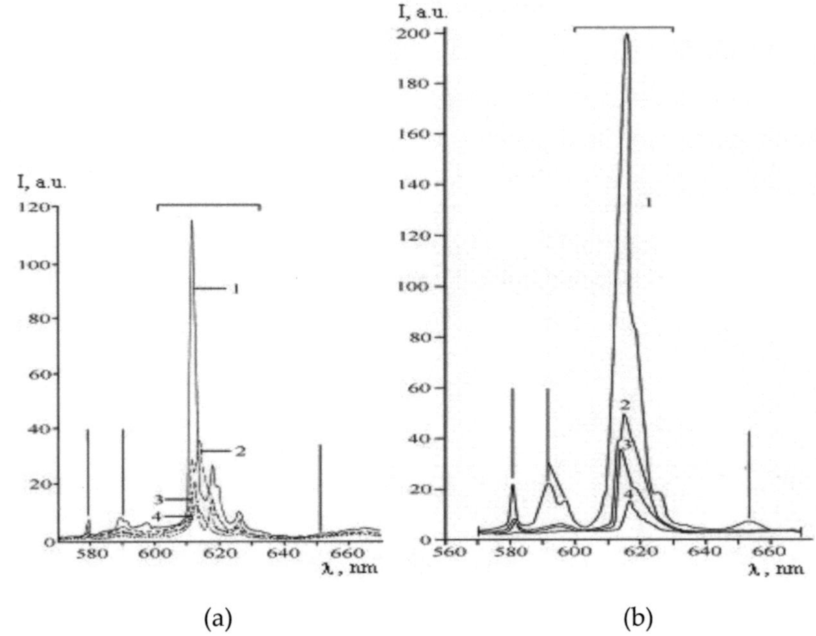

Figure 10. The luminescence spectra ($\lambda_{ex.}$ = 365 nm) of the europium (3+) chelate compounds (a) and of CPU films with 1%wt. of these chelate compounds (b): Eu(TTA)$_3$phen (1); Eu(TTA)$_3$TPPO (2); Eu(TTA)$_3$ (3); Eu(fod)$_3$ (4)

As it can be seen the luminescence of Eu (3+) chelate compounds introduced into PU matrix (only 1%wt.) is more intensive than luminescence of isolated metal chelate compounds (100% wt). In addition, the intensity of luminescence of tetra coordinated Eu(3+) chelate compounds (Eu(TTA)$_3$phen and Eu(TTA)$_3$TPPO) both isolated and introduced into CPU matrix, considerably exceeds such intensity for 3-coordinated Eu(3+) chelate compound Eu(TTA)$_3$ that does not contain additional ligands The intensity of photoluminescence of Eu(fod)$_3$ also is considerably lower.

Estimation of Eu (3+) environment symmetry in various complexes via the coefficient of asymmetry (η) defined as ratio of intensity of $^5D_0 \rightarrow {}^7F_2$ –transition to intensity of $^5D_0 \rightarrow {}^7F_1$ transition. (Haopeng et al., 2008) shows that the greatest coefficient of asymmetry (η = 9) has CPU-1%Eu(TTA)$_3$phen characterized by the greatest intensity of luminescence. Consequently, the presence of additional ligand in the external coordination sphere of Eu (3+) favours increasing of luminescence intensity. Then increasing of Eu-chelate compounds luminescence intensity in PU can be explained in particular by additional coordination of lanthanide ion with the functional groups of PU and/or by formation of donor-acceptor complexes between aromatic fragments of PU and quasi-aromatic chelate rings of chelate compounds of rare-earth metals.

5. Dielectric relaxation and conductivity

The dielectric properties of PU were studied by broad band DRS measurements in wide range of temperature (-40 to 120 °C). The data are analyzed within the various formalisms. The direct current conductivity was both measured using two-electrode method and was estimated using DRS complex electric resistance $\sigma_{dc} = d/(AR_{dc})$ and Z″(Z′) isotherms (Cole-Cole diagram). Figure 11-13 illustrate obtained dielectric spectra. Calculated conductivity values are listed in Table 5.

According to two-electrode method the direct current conductivity of PU can drastically change in the presence of some metal chelate compounds. At the room temperature σ_{dc} for the CPU-5%Eu increases by one order as compared with CPU-0. In the presence of polyheteronuclear metal chelate compounds σ_d enlarges from 2 to 3 orders (fig. 13, table 5). DRS analysis of complex dielectric permittivity as well as complex admittance σ*, complex electrical modulus M* and impedances Z', Z" allows reveal the nature of the observed conductivity.

System	a) σ_{dc}, Sm/cm 20°C	b) σ_{dc}, Sm/cm 40°C	System	a) σ_{dc}, Sm/cm 20°C	b) σ_{dc}, Sm/cm 20°C
CPU-0	1,78 10^{-12}	$1{,}3\cdot10^{-11}$	LPU-0	$4{,}65\cdot10^{-12}$	$4.6\cdot10^{-12}$
CPU-Cu	2,86 10^{-11}	$2\cdot10^{-9}$	LPU-Cu	$4{,}25\cdot10^{-11}$	$3.8\cdot10^{-11}$
CPU-Cu_2Zn	2,47 10^{-9}	$0{,}7\cdot10^{-8}$	LPU-Cu_2Zn	$1{,}51\cdot10^{-9}$	$1.2\cdot10^{-10}$
*CPU^{2000}-0	-	$1*10^{-10}$	*CPU^{2000}-Cu_2Zn	-	$1*10^{-7}$
*CPU^{2000}-Cu	-	$1*10^{-9}$			

* The PU films synthesized with PPG-2000

a) σ_{dc} measured using two-electrode method and b) σ_{dc} obtained using DRS data

Table 6. PUs conductivity at a direct current

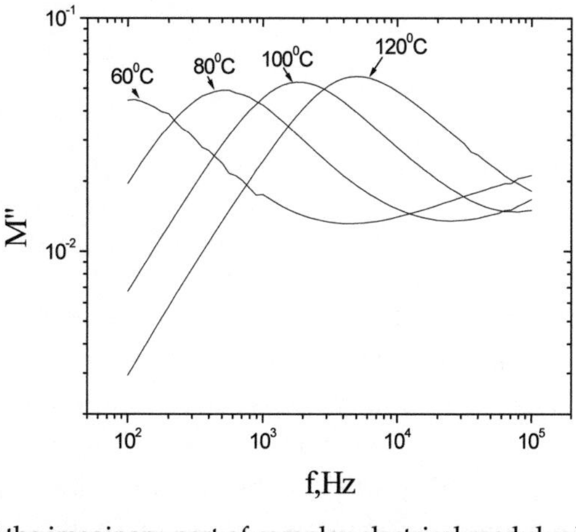

Figure 11. Log-log plots of the imaginary part of complex electrical modulus M″ *vs.* frequency for CPU–5%Eu at several temperatures

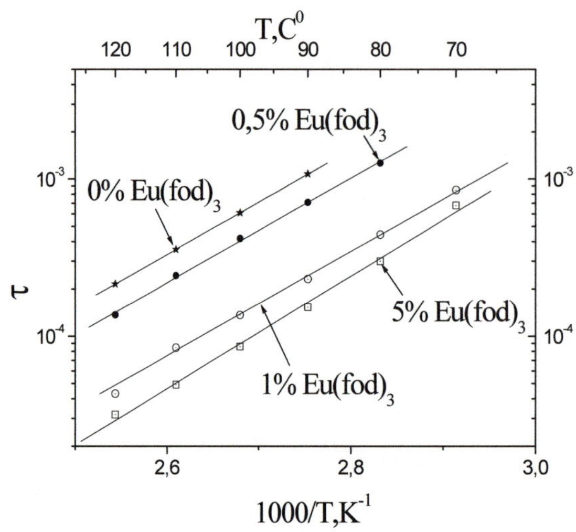

Figure 12. The thermal dependence of the relaxation time (τmax) for the CPU with various content of Eu (3+) chelate compound.

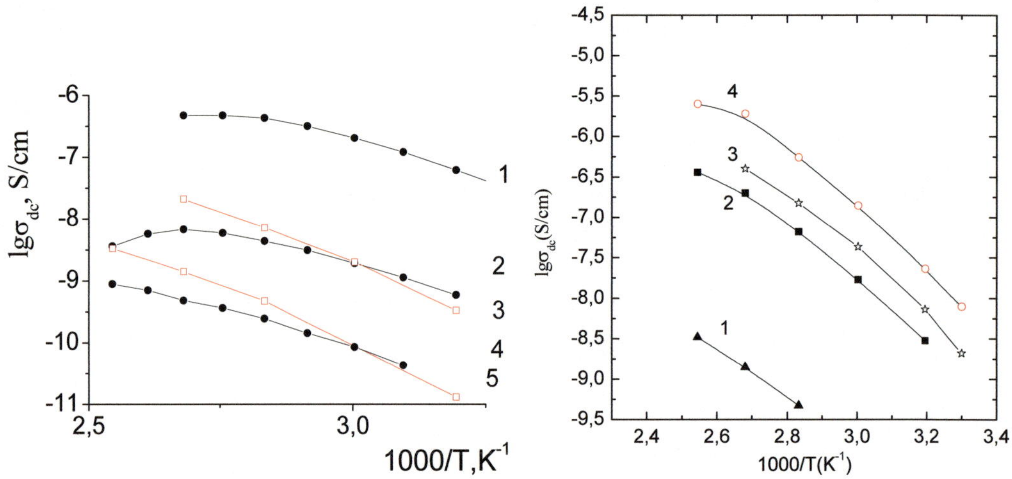

Figure 13. The $\log\sigma_{dc}$ vs. 1/T for CPU: (a) CPU with various length of flexible component: CPU (PPG-2000) – 1%Cu_2Zn (1); CPU (PPG-1000) – 1% $Cu(eacac)_2$ (2); CPU (PPG-2000) – 1% $Cu(eacac)_2$ (3); CPU (PPG-2000) – 0 (4); CPU (PPG-1000) – 0 (5) and (b) CPU-0 (1) and CPU-Cu_2Zn formed in the presence of various solvents: 1, 4-dioxane (2); dichloromethane (3) and DMFA (4).

The curves on the fig. 11 have well defined maxima in temperature region of 60 to 120°C. According to (Pathmanatham & Johari, 1990; Kyritsis & Pissis, 1997) these maxima correspond to conductivity relaxation. Increasing of temperature is accompanied with shift of conductivity relaxation maxima to higher frequencies (fig. 11). The fact concerned to increasing of segmental mobility in PU. The metal chelate compounds introduction and increasing of their content in the system result in increasing of PU segmental mobility.

Experimental dependences of relaxation time ($\tau_{\max} = \dfrac{1}{2\pi f_{\max}}$) in log scale *vs.* 1/T (fig. 12) for metal-containing PU are linear indicating the Arrhenius-type of temperature dependence of τ_{max}.

Decreasing of τ_{max} value with the increasing of Eu(3+) content in PU confirm increasing of macro chains mobility in metal-containing CPU (Kozak et al., 2006). The activation energy of conductivity relaxation for CPU-0, CPU-0,5%Eu, CPU-1%, CPU-5% are approximately similar. Experimental dependences of logσ_{dc} vs. 1/T are non-Arrenius both for PU-0 and metal containing PUs. It fit the theoretical curves of Vogel–Tamman-Fulcher (VTF) equation $\sigma_{dc}=\sigma_0\exp(-B/(T-T_0)$ (fig. 13) indicating influence of the PU free volume on charge transport. The results obtained give evidence of significant influence of structural organization in the modified PU on its conductivity level.

As it can be seen PU's direct current (σ_{dc}) conductivity grows with increasing of temperature and that is characteristic to ionic conductivity. The metal ion participation as current carrier is unlikely because to small amounts of metal ion in the modified PUs (~ 0.025-0.25% wt.). That fact and coordination immobilization of the modifiers in polymer makes unlikely increasing of the conductivity due to the metal chelate compound conductive properties. On the other hand ionic mechanism of conductivity and adequate amount of protons presented in PU matrix as well as observed increasing of polymer chain mobility in modified PU allows us to suppose proton participation in the process of charge transport.

Comparison of direct current conductivity of CPU based on PPG-2000 with conductivity of CPU based on PPG-1000 shows increasing of conductivity at the direct current of such system up to 10^{-7}Sm/sm at the 40°C due "softening" of PU. Nevertheless it can be seen that conductivity level of maximum soft LPU is at least one order lower then conductivity of CPU.

6. "Metal chelate compound - polymer" complexing and formation of the additional network of coordination bonds in metal containing PU

Mutual influence of metal chelate compound and polymer matrix due to complex formation is a decisive reason of observed changes of structural, dynamic, relaxation etc. characteristics of the metal contained PU. The complexing of metal chelate compounds with PU matrix was analysed using electron spectroscopy and EPR.

6.1. The complexing of the metal chelate compound with PU matrix according to the electron spectroscopy

The electron spectroscopy allows analyse both character of complexing of metal chelate compound with the polymer matrix and state of metal chelate compound in PU. The electron spectra of transition and rare-earth metal chelate compound in PU indicate presence of the band of *d-d*-transitions for the transition metal chelate compound introduced into PU (fig.14) and band of π-π-transitions for the rare-earth metal chelate compounds introduced into PU. That points on saving of chelate structure of the complexes in polymer matrix. While the change the band intensity, its broadening and shift to a long-wave region testifies their participation in complexing with PU.

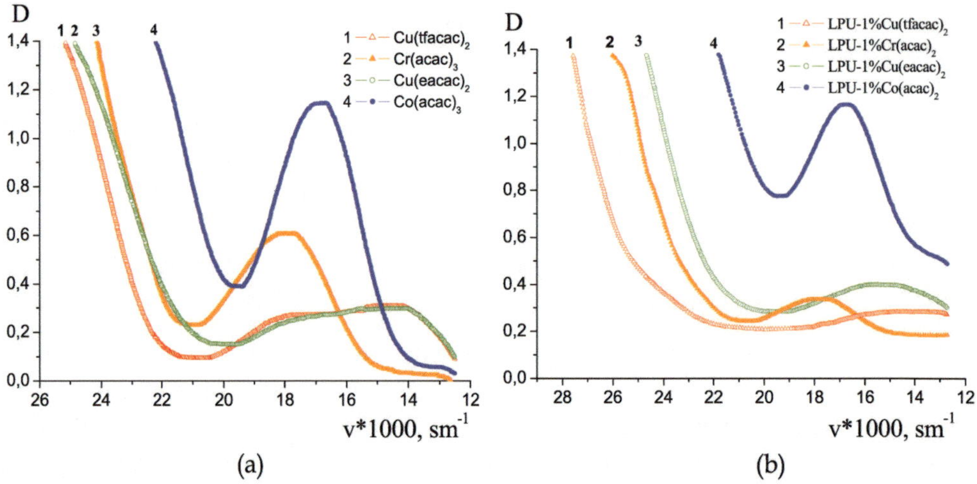

Figure 14. The electron spectra of transition metal chelate compounds in dichloromethane (a) and LPU (b): Cu(tfacac)$_2$ (1), 1%Cr(acac)$_3$ (2), Cu(eacac)$_2$ (3); Co(acac)$_3$ (4); LPU Cu(tfacac)$_2$ (ε = 28 l/mol×sm) (1); LPU-1%Cr(acac)$_3$ (ε = 33 l/mol×sm) (2); LPU-1%Cu(eacac)$_2$ (ε = 40 л/ l/mol×sm) (3); LPU-1%Co(acac)$_3$(ε = 117 l/mol×sm) (4)

Figure 14 represents comparison of electron spectra in the visible region of the LPU films with 1%wt. of transition metal (copper, chrome, cobalt) chelate compounds (fig. 14, a) and the spectra of this metal chelate compounds dissolved (c= 10^{-2} M) in dichloromethane (CH_2Cl_2) (fig. 14, b).

In addition to described above changes in electron spectra the influence of fluorine on complex ability of metal chelate compound in PU is evident due to, the rise of absorption level (ε = 40 l/mol×cm) and hypsohromic shift of maxima of band of d-d-transitions for LPU - 1% Cu(eacac)$_2$ as compared with LPU-1% Cu(tfacac)$_2$ that have fluorine in ligand (ε = 28 l/mol×cm). Fig. 15 illustrates the detailed analysis of electron transitions of copper ion in β-diketonates (4 transitions for D_{2h} symmetry). Calculated maxima positions of Gaussian components of adsorption band corresponding to electron d-d-transitions of copper ion for Cu(tfacac)$_2$ and Cu(eacac)$_2$ in solution and in PU are listed in the table 6.

System	υ, cm^{-1}			
	$d_{чн} \rightarrow d_{z^2}$	$d_{чн} \rightarrow d_{x^2-y^2}$	$d_{чн} \rightarrow d_{xz}$	$d_{чн} \rightarrow d_{yz}$
Cu(tfacac)$_2$ in CH_2Cl_2	13658	15386	17930	19579
LPU -1% Cu(tfacac)$_2$	13435	15179	16322	17628
Cu(eacac)$_2$ in CH_2Cl_2	13715	15376	16686	18261
LPU-1% Cu(eacac)$_2$	12771	14497	16059	17738

Table 7. The allocation of Gaussian components of copper chelate compounds adsorption band.

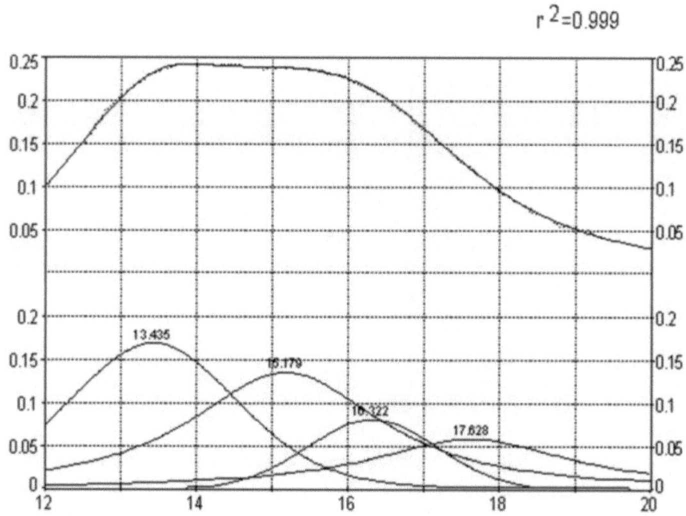

Figure 15. The adsorption band of LPU+1%Cu(tfacac)2 with Gaussian components allocation.

The maxima of d_{xz} and d_{yz} transitions of Cu(2+) ion in copper chelate compounds, immobilized in PU matrix are shifted to the long-wave region, as compared to corresponding transitions of copper chelate compounds in solution (table 6). Visible broadening of the d_{z^2} component of absorption band of Cu (2+) chelate compounds in PU indicates the prevalence of axial coordination of macro ligand in "PU-metal chelate compound" complexes.

6.2. Complex formation in the "polymer-modifier" system according to EPR data

The EPR data confirm the complexing between the metal chelate compound and PU functional groups. The state of paramagnetic Cu(2+) containing chelate compounds in PU can be directly analyzed using EPR due to sensitivity of spin-electron parameters A_{II} and g_{II} (see table 7) of the tetragonal copper chelate compounds to symmetry and chemical nature of the copper nearest environment.

Decreasing of A_{II} and increasing g_{II} of $Cu(acac)_2$, $Cu(tfacac)_2$, $Cu(eacac)_2$ and $(Cu_2Zn_2(NH_3)_2Br_2(HDea)_4)Br_2$ immobilized in PU as compared with undisturbed compounds indicate the participation of the modifiers in the complexing with PU electron donor groups.

Figure 16 illustrates the representative EPR spectra of some copper containing modifiers both isolated and immobilized in PU with different topology (linear and cross-linked). The EPR spectra of polyheteronuclear powdered crystalline samples have anisotropic shape with weakly resolved HFS due to broadening of the spectrum components and possible tetrahedral distortion of the copper ion surrounding in the polyatomic complex. Immobilization of such chelate compounds in a PU network resulted in decreasing of the EPR signal intensity.

System	g_{II}	$A_{II}\times10^{-4}$ cm^{-1}	$g_{\perp}$	$A_{\perp}\times10^{-4}$ cm^{-1}	g_0	a_0*10^{-4} cm^{-1}
[1]$Cu(tfacac)_2$ *	2,271	187	2,052	23	-	-
CPU-1% $Cu(tfacac)_2$	2,275 2,290	172 151	2,059	21	2,131 2,136	71 64
LPU-1% $Cu(tfacac)_2$	2,283 2,302	162 141	2,049	19	2,127 2,133	67 60
[1]$Cu(eacac)_2$ *	2,276	187	2,055	22	2,128	68
CPU-1% $Cu(eacac)_2$	2,249	150	-	-	-	-
LPU-1% $Cu(eacac)_2$	2,298	173	-	-	-	-
[1]$Cu(acac)_2$ *	2,250	189	2,052	24	2,118	75
CPU-1% $Cu(acac)_2$	2,269	182	-	-	-	-
LPU-1% $Cu(acac)_2$	2,254	188	2,052	29	2,119	82
[2]$(Cu_2Zn_2(NH_3)_2Br_2(HDea)_4)Br_2$	2,370	122	-	-	-	-
CPU-1% CuZn	2,370 2,300	132 142	-	-	-	-
LPU-1%CuZn	-	-	-	-	-	-

[1)]undisturbed Cu(2+) complex in glassy matrix chloroform/toluene (40/60) (at -196°C)
[2)] powder of polycrystalline sample
* (Lipatova & Nizelskii, 1972)

Table 8. Electron-spin parameters of isolated and polymer immobilized copper complexes.

The most reasonable explanation for this effect is distortion of the modifier's symmetry or geometry in PU-CuZn. The shape of EPR signal in PU network modified with $(Cu_2Zn_2(NH_3)_2Br_2(HDe)_4)Br_2$ indicates formation of complexes of various content and structure.

6.3. Low molecular probes dynamic and formation of additional network of the coordination bonds in metal containing polyurethanes

Using QENS and EPR with paramagnetic probes of various natures it was shown that complex formation of metal containing modifier with macro chains results in appearance of additional spatial obstacles for probe diffusion as compared with metal free network. The dynamic of low molecular probes and complex formation in the nanostructured polyurethane network containing Co (3+) chelate compounds immobilized *in situ* were analyzed.

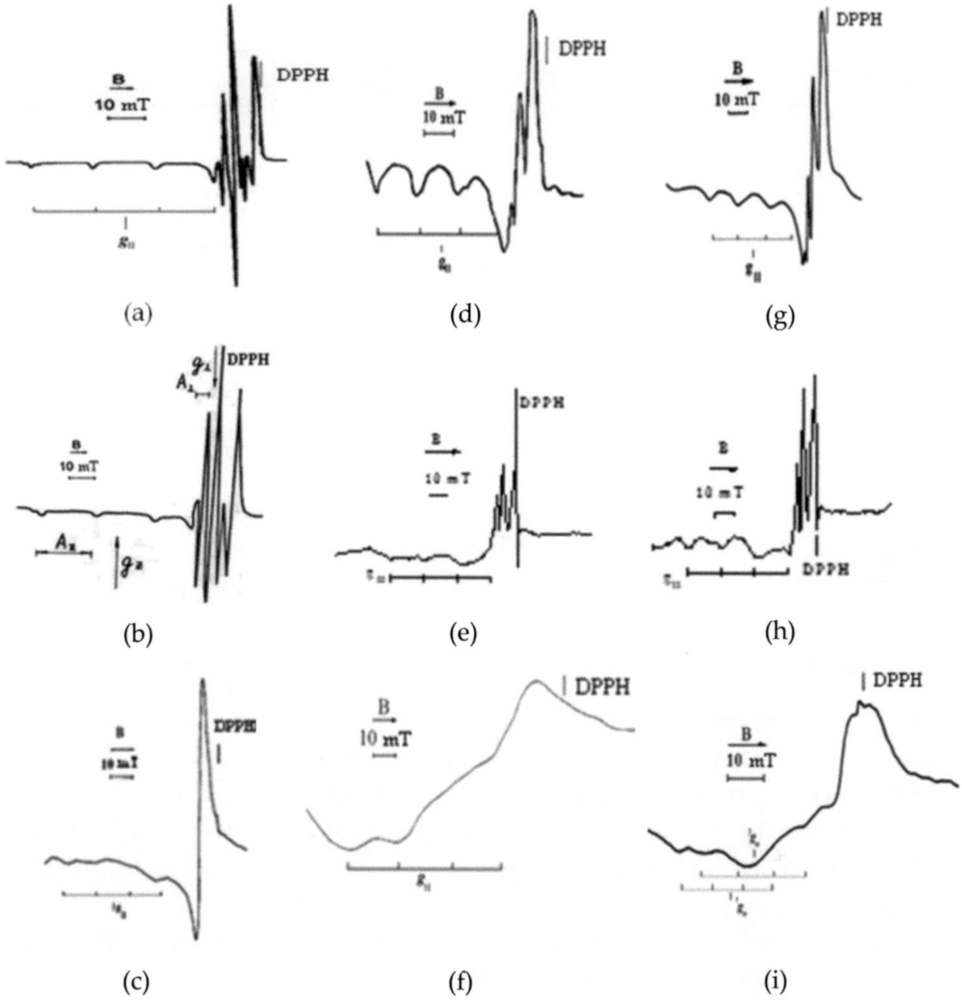

Figure 16. EPR – spectra of matrix isolated $Cu(eacac)_2$ (a), $Cu(acac)_2$ (b), $(Cu_2Zn_2(NH_3)_2Br_2(HDea)_4)Br_2$ (c) in chloroform-toluene at -196°C; LPU with 1%wt. of $Cu(eacac)_2$ (d), $Cu(acac)_2$ (e), $(Cu_2Zn_2(NH_3)_2Br_2(HDea)_4)Br_2$ (f) and CPU with 1%wt. of $Cu(eacac)_2$ (g), $Cu(acac)_2$ (h), $(Cu_2Zn_2(NH_3)_2Br_2(HDea)_4)Br_2$ (i).

According to EPR data obtained using complex spin probe (Kozak et al., 2006) it was demonstrated that in cobalt containing CPU the complexes "polymer-metal chelate compound" of two types are formed. The analysis of rotational diffusion of nitroxyl spin probe TEMPO (see table 2) reveals the decreasing of PU segmental mobility due metal chelate compound introduction and/or it content increasing. The dynamic of solvent molecules diffusion in swelled CPU-0, CPU-5%Co films and in probe liquid was analysed to compare the ratio of one-particle and collective modes of the solvent molecules motion.

System	$D \cdot 10^{-6}$, cm^2/c	$D^F \cdot 10^{-6}$, cm^2/c	$D^L \cdot 10^{-6}$, cm^2/c	D^L/D, %
Probe liquid	3,52	3,18	0,35	8,0
The solution of $Co(acaca)_3$ 5%wt. in the probe liquid	3,33	3,04	0,29	8,7
CPU-0 (CH_2Cl_2)	2,71	2,40	0,31	11,5
CPU-Co5% (CH_2Cl_2)	1,85	1,32	0,53	28,6
CPU-Co5% (DMF)	1,25	0,89	0,36	28,8

where D^F – one-particle ("Frenkel ") diffusion coefficient ;
D^L – collective ("Lagrangian") diffusion coefficient.

Table 9. The diffusion parameters of the probe molecules in the swelled PU films

The sharp decreasing of both the general and one-particle component of diffusion coefficient for the metal containing PU as compared with the metal-free PU indicates the appearance of the spatial hindrances for the liquid molecule dynamics.

7. Conclusion

Immobilization *in situ* mono- and polyheteronuclear chelate compounds of transition and rare-earth metal in linear and cross-linked polyurethanes results in nanoscale structuring of forming polymer and is accompanied with polymer matrix enrichment by the nanosize heteroligand macro complexes of metal formed simultaneously with organic nanosize structures characteristic for metal-free polymer. Nanostructuring of formed in this way polyurethane favours creation of a new hierarchy in structural organization of the polymer as compared with metal free system as well as changes in dynamic, relaxation, optical, dielectric, surface etc. properties of the modified polyurethane.

Analysis of structural heterogeneity of metal-modified polymer indicates existence of two types of nanosize heterogeneities in the bulk of polyurethane. One of them is inherent to segmented PU and another is generated in the presence of transition metal chelate compound. The structural heterogeneity of PU influences the local segmental mobility of macro chains, resulting in "dynamic heterogeneity" as well as in "thermodynamic heterogeneity" of the systems.

The possible origin of the formation of the ordered micro regions is segmental structure of PU containing the soft and hard blocks with different complex ability relative to metal chelate compound. The PU's surface structure depends on the boundary "polymer-support" or "polymer-air". Concentration of PU less polar groups that form the weak complexes with metal chelate compound at the „polymer-support" boundary can facilitate the partial segregation of metal containing centres at this boundary.

The essential increasing of luminescence intensity of the rare-earth metal in the polyurethane environmental is a way for creation of new optically active materials. The intensity of PU-Eu luminescence depends both on the europium chelate compound content and polymer topology. Contrary to LPU the relationship of luminescence intensity *vs.* modifier percentage in CPUs is linear.

Increasing of the polyurethane conductivity to semi-conducting level is caused by the drastic increasing of macro chain mobility in the presence of polyheteronuclear modifiers. Conductivity level of LPU is at least one order lower then conductivity of CPU.

The results obtained indicate significant influence of structural organization of the modified polyurethane on its properties. The effect is caused by complex formation between metal chelate compound and functional groups of the forming polymer. The analysis of dynamic of low molecular probes and complex formation in the nanostructured polyurethane gives experimental evidence of existence of additional coordination bond network in metal-contained polyurethanes.

Author details

Nataly Kozak and Eugenia Lobko
Institute of Macromolecular Chemistry National Academy of Sciences of Ukraine, Ukraine

Acknowledgement

We would like to express our sincere gratitude to Professor Svetlana B. Meshkova (A. V. Bogatsky Physic-Chemical Institute of National Academy of Sciences of Ukraine, Odessa) and Professor Vladimir N. Kokozay (Taras Shevchenko Kyiv University) for synthesized heteroligand rare-earth metal's and polyheteronuclear chelate compounds, respectively.

8. References

Bershein, V.A. & Yegorov, V.M. (1990). *Differential Scanning Calorimetry in Polymer Physic.* (in Russ.). Chemistry, ISBN 5-7245-0555-X, St. Petersburg

Buchachenko, A. L., Wasserman, A.M., Alexandrova, T.A. et al. (1980). Spin Probe Studies in Polymer Solids, In: *Molecular Motions in Polymers by ESR,* R.F. Boyer & S.E. Keinath, (Eds.), 33-42, ISBN 3718600129, Harwood Academic Publisher, Chur, Switzerland

Bulavin, L.A.; Karmazina, T.V.; Klepko, V.V.; Slisenko, V.I. (2005). Neutron spectroscopy of condensed medium . (in Russ.). Akademperiodica, ISBN 966-360-009-8, Kyiv

Kovarskii, A.L. (1996). Spin Probes and Labels. A Quarter of a Century of Application to Polymer Studies, In: *Polymer Yearbook,* R.A. Pethrick, (Ed.), 113-139, ISBN 9783718659142, Harwood, Switzerland

Kozak, N.V.; Kosyanchuk, L.F.; Lipatov, Yu.S. et al. (2000). Effect of Zn2+, Cr(2+) and Ni2+ Ions of Cross-linked Segmented Polyurethanes. *Polymer Science,* Ser. A, Vol. 42, № 12, (June-August 2000), pp. 1304-1309, ISSN 0965-545X

Kozak, N.; Nizelskii, Y.; Mnikh, N. et al. (2006). Formation of Nanostructures in Multicomponent Systems Based on Organic Polymer and Coordination Metal Compound. *Macromolecular Symposia,* Vol. 243, (November 2006), pp. 243-262, ISSN 1521-3900

Kozak, N.V.; Lobko, Eu.V; Perepelitsina, L.M. et al. (2010). The Surface Characteristics of Polyurethanes, Contained β-diketonate Europium (3+). (in Ukr.). *Ukrainian Chemistry Journal,* Vol. 76, № 8, (June-August 2010), pp. 121-126, ISSN 0041-6045

Kratky, O.; Pilz, I.; Schmitz, P.J. (1966). Absolute intensity measurement of small-angle x-ray scattering by means of a standard sample. *Journal of Colloid Interface Science,* Vol. 21, №1, pp. 24-34, ISSN 1095-7103

Kyritsis, A. & Pissis, P. (1997). Dielectric studies of polymer-water interactions and water organization in PEO/water systems. *Journal of Polymer Science*: Ser. B, Polymer Physics, Vol. 35, (July 1997), pp. 1545-1560, ISSN 0965-545X

Lipatova, T. E. & Nizelskii, Yu.M. (1972). Complex Forming and Mechanism of the Catalysis of the Reaction Urethane Formation in situ β-diketonate of metals. (in Russ.). In: *The Successes of the Polyurethane Chemistry,* Yu. S. Lipatov, pp. 214-244, Naukova Dumka, Kyiv

Lipatova, *T.E.;* Shylow, V.V., Alexeeva, *T.T.;* et al. (1987), The Influence of the Support Character on the Structure of the Linear Polyurethanes Surface. (in Russ.). *Polymer Journal,* Seria B, Vol. 29, № 4, (December 1987), pp. 55-260, ISSN 1560- 0904

Lipatova, T.E. & Alexeeva, *T.T.* (1988). The spectroscopy investigated of the surface of segmented polyurethanes. (in Ukr.). *The Ukrainian Chemistry Journal,* Vol. 54, № 6, (June 1988) , pp. 324 – 628, ISSN 0041-6045

Lipatov, Yu. S.; Kosyanchuck, L. F.; Kozak, N. V et al. (1997). Effect of Metal Compounds on the Surface Properties of the Solid Polyurethanes being Formed in their presence. *Journal of Polymer Materials,* Vol. 14, № 3, (December 1997), pp. 63-268, ISSN 0970-0838

Lipatov, Yu. S.; Kozak, N. V.; Nizelskii, Yu. M. et al. (1999). Effect of Metal Compounds on the Physical Aging of Poly (urethanes). *Polymer Science,* Series A, Vol. 41, №8, (September 1999), pp. 1308-1315, ISSN 0965-545

Lobko, Eu.V; Kozak, N.V.; Meshkova, S.B. et al. (2010). The Photoluminescence of Polyurethanes, which Modified by the Chelate Compounds of threes- and tetra coordinated europium (3+). (in Ukr.). *Polymer Journal,* Vol. 32, № 5, (December 2010), pp. 410-415, ISSN 1818-1724

Nizelskii, Yu. N.; Shtompel, V.I.; Kozak, N.V. et al. (2005) The Nanostructure Heterogeneity of Polyurethane Films, Formed in the Presence of β-diketonate of Metals. *Report of National Academy of Science of Ukraine*, № 10, (October 2005), pp. 142-148, ISSN 1025-6415

Nizelskii, Yu. & Kozak N. (2006). In Situ Nanostructured Polyurethanes with Immobilized Transition Metal Coordination Complexes. *Journal of Macromolecular Science*, Part B, Vol. 46, (April 2007), pp. 97-110, ISSN 0022-2348

Pathmanatham, K. & Johari, G.P. (1990) Dielectric and Conductivity Relaxations in Poly (hema) and of Water in its Hydrogel. *Journal of Polymer Science*: B, Vol. 28., Issue 6, (May 1990), pp. 675-689, ISSN 0887-6266

Pissis, P.; Kanapitsas, A. (1996). Broadband dielectric relaxation spectroscopy at 10^{-4} – 10^{10} Hz. *Journal of the Serbian Chemical Society*, Vol. 61, № 9, pp.703-715, ISSN0352-5139

Poluectov, N.S.; Kononenko, L.I.; Yefrushyna, N.P. et al. (1989). Spectrometric and Luminescence Methods of lanthanides detection, (in Russ.), Naukova Dumka, ISBN 512000749X, Kyiv

Porod G. (1982). General theory. In: *Small-angle X-ray Scattering*. O.Glatter & O.Kratky (eds.)., 17-51, Academic Press, ISBN 0-12-286280-5, London

Saunders, J.H. & Frish, K.C., (1962). *Polyurethanes. Chemistry and Technology*. Part 1, Chemistry, Interscience Publisher, John Wiley&Sons, ISBN 0898745616, New York-London

Skopenko, V.V.; Garnovsky, A.D.; Kokozay, V.M.; et al. (1997). *The direct synthesis of coordination compounds*, (in Russ.), Ventury, ISBN 966-570-025-1, Kiev

Stompel, V.I. & Kercha, Yu.Yu. (2008) *The Structure of Linear Polyurethanes*, (in Ukr.), Naukova Dumka, ISBN 978-966-00-0699-7, Kyiv, Ukraine

Schmidt, P.W.; Hight, R.J. (1960). Slit height corrections in small angle x-ray scattering. *Acta Crystalographia*, Vol. 13, (June 1960), pp.480-483, ISSN 0365-110X

Tavana, H.; Gitiafroz, R.; Hair, M.L. et al. (2004). Determination of solid surface tension from contact angels: the role of shape and size of liquid molecules. *The Journal of Adhesion*, Vol. 80, (August 2004), pp. 705-725, ISSN 0021-8464

Vasserman, A.M.; Kovarskii, A.L. (1986). *Spin Labels and probes in physical chemistry of polymers*", (in Russ.), Nauka, Moscow, Russia)

Vinogradova, E.A.; Vassilyeva, O.Yu.; Kokozay, V.N. (2002). An incomplete cube-like array of Cu, Zn, I with O atoms from 2-dimethylaminoethanol formed directly from zero-valent metal powders. *Inorganic Chemistry Communications*, Vol. 5, (May 2001), pp. 19-22, ISSN 1387-7003

Wirpsza, Z. (1993). *Polyurethanes: chemistry, technology and applications*. Elis Horwood Limited, ISBN 0-13-683186-9, London

Yang, H.; Zhang, D.; Shi, L. et al. (2008). Synthesis and strong red photoluminescence of europium oxide nanotubes and nanowires using carbon nanotubes as templates. *Acta materialia*, Vol. 56, (March 2008), pp. 955-967, ISSN 13596454

Ying, J. (Ed.) (2002). *Nanostructured Materials: Advances in Chemical Engineering,* Massachusetts Institute of Technology, ISBN 0-12-008527-5, Cambridge

Chapter 5

Thermal Analysis of Polyurethane Dispersions Based on Different Polyols

Suzana M. Cakić, Ivan S. Ristić and Olivera Z. Ristić

Additional information is available at the end of the chapter

http://dx.doi.org/10.5772/35800

1. Introduction

1.1. Water-based polyurethane dispersions

Water-based polyurethane dispersions (PUD) are a rapidly growing segment of polyurethane (PU) coatings industry due to environmental legislations such as the clean air act and also due to technological advances that made them an effective substitute for the solvent-based analogs. Water-based or waterborne PUD have gained increasing importance in a range of applications, due in large part to properties such as adhesion to a range of substrates, resistance to chemicals, solvents and water, abrasion resistance and flexibility. Water-based PUD show very good mechanical and chemical properties and match the regulatory pressures for low volatile organic compound (VOC) containing raw paints. The continuous reduction in costs and the control of VOC emissions are increasing the use of water-based resins, motivating the development of PU dispersed in water. PU obtained from water-based PUD have superior properties when compared with similar materials obtained from organic media. Water-based PUD are used in many application areas to coat a wide range of substrates - for example footwear adhesives, wood lacquers for flooring and furniture, leather finishings, plastic coatings, printing inks and automotive base coats (Rothause et al., 1987; Kim et al., 1994; Ramesh et al., 1994).

Regarding the chemical nature of PU, the water based PU are applied with higher solids content, compared to the solvent based PU, because their viscosity does not depend on the molecular weight of PU, as is the case for solvent based PU (Gunduz & Kisakurek, 2004). Thus waterborne PUD can be prepared at high solid contents with a molecular weight enough to form films with excellent performance resulting solely upon "physical drying". This means that the film formation occurs by simple evaporation of water even at room temperature.

Waterborne PUD are fully-reacted PU systems produced as small discrete particles. 0.1 to 3.0 micron, dispersed in water to provide a product that is both chemically and colloidally stable, which only contains minor amounts of solvents and does not emit VOC. Polymeric structure of waterborne PUD is formed by usually reacting an excess of aliphatic isocyanates (mainly IPDI or HDI based), with a polyol or a mixture of polyols to form a prepolymer containing the so called soft segment. The polyols are generally polyesters, polyethers, or polycarbonates. The hard segment is generally formed by chain extending the prepolymer with short chain diamines and from the short chains containing ions. Due to incompatibility between the two segments of the polymer chain, the hard segment separates and aggregates into domains that act as reinforcing fillers to the soft segment. The degree of phase separation as well as the the concentration of the hard segments are contributing factors to the good properties of PUD. PU backbone with a minority of the repeat units contains pendant acid or tertiary nitrogen groups, which are completely neutralized or quarternized, respectively, to form salts. Such ionomeric groups are absolutely necessary for the formation of dispersions, because they act as internal surfactants, and are not incorporated in the chain of the solvent-based PU. Ionic centers in the hard segment generally favor segregation and cohesion within the hard segment domains due to their strong electrostatic forces and thermodynamic incompatibility with the polymer matrix. Water-based PUD can be divided into two classes. The first group consists of polymers stabilized by external emulsifiers, and second one achieves stabilization by including hydrophilic centers in the polymer. Such hydrophilic centers may be one of the three types: non-ionic, cationic groups and anionic groups. These hydrophilic groups fulfill the function as internal emulsifiers and make possible to produce stable water/based emulsions. Water–based PUD can be classified into anionic, cationic and nonionic systems (Rothause & Nechtkamp, 1987; Kim et al., 1996).

Several processes have been developed for the synthesis of PUD. All of these have in common the first step, in which a medium molecular weight polymer (the prepolymer) is formed by the reaction of suitable diols or polyols (usually macrodiols such as polyether or polyester) with a molar excess of diisocyanates or polyisocyanates. In this reaction mixture, an internal emulsifier is added to allow the dispersion of the polymer in water; this emulsifier is usually a diol with an ionic group [carboxylate, sulfonate, or quaternary ammonium salt) or a nonionic group poly(ethylene oxide)]. The internal emulsifier becomes part of the main chain of the polymer. The critical step in which the various synthetic pathways differ is the dispersion of the prepolymer in water and the molecular weight build-up. The most important dispersions are emulsifier-free ionomer dispersions. The resulting dispersions are mainly anionic or non-ionic, that have the potential for wide variations in composition and property level. They can be obtained by different processes, however, the earliest process to prepare the aqueous PUD is known as acetone process and this process has remained technically important so far (Hepburn, 1992; Oertel, 1985). Within the last three decades several new processes have been developed such as prepolymer mixing process, hot melt process and ketamine/ketazine process.

The facts that aqueous/water PUD have become increasingly important for industrial and academic research in recent years is due to the following reasons:

1. the environmental law requires for the development of ecological-friendly products for which the emissions of volatile organic compounds (VOC) have been reduced to a minimum,
2. the economic reasons (substitution of expensive organic solvents in conventional PU with water),
3. the water PU surpasses performance of conventional isocyanate- and/or solvent-containing PU,
4. continuous increase in solvent prices, low raw material costs and easy to clean up the reactor system made waterborne PU system more popular in the industry.

2. Ingredients for water–based PUD

The basic components used to build up PUD include long–chain polyether, polyester or polycarbonate polyol, diisocyanate, aromatic or (cyclo)aliphatic, low molecular weight glycol and /or diamine, bis–hydroxycarboxylic acid and a neutralization base. In general, an excess of diisocyanate is treated with a long-chain linear polyol, bis-hydroxycarboxylic acid and other low-molecular-weight glycol to form an isocyanate-terminated prepolymer with a segmented structure. In this polymer, the long-chain polyol units form soft segments, and the urethane units-built up from diisocyanate, glycol and bis-hydroxycarboxylic acid form hard segments. The pendant carboxylic acid groups are neutralized with base to form internal salt group containing prepolymers that can be easily dispersed in water. The microphase separation between the incompatible soft- and hard-segment sequences contributes to the unique properties of PUD. The PU chains with NCO terminating groups can be extended with glycol forming urethane groups. Chain extenders are low molecular weight, hydroxyl and amine terminated. Aliphatic isocyanates: hexamethylene diisocyanate (HDI), isophorone diisocyanate (IPDI) and (4,4'–diisocyanatodicyclohexylmethane ($H_{12}MDI$), improve thermal and hydrolytic stability, resistance to UV degradation and they do not yellow (Bechara, 1998).

Aliphatic diisocyanates are less reactive than aromatic ones and they must be used with certain catalysts. Aromatic isocyanates: methylene diphenyl diisocyanate (MDI), toluene diisocyanate (TDI) and 1,5–naphthalenediisocyanate (NDI) on the other hand, provide for toughness but yellow upon exposure to UV light. Although early water dispersed PU resins heavily utilized TDI, there is a high tendency to shift to aliphatic diisocyanates or to the aromatic diisocyanates with NCO groups not directly attached to an aromatic nucleus (Gunduz & Kisakurek, 2004).

The two key classes of polyols are polyethers and polyesters. Polyester polyols have been largely used in PUD paints as they exhibit outstanding resistance to light and aging. There are four main classes of polyester polyols: linear or lightly branched aliphatic polyester polyols (mainly adipates) with terminal hydroxyl groups, low molecular weight aromatic polyesters for rigid foam applications, polycaprolactones, polycarbonate polyols. Polyether polyols are susceptible to light and oxygen when hot, however, they improve water dispersion, and impart chain flexibility. These are made by the addition of alkylene oxides, usually propylene oxide, onto alcohols or amines which are usually called starters or 'initiators'. Polyether based on propylene oxide contains predominantly secondary hydroxyl end–groups. Secondary hydroxyl end–groups are several times less reactive with isocyanates than primary hydroxyl groups and for some applications, polyether based only on propylene oxide may have inconveniently low reactivity. The primary hydroxyl content may be increased by a separate reaction of the polyoxypropylene polyols with ethylene oxide to form a block copolymer with an oxyethylene tip.

In the choice of polyol for PU application, selected polyols must be competitive with other polyols and also enable the final PU product to be cost competitive with other materials in the end application.

In a typical anionic PUD process, anionic groups (carboxylic and sulfonic) are introduced along the length of the polymer chain by using hydrophilic monomers or internal emulsifiers. As the hydrophilic monomer, dimethylol propionic acid (DMPA) is the most widely used acid, which has two hydroxyl groups, therefore, it can be one of the main constituents of the PU backbone. DMPA improves the hydrophilic property by serving as the potential ionic center with N–methyl pyrrolidone as the co–solvent. Tartaric acid (TA) can also be used, but it usually results in branching. However, TA improves the mechanical properties of PU paint. Study has shown that the particle size of dispersion depends on the content of DMPA. Therefore, increased amount of DMPA leads to more ionic centres in the PUD backbone and thereby increasing hydrophilicity of the polymer and hence reductions in particle size (Dieterich, 1981; Jacobs & Yu, 1993; Rosthauser & Nachkamp, 1986).

The chain extension step also has a high influence on the properties of the resin produced, not only due to the structure and concentration of the extender but also due to process variables that influence the particle size distribution. Chain extenders are difunctional glycols, diamines or hydroxyl amines. If a diol of low molecular weight reacts with the NCO terminated PU chains in the chain extension reaction step, urethane linkages will be formed but if a diamine is used as chain extender, the NCO terminated PU chains will form urea linkages. The higher density of hydrogen bonds of polyurea hard segments is responsible for the improved mechanical properties of polyurea and PU/urea products. Typical chain–extending agents are as follows: water, diethylene glycol, hydroquinone dihydroxyethyl ether, bisphenol A, bis(hydroxyethylether), ethanolamine, hydrazine, ethylene diamine. Aliphatic diamines such as hydrazine or ethylene diamine are used as chain extenders in processes directed to preparing waterborne PUD. In the chain extension step, it is most important to control the extremely fast reaction between NCO groups and NH_2 groups accompanied by the viscosity rise. Molecular weight of PUD increases by the formation of urea linkages through the chain extension step and it is the most important step to determine the molecular weight of PUD (Kim, 1996; Delpech & Coutinho, 2003).

In the synthesis of PUD, to neutralize the carboxyl and/or sulfo groups, are used agents that contain one or more bases and to form internal salt groups containing prepolymers that can be easily dispersed in water. During the neutralization, carboxyl and/or sulfo groups serve for anionic modification or stabilization of the PUD. Tertiary amines and in particular triethylamine are preferably used. The structure of waterborne anionic PUD is illustrated in Scheme 1.

Most commonly used catalysts in PU chemistry are tertiary amine catalysts and metal catalysts, especially tin catalysts. Tertiary amines are catalysts for the isocyanate–hydroxyl and the isocyanate–water reactions. Organotins are the most widely used, however organomercury and organolead catalysts are also used but have unfavourable hazardous properties.

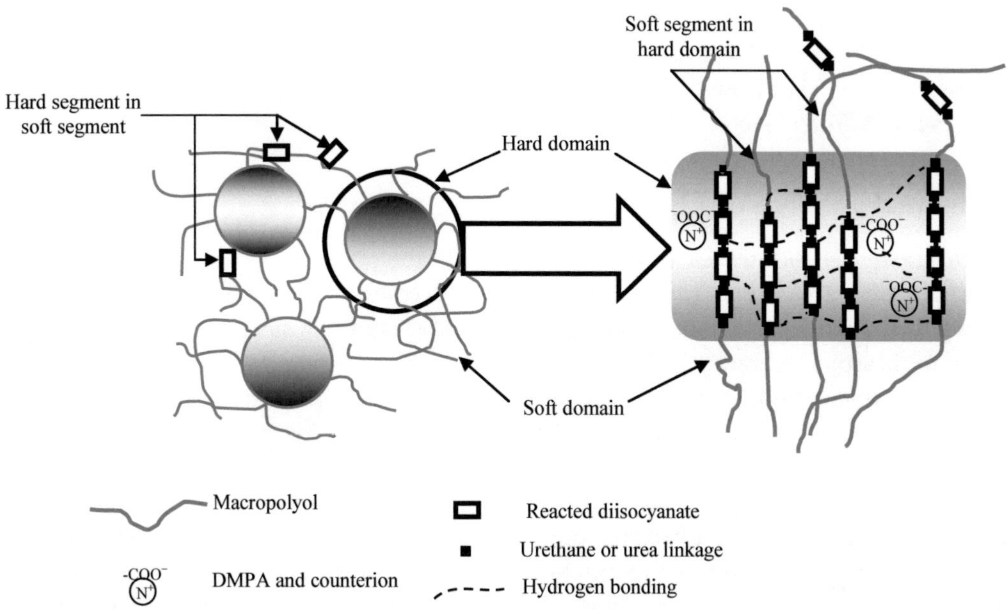

Scheme 1. Structure of waterborne anionic PUD (Tawa & Ito, 2006)

2.1. Various methods for preparing water–based PUD

The most important process is the prepolymer mixing process that has the advantage of avoiding the use of a large amount of organic solvent. In this process hydrophilically (carboxylate molecule) modified prepolymer is directly mixed with water. If the mixture viscosity is too high, a small amount of a solvent such as N–methyl pyrrolidone can be added before the dispersion step. Chain extension is accomplished by the addition of di– or polyamines to the water–based prepolymer dispersion.

The acetone process can be considered the link between the solvent synthesis and the prepolymer mixing process. In effect, the prepolymer is synthesized in a hydrophilic organic solvent, for example acetone solution and afterwards it is subsequently mixed with water.

The hot melt process explains the process of obtaining a PUD by the reaction of NCO–terminated ionic modified prepolymer with, for example ammonia or urea resulting in a prepolymer with terminal urea or biruet groups, respectively. The terminal urea or biruet prepolymer is methylolated with formaldehyde and mixed with water, forming dispersion spontaneously. By polycondensation (lowering the pH, increasing the temperature), chain–extension or cross–linking was obtained.

Ketamine and ketazine process explains the process of obtaining a PUD by reaction of NCO–prepolymers containing ionic groups mixing with a blocked amine (ketamine) or hydrazine (ketazine) without premature chain extension. These mixtures can be emulsified with water even in absence of co–solvents. The reaction with water liberates the diamine or hydrazine, which then reacts with the prepolymer.

Non–ionic dispersions are obtained similar to ionomer dispersions if the ionic centre is replaced by lateral or terminal hydrophilic ether chain. The temperature of dispersing process has to be kept below 60 °C. Non–ionic dispersions are stable towards freezing, pH changes and addition of electrolytes.

3. Thermal analysis of PU

Thermal analysis techniques have been used for many years in many scientific and industrial laboratories for studying the thermal decomposition of polymeric materials. Among them thermogravimetry (TG) is one of the most common since the mass of a sample is easy to measure accurately and valuable information regarding the nature of the process can be extracted from a mass loss against time or temperature plot. The thermal decomposition of PU (their degradation attributed to absorbed thermal energy) is important phenomenon from both fundamental and industrial applications (Pielichowski et al., 2004). The understanding of degradation processes allows determination of optimum conditions for designing PU in order to obtain high-performance polymer materials. Fundamental research has established that the thermal decomposition of PU is a complex heterogeneous process and consists of several partial decomposition reactions (Scaiano, 1989). The study of the decomposition of PU is particularly difficult since they degrade with the formation of various gaseous products and a number of decomposition steps are typically observed in thermogravimetric analysis (TGA) experiments. Some authors claim that the study of the thermo-degradation behavior of PU at high temperatures provides a fingerprint of the material that has to do not only with the characteristics of the original material, but also with its processing and the final quality of the end use products (Prime et al., 1988). The thermal stability of a material is defined by the specific temperature or temperature-time limit within which the material can be used without excessive loss of properties (Chattopadhyay & Webster, 2009). With respect to commercial applications, the investigation of thermal decomposition processes has two important aspects. The first concerns the stabilization of a polymer to obtain novel materials with a desired level of thermal stability that will be able to fulfill the demands of contemporary materials engineering. The second is to understand material behavior at higher temperature as well as to obtain characteristic thermal decomposition data.

A waterborne PUD can be defined as a binary colloidal system in which PU particles are dispersed in a continuous aqueous medium. PUD are usually prepared as low molecular weight NCO-terminated prepolymers for ease of dispersion. Then, diamines are generally used to increase the molecular weight by reaction with the terminated NCO end groups (chain extension). The presence of ionic species in PUD has a considerable effect on the physical properties. PUD are now one of the most rapid developing and active branches of PU chemistry.

PU are synthesized by the prepolymer reaction of a diisocyanate and a polyol (mainly polyethers and polyesters). If a diol of low molecular weight reacts with –NCO-terminated prepolymers in the chain extension reaction step, urethane linkages will also be formed but if a diamine is used as chain extender, the reaction between the $–NH_2$ groups and the –NCO terminated prepolymers will form urea linkages. In this case, poly(urethane-urea)s, which are the most important class of polyureas, are produced. These copolymers show reduced plasticity in comparison to homopolyurethanes. The resulting PU or poly(urethane-urea) chains consist of alternating short sequences forming soft (flexible) and hard (rigid) segments. The soft segments, originated from the polyol, impart elastomeric characteristics to the polymer. The hard segments are mainly produced by reacting the isocyanate and the chain extender. They are polar and impart mechanical properties to PU. The hard segments contain the highly polar urethane linkages. Due primarily to interurethane and urea hydrogen bonding, the two segment types tend to phase-separate in the bulk, forming microdomains. The hard segments act as physical crosslinks and, as a consequence, the physical, mechanical and adhesive properties depend strongly on the degree of phase separation between hard and soft segments and interconnectivity of the hard domains. The urethane linkages in PU can serve as H-bond acceptor and donor. In polyether-based PU, the urethane –NH can bond to either the polyether –O– linkage or the urethane –C=O groups. In the case of poly(urethane-urea) formation, there is an additional –NH from urea linkage participating in the interactions (Delpech & Coutinho, 2000). The degradation of thermoplastic PU has been extensively studied, and a number of reviews are available (Lu et al., 2002; Fambri et al., 2000). Thermal degradation of ester- and ether-based thermoplastic PU is performed under vacuum, air and nitrogen, allowing investigators to determine the mode of degradation (Dulog & Storck, 1996).

Polyester-based thermoplastic PU exibit rapid degradation in air and nitrogen, indicating that a nonoxidative mechanism is involved. In contrast, the significantly improved thermal stability of ether-based PU under vacuum and nitrogen indicates that the oxidative process plays a major role in the decomposition of ether-based thermoplastic PU. In general, the ester-based PU normally exibit better thermal and oxidative stabilities than the ether-based ones. The mechanism of thermal degradation of PU is very complex due to the variety of products formed.

It is proposed that the thermal degradation of thermoplastic PU is primarily a depolycondensation process, which starts at about 200 °C (Cakić et al., 2006 a). The first stage of decomposition is because of degradation of hard segments and starts at about 200 °C and at ~ 360-380 °C, while the second step of degradation is because of degradation of soft segments and ends above 480 °C. Waterborne PU should exhibit some different features in thermal degradation due to their unique chain structure, for example, salt-forming groups. Therefore, it is necessary to analyses their thermal degradation behavior to understand the structure-property relationship. The properties of PUD are mainly determined by the interactions between the hard and soft segments, and by the interactions between the ionic groups. The ionic group content, solids content, segmented structure, molecular weight of the macroglycol, the type of chain extender and the hard/soft segments ratio, determine the properties of PUD.

In the following sections, we will review typical results to demonstrate the utility of TGA in deducing the structural and bonding information about waterborne PUD based on different polyols. The thermal stability of PU and poly(urethane –urea)s cast films with anionomer character, obtained from waterborne dispersions and based on isophorone diisocyanate (IPDI), dimethylolpropionic acid (DMPA), poly(propylene glycol) (PPG), polycarbonate diol (PCD) and glycolized products obtained from recycled poly(ethylene terephthalate)(PET) is also compared. Three types of chain extenders were used: ethylene glycol (EG), propylene glycol (PG) and ethylene diamine (EDA). The effect of type of polyols, chain extender, type of catalyst, ionic content, length of soft segment, hard segment content and the presence of urea or urethane linkages on the thermal stability of the waterborne anionic PUD are discussed.

4. Water-based PUD based on poly(propylene glycol) and selective catalyst

One of the inherent drawbacks of waterborne PU technology is the formation of carbon dioxide due to the side reactions of isocyanate with water. When an isocyanate reacts with water, the products are a urea linkage (via an amine intermediate) and carbon dioxide. The carbon dioxide formation is problematic in that it causes imperfections in the coating during cure, such as blistering and pin-hole formation. The main aspect in the development of waterborne PU is in the first place to find methods for preventing the undesired secondary reactions with water and achieving the best crosslinking. This reaction is reduced to a minimum by the use of non-tin catalysts. One novel approach to control the water side reaction is the use of catalysts which selectively catalyze the isocyanate-polyol reaction and not the isocyanate-water reaction (Colling et al., 2002; Blank &Tramontano, 1996).

The relative selectivity (S) obtained from equation $S = P_{urethane}/P_{urea}$, was measured as urethane IR peak area ($P_{urethane}$)/ urea IR peak area (P_{urea}) ratio, by method given by Blank (Blank et al., 1999). After the integration of characteristic absorption max of urethane (1700 cm^{-1}, 1540 cm^{-1}) and urea (1640 cm^{-1}, 1570 cm^{-1}) was done, the relative selectivity was calculated. The manganese catalyst, a complex of Mn(III)–diacetylacetonatomaleate with various ligands based on acetylacetonate and maleic acid, used in some of the experiments (Stamenković et al., 2003; Cakić et al., 2006), has shown a high selectivity for the isocyanate–hydroxyl reaction in comparison to the commercially available zirconium catalyst (Blank et al., 1999). Zirconium catalyst is a proprietary zirconium tetra-dionato complex in the reactive solvent with the metal content of 0.4%.

TG is a suitable method to evaluate the thermal properties of several types of PU elastomers. The thermal stability of PU has been studied extensively because of the great importance of this group of materials (Chang et al., 1995). These thermoplastic elastomers generally are not very thermally stable, especially above their softening temperatures (Wang & Hsieh, 1997), and their mechanism of thermal degradation is very complex due to the variety of products formed. Commonly, it presents a bimodal profile where the first mode is related to the hard segments of PU. Usually, at a low heating rate, the degradation process results in differential weight loss (DTG) curves with several peaks, which is an indication of the complexity of the degradation (Delpech & Coutinho, 2000).

4.1. Experimental

Poly(propylene glycol) (PPG)), (M_n = 1000, hydroxyl value 111 mg KOH/g, dried under vaccuum, at 120 °C), isophorone diisocyanate (IPDI) and dimethylolpropionic acid (DMPA), were obtained from Aldrich Chemical Co. 1-methyl-2-pyrrolidone (NMP), dimethyl formamide (DMF) and triethyl amine (TEA) were received from Merck (Darmstadt, Germany). Ethylene glycol (EG) and propylene glycol (PG) obtained from Zorka Co.(Šabac Serbia). Dibutyltin dilaurate (DBTDL), was supplied by Bayer AG. Zirconium catalyst (ZrCat) was supplied by King Industries Inc., Norwalk, CT, USA. Manganese catalyst (MnCat) has been used in the reactive diluent with a metal content of 0.4% (Stamenković et al., 2003; Cakić et al., 2006 b). Water–based PUD from PPG with selective catalyst were prepared using the prepolymer method has been described in detail in our previously work (Cakić et al., 2009).

In the first step, PPG and DMPA were dispersed in DMF to obtain a homogeneous mixture and heated at 70 °C. IPDI and DBTDL were added to the homogenized mixture at 80 °C. An NCO/OH equivalent ratio of 3.0 was used. Hard : Soft segment ratio was defined as a ratio between IPDI weight and polyol weight in the starting formulation and is calculated as 1.5. The reaction times were determined by the dibutyl amine back titration method. After obtaining completely NCO terminated prepolymer, the mixture was cooled to 60 °C and the carboxylic groups were neutralized with TEA (DMPA equiv) dissolved in NMP. In second step the chain extension was carried out with EG or PG. The selective catalysts, ZrCat or MnCat, at concentration of 2% relative to the resin solids, have been added to the reaction solution. Water was added to the mixture and stirred to obtain dispersion of organic phase in water. The waterborne PUD contains 40 wt% solids (Cakić et al., 2009). Films were prepared by casting the waterborne dispersions on leveled surfaces and allowing them to dry at room temperature, for 7 days, and then at 60 °C, for 12h (Coutinho, 1996, 2003). When chain extenders were EG and PG, the PUD had to be cast in Teflon surfaces due to the high adhesiveness observed on the glass surface, making demoulding impossible. After demoulding, the films were kept into a desiccator to avoid moisture.and polyurethanes were formed.

4.2. Results and discussion

Degradation profile of waterborne PUD is influenced by the variation of chain extender presented in Fig.1. It was verified that the thermal stability was influenced by chain extender type. In a general way, thermal stability was higher when EG chain extender was used, in comparison to PG, probably because of the higher reactivity of the primary hydroxyl groups. The onset temperatures calculated for the first stage for PU chain-extended with EG were about 234 °C and 140 °C for PG. Unsymmetrical structure of IPDI enables easier diffusion of EG (Gunduz & Kisakurek, 2004).

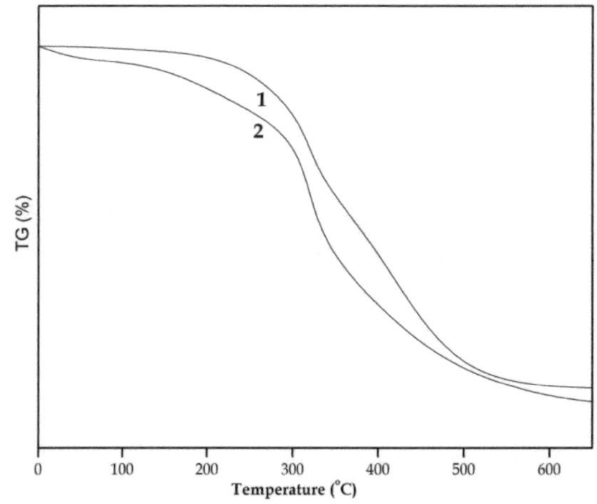

Figure 1. TG curves of PUD without catalyst with EG (1) and PG (2) as chain extender

Fig. 2 shows the degradation profile of PUD with variation of catalyst using catalysts of different selectivity. EG and PG formed urethane linkages by reaction with terminal NCO groups. The initial onsets observed are: 234 °C when EG was employed, 275 °C and 311 °C, when the chain extender was EG with ZrCat and EG with MnCat, respectively. The initial onsets observed are: 140 °C when PG was employed, 290 °C and 305 °C, when the chain extender was PG with ZrCat and PG with MnCat, respectively. The thermal stability was

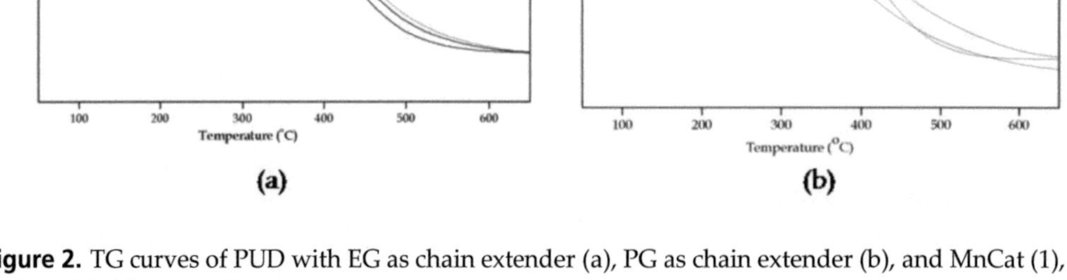

Figure 2. TG curves of PUD with EG as chain extender (a), PG as chain extender (b), and MnCat (1), ZrCat (2), without catalyst (3)

higher when MnCat was used, in comparison with the use of ZrCat. This result suggested that all the residual NCO groups in PU particles did not react with the chain extender completely. Because the viscosity of particle is high at low temperature in the chain extension step, it would take long time for chain extenders to diffuse into the particle. Therefore, the efficiency of chain extension increased as total surface area of particles increased (Jang et al., 2002; Cakić et al., 2007 a). In general, the presence of more selective catalyst has also been found to have a stabilizing effect on the resultant PU, as can be observed in the curves obtained for the samples chain-extended with EG and PG, in comparison to the samples obtained without selective catalyst probably due to favoring of isocyanate-polyol reaction and not the isocyanate-water reaction. The PUD was formed with higher hard segment proportions.

The DTG curves show that there are different stages of degradation which are not perceptible in TG curves, showing the close relation and mutual influence between the degradation process of hard and soft segments.

Fig. 3a shows the DTG curves corresponding to the TG degradation profiles presented in Fig. 2a, in which the catalysts were varied (MnCat or ZrCat). The chain extender employed was EG. Two peaks are observed. The first group of peaks, corresponding to the degradation of rigid segments formed by urethane and urea linkages, presents maximum of the peak from 200 to 250 °C. The second group, related to the degradation of PPG soft segment, varying maximum of the peak from 375 to 418 °C . The peaks shifting towards higher temperatures resulting from addition of more selective catalyst confirm the assumption that all isocyanate groups had not reacted with the added chain extender. Selective catalyst isocyanate-polyol reaction causes greater incorporation of chain extender in hard segments, which is reflected on higher thermal stability of hard segments (Jang et al., 2002; Lee et al., 1995; Cakić et al., 2007 b).

Fig. 3b depicts DTG curves related to TG profiles observed in Fig. 2b, in which the catalyst were varied (MnCat or ZrCat). The chain extender employed was PG. The first group of peaks presents maximum of the peak appearing in the range from 160 °C to 195 °C. The

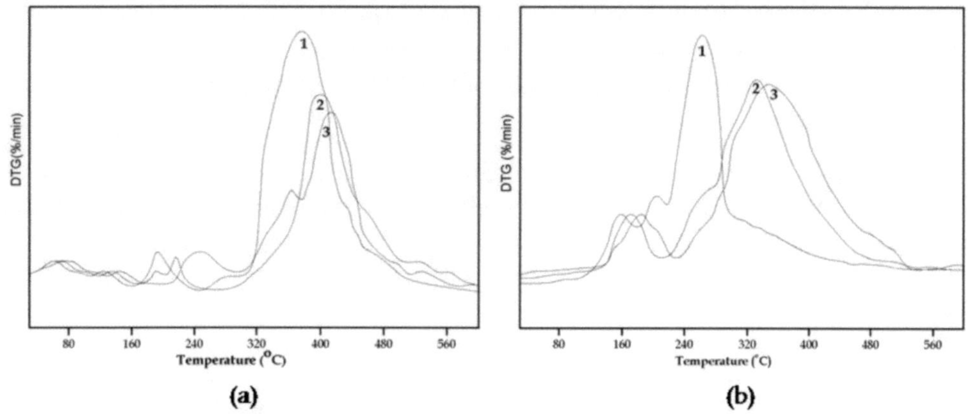

Figure 3. DTG curves of PUD with EG as chain extender (a), PG as chain extender (b), and without catalyst (1), ZrCat (2), (3) MnCat

second group can be observed in the range, varying maximum of the peak, from 267 to 347 °C, for PPG soft segments. A marked difference can be observed, promoted by changing the type of chain extender in DTG profiles, especially in the first stage of weight loss, corresponding just to urethane (EG or PG as chain extender) linkage degradation. The soft segment, formed only by PPG degradation step seemed to be also affected. The rigid segment formed from EG retarded the weight loss of PPG chains (peak at 375 °C), while PG showing peaks at 267 °C (Cakić et al., 2006, 2007 c). All DTG curves showed that there are different stages of degradation which are not perceptible in TG curves, showing the close relation and mutual influence between degradation of hard and soft segments.

The degradation profiles of PU cast films obtained from water-based dispersions were influenced by the type of chain extender, length of the hard segment and type of catalysts. The presence of more selective catalysts, which formed urethane linkages with higher hard segment proportions, had a marked influence on the degradation of the polymers, especially in elevated quantities, improving the thermal stability of the materials. The DTG curves showed that the length of the hard segment had a strong influence on the thermal profile of the samples as a whole. The type of chain extender, forming urethane linkages, affected the whole process of degradation and the presence of more selective catalyst improved the thermal resistance of the chains.

5. Water-based PUD based on glycolized products obtained from recycled poly (ethylene terephthalate)

Recycling of polymers has received a great deal of attention (Atta et al., 2006, 2007). Although several methods have been proposed for recycling waste poly(ethylene terephthalate) (PET), it is suggested that the most attractive method is glycolysis of chemicals into the corresponding monomers or raw chemicals that could be reused for the production of plastics or other advanced meterials (Patel et al., 2007).

Two-stage PUD synthesis was applied: the first, glycolysis of PET using different types of glycols (PG), triethylene glycol(TEG) and poly(ethylene glycol) (PEG 400), with different molar ratio of PET repeating unit to glycol (1:2 and 1:10); the second, preparation of PUD of the products formed.

PUD are prepared by anionic dispersion process (Cakić et al., 2011), using IPDI, glycolyzed products, DMPA as potential ionic center which allow water dispersibility and ethylene dimine (EDA) as chain extender.

5.1. Experimental

Example 1 of glycolysis reaction: Small pieces of PET waste (100 g), equivalent to 0.5 mol repeating unit (mol.wt. 192 $gmol^{-1}$) were added to 88.64 g PG (mol.wt. 76.09 $gmol^{-1}$), 173.07 g TEG (mol.wt.150 $gmol^{-1}$) or 461.5 g PEG 400 (mol.wt. 400 $gmol^{-1}$), such that the molar ratio of PET repeating unit to glycol was 1:2.

Example 2 of glycolysis reaction: In the second experimental runs of depolymerisation, appropriate amount of PET waste were added to 396.1 g PG, 750 g TEG or 2173 g PEG 400, so that molar ratio of PET repeating unit to glycol was 1:10. These mixtures (with different molar ratio PET/glycol) and 0.5 wt.% zinc acetate (based on the weight of PET as transesterification catalyst) were charged to a glass reactor, which was fitted with stirrer, reflux condenser, nitrogen inlet and temperature controller. This reactor was immersed in an oil bath and the content of the reaction kettle was heated at 190 °C for 2 h, subsequently the temperature was raised to 210 °C until all the solids disappeared.

The obtained glycolyzed oligoester polyols were analysed by the hydroxyl value (HV) determination according to the conventional acetic anhydride/pyridine method (Cakić et al., 2011). The hydroxyl value of the oligoester polyol obtained in the glycolysis reaction based on molar ratio of PET repeating unit to glycol, 1:2, with PG was HB_{PG}=490 mg KOH/g, TEG HB_{TEG}=370 mg KOH/g and PEG HB_{PEG400}=297 mg KOH/g.

The hydroxyl value of the oligoester polyol obtained in the glycolysis reaction based on molar ratio of PET repeating unit to glycol, 1:10, with PG was HB_{PG}=201 mg KOH/g, TEG HB_{TEG}=209 mg KOH/g and PEG HB_{PEG400}=192 mg KOH/g.

5.2. Synthesis of PUD based on glycolyzed products with molar ratio PET/glycol, 1:2

Anionic PUD based on glycolyzed products with molar ratio PET/glycol, 1:2, were prepared by modified acetone process. Acetone was added to the prepolymer and the dispersion is formed by the addition of water to this solution. Procedure for synthesis of anionic PUD has been developed adjusting the molar ratio of DMPA as a hydrophilic monomer to IPDI as 1:3.3. Mass of oligoester polyol, obtained by PET glycolysis, according to example 1, with a hydroxyl number which is equivalent to the hydroxyl number of 0.06 mol of poly(propylene glycol) PPG_{1000} (110 mg KOH/g), was for PG 15 g, TEG 20 g, poly (ethylene glycol) (PEG 400) 25 g.

The oligoester polyol and hydrophilic monomer (8 g, equ. 0.06 mol) was mixed in the cosolvent DMF (50:50 w/w), in a 250-ml round four-neck glass reactor connected to a stirrer, a thermometer, a reflux condenser and a nitrogen gas inlet. The reaction was carried out at 70 °C for 30 min to obtain a homogeneous mixture and the uniform distribution of hydrophilic monomer to PU backbone. IPDI (44.4 g, equ.0.2 mol) and catalyst DBTDL (0.03% of the total solid) were added to the homogenized mixture at 80 °C for about 4h until the amount of residual NCO groups reached a theoretical value, as determined by the dibutyl amine back-titration method. To reduce the viscosity and obtain a homogenous mixture of NCO prepolymers, acetone was added 50 wt% to the solid reaction mass. The theoretical value of NCO groups for PUD based on oligoester polyol obtained from the glycolysis with molar ratio of PET repeating unit to glycol, 1:2, was 19.2% for PG, 17.7%, for TEG and 16.1% for PEG.

After obtaining completely NCO terminated prepolymer, the mixture was cooled to 60 °C and the carboxylic groups in hydrophilic monomer were neutralized with TEA (DMPA equ). TEA was dissolved in NMP by stirring the solution for 60 min.

PU anionomer was cooled to 30 °C then dispersed in water (50% of total mass) under high speed stirring for 30 min. The rate of water addition to the mixture was carefully controlled, to obtain a stable inversion. Upon completing the phase inversion, EDA (0.03 mol) was added for 60 min. at 35 °C. PUD of about 30 wt% solids was obtained upon removal of acetone by rotary vacuum evaporation at 35 °C.

5.3. Synthesis of PUD based on glycolyzed products with molar ratio PET/glycol, 1:10

Anionic PUD based on glycolyzed products with molar ratio PET/glycol, 1:10, were prepared by prepolymer mixing method in two steps: synthesis of NCO-terminated prepolymers and the preparation of dispersions by introducing anionic centers to aid dispersions (Athawale & Kulkarni, 2009). The prepolymer mixing method has been developed adjusting the molar ratio of DMPA to IPDI as 1:3, which was increased compared to the previous procedure. Mass of oligoester polyol, obtained by PET glycolysis, according to example 2, with a hydroxyl number which is equivalent to the hydroxyl number of 0.1 mol of poly(propylene glycol) PPG_{1000} (110 mg KOH/g), was for PG 54.5 g, TEG 52.6 g, poly (ethylene glycol) (PEG 400) 57.3 g. The mass of oligoester polyols and hydrophilic monomer was the same as in the previous procedure, but the mass of IPDI was (66.6 g, equ.0.3 mol). determined by molar ratio of a hydrophilic monomer to IPDI as 1:3. In order to obtain NCO terminated prepolymer, synthesis was controlled by determing the NCO groups by the dibutyl amine back-titration method until a theoretical value was achieved. The theoretical value of NCO groups for PUD based on oligoester polyol obtained from glycolysis with molar ratio of PET repeating unit to glycol, 1:10, was 14.2% for PG, 13.7%, for TEG and 13.9% for poly(ethylene glycol). The neutralization of the carboxylic groups in hydrophilic monomer, extension chain with EDA and the preparation of stable dispersion was done as in the previous modified acetone process. The solid content in this dispersion was 30%.

Films were prepared by casting the aqueous dispersions on the glass surface and allowing them to dry at room temperature for 7 days and then at 60 °C, for 12h (Coutinho, 1996, 2003). Films were cast by 100 μm applicators from the solutions onto glass surface (7cm x 2 cm) to obtain dry film thickness of 25-30 μm, making demoulding impossible. After demoulding, the films were kept into a desicator to avoid moisture. TG experiments were performed in a Perkin-Elmer TG-7 analyser. Film samples about 20 mg were placed in a platinum sample pan and heated from 30 to 600 °C, with an air flow of 200 mL min^{-1} and heating rates of 10 °C min^{-1}. During the heating period, the weight loss and temperature difference were recorded as a function of temperature.

TGA was used to analyze decomposition behavior of cured films of PUD synthesized with glycolized products obtained from PET waste. TG curves are depicted in Figs. 4a,4c and 4e (curve marked as 1 a show lower molar ratio of PET / glycol (1:2) in the glycolized oligoester) represents the degradation of PUD influenced by the variation of oligoester polyols. It was verified that the thermal stability was influenced by glycol type and different molar ratio of PET repeating unit to glycol in glycolysis reaction. Figs. 4b, 4d and 4f depict the behavior of corresponding differential weight loss (DTG) curves.

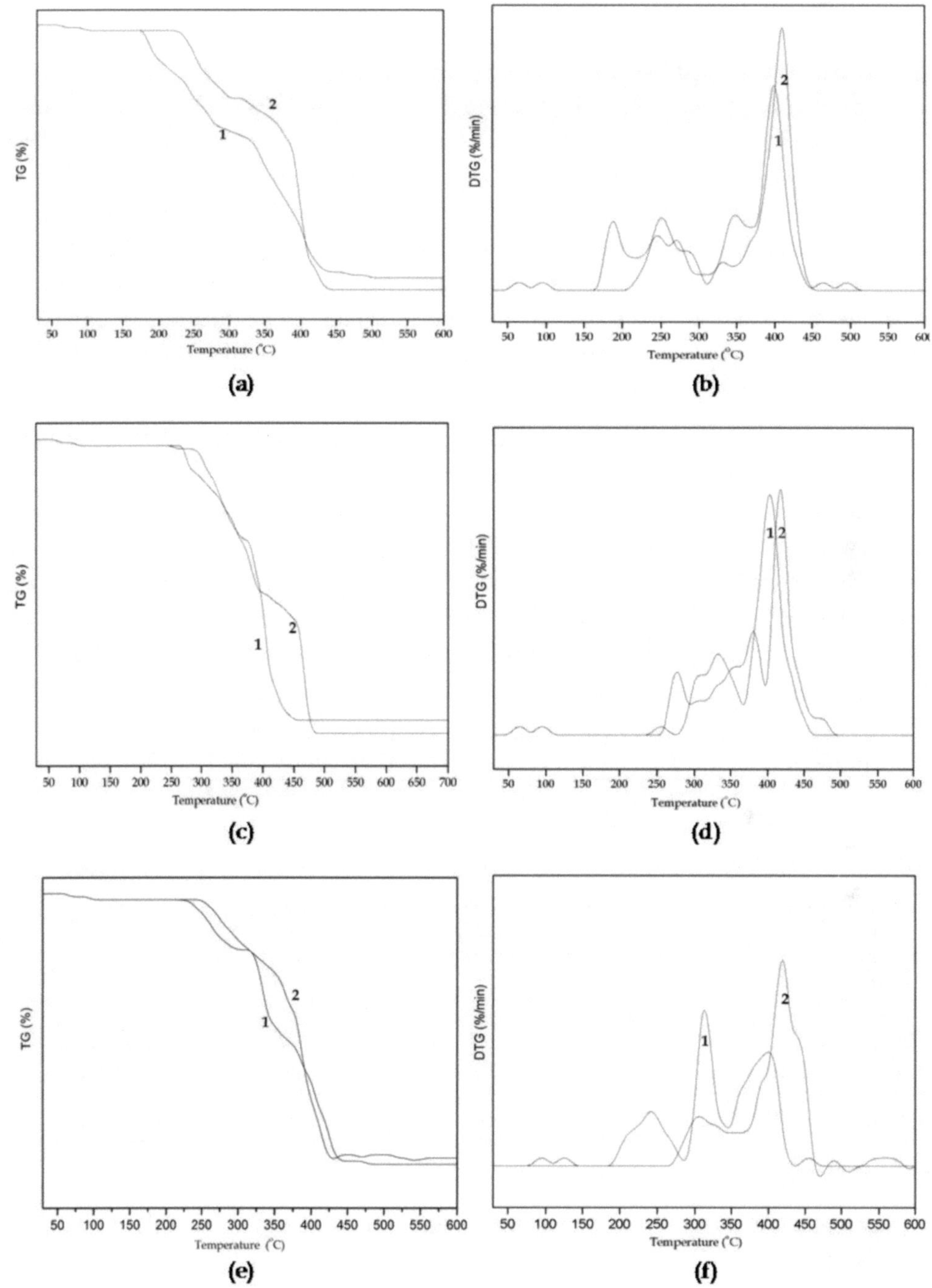

Figure 4. TG curves (a) and DTG curves (b) of PUD synthesized from glycolized oligoester PET/PG with molar ratio 1:2 (1) and 1:10 (2). TG curves (c) and DTG curves (d) of PUD synthesized from glycolized oligoester PET/TEG with molar ratio 1:2 (1) and 1:10 (2). TG curves (e) and DTG curves (f) of PUD synthesized from glycolized oligoester PET/poly(ethylene) glycol 400 with molar ratio 1:2 (1) and 1:10 (2).

PUD (synthesized from glycolized oligoester PET/PG (1:2), Fig. 4b) synthesized from depolymerised oligoesters with lower molar ratio of PET repeating unit to glycol in glycolysis reaction showed lower thermal stability in the initial stage of degradation may be due to the presence of greater amount of aromaticity in polyester backbone which makes the PU chains susceptible to scission and relieves the structure crowing (Athawale & Kulkarni, 2010). In later stage (above 300 °C), it showed enhanced thermal stability. It has also been proved that two or three peaks of first decomposition were well correlated with the higher value of polydispersity of GPC results (1.65), for oligoester polyols PET/PG (1:2) compared to the values of polydispersity (1.20), for oligoester polyols PET/PG(1:10). Curve marked as 2, in Figures 4b, 4d and 4f, which shifted third decomposition step temperature to the higher values, shows that glycolized oligoester obtained with higher molar ratio of PET/glycol of 1:10 have better thermal stability of obtained PUD.

Because of the presence of oligoester polyols wich lower molecular weight in glycolysis reaction and a diamine were used in the synthesis of PUD, two kinds of hard segments are formed, urethane and urea. It has been estabilished that the urethanes have lower thermal resistance than urea and therefore the first decomposition process at about 190 - 250 °C and the second at about 270 – 290 °C of PUD should correspond to the urethane and urea hard segments, respectively. The decomposition temperature of the soft segment is observed at 400-430 °C. The decomposition temperature for investigated samples are listed in Table 1.

Sample	First decomposition		Second decomposition	Third decomposition
	T_{1on} (°C)	T_{1max} (°C)	T_2 (°C)	T_3 (°C)
PET/PG (1:2)	190.2 247.0	270.7	349.7	400.3
PET/TEG (1:2)	-	288.0	335.1	401.9
PET/PEG$_{400}$ (1:2)	248.6	-	316.5	403.4
PET/PG (1:10)	251.7	288.1	327.0	411.3
PET/TEG (1:10)	-	277.6	379.6	416.4
PET/PEG$_{400}$ (1:10)	-	-	305.4	422.4

Table 1. Temperature of decomposition of PUD

The degradation profile of PUD was dependent on mole ratios of PET to glycol in glycolyzed products.

The samples based on PET/glycol, at molar ratio of 1:10, had better thermal stability than samples based on PET/glycol, at molar ratio of 1:2. The higher values of temperature for third decomposition stage, for samples with molar ratio of 1:10, probably is due to the increased length of glycol in glycolyzed oligoester polyol.

6. Waterborne PUD based on polycarbonate diols (PCD)

The polyols used in PUD synthesis are of polyether-, polyester-, polycaprolactone- and polycarbonate- origin. The use of individual types of polyol chains and their functionalities depend on the purpose of the potential use, e.g.; PUD made from polyesters can have slightly elevated strength and oil resistance compared to polyether-based PUD and have been largely used in PU paints as they exhibit outstanding resistance to light and aging. Polyether polyols are susceptible to light and oxygen when hot, however, they improve water dispersion, and impart chain flexibility (Gunduz & Kisakurek, 2004). The use of PCD in PUD, as compared to other polyols, imparts better hydrolysis resistance, improved ageing and oil resistance, excellent elastomeric properties even at low temperature, improved mechanical properties, good weathering and fungi resistance (Garcia et al., 2010). PCD used as the soft segment component in PUD synthesis are usually obtained from dimethylcarbonate or ethylene carbonate and a linear aliphatic diol (Foy et al., 2009). The properties of PUD are related to their chemical structure (Cakić et al., 2009; Athawale & Kulkarni, 2010) and are mainly determined by the interactions between the hard and soft segments, and the interactions between the ionic groups (Garcia et al., 2011). The properties of PUD are strongly influenced by composition and ionic content, an important target in an investigation of the role of the composition (Lee at al., 2004, 2006).

6.1. Experimental

Water-based PUD derived from IPDI, with different molar ratio PCD to DMPA, were prepared by the modified dispersing process. The ionic group content in PU-ionomer structure was varied by changing the amount of the internal emulsifier, DMPA (4.5, 7.5 and 10 wt% to the prepolymer weight).

Three waterborne PUD were prepared using NCO/OH = 1.5 by method in which the dispersing procedure was modified (Lee et al., 2006). In the modified procedure only the dispersing stage was varied compared to the standard procedure. The prepolymer solution was mixed with a small amount of deionized water for dispersion of polymer in water. Solvent was added for reducing the viscosity, if necessary.

Into a 250 ml glass reaction kettle, equipped with a mechanical stirrer containing a torque meter, a thermometer, a condenser for reflux and nitrogen gas inlet, was added 60 g (0.03 mol) of PCD (dried under vacuum at 120 °C); and 4, 8 or 12 g (0.03, 0.06 or 0.09 mol) of DMPA dispersed in 30 ml DMF. The reaction mixture was heated at 70 °C for 0.5 h to obtain a homogeneous mixture. This step is important for the resulting equal uniform distribution of hydrophilic monomer, DMPA, on PU backbone. After that 20, 32 or 40 g (0.09, 0.15 or 0.18 mol) of IPDI and DBTDL (0.03 wt. % of the total solid) were added to the homogenized mixture and stirred at 80 °C for 2.5 h. Dibutyl amine back titration method was used for the determination of the reaction time necessary to obtain completely NCO-terminated prepolymer. Then the mixture was cooled down to 60 °C and carboxylic groups (DMPA equiv) were neutralized with TEA dissolved in NMP (2 wt % of the total reaction mass) by stirring the reaction mixture at 60 °C for 1h.

Subsequently, the prepolymer solution was mixed with 0.3 ml of deionised water for dispersion step-by-step. Stirring was increased during the addition of water, and the mixture was diluted with NMP. Waterborne PUD was obtained by drop-wise addition of water to the mixture in order to obtain PUD with solid content of about 30% at 30 °C for 1h. The chain extension was carried out with solution of 0.9 or 1.8 g (0.015 or 0.03 mol) of EDA in 2 ml of deionised water at 35 °C for 1h. The mixture was heated to 80 °C under vacuum in order to remove NMP and to obtain PUD with solid content of about 30%.

The thermal stability of PU was determined by TG. The DTG thermogram of cured films of PUD based on PCD showed several degradation steps (Fig.5b). Detailed analysis of the thermogram is represented in Table 2. The decrease in the DMPA content produces a slight increase in the decomposition temperature (Fig.5a). However, the degradation mechanism was very complex due to the different stability of the hard and soft segments.

The removal of residual water due to incomplete drying of PU was produced at around 130 °C. The DTA thermogram of the used aliphatic PCD shows the main degradation at 350 °C and other less important at 265 °C (Garcia, 2010, 2011). Because this diol and EDA were used in the synthesis of PU, two kinds of groups in hard segments have been formed, i.e., urethane and urea ones. The decomposition temperature of PUD is mostly influenced by the chemical structure of the component having the lowest bond energy (Coutinho et al., 2003; Cakic et al.,2009). The urethane bond has lower thermal resistance than the urea bond and thus the first decomposition process at about 280 °C corresponds to the beginning of the urethane part of hard segment degradation and second at about 300 °C to the degradation of the urea part of hard segment. The degradation in PU at 264–268 °C is characteristic of the polyol. The degradation of soft segment (mainly composed of polyol) is produced at 329–338 °C. The soft and hard segments content were quantified from the weight loss at above mentioned temperatures (265 °C from PCD degradation, 280 °C and 300 °C correspond to the urethane and urea hard segment degradation, and 330 °C from degradation of soft segment).

According to Table 2, the decrease in DMPA content produced a slight increase in the decomposition temperature and a decrease in the weight loss for the decomposition of urethane and urea hard segments, which can be ascribed to a decrease in the amount of hard segment. The decomposition temperature of the soft segments is produced at 329–338 °C and the weight loss increase by decreasing DMPA content 20.9 and 23.5 to 6.2wt.% in PU ionomers.

Sample PUD	Residual water		Soft segment		Hard segment	
	T (°C)	Weight loss (wt%)	T (°C)	Weight loss (wt%)	T (°C)	Weight loss (wt%)
4.5%DMPA	129.8	1.1	268; 329	10.1; 20.9	282; 301	23.5; 45.4
7.5%DMPA	137.3	1.0	264; 338	3.8; 23.5	290; 304	17.5; 53.7
10%DMPA	128.8	0.9	267; 331	10.5; 6.2	280; 301	22.9; 59.3

Table 2. Temperature of decomposition and weight loss of PUD (obtained by TG measurements)

The decrease in DMPA content produces a decrease in the hard segment content of PU ionomers. The resistance to thermal degradation of PU ionomer increased by decreasing the DMPA content due to the lower hard segment content.

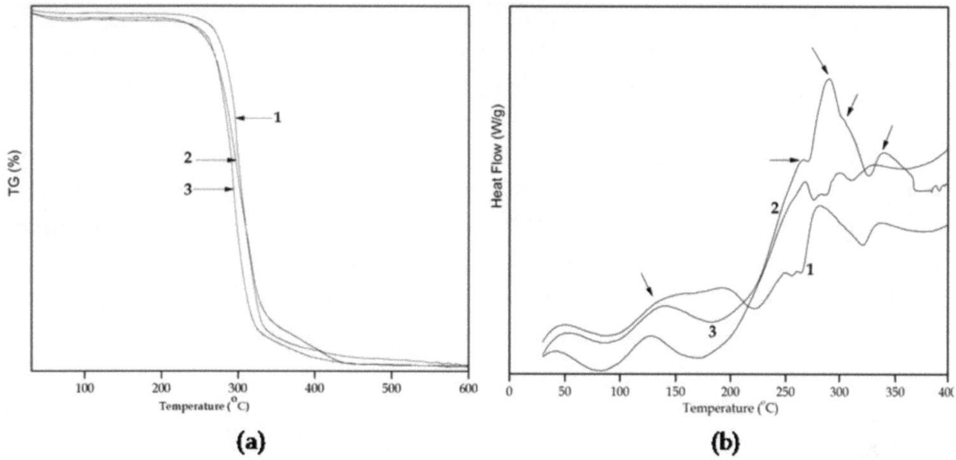

Figure 5. TGA curves (a) and DTA curves (b) of cured films of PUD based on PCD with 4.5% DMPA (1), 7.5% DMPA (2), 10% DMPA (3).

7. Conclusions

The wide application of PUD makes necessary better understanding of the chemistry-structure relationship that improves the thermal stability as this is important prerequisite to obtain tailor-made products for high performance applications. In this work, the investigation on thermal degradation of PUD with well-defined architectures indicated that diol types and DMPA content had great influence on thermal stability. PUD with lower DMPA content has shown enhanced thermal stability. The degradation profiles of PU aqueous dispersions were influenced by the type of chain extender, length of the hard segment and type of catalysts. The TG curves showed that the length of the hard segment had a strong influence on the thermal profile of the samples as a whole. The possibility for using glycolysis products of waste PET in PUD manufacturing was confirmed. The effects of glycol type and the different mole ratios of PET to glycol on thermal properties of PUD have been described. The degradation profile of the dispersions was dependent on mole ratios of PET to glycol in glycolyzed products. The samples based on (PET/glycol molar ratio 1:10) have shown enhanced thermal properties, which can be ascribed to increased length of glycol in glycolyzed oligoester polyol.

Author details

Suzana M. Cakić and Olivera Z. Ristić
University of Niš, Faculty of Technology, Leskovac, Serbia

Ivan S. Ristić
University of Novi Sad, Faculty of Technology, Novi Sad, Serbia

8. References

[1] Atta A.M.; Abdel-Raouf M.E.; Elsaeed S.M.; Abdel-Azim A.A. (2006) Curable resins based on recycled poly(ethylene terephthalate) for coating applications. *Prog. Org. Coat.,* 55, pp. 50–59.

[2] Atta A.M.; El-Kafrawy A.F.; Aly M.H.; Abdel-Azim A.A. (2007) New epoxy resins based on recycled poly(ethylene terephthalate) as organic coatings. *Prog. Org. Coat.,* 58, pp. 13–22.

[3] Athawale V.D. & Kulkarni M.A. (2009) Preparation and properties of urethane/acrylate composite by emulsion polymerization technique. *Prog. Org. Coat.,* 65, pp. 392–400.

[4] Athawale V.D. & Kulkarni M.A. (2010) Polyester polyols for waterborne polyurethanes and hybrid dispersions. *Prog. Org. Coat.* 67, pp. 44–54.

[5] Bechara I. (1998) Formulating with polyurethane dispersions. *Eur. Coat. J.,* 4, pp. 236–243.

[6] Blank W.J.; He Z.A. & Hessell E.T. (1999) Catalysis of the isocyanate-hydroxyl reaction by non-tin catalysts. *Prog. Org. Coat.,* 35, pp. 19-29.

[7] Blank W.J. & Tramontano V.J. (1996) Properties of crosslinked polyurethane dispersions. *Prog. Org. Coat.,* 27(1), pp. 1-15.

[8] Cakić S. M.; Lačnjevac Č.; Stamenković J.; Ristić N.; Takić Lj.; Barać M.; Gligorić M. (2007 a) Effects of the acrylic polyol structure and the selectivity of the employed catalyst on the performance of two-component aqueous polyurethane coatings. *Sensors,* 7(3), pp. 308-318.

[9] Cakić S.M.; Nikolić G.S.; Stamenković J.V. (2007 b) Thermo-oxidative stability of waterborne polyurethanes with catalysts of different selectivity evaluated by non-isothermal thermogravimetry. *J. Serb. Chem. Soc.,* 72(7), pp. 723-735.

[10] Cakić S.M.; Nikolić G.S.; Lačnjevac Č.; Gligorić M.; Rajković M. (2007 c) The thermal degradation of aqueous polyurethane with catalysts of different selectivity. *Prog. Org. Coat., 60(2),* pp. 112-116.

[11] Cakic S.; Lačnjevac Č.; Rajković M.B.; Rašković Lj.; Stamenković J. (2006 a) Reticulation of Aqueous Polyurethane Systems Controlled by DSC Method. *Sensors,* Vol. 6, No. 5, pp. 536-545.

[12] Cakic S.; Lačnjevac Č.; Nikolić G.; Stamenković J.; Rajković M.B.; Gligorić M.; Barać M. (2006 b) Spectroscopic Characteristics of Highly Selective Manganese Catalysts in Acqueous Polyurethane Systems. *Sensors,* 6, pp. 1708-1720.

[13] Cakić S.; Ristić I.; Djordjević D.; Stamenković J.; Stojiljković D. (2010) Effect of the chain extender and selective catalyst on thermooxidative stability of aqueous polyurethane dispersions. *Prog. Org. Coat.,* 67, pp. 274–280.

[14] Cakić S.; Ristić I.; M-Cincović M.; Nikolić N.; Ilić O.; Stojiljković D.; B-Simendić J. (2011) Glycolyzed products from PET waste and their application in synthesis of polyurethane dispersions. *Prog. Org. Coat.,* In press, doi: 10.1016/j. porgcoat.2011.11.024.

[15] Cakić S.; Stamenković J.; Djordjević D.; Ristić I. (2009) Synthesis and degradation profile of cast films of PPG-DMPA-IPDI aqueous polyurethane dispersions based on selective catalysts. *Polym. Degrad. Stab.,* 94, pp. 2015–2022.

[16] Chang T.C.; Chiu Y.S.; Chen H.B.; Ho S.Y. (1995) Degradation of phosphorus-containing polyurethanes. *Polym. Degrad. Stab.*, 47, pp. 375-381.

[17] Chattopadhyay D.K. & Webster D.C. (2009) Thermal stability and flame retardancy of polyurethanes. *Prog. Polym. Sci.* 34, pp. 1068-1133.

[18] Collong W.; Gobel A.; Kleuser B.; Lenhard W.; Sonntag M. (2002) 2K waterborne clearcoat-a competition between crosslinking and side reactions. *Prog. Org. Coat.*, 45, pp. 205-209.

[19] Coutinho F.M.B. & Delpech M.C. (1996) Some properties of films cast from polyurethane aqueous dispersions of polyether-based anionomer extended with hydrazine. *Polym. Test.*,15, pp. 103-113.

[20] Coutinho F.M.B.; Delpech M.C.; Alves T.L.; Ferreira A.A. (2003) Degradation profiles of cast films of polyurethane and poly(urethane-urea) aqueous dispersions based on hydroxyterminated polybutadiene and different diisocyanates. *Polym. Degrad. Stab.*, 81, pp. 19-27.

[21] Delpech M.C. & Coutinho F.M.B. (2000) Waterborne anionic polyurethanes and poly(urethane-urea)s: influence of the chain extender on mechanical and adhesive properties. *Polym. Test.*, 19, pp. 939-952.

[22] Dieterich D. (1981) Aqueous Emulsions, Dispersions and Solutions of Polyurethanes; Synthesis and. Properties. *Prog. Org. Coat.*, 9, pp. 281-340.

[23] Dulog L. & Storck G. (1966) Die oxydation von polyepoxiden mit molekularem sauerstoff, *Macomol. Chem.*, 91, pp. 50-73.

[24] Fambri L.; Pegoretti A.; Gavazza C.; Penati A. (2000) Thermooxidative Stability of Different Polyurethanes Evaluated by Isothermal and Dynamic Methods. *J. Appl. Polym. Sci.*, 81, pp. 1216–1225.

[25] Foy E.; Farrell J.B.; Higginbotham C.L. (2009) Synthesis of Liner Aliphatic Polycarbonate Macroglycols Using Dimethylcarbonate. *J. Appl. Polym. Sci.*, 111, pp. 217–227.

[26] Garcia-Pacios V.; Costa V.; Colera M.; J. Martin-Martinez M. (2010) Affect of polydispersity on the properties of waterborne polyurethane dispersions based on polycarbonate polyol. *Int. J. Adhes. Adhes.*, 30, pp. 456–465.

[27] Garcia-Pacios V.; Costa V.; Colera M.; Martin-Martinez J.M. (2011) Waterborne polyurethane dispersions obtained with polycarbonate of hexanediol intended for use as coatings. *Prog. Org. Coat.*, 71, pp. 136–146.

[28] George Woods, (1987). *The ICI Polyurethanes Book*, 2nd Edition, Wiley, New York.

[29] Gunduz G. & Kisakurek R.R. (2004) Structure–Property study of waterborne polyurethane dispersions with Different hydrophilic content and polyols. *J. Disper. Sci.Technol.*, 25 (2), pp. 217-228.

[30] Hepburn C. (1992) *Polyurethane Elastomers*, Second ed., Elsevier, New York.

[31] Jacobs, P.B. & Yu, P.C. (1993) Two-Component Waterborne Polyurethane Coatings. *J.Coat. Tech.*, 65 (822), pp. 45-50.

[32] Jang J. Y.; Jhon Y.K.; Cheong I.W.; Kim J.H. (2002) Effect of process variables on molecular weight and mechanical properties of water-based polyurethane dispersion. *Colloids Surf. A- Physicochem. Eng. Aspects*, 196, pp. 135-143.

[33] Kim B.K. & Min L.Y. (1994) Aqueous dispersion of polyurethanes containing ionic and nonionic hydrophilic segments. *J. Appl. Polym. Sci.*, 54, pp. 1809-1815.
[34] Kim B. K. (1996) Aqueous polyurethane dispersions. *Coll. Polym. Sci.*, 274, pp. 559-611
[35] Lee D.K.; Tsai H.B.; Tsai R.S. (2006) Effect of Composition on Agueous polyurethane Dispersions Derived from polycarbonatediols. *J. Appl. Polym. Sci.*, 102, pp. 4419–4424.
[36] Lee D.K.; Tsai H.B.; Wang H.H.; Tsai R.S. (2004) Aqueous Polyurethane Dispersions Derived from Polycarbonatediols. *J. Appl. Polym. Sci.*, 94, pp. 1723–1729.
[37] Lee H.T.; Hwang Y.T.; Chang N.S.; Huang C.C.T.; Li H.C. (1995) Waterborne, High-Solids and Powder Coatings Symposium, New Orleans, pp. 224.
[38] Lu M.G.; Lee J.Y.; Shim M.J.; Kim S.W. (2002) Thermal Degradation of Film Cast from Aqueous Polyurethane Dispersions. *J. Appl. Polym. Sci.*, 85, pp. 2552–2558.
[39] Oertel G. (1985) *Polyurethane Handbook*, Hanser Publishers, Munich.
[40] Patel M.R.; Patel J.V.; Mishra D.; Sinha V.K. (2007) Synthesis and Characterization of Low Volatile Content Polyurethane Dispersion from Depolymerised Polyethylene Terphthalate. *J. Polym. Eviron.*, 15, pp. 97–105.
[41] Pielichowski K.; Slotwinska D.; Dziwinski E. (2004) Segmented MDI/HMDI based polyurethanes with lowered flammability. *J. Appl. Polym. Sci.*, 91, pp. 3214-3224.
[42] Prime R.B.; Burns J.M.; Karmin M.L.; Moy C.H.; Tu H.B. (1988) Applications of DMA and TGA to quality and process control in the manufacture of magnetic coatings. *J. Coat. Technol.*, 60, pp. 55–60.
[43] Ramesh S. & Radhakrishna G. (1994) Synthesis and characterization of polyurethane ionomers. *Polym. Sci.*, 1, pp. 418-423
[44] Randell D. & Lee S. (2000) *Polyurethane Book*, 2nd Editors, John Wiley & Sons, New York.
[45] Rothause J. W. & Nechtkam K. (1987) Advances in Urethane Science and Technology, 10, pp. 121
[46] Rosthauser J.W. & Nachtkamp K.J. (1986) Waterborne polyurethanes. *J.Coat. Fabrics.*, 16, pp. 39-79.
[47] Scaiano J. C. (1989) Laser Photolysis in Polymer Chemistry. Degradation and Stabilization of Polymers. Elsevier, Amsterdam.
[48] Stamenković J.; Cakić S.; Nikolić G. (2005) Study of the catalytic selectivity of an aqueous two-component polyurethane system by FTIR spectroscopy. *Chem. Ind.*, 57, pp. 559-562.
[49] Tawa T. & Ito S. (2006) The Role of Hard Segments of Aqueous Polyurethane-urea Dispersion in Determining the Colloidal Characteristics and Physical Properties. *Polym. J.*, 38(7), pp. 686-693.
[50] Wang T.L. & Hsieh T.H. (1997) Effect of polyol structure and molecular weight on the thermal stability of segmented poly(urethaneureas). *Polym. Degrad. Stab.*, 55, pp. 95-102.

Chapter 6

Polyurethane Flexible Foam Fire Behavior

Ahmadreza Gharehbagh and Zahed Ahmadi

Additional information is available at the end of the chapter

http://dx.doi.org/10.5772/47965

1. Introduction

Polyurethanes are a broad range of polymers, which are formed from the reaction between diisocyanates or polyisocyanates with diols or polyols. According to **the types and amounts of, polyols, isocyanate, ingredients** and the overall reaction circumstances, a broad range of products **like flexible foams, rigid foams**, elastomers, coatings and adhesives are produced.

Since the polyurethane products specially foams are playing an indispensable rule in our daily life **because of wide range of applications in automotive, household, refrigerators, insulations,** reducing of the fire risk of such a products are become more vital.

Conventional polyurethane **flexible** foams are easily ignited by a small flame source and burn rapidly with a high rate of heat release and smoke and toxic gases. This high flammability of polyurethane flexible foam is related to its cellular and open cell structure and low density of such foams. Oxygen can easily pass through the cells of the combustible material and in subjecting with an accident, smoldering cigarette or an electrical shock, foam catch fire [1].

Polyurethanes can be resisted against fire by different ways. **Depends on the types and applications of them, fire resisting could be done by the flame retardants using or by changing in** the polymer structure. In the whole picture the polymer ignition can be controlled by the following **factors**.

1. Extinguishing material reduction
2. Air supplying source reduction
3. Fire diffusion and heat generation reduction4. Increasing of the energy needed for entire combustion process

Different types of the fire retardants could be used according to one of the above mentioned categories. The flame retardants are acting according to one of the following mechanisms.

1. Reaction with the flame and preventing of the spreading of the fire by the created free radical blocks.
2. Preventing of the oxygen diffusion into the polymer
3. Lowering of the flame temperature with **removing energy from the system**
4. Char creation and creating a free place between the solid polymer and the disposed area.
5. Polymer expansion and making a free place between the fire and the decomposed polymer.

There are lots of materials which are known as the fire retardants according to the following groups.

- Halogenated flame retardants which are acting in the gas phase with disturbing the hydrogen-oxygen reaction. They react with hydrogen and create the halogen free radicals then they block free radicals of decomposed polymer.
- Metal oxides which act in solid or gas phase and some of the members of this group cause to reduce the flame temperature.
- Phosphorous containing compounds which create char on the extinguishing area of the polymer and prevent the oxygen feeding to the flame.
- Halogen free FR which the two main candidates are EG and Melamine. The heat stability of the polyurethanes especially the rigid foams at high temperatures depend on the isocyanurate to allophanates and biurets cross-linked bond ratio. Carbodiimide is produced by the condensation reaction of isocyanate with lose of CO2 (Fig.1). This reaction can be catalyzed by the cyclic phosphine-oxide. Generated carbodiimide is used as an unti-hydrolyze agent in the polyurethanes. The heat stability of the diverse products of polyurethanes is classified in Table (1).

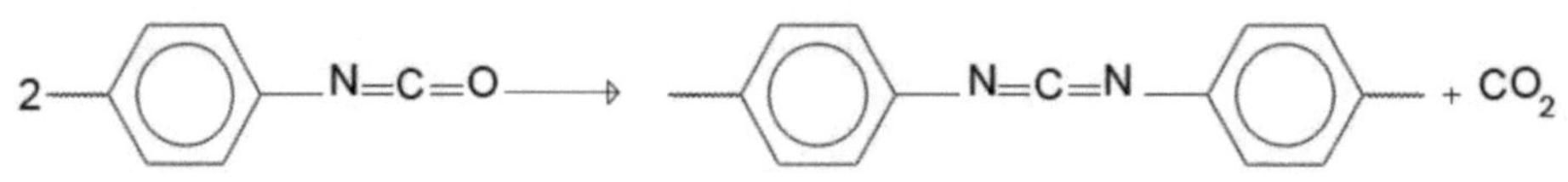

Figure 1. Condensation reaction of isocyanate to make Carbodiimide

Bonds	Decomposition Temperature(ºC)
allophanate	100-200
biuret	115-125
Urea	160-200
Urethane	180-200
Substituted Urea	235-250
carbodiimide	250-280
Isocyanurate	270-300

Table 1. Heat stability of the diverse products of polyurethanes

2. Polyurethane foam morphology

Polyurethane morphology plays a vital rule on the fire properties of the polyurethane foam. The porous structure of the foam helps to diffuse the oxygen easily in to the foam and accelerate the ignition process. Fig.2 shows the SEM Picture of the polyurethane flexible foam with no filler inside. As it is clear, the cell structure of the foam includes Cell window, Strut and Strut joint [1].

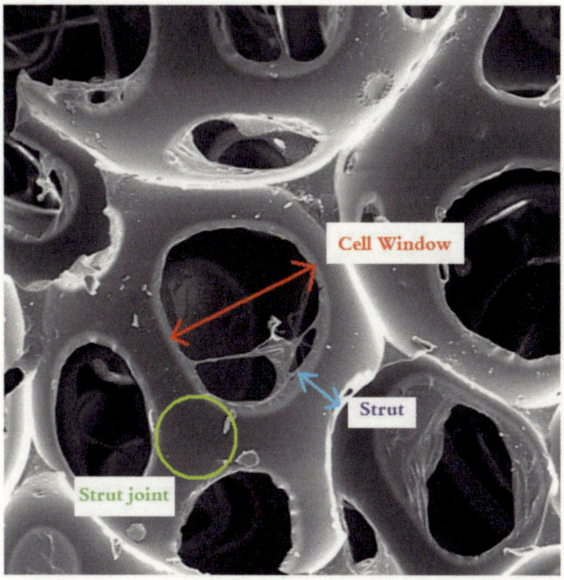

Figure 2. Fundamental concepts of polyurethane foam Cell Structure (SEM×200)

3. Polyurethane flexible foam fire retardants

3.1. Halogenated phosphorous flame retardants

In recent years the phasing out of some types of halogenated FR (flame retardant) due to persistence at environment and bioaccumulation and toxicity has been more investigated.

TMCP (Tris (2-chloroisopropyl) phosphate) and TDCP (Tris (1, 3-dichloroisopropyl phosphate) are two well-known liquid FR which are used in polyurethane flexible foam to make fire resisted **(Figure3-4). Table (2) illustrate some important parameters of the mentioned fire retardants.** [2]

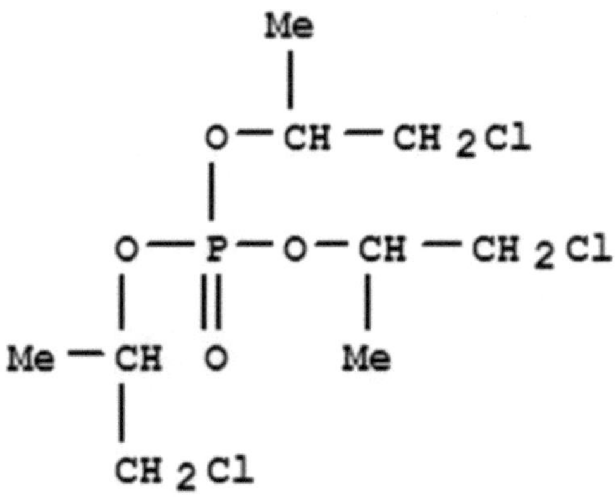

Figure 3. Chemical structure of TMCP

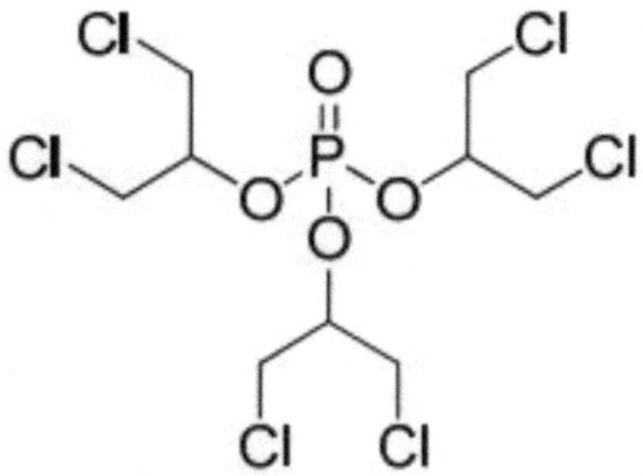

Figure 4. Chemical structure of TDCP

Property	TMCP	TDCP
CL content (%)	32.5	49
Mw(g/mol)	327.55	430.91
P content (%)	9.5	7.1
physical state at 25°C	clear liquid	clear liquid
Water solubility(%)	< 0.05	< 0.05

Table 2. Properties of TMCP and TDCP

Studies show that in the foams with only liquid FR (TMCP , TDCP) a very divergent combustion behaviour has been indicated. TMCP containing foams show lower TWL(total weight loss) and shorter burn time compared to TDCP containing foams.Moreover, TMCP containing foams didn't show any significant dripping and subsequent hole formation, a phenomenon seen at all levels of TDCP addition. TMCP and TDCP addition leads to decrease in the THE(total heat evolved) but an increase in the amount of smoke and carbon monoxide produced and this is why normally some amount of other FR such as melamine is added to the TMCP and TDCP containing foams to decrease total heat evolved, total smoke produced and CO emission significantly[2,3].

3.2. Halogen-free flame retardants

Due to the above mentioned reasons it has been a driving force to move toward the halogen free FR to compensate those weakness of halogenated FR, **despite of some disadvantages that the halogen free FR have e.g. they are mostly in solid state and they show process difficulties.** There are different types of halogen-free flame retardants which are behaving with different mechanisms. First group acts according to the expansion inside the polymer and oxygen-diffusion prevention and second group does by the cooling of the ignited surface of the polymer. One important example of the above mentioned groups are leaded by expandable graphite (EG) and Melamine powder respectively.

3.2.1. *Expandable graphite*

EG is a graphite intercalation compound in which some oxidants, such as sulfuric acid, potassium permanganate, etc. are inserted between the carbon layers of the graphite [4]. Fig.5 illustrates the chemical structure of Graphite, diamond and C60 [1].

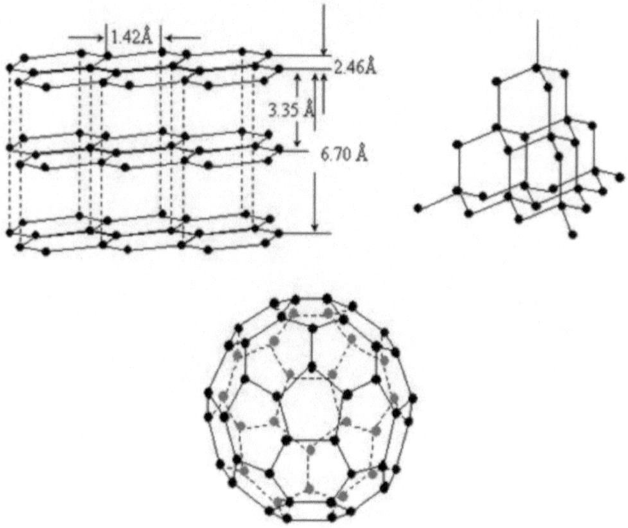

Figure 5. Comparison of Lattice graphite, Diamond and C60

When EG is subjected to a heat source, it expands to hundreds of times of its initial volume and creates voluminous, stable carbonaceous layer on the surface of the materials. This layer limits **the heat transfer from the heat source to the substrate and the mass transfer from the substrate to the heat source resulting in protection of the underlying material [5, 6]. The redox process [7]** between Sulfuric acid and graphite generates the blowing gases according to the reaction:

$$C + 2H_2SO_4 = CO_2\uparrow + 2H_2O\uparrow + 2SO_2\uparrow$$

The fire retardancy of EG is done by two steps [1]:

- The EG expands under the impact of Heat up to about 500 times of its original volume and creates a very large surface. It allows a quick oxidation of the carbon. The oxygen is taken out of the air and makes the air almost inert. This inert air extinguishes the fire.
- EG doesn't create flames while oxidation and will extinguish if no more heat will be applied to the glowing graphite. Therefore, no source of fire will be generated by the oxidizing graphite.

The more characteristic factors for EG which should be considered are:

- SET (start expansion temperature)
- Expansion volume
- Strength

Figure (6, 7) show particle size and distribution of two types of EG with different sizes (0.18mm, 0.25mm)

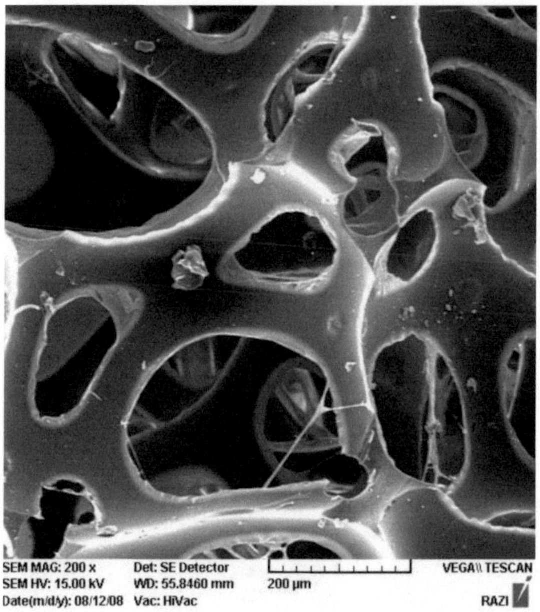

Figure 6. Particle size and distribution of 8% of EG (0.18mm) (SEM ×200)

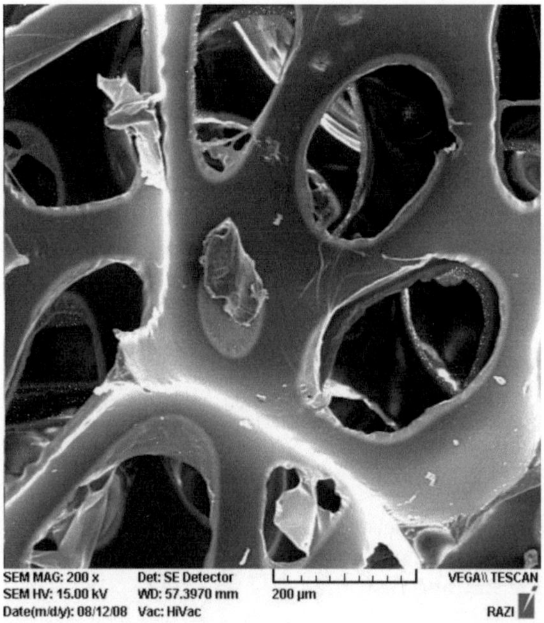

Figure 7. Particle size and distribution of 8% of EG (0.25mm) (SEM×200)

3.2.2. Melamine

Melamine acts as fire retardant and smoke-suppressant according to the following combined mechanisms [8].

- Melamine is believed to act as a heat sink, increasing the heat capacity of the combustion system and lowering the surface temperature of the foam. Thus the rates of combustible gas evolution and burning are reduced.
- The nitrogen content of the melamine may partly end up as nitrogen gas when melamine burns, providing both a heat sink and inert diluents in the flame. The presence of melamine in the foam results in less heat generated by the flame, consequently less heat fed back to the foam and the rate of foam pyrolysis, i.e. generating of volatile fuel is reduced.
- Due to a chemical interaction between melamine and the evolved isocyanate fraction creating from degradation of polyurethane foam. This interaction reduces the amount of diisocyanate the main contributor to the smoke and CO release (Fig.8).

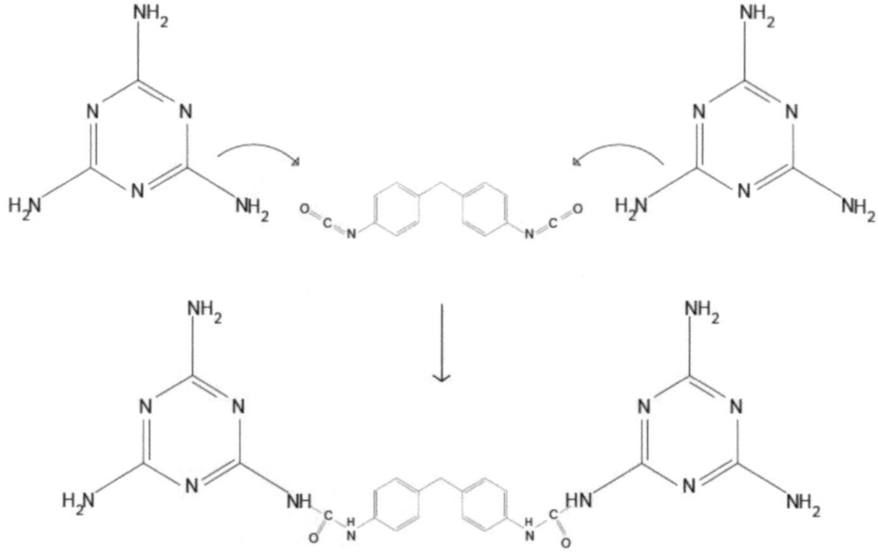

Figure 8. Chemical reaction between melamine and diisocyanate (MDI)

Figure (9) shows particle size and distribution of melamine powder inside the flexible foam.

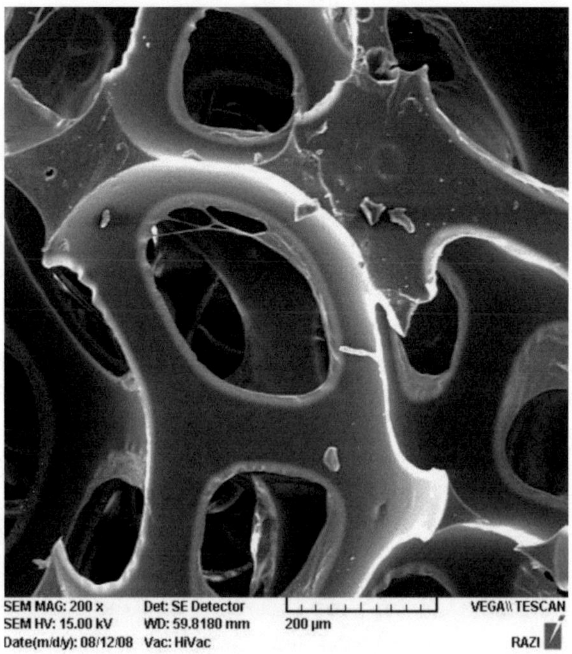

Figure 9. Particle size and distribution of 8% of melamine (SEM ×200)

4. Properties of the polyurethane flexible foam with different types of fire retardants

Comparison between halogenated flame retardants which are mainly liquid with halogen free flame retardants (expandable graphite and melamine powder) which both are solid can be categorized as **four** items.

- Processing
- Reactivity
- Fire properties
- Physical properties.

4.1. Foaming process

The best choice for the processing as it is clear will be the liquid grade which has a good dispersion inside the polyol and less side effects.

Expandable Graphite has a limit pot life (3-4 hours) and when it subjects with the polyol component, it attacks to the catalyst of the polyol and destroys the catalyst during the foaming process. **For the foam producing, the highly recommendation is the EG containing polyol should react with proper isocyanate before the EG pot life reaches or the EG should be injected by an individual stream and mix with the polyol stream in the mixing head instead of pre-mixing with polyol.** Otherwise the produced foams will be collapsed. The other disadvantage of this technology is related to the fact that EG is very corrosive and make the mixing head to be damaged and it is preferred to be used a hardened grade of mixing head, a damaged mixing head needle picture is showed in (Fig.10)[9].

The advantages of this technology is the good homogeneity of the EG particles inside the polyol.

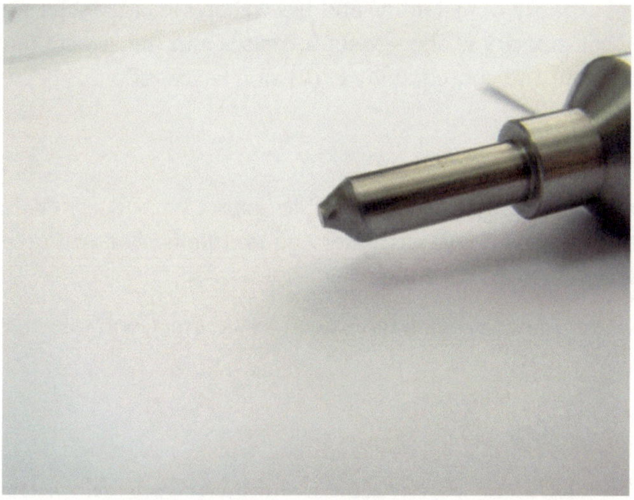

Figure 10. Mixing head needle corrosion by EG

Despite the EG, melamine has the longer pot life inside the polyol, which is around 24 hours but the fast sedimentation of the melamine powder in the polyol will be the main disadvantages so we need a suitable method to disperse the melamine powder in the polyol very well to achieve a homogeneous mixture.

4.2. Reactivity

Foam reactivity is determined by the following parameters:

- Cream time (sec): Cream time is the time when the polyol and isocyanate mixture begins to change from the liquid state to a creamy and starts to expansion subsequently.
- Gel time (sec): Gel time is the time the foam start to stiffen
- Rise time (sec): rise time is the time the foam reach to its maximum height
- Recession factor (%): the height percentage the foam is settled after 5 min after the rise time
- Expansion factor (cm/kg): the proportion of the maximum height of the foam to foam weight.

The flame retardants would affect on the foam reactivity depend on the types of them, whether they are solid or liquid. Because they make changes in cell structure and total system heat capacity. The recession factor goes up with addition of EG and melamine but with different slopes. This is due to the increase in the average cell size of the foam. The bigger the flake size, the larger the cells and higher the recession factor. On the other hand melamine powders with small sizes are embedded on struts and joints and increase the viscosity and reduce the drainage rate which consequently, decreases the number of cells with bigger sizes [10]. Melamine powders with bigger size (bigger than struts and joints) are embedded inside the cell walls and open the cells.

Expansion factor which is related to the free rise density (FRD) reduces with addition of the EG and melamine in the foam. This is due to the increase the heat capacity of the entire system because of high heat capacity of melamine and EG. When melamine and EG content increases in the system, the heat capacity of the system increases and the system temperature reduces, therefore, the foam height and consequently expansion factor reduces [11].

4.3. Fire properties

The fire properties of the polyurethane flexible foams have been evaluated by different types of methods depends on the customer requirements. For example, the automotive, railway and airplane industries have their own standards.

The most important parameters which have been tested are: Cone calorimetry, flammability, smoke density and toxicity.

4.3.1. Cone calorimetry ISO 5660

The principle of the calorimetry by oxygen consumption (cone calorimeter) is based on the relation between the oxygen consumption and the heat release during the combustion. The ratio between the heat release and the weight of oxygen consumed is a constant (Huggett constant) equal to 13100 kJ/kg. It has been previously demonstrated that cone calorimeter results are in good correlation with results obtained in full scale fire test on upholstered furniture [3].

Samples of flexible foams (10*10*5cm) were exposed in a Stanton Redcroft Cone Calorimeter according to ASTM 1356-90 under a heat flux of 35kW/m2 (case of fully involved real firs). This flux was chosen because it corresponds to the evolved heat during a fire. An electrical spark igniter ignited volatile gases from the heated specimen. At least three specimens have been tested for each formulation. Data were recorded with a computer connected to the cone calorimeter. The test gives the opportunity to evaluate:

- RHR: Rate of Heat Release (kW/m2)
- Figra: fire growth rate: RHR/time (kW/m2/s)
- Weight loss (wt. %)
- Emission of carbon monoxide (ppm)
- TVSP: Total volume of smoke production (m3)
- THE: total heat evolved (kJ/cm2/g)

The combustion of flexible polyurethane foams is a two steps process (Fig.11).

The first step corresponds to the melting of the foam into a tar and the second to the combustion of the tar previously produced. [3]

These two degradation steps lead to two distinct peaks of rate of heat released.

- The RHR1 values (the values of RHR of the first and second RHR peaks).
- The T1 and T2 values (times at which RHR1 and RHR2 occur).
- The Figra2 values (the two maximum peaks on the Figra curve).

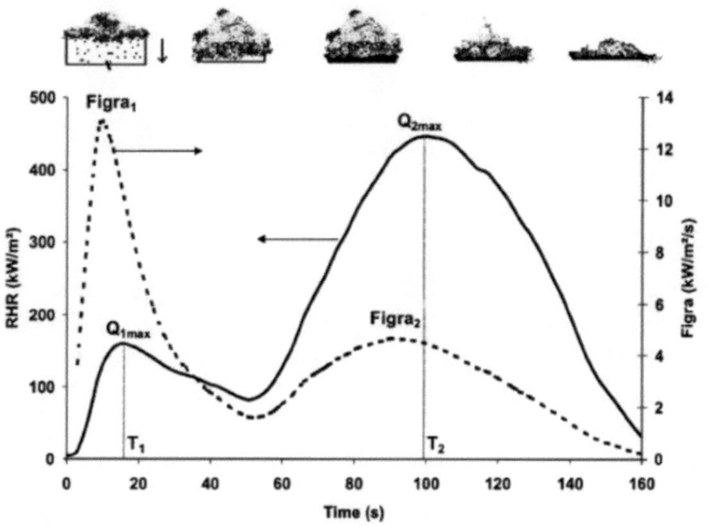

Figure 11. Combustion of flexible polyurethane foams: a two-stage process

4.3.2. Flammability

Flammability of the polyurethane foam is running with wide range of test methods depends on the applications and customers specification. Some fire tests standards include: FMVSS NO.302, British Standard 5852, ISO 9772 and FAA/JAA 25.853 Appendix F. as an example the airplane seat foam fire tests according to FAA/JAA 25.853 Appendix F have been investigated.

In this test 5 samples with 75mm*305mm*13mm dimension have been subjected with flame vertically for 12 sec and the following parameters have been investigated.

Burning time (the time that burning is continuing after removing the flame source) Burned length (the length of the foam which is damaged by the burning process) Time of dripping (the time which droplet continues to burn).

Synergetic effect

The synergetic effect of different types of FR has been observed. for instance, the fire properties of the EG loaded foams is much worse than when it is used by mixing with a liquid FR such halogenated phosphorous flame retardant. **Also when some amount of melamine is added to the TMCP and TDCP containing foams it helps to decrease total heat evolved, total smoke produced and CO emission significantly[2].**

Also the mixing of the liquid FR could boost the fire properties of the melamine loaded foam considerably.

4.3.3. Smoke density and toxicity

Smoke density and Toxicity are measured according to Airbus Directive ABD0031 (2005) on two categories:

1. Flaming
2. Non-flaming

Samples with 76mm*76mm*13mm are chosen to do the above mention tests against them in flaming and non-flaming status.

4.4. Physical properties

Physical and mechanical properties of the flexible polyurethane foams are evaluated by different types of tests in order to make an entire picture from the foam part performance during the consuming by the customer. For instance, flexible polyurethane foam is widely used as car seat foam and it has to keep its shape and other properties such as hardness and compression set during the time which is used. The most important properties of the polyurethane flexible foam as car seat foam according to RENAULT specifications are viewing as below.

- Core Density

- Compression Load Deflection (CLD: P25/5) and Sag-Factor according to D411003
- Compression Set according to D451046
- Tensile strength & Elongation at break according to
- Tear Strength according to D411048
- Resilience in 1st and 5th cycle according to D455128

When the polyurethane flexible foam is going to be fire resisted, some fire retardants in liquid or solid forms are entered in to the foam structure and make some changes in the physical properties of the final foam part. Mostly the valuable changes have been observed by the solid FR addition rather than the liquid one.

Depending on the fire retardant nature, shape and size, their addition may have some positive or negative effect on foam physical-mechanical properties. By loading the solid FR with the same amount, the foams become softer, because both additives have a similar size as cell windows and make the foam inhomogeneous. With EG, the homogeneity would be less than the foam loaded by melamine, because of its bigger size and flake shape which makes the foam much softer [1].

Sag-factor or the comfort index [12] increasing when the percentage of EG and melamine increases. It means that by adding solid FR, the comfort index would change considerably. Compression set, which is another very important factor, has increased by rising the EG percentage, but there was almost no changes in CS by increasing the melamine content. This effect is due to destroying effect of the cells structure by both additives but mainly by the EG.

Tear strensgth of the foams has improved by increasing the EG which could be related to the rigidity of EG flakes but deteriorates when melamine is added.

Finally, the resilience in 1st cycle is decreased for all additives but it is recovered in 5th cycle, because in 1st cycle the polymer chains have lost their flexibility due to rigid particles but after 5 cyclic movements the particles are embedded in struts and joints and the foam restores its flexibility.

5. Statistical method

Principle component analysis (PCA) is a useful method to illustrate relations between different parameters by using STAT-BOX-ITCF [13, 14].

Interpretation of the results consists first in the checking the representation of the variables in the circles of correlation. The correlations between variables are deduced from the relative position and the length of their corresponding vectors on the circle of correlation. An example of interpretation is done in (Fig.12); the angle between two vectors defines the intensity of the correlation (vectors 1 and 5). If α is=90^0, no relation exists between the variables. The strength of the correlation is higher when the angle is close to 0^0 or 180^0. So, orthogonal vectors (vectors 1 and 2) mean no correlation between the variables. Data are strongly correlated if their vectors are collinear (vectors 1 and 3, and vectors 1 and 4). The nature of the correlation also depends on the direction of the vectors: if vectors have the same direction (vectors 1 and 4) the variables are correlated, i.e. an increase in the variable linked to the vector 1 corresponds to an increase in the variable linked to the vector 4. Inversely, if vectors are opposite (vectors 1 and 3), the variables are anti-correlated.

The correlation between two variables is also a function of the length of the vectors. As example, vectors 2 and 6 are co-linear and so should be anti-correlated. But the weak length of the vector 6 means that its corresponding variable does not influence the variable linked to vector 2 [2].

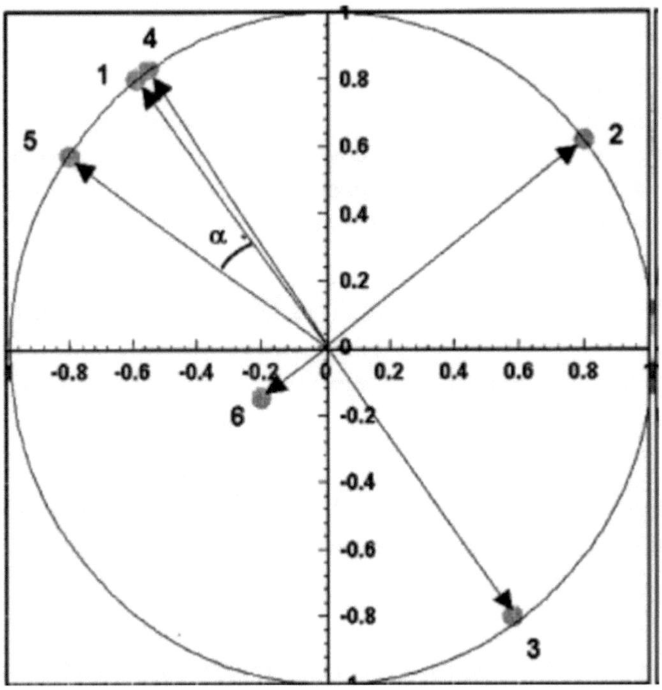

Figure 12. Interpretation of principal components analysis

5.1. Cone calorimeter–FMVSS 302

The principal components analysis from cone calorimeter and FMVSS 302 data shows the following correlations (Fig.13)

- *RHR1* is moderately correlated with *Figra1*: Figra1 is a variable that depends on the first peak of HRR (also called q1max), *d* the time it occurs. So, it seems quite coherent to find this kind of relation if the relative variation of the time is low.
- *RHR1* is correlated with *Figra2*. In the cone calorimeter, the foam degradation occurs in two main steps. It is obvious that an important consumption of fuel in the first step leads to a lower Figra2.
- FMVSS is strongly correlated with Figra1 and Figra2 and inversely correlated with *RHR2*. The lower Figra1 and Figra2, the slower the flame spread. A high RHR2 means loss of heat by dripping.

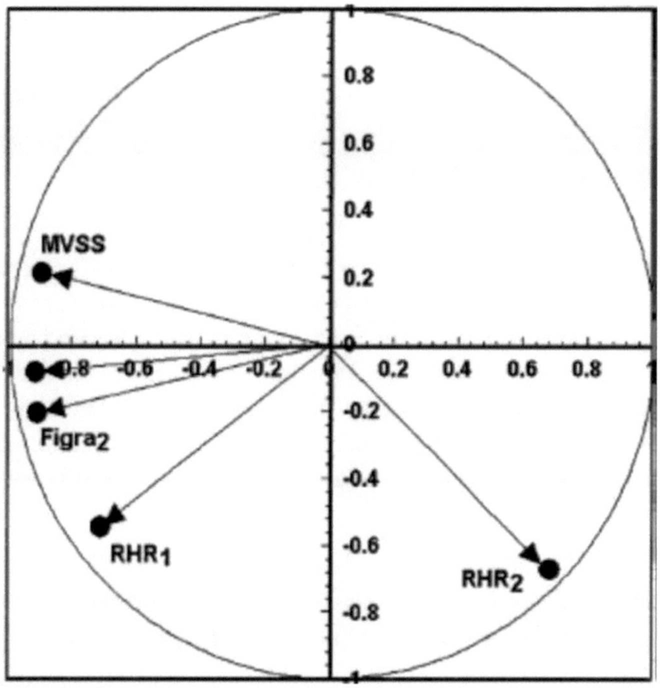

Figure 13. Correlation circle—relationship: cone calorimeter/FMVSS.

From the energy assessment of the foam consumption during 1s, we can find a relation between the propagation speed of the flame and the energy of the tar produced by the combustion (Fig.14).

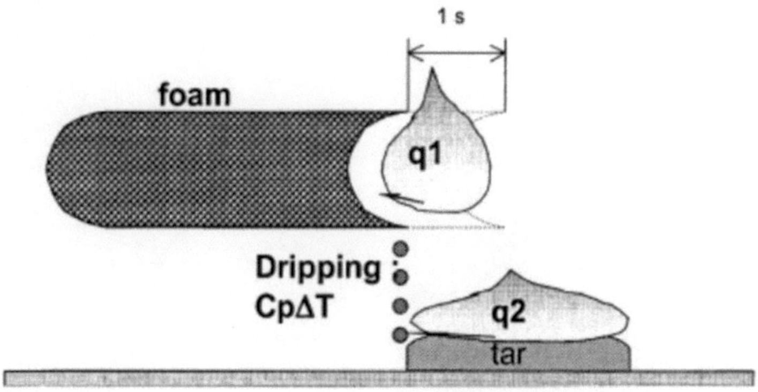

Figure 14. Principle of FMVSS.

As a first hypothesis, we can consider the following relation:

$$q1 + q2 - \Delta Q = Q = \text{constant}$$

- ΔQ corresponds to the part of heat used to melt the polymeric matrix leading to dripping.

- *q*1 corresponds to the energy released during the first stage of the combustion that leads to the formation of the tar (Figra1).
- *q*2 corresponds to the energy released by the combustion of the tar (Figra2).

This relation indicates the different strategies to decrease the value of RHR1 (and so Figra1), that is to say the flame spread in the FMVSS 302 tests:

- To decrease the total heat evolved *Q* using specific FR additives.
- To decrease the heat released during the first stage of degradation of the foam and as a consequence to decrease the heat fed back to the virgin polymer (decrease in Figra1).
- To increase RHR2, that is to say to reduce the energy of combustion by dripping.
- To delay the heat released by the tar. When the foam is molten, the tar starts to burn and this tar is not immediately lost by dripping. Hence, it is of interest to delay the combustion of this tar to enable it to drip (decrease in Figra2). An increase in RHR2 is not sufficient to reduce the flame spread and it is important that the high energy tar degrades at a later stage.

Hence, we may propose that the flame propagation rate in FMVSS 302 testing is much lower when easy melting and dripping allows heat reduction and tar dripping. It may be proposed that *q*2 corresponds in fact to the almost complete combustion of the tar.

Comparing the RHR curves of foams processed with variable water level, we note that the density of the foam strongly influences the first RHR peak (Fig.15). The higher the water content (the lower the density) the faster the step of melting under cone calorimeter conditions.

The effect of density on RHR1 may explain the previous correlation found between density and FMVSS 302. Low density leads to rapid melting and to a high flame propagation rate.

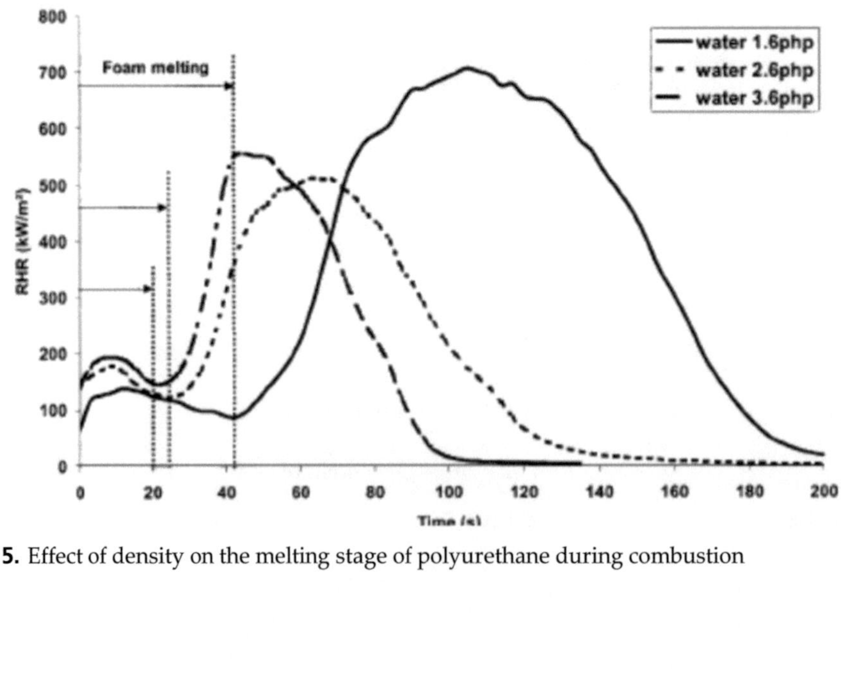

Figure 15. Effect of density on the melting stage of polyurethane during combustion

5.2. Cone calorimeter–British Standard

Ignition Source Crib 5 test to SI 1324 Sch. 1 Pt. 1

The statistical computation was made considering the two different sets of foams: the foams containing TMCP–melamine and the ones containing TDCP–melamine. The level of fire retardant additives has been included in the computation but is not shown on the circles of correlation.

Considering the TMCP–melamine foams (Fig.16) it is of interest to note that the lower are Figra1 and Figra2, the lower are the burn times, TWL and DWL. We also note that *T*2 is strongly inversely correlated with the data of SI 1324 Sch. 1 Pt. 1, that is to say the higher *T*2 the

Better results under the SI 1324 test (lower TWL, DWL and burn time).

The statistical computation of the data from the formulations TDCP–melamine clearly shows that the fire behavior of these foams in the SI 1324 test is linked to the second stage of degradation of the foam in the cone calorimeter (Figra2 and *T*2). Indeed, the Figra curve represents the fire growth rate of foam during combustion.

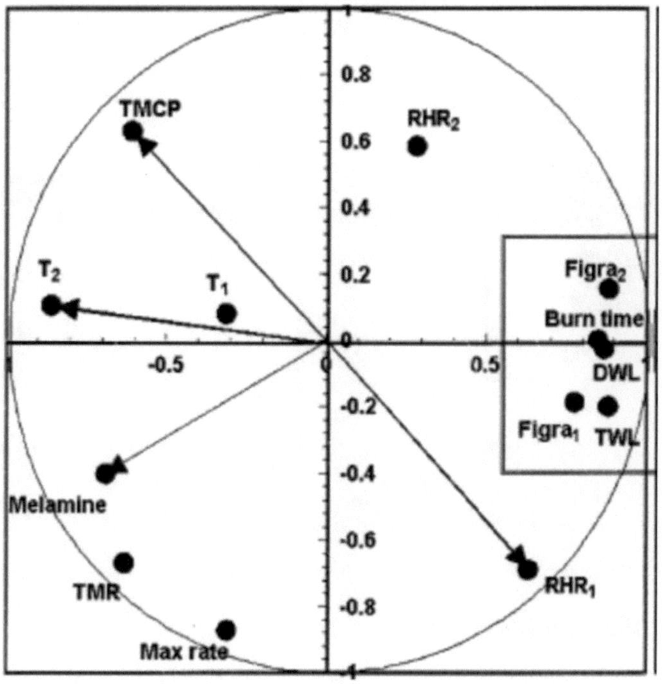

Figure 16. Correlation circle—relationship: cone calorimeter/SI

1324, TMCP–melamine formulations

A high Figra means a high rate of flame propagation and so leads to a high weight loss of the material. Hence, it is not surprising that Figra curves are strongly linked to the BS5852 results. The combustion of PU foam occurs in two steps: the "melting" of the foam and the combustion of the tar. The tar combustion is the most exothermic part of the combustion. A decrease in the heat released by the tar reduces the flame propagation and leads to a decrease in the weight loss of the foam (Fig. 17).

The TDCP and TMCP additives differ in their chlorine and phosphorus content and also in their temperature of degradation. TMCP degrades earlier than TDCP (150 ^{0}C and 210 ^{0}C, respectively); this temperature corresponds to a 5 wt. % weight loss under thermo gravimetric analysis conditions). A previous study [15] has clearly shown that TMCP is efficient in the early stage of combustion but no interaction with melamine is observed (temperature of 5 wt. % weight loss of melamine is 290 ^{0}C). TDCP acts later and when melamine starts to degrade about 50 wt. % of TDCP is available in the system, so a strong TDCP–melamine synergy is observed. The use of TDCP or TMCP in combination or not with melamine leads to very distinctive fire properties of the foams.

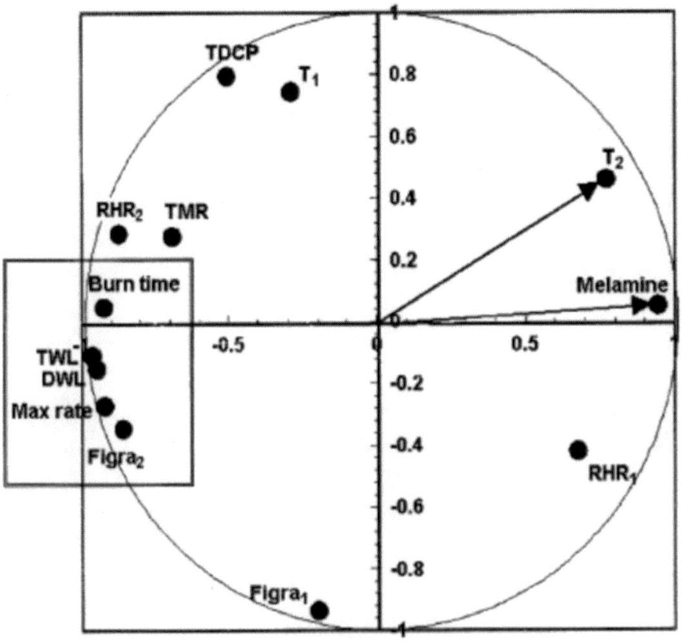

Figure 17. Correlation circle—relationship: cone calorimeter/SI

1324, TDCP–melamine formulations

Considering the TMCP–melamine foams, it is of interest to note that the higher the TMCP content the lower is RHR1. That confirms the early effect of TMCP that acts by decreasing the heat released by the foam in the first stage of the combustion. Secondly, the melamine content is inversely correlated with RHR2. As described previously, the temperature of decomposition of melamine is high (290 ^{0}C) and this inverse correlation indicates an efficiency of melamine during the combustion of the tar.

Regarding the TDCP–melamine formulations, we note a positive effect of the TDCP amount on the RHR1 peak. Even if TDCP degrades later than TMCP, a part of the TDCP is efficient in the first stage of the combustion.

The melamine content is strongly correlated with the SI 1324 data. High melamine content leads to a decrease in TWL, DWL burn time and maximum rate of weight loss.

The Figra2 and RHR2 peaks are also correlated with these data.

5.3. Properties–FMVSS 302

The statistical treatment shows that the FMVSS 302 rating is an inverse function of the density of the foam which is itself a function of the water index (Fig. 18). No significant relations may be proposed between FMVSS 302 and porosity or TDI index because data did not show any variation of the porosity (same SnOct content) and only a low variation of the TDI index.

The porosity index of the foam is strongly correlated with the SnOct range used in the foam manufacturing.

The previous study of conventional foams has revealed correlations between the FMVSS 302 testing and these parameters. The PCA study shows the absence of correlation between the EO content, the porosity (and so the SnOct range) and the index of the foam with the FMVSS 302 testing. However, it clearly shows that FMVSS 302 is strongly and inversely correlated with the density of the foam as it has been previously supposed.

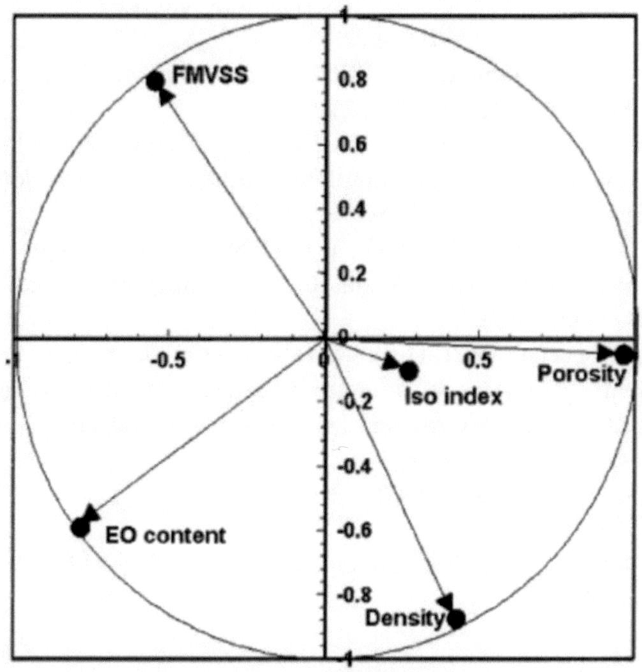

Figure 18. Correlation circle—relationship: physical properties/FMVSS.

Author details

Ahmadreza Gharehbagh
Iran Polyurethane Mfg.Co. NO.30, Tehran, Iran

Zahed Ahmadi
Color and Polymer Research Center, Amirkabir University of Technology, Tehran, Iran

6. References

[1] R. Bashirzadeh, A. Gharehbaghi, Journal of Cellular Plastics December 30, 2009, An Investigation on Reactivity, Mechanical and Fire Properties of Pu Flexible Foam
[2] Jerome Lefebvre and Michel Le Bras, Benoit Bastin and Rakesh Paleja, Rene Delobel, Journal fire sciences, Vol.21-september 2003
[3] Jerome Lefebvre, Benoit Bastin, Michel Le B, Sophie Duquesne,Christian Ritter, Rakesh Paleja, Franck Poutch , Flame spread of flexible polyurethane foam: comprehensive study, Polymer testing, 23(2004) 281-290
[4] Lei Shi, Zhong-Ming Li, Wei Yang, Ming-Bo yang, qiu-Ming Zhou, Rui Huang, Powder Technology 170(2006) P.178-184.
[5] Bourbigot, S.; Le, B.M.; Decressain, R.; Amoureux, J.P. J.Chem.Soc.-Faraday T.1996, 92(1), 149-158.
[6] Delobel R, Lebras M, Ouassou N, alistiqsa F. JFire Sci 1990; 8:85-108.
[7] Camino G,Duquesne S, Delobel R, Eling B,Lindsay C,Roels T.Fires and polymers. In: Nelson GL, Wilkie CA, editors.Materials and solutions for hazard prevention. Washington DC: ACS Pub; 2001.P.90.
[8] Dennis Price, Yan Liu, G.John Milnes, Richard Hull, Baljinder K.Kandola and A.Richard Horrocks Fire and Materials .2002; 26:201-206.
[9] Gharehbaghi, R. Bashirzadeh, and Z. Ahmadi, Polyurethane flexible foam fire resisting by melamine and expandable graphite: Industrial approach, Journal of Cellular Plastics, online published on 19 September 2011
[10] Turner, R.B., Nichols, J.B. and Kuklies, R.A. (1988). The Influence of viscosity in Cell opening of Flexible Molded Foams, In: Proceedings of the SPI, 31st Conference, Technomic, Lancaster, PA.
[11] A.Konig, U.Fehrenbacher and T. Hirth, E.Kroke, J.of Cellular plastics Vol.44-Nov.2008
[12] Kaneyoshi Ashida, Polyurethane and Related Foams, Chemistry and Technology (2007)
[13] G. Philippeau, Comment interpre´ter les re´sultants d'une analyse en composantes principales, I.T.C.F, Paris, 1986.
[14] H. Harman, Modern factor analysis, The University of
[15] Chicago Press, Chicago, 1976.
[16] B.Bastin, R. Paleja, J.Lefebvre, in: Polyurethanes EXPO 2002, API Conference, Salt Lake City, Utah,2002,p.244.

Section 3

Applications

Chapter 7

Polyurethane in Urological Practice

Valentina Cauda and Furio Cauda

Additional information is available at the end of the chapter

http://dx.doi.org/10.5772/36408

1. Introduction

Polyurethane (PU) is one of the most bio- and blood-compatible materials currently used for fabrication of various medical devices, e.g. blood bags, vascular/ureteral catheters and artificial heart. Originally, PU was conceived with other copolymers, aiming at similar goals, i.e. enough versatility to successfully meet biomedical devices constraints, such as biocompatibility, resistance to sterilization, physical features invariance over time and infection resistance during indwelling. PU in particular, has also other key properties particularly suited to biomedical industry, including strength, versatility and low cost. The surface of PU can be chemically functionalized with organic and biologically active molecules, resulting in improved durability, compliance, acceptance and tolerance in the human body during implantation. These features additionally strengthen PU as an appealing candidate for biomedical applications.

Normally, several polymers such as natural rubber, polyethylene, polyvinylchloride, fluoropolymers, hydrogels and silicon are used in biomedical applications. Despite the widespread use of these materials, PU still covers a relevant and dominant role, thanks to its high blood and tissue biocompatibility for improving the quality of patient's life. It conjugates a good stability over long implantation times, excellent physico-mechanical and surface tuning properties via anchoring of molecules. PU has simply a unique mix of features, highly required for almost any medical device.

PU is also widely used in cardiovascular applications, in particular for the preparation of venous and intravenous catheters and balloons for angioplasty and angiography. It has also been successfully used for tissue replacement and augmentation in breast implants, facial reconstruction and body joints. Artificial organs based on PU such as heart, kidneys and lungs have already been developed. Thanks to its above described unique features, PU has recently been proposed for new promising application fields, e.g. controlled drug delivery devices. A deep review on the use of PU in medicine and medical devices is also available (Zdrahala and Zdrahala 1999).

Given its large use, in this chapter we will focus on the use of PU in urological applications, in particular as ureteral catheter, or stent, in endourology routines.

In urology, catheterization is defined as the insertion of tubes (stents), such as urinary catheters, into the patient's bladder through the urethra. A stent is usually a tube with ending coils at both sides (pig-tails), or with lateral holes to further improve urine drainage (JJ or double-J). Urethral stents are usually in latex, silicone or polyurethane, allowing patients' urine to drain freely from the bladder. Stents can also be used to inject liquids for treatment or bladder conditions diagnosis. A patient is typically catheterized in the case of acute or chronic urinary retention, orthopedic procedures that may limit movement, benign prostatic hyperplasia, incontinence, and effects of various surgical interventions involving bladder and prostrate.

In endourology, urine is drained by indwelling a catheter between the kidney and the patient's bladder, hence inserting the stent into the ureter. Nowadays, ureteral stenting has become a common procedure for safe urine drainage. It is effective for managing several diseases, such as ureteral obstruction by stones or clots, benign or malignant ureteral obstruction, or post-surgical treatments, i.e. ureteroscopy and ureteral surgery. Ureteral stents almost painlessly keep the ureteral lumen open, ensuring that urine flows while maintaining the correct renal function. Stents also promote ureteral healing and prevent strictures formation.

From a general urological viewpoint, these devices must be easily maneuverable, affordable, and radiopaque for a correct positioning under fluoroscopic guidance. Hence, ureteral stents are mainly fabricated in PU or silicone since some patients can be allergic or sensitive to latex after long-term use. Radiopacity is ensured by adding metallic salts, e.g. based on barium. According to a critical study on materials (Mardis et al. 1993), the use of strong materials, e.g. PU, permits the reduction of the stent wall thickness, enlarging the inner lumen and the size or number of lateral holes and increasing urine flow. Silicone stents are weaker than PU, and need to be fabricated with smaller inner diameters, compromising urine flow and increasing the lateral compressibility.

Despite the numerous advantages enabled by PU, some complications and challenges remain. Indeed, PU is not perfectly biocompatible, in the sense that it somehow affects the epithelial cells of the ureter (urothelium): urothelial ulceration and erosion may occur. Other complications are related to stent migration or fracture, erosion, development of uretero-arterial fistula, fever, infection, voiding symptoms including dysuria and hematuria (Arshad et al. 2006). Among these, the stent encrustation represents one of the most serious complications resulting from the use of PU double-J stents: a stent can be encrusted by inorganic salts flowing with urine, and bacterial colonies can grow on the surface. These infections are very common among the general population, sometimes leading to death. During infections, the bacteria grow in the internal lumen of the stent, forming the so-called "biofilm", normally an aggregation of bacteria with their extracellular products and several inorganic salts (Costerton 2007). This matrix covers the cells, leading to a reduced susceptibility to prophylactic antibiotics (Tenke et al. 2004). Moreover, the development of these encrustations can obstruct the device and impair the urinary flow, compromising with patient care and leading to kidney infections, sepsis and shock (Warren et al. 1994).

Temporary prevention from encrustation includes stent replacement at regular intervals, modification of the type or size of catheters, washing the catheter and bladder with acidic, antiseptic or saline solutions (Arshad, Shah and Abbasi 2006). Antibiotics are still orally administered whenever stent is replaced or inserted for preventing infections (Reid 2001).

However all these approaches are mostly ineffective. For this reason stent surface modifications have been proposed to prevent bacterial and inorganic molecule adhesion. Various strategies have been conceived, using silver-coated surfaces (Leung *et al.* 1992; Multanen *et al.* 2000), surface modification towards hydrophobicity (Jansen *et al.* 1993) or functional groups creation with intrinsic antimicrobial activity. Heparin is a good candidate for solving these problems. In previous *in vitro* and *in vivo* studies, heparinization of medical devices showed a reduction of microbial colonization (Appelgren *et al.* 1996; Cauda *et al.* 2008; Ruggieri *et al.* 1987). Heparin is a highly sulfated, anionic polysaccharide known for its anti-coagulant and anti-thrombogenic properties (Piper 1946). It has a strong negative electrical charge able to prevent cells adhesion since bacterial' cell membrane surface is also negatively charged. Coating stents with heparin can be practical and low-cost. Over the last three decades several studies, especially in vascular medicine, were indeed reported, showing that this approach is effective (Appelgren, Ransjo, Bindslev et al. 1996; Hildebrandt *et al.* 1999; Ruggieri, Hanno and Levin 1987).

Another successful approach preventing biofilm formation comprises the use of Diamond-Like Carbon (DLC) coatings on the ureteral stent. DLC is a thermodynamically meta-stable state of carbon where diamond-like (sp3-hybridized) and graphite-like (sp2-hybridized) bonding coexist with a large fraction of sp3 bonds. Coatings can be prepared by miscellaneous deposition methods, e.g. ion deposition, sputtering, pulsed laser deposition and plasma-enhanced chemical vapor deposition, using accelerated hydrocarbon ions as film forming particles (Grill 1999). In general, they are characterized by high mechanical hardness and chemical inertness. Depending on the deposition conditions, the properties of DLC films can be adjusted depending on the applications, e.g. allowing enhancement of the wear and corrosion resistance of precision cutting and machining tools. These films are already used as protective coatings on magnetic hard disks and optical glasses. They showed excellent tribological and mechanical properties, corrosion resistance, biocompatibility, and hemocompatibility (Anne Thomson *et al.* 1991; Roy and Lee 2007; Voevodin and Donley 1996). Recent studies have been focused on their ability to decrease the formation of crystalline bacterial biofilm as well as stent related side effects and discomfort (Laube *et al.* 2006; Laube *et al.* 2007).

The evaluation of the *in vivo* efficacy of both heparin and plasma deposited DLC-coated ureteral stent have been reported. Recent works show the superiority of both coated stents for preventing biofilm adhesion and encrustation compared to uncoated PU catheters (Cauda *et al.* 2009).

Despite these recent advances, oral administration of antibiotics (e.g. ciproflaxin) cannot be avoided. Papers in this respect report that bacteria proliferation has been addressed with a local release of antibiotics or antiseptics (Cormio *et al.* 2001; John *et al.* 2007; Leung *et al.* 2001; Raad *et al.* 1997). Drug-Eluting Stent (DES) has the advantage of maximizing the local tissue levels of therapeutic agents while minimizing systemic toxicity. The problem has been faced with antibiotics incorporation (Gorman and Woolfson 2002) using novel biomimetic and bioactive silicones. For example, with a gentamicin-releasing urethral catheter, encrustation inhibition in a rabbit model has been shown in the short term (Cho *et al.* 2001). Other authors (Cadieux *et al.* 2006) reported on a triclosan-loaded ureteral stent implanted in rabbit bladders with bacterial infection. The study showed a significant decrease of urinary tract infection rate. However, a very high control on the delivery kinetics has been not achieved yet.

Another important problem with stents is that they require an additional cytoscopic procedure for their removal. Stents removal may be uncomfortable for the patient, in particular when these are encrusted, therefore requiring hospitalization and anaesthesia. Considering this big concern, the development of time bio-degradable stent materials has become a major keypoint. In principle, bio-adsorbable stents shall be designed to maintain their integrity for a given period of time, and undergo a dissolution process followed by spontaneous expulsion by the patient at the same time.

So far, the bio-degradable stents degrade very fast (typically 48h) or leave fragments removable by lithotripsy and ureteroscopy (Lingeman *et al.* 2003). Some tests on pigs with degradable poly-L-lactic-L-glycolic acid (PLGA) devices showed fragments embedded in the ureteral walls within cystic sacs, normally leading to fibrosis, inflammation and a large foreign body giant cell reaction (Olweny *et al.* 2002). These results show that further enhancement on their degradable characteristics is highly required. It is straightforward that the future of ureteral stents will head towards this direction, i.e. to chemically controlled biodegradation, combined to a concomitant release of biologically active molecules on target sites and with minimally invasive surgical procedures. Moreover, highly engineered stents can possibly work as a "scaffolds" for tissue regeneration via cell attachment and proliferation, finally controlling the local inflammation and healing (Zdrahala and Zdrahala 1999).

This chapter focuses in details on the use of PU catheters for endourological applications, providing new insights on the *in vivo* performances of PU stent. Given to our past expertise, we report on the influence of the PU surface treatments in preventing the encrustation and the formation of bacterial biofilm when implanted into the ureter.

2. Application

As an application of PU in biomedical devices, in this paragraph we present our clinical experience with ureteral stents. In particular the focus is driven on the *in vivo* efficacy of PU stents and their surface modification with heparin or with diamond-like carbon. Parameters like biofilm formation, inorganic encrustation extent, stiffness, brittleness or failure of the PU material depending on the indwelling time will be examined. The surface chemistry and morphology characterization of the indwelled stents will also be evaluated according to the surface coating and biofilm formation. The characterization techniques used in this work included Field Emission Scanning Electron Microscopy (FESEM), Energy Dispersive Spectroscopy (EDS), and Infrared (IR) spectroscopy.

2.1. Experimental part: Patients, stents and characterization methods

We review here the results collected from 2006 to 2010, concerning the characterization studies on uncoated, heparin- and DLC-coated PU double-J ureteral stents (all provided by Cook Ireland LTD) after indwelling. We enrolled 59 patients (from 45 to 75 years old). 49 patients showed unilateral ureteral obstruction, thus requiring ureteral stenting in the affected ureter. The patients with unilateral obstruction received an heparin-coated, DLC-coated or an uncoated PU stent in the ureter to be treated.

10 patients suffered from bilateral obstruction, therefore the coated and uncoated PU stents were indwelled at the same time in both ureters, respectively. Each patient with bilateral obstruction received randomly both the coated stent (with heparin or with a diamond-like carbon coating) in one ureter, and the uncoated one, thus the pure PU stent, in the other ureter.

Stents indwelling was also studied for different periods of time. The stent types and indwelling times are reported in Table 1.

Number of stents	Coating	Indwelling time	Type of indwelling
14	Heparin	1-3 months	Unilateral
9	Heparin	> 3 months	Unilateral
19	None[a]	1-3 months	Unilateral
3	None	> 3 months	Unilateral
4	DLC	1 month	Unilateral
5[b]	None	1 month	Bilateral
5[b]	Heparin	1 month	Bilateral
5[b]	None	1 month	Bilateral
5[b]	DLC	1 month	Bilateral

[a] None indicates that the stent surface is of pure PU.

[b] These stents were implanted bilaterally, thus inserting both the uncoated and the coated stents into both ureters respectively of the same patient.

Table 1. Stents types and indwelling periods of the PU, heparin-coated and diamond-like carbon coated PU stents.

The criteria used for patient selection for indwelling time of 1 month were mainly post–endoscopic stone treatment, ureteropelvic junction (UPJ) obstruction (awaiting surgery). In the long term indwelling (thus 3 or more months) stenting was performed in case of hydronephrosis due to extrinsic compression in patients with multi-cystic disease (no surgical indications) and UPJ obstruction after failure of endoscopic and surgical treatments (patient refused re-intervention). Pregnant women, patients with dermatitis or a burn over the insertion site were excluded. All patients enrolled in this study gave their informed consent.

In all cases, double-J stents were placed using a retrograde uretero-pielography approach during cystoscopy, as suggested by the stent producer, to evaluate the excretory system.

Ciprofloxacin (500 mg twice a day) was administered for prophylaxis for the first 4 days after the procedure. Any occurrence of technical problems and violations of aseptic conditions during the procedures were recorded. The antibiotic therapy administrations, other therapeutic interventions administered during the period of indwelling, the presence of fever, infections or urinary symptoms were also recorded. After the procedure, the follow-up check included urine analysis and culture on day 15 and afterwards every three months. In addition, a complete blood count, serum creatinine levels, and ultrasonography were performed on day 30, and every 3 months thereafter. All patients received instructions to present themselves at the institution in the event of experiencing side pain, fever, dysuria, hematuria, or vomiting.

The stents were removed after different times (see Table 1) during cystoscopy. Characterization of the stents after the indwelling period was carried out using morphological and compositional analysis and the results obtained were compared with the reference stent before use. In particular they were cut both perpendicularly and parallel to the stent axis in order to obtain cross sections or plan views of the inner and outer surfaces (Figure 1). The samples were then characterized by three different techniques:

a. FESEM (JEOL JSM 6500F) for the morphological characterization of the stent surface and the formed encrustations;
b. EDS (INCA) to collect data about the chemical composition of the stent material and the deposited encrustation at the inner and outer surface of the samples;
c. IR spectroscopy on Attenuated Total Reflectance mode (ATR, Bruker Equinox 55) for the identification of the chemical compounds deposited on the inner and outer surface of the samples.

Measurements (a) and (b) were carried out at the same time. Since the stent are made of a polymeric material (polyurethane), i.e. constituted by carbon, oxygen, hydrogen, and nitrogen, these elements are excluded from the elemental analysis on the encrustation. Moreover, the specimen is fixed to the sample holder for the FESEM and EDS characterizations by a carbon sticker, the composition of which also interferes with the elemental analysis of the stent itself. For these reasons, the detection of bacterial biofilm is not possible with the EDS technique and therefore IR spectroscopy is required.

IR spectroscopy was carried out directly on the inner and outer surfaces of the cut stent in Attenuated Total Reflectance (ATR) mode by means of a diamond immersion probe (as shown in Figure 1, scheme a).

All the described measurements were also performed on reference stents without *in-vivo* indwelling.

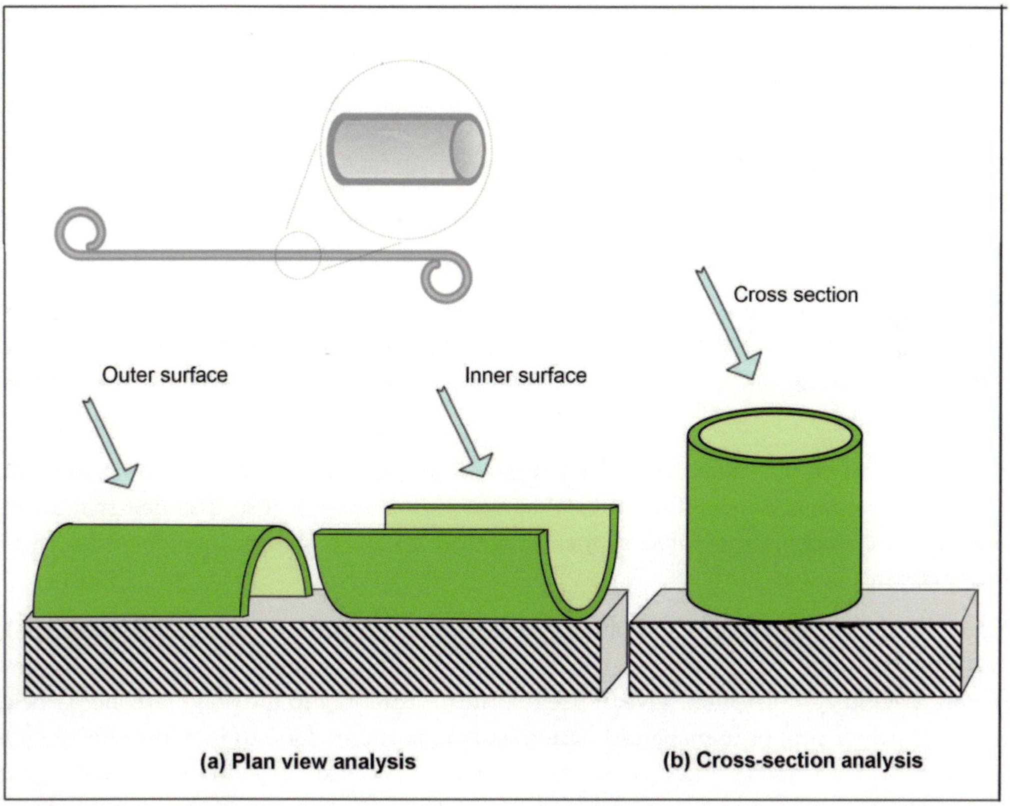

Figure 1. Scheme of the analyzed surfaces for the polyurethane ureteral stents for the characterization with FESEM and IR spectroscopy IR. (a) For the plan view analysis, a cut along the stent axis was carried out for characterizing the outer (left) and inner (right) surface; (b) Cross-sections were obtained by cutting the stent perpendicular to its axis.

2.2. Clinical results

No technical problems or violations of aseptic conditions during endoscopic procedures were recorded. In all cases retrograde uretero-pielography, performed at the start of the procedure, showed unilateral or bilateral ureteral obstruction with various degrees of excretory system dilatation.

None of the patients reported fever, side pain or voiding symptoms during the period of study. Urine culture was negative in all cases. In three patients, groups frequency/urgency symptoms were recorded. Until stents removal, these patients were successfully treated with antimuscarinics. Follow-up examination revealed no diferences in blood count, serum creatinine and ultrasonographic features of kidney and ureter with respect to the baseline values. During indwelling, all the stents were well tolerated.

Stents were removed without technical difficulties in all cases. No technical problems occurred during the endoscopic procedures on ten patients showing chronic unilateral obstruction. In these cases, uretero-pielography showed ureteral obstruction with severe excretory system dilatation.

2.3. Characterization results on starting stents surfaces

The starting stent surfaces were analyzed prior to indwelling, in order to evaluate the differences between the pure PU surface, the heparin- and DLC-coatings.

To understand the performances of both coatings in preventing bacterial adhesion and encrustation with respect to the PU material, the stents were also characterized after different indwelling times (see Table 1 for details). These results will be discussed in the next paragraphs.

Here we report on the results from the different characterization techniques concerning the stents before patient's indwelling (hereafter "reference-stents"). It is intended that all the results and comments presented in the Paragraph 2.4 are based on the comparison with these reference stents.

Figure 2 shows the surface morphology (measured by FESEM) of the PU (uncoated), heparin-coated and DLC-coated stents. The internal and external surfaces of the PU stent (Figure 2.a and 2.b) presented several irregularities, attributed to the polyurethane particles (of about 0,5 – 1 μm) or undispersed barium sulfate particles, used to impart radiopacity to the stent. The elemental analysis by means of EDS showed no additional results, since carbon, oxygen and nitrogen are excluded due to the reasons mentioned in the Paragraph 2.1. For this reason the results of the EDS analysis are not shown.

At the edge of heparin-coated reference-stent it was possible to observe the smooth heparin layer on the polyurethane substrate (Figure 2.d), with a thickness of about 5 μm.

The EDS shows the presence of sulphur, an element present in the heparin chemical formula together with C, O and N. Barium was present in the black paint line, used as fluoroscopic marker, at the outer surface of the catheter.

Figure 2.e shows the external surface, close to a lateral drainage hole, of the DLC-coated reference-stent. At the internal surface of this stent (Figure 2.f) several grains and thin cracks were observed, possibly attributed to the carbon coating. The elemental analysis on DLC-coated stent surface detected carbon, oxygen and nitrogen, attributed to PU.

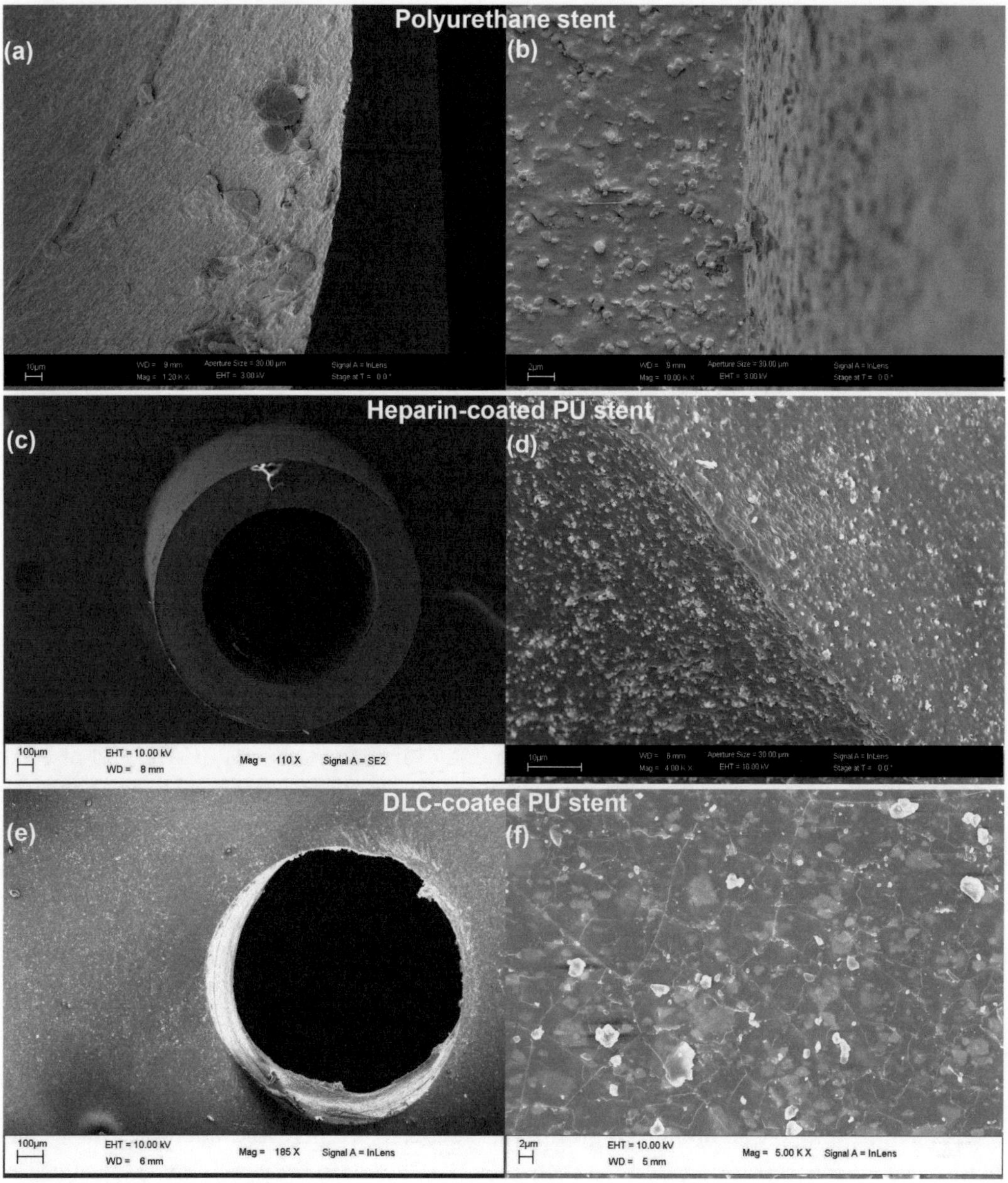

Figure 2. FESEM results of the stents surfaces of PU, Heparin-coated PU and carbon-coated PU.

The IR spectra of the reference-stents are shown in Figure 3 and were recorded at the internal surface, showing the functional groups of PU, heparin and DLC.

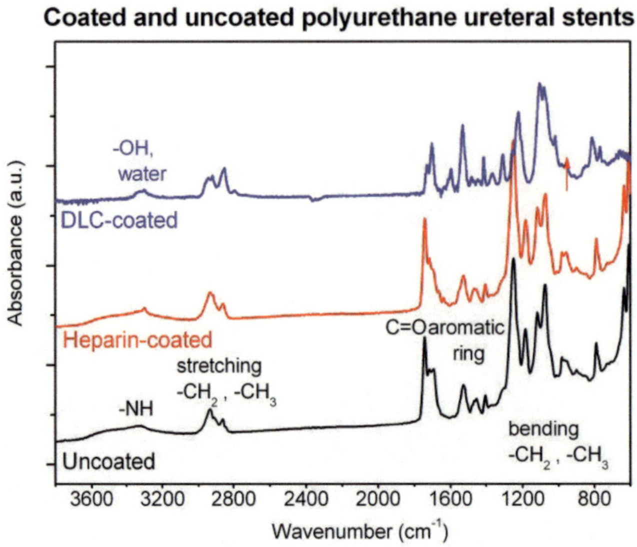

Figure 3. IR of the stents surface: comparison between PU stents without coating and with heparin and diamond-like carbon coatings.

The PU spectrum (in black) in the 3800-1200 cm^{-1} region reflected the vibration modes of its functional groups. Hydroxyl groups (-OH) and physisorbed hydration water were responsible for the bands from 3700 to 3100 cm^{-1}. The stretching vibration of the –NH group was associated to the band at 3300 cm^{-1}, while peaks at 2900 and 2860 cm^{-1} are the stretching vibrations of alkyl $-CH_2$ and $-CH_3$ groups. The peak at 1690 cm^{-1} represents another vibration mode (bending) of clustered water. The band at 1740 cm^{-1} indicated the carboxyl group C=O vibration mode, at 1450 cm^{-1} the mode of the aromatic ring (C_6H_6) and at 1400 cm^{-1} the bending modes of alkyl $-CH_2$ and $-CH_3$ groups. In the spectral zone below 1200 cm^{-1} only collective vibrations of the single bonds were observed since the bands in this range are typical of the polymer chain and constitute its "fingerprint".

In the DLC-coated stent spectrum (in blue), no significant differences were appreciable with respect to the previous PU spectrum, however several changes in the peak intensities were observed. In particular, the band from 3400 to 3250 cm^{-1}, representing water, –OH groups and amine group (-NH) were more intense. A similar increase was observed for to the bending peak of water at 1690 cm^{-1}, concluding that the DLC-coated stent surface was more hydrophilic than the untreated polyurethane catheter. Stretching peaks at 2940 and 2850 cm^{-1}, representing alkyl $-CH_2$ and $-CH_3$ groups, also showed an increased intensity, due to the plasma treatment implanting hydrocarbon ions. For the same reason, similar changes in intensity were observed at 1400 cm^{-1} for the peak related to $-CH_2$ and $-CH_3$ bending vibrations and an additional peak at 1597 cm^{-1} attributed to the C=C group. The IR light beam penetrated under the thin DLC-modified surface and also detected the polyurethane surface. For example, the aromatic ring belonging to PU was also slightly observed in the range between 1430 and 1290 cm^{-1}.

The spectrum of the heparin-coated PU surface (in red) shows some new features with respect to the PU reference-stent (in black). In particular, the broad band from 3700 to 3100 cm^{-1}, representing water and –OH groups, showed a lower intensity. A similar decrease was observed for the bending peak of water at 1690 cm^{-1}. It was than concluded that the heparin coated stent surface showed lower hydration than the uncoated one. In contrast, stretching peaks at 3300 cm^{-1}, belonging to –NH groups, and peaks at 2900 and 2860 cm^{-1}, representing alkyl $-CH_2$ and $-CH_3$ groups, increased, since both functional groups were also present in the heparin chemical formula. For the same reason, similar changes in intensity were detected at 1400 cm^{-1} for the peak related to $-CH_2$ and $-CH_3$ bending vibrations. Additional peaks appeared in the range from 1650 to 1600 cm^{-1}, due to the presence of carboxylate (-COOH) and sulphate ($-SO_4$) groups belonging to heparin. The other peaks of the red spectrum are no longer discussed, since they belonged to the PU substrate.

2.4. Characterization results on the indwelled stents

In the following sections, the results on the indwelled stents are reported and divided according to the indwelling time (from 1 to 3 months, more than 3 months) and the adopted approach (unilateral or bilateral indwelling). In addition, some example will be given in order to show how the patient pathology, such as recidivist calculosis, would affect the stent surface.

2.4.1. Stents indwelled unilaterally for 1 - 3 months

In this section we will examine the surface characterization of stents indwelled unilaterally and for a period ranging from 1 to 3 months. We have examined (see also Table 1):

i. 19 PU stents;
ii. 14 heparin-coated stents;
iii. 4 DLC-coated stents.

The samples were analyzed by means of the three techniques described above. The characterization aimed to evaluate the presence of bacterial biofilm and inorganic encrustations, such as calcium oxalate (CaC_2O_4), sodium chloride (NaCl), brushite ($CaHPO_4 \bullet 2H_2O$) and other salts, such as silica (SiO_2) and compounds of magnesium (Mg) and potassium (K). The obtained data were used to evaluate the behavior of PU ureteral stent *in vivo* according to the indwelling time and the surface treatment.

Figure 4 shows the comparison between the surfaces of PU, heparin-coated and DLC-coated stents, indwelled unilaterally into three different patients. They were all indwelled into the patient's ureter after the stone removal from the kidney by endoscopic lithotripsy with holmium laser. In the cases of both heparin-coated and DLC-coated stents almost clean and encrustation-free surfaces were observed (Figures 4.e-4.l). In contrast, higher levels of encrustation were detected at both the inner and outer surfaces of the PU stent (Figures 4.a-4.d).

The elements found at both inner and outer stent surfaces by means of Energy Dispersive Spectroscopy (EDS) are reported in Table 2. Some inorganic salts were detected at all the stent surfaces, such as sodium chloride (NaCl, evidenced by the presence of both Na and Cl, Table

2). In addition, oxides of calcium (evidenced only from the presence of Ca, whereas oxygen and carbon were both not taken into account by EDS analysis), silica (revealed by Si element), phosphorous (P) and potassium (K) were collected. In the case of the heparin-coated stent, the sulphur (S) element present at both inner and outer surfaces clearly derived from the heparin layer.

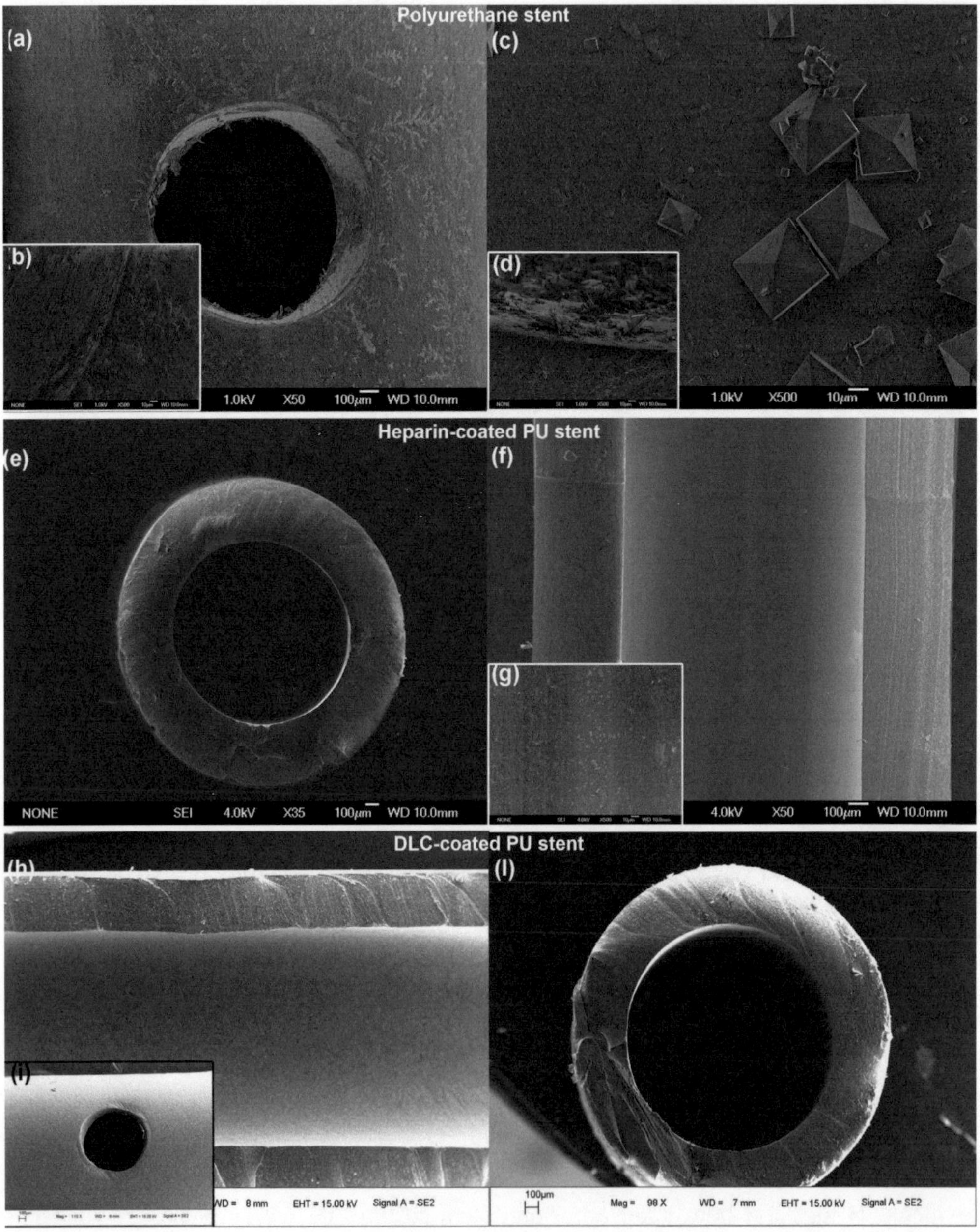

Figure 4. FESEM results of PU, heparin-coated and DLC-coated polyurethane stents surfaces after unilateral indwelling for a period ranging from 1 to 3 months.

PU stent				Heparin-coated stent				DLC-coated stent			
Inner Surface		Outer Surface		Inner Surface		Outer Surface		Inner Surface		Outer Surface	
Element	Atom %	Element	Atom %	Element	Atom %	Element	Atom %	Element	Atom %	Element	Atom %
Na K	29.17	Na K	37.71	Na K	22.79	Na K	0.00	Na K	0.00	Na K	10.70
Mg K	0.00	Mg K	0.00	Mg K	0.00	Mg K	0.00	Mg K	0.00	Mg K	0.00
Si K	3.55	Si K	0.00	Si K	0.00	Si K	0.00	Si K	0.00	Si K	0.62
P K	2.14	P K	2.48	P K	0.00	P K	0.00	P K	0.00	P K	0.00
S K	1.89	S K	0.00	S K	38.40	S K	61.01	S K	0.00	S K	0.00
Cl K	20.45	Cl K	38.09	Cl K	10.07	Cl K	0.00	Cl K	100.00	Cl K	88.78
K K	2.46	K K	2.06	K K	0.00	K K	0.00	K K	0.00	K K	0.00
Ca K	34.99	Ca K	0.00	Ca K	2.61	Ca K	0.00	Ca K	0.00	Ca K	0.00
Bi K	5.35	Bi K	19.66	Ba L	26.13	Ba L	38.99	Ba L	0.00	Ba L	0.00

Table 2. EDS analysis on PU, heparin-coated and DLC-coated stent surfaces after 1-3 months of unilateral indwelling.

IR spectroscopy (here only the spectra carried out at the internal surfaces are shown, see Figure 5) confirmed the previous findings (a higher content of water was observed in the case of DLC-coated stent, blue spectrum). It was therefore concluded that the stent surfaces were clean and the bacterial biofilm was not detected.

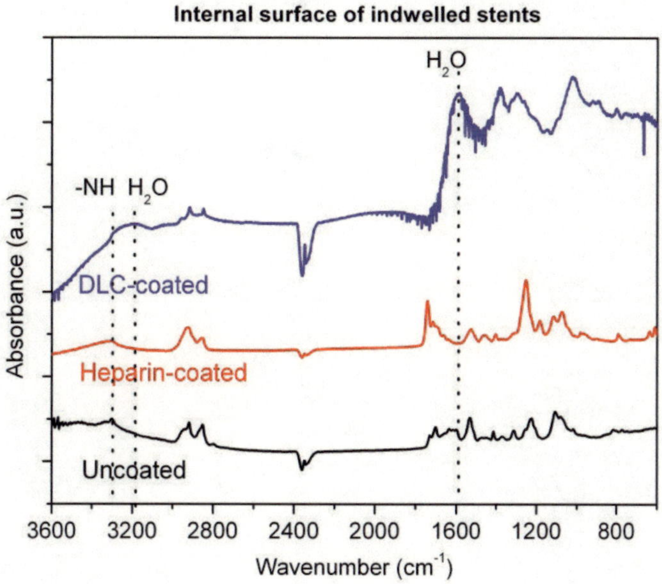

Figure 5. IR of the internal stent surfaces.

By combining the results of the three characterization techniques, the following considerations can be drawn: (i) the inner surface of the stents, independently from the surface treatment, was more encrusted than the outer one; (ii) the PU stent showed higher level of inorganic encrustation with respect to both surface-modified stents.

To see how the patient's pathology and conditions affected the ureteral stents, in Figure 6 we compare the surfaces of three PU, heparin-coated and DLC-coated stents respectively, indwelled into stone-former patients. All the catheter surfaces were heavily encrusted; however lower level of depositions were observed at both the coated stent surfaces (Figures 6.d-6.i) with respect to the PU stent (Figures 6.a-6.c).

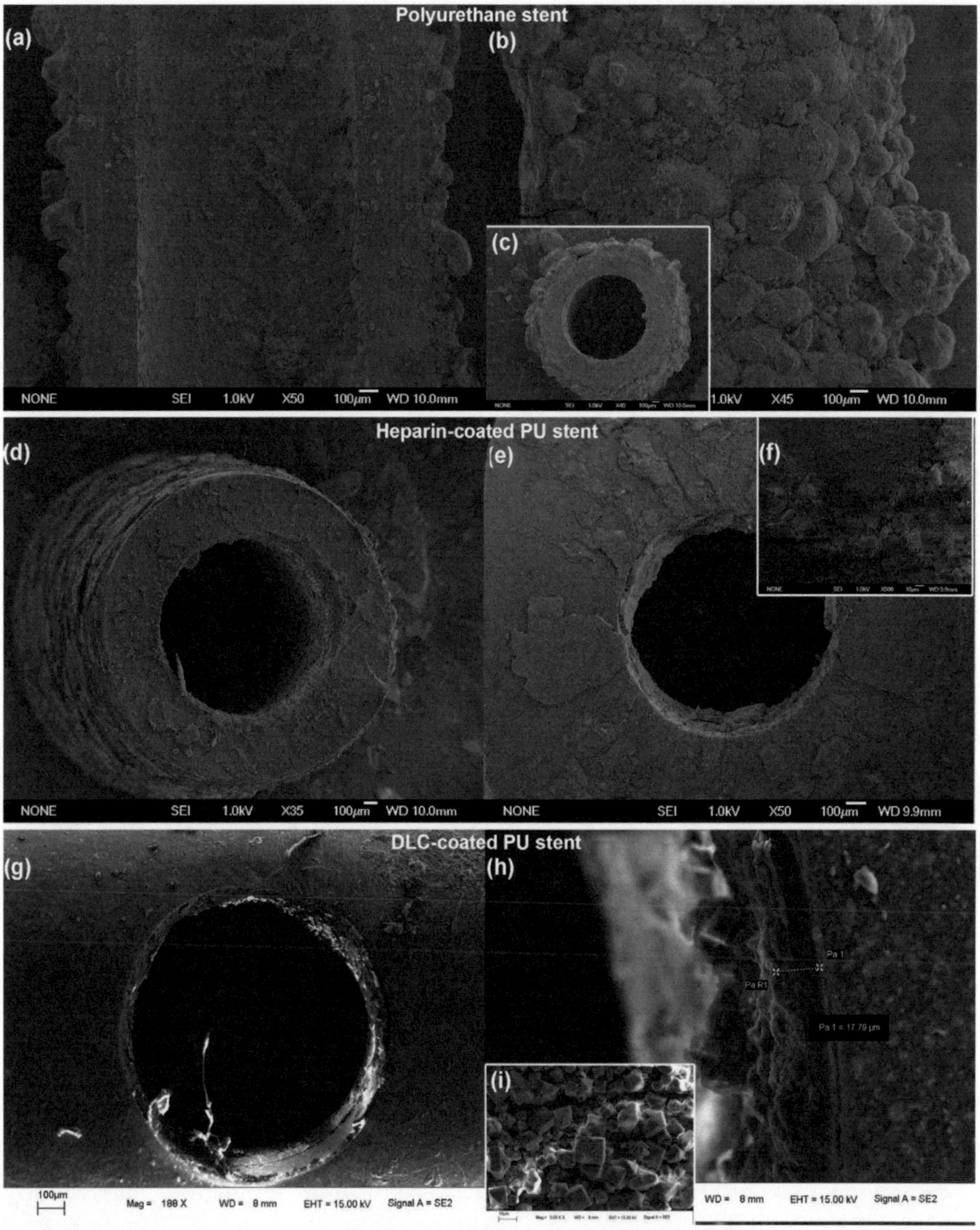

Figure 6. FESEM images of PU, heparin-coated and DLC-coated stents after indwelling of 1-3 months into stone-forming patients.

The encrustation of the three stents was mainly composed of calcium oxalate, as clearly detected by the IR spectroscopy (Figure 7) and the EDS analysis (here not shown). In particular, IR spectrum of the oxalate crystals was very well defined, with the characteristic peaks at 1706 and 1313 cm^{-1} representing the vibration modes of C=O and C-O groups, respectively. These spectra showed the high purity of the isolated bio-mineral on the stent surfaces. In addition, the vibration modes of the polymeric substrate (heparin, DLC and polyurethane) were no longer recognizable. One can then conclude that the encrustation was thicker than the depth of the analysis (about 1 μm using the Attenuated Total Reflection (ATR) detection mode).

These findings do not allow to conclude which stent is more or less encrusted with respect to the surface treatment. Indeed, in stone former patients with recidivist calculosis, whatever stent is applied, the urologist has to plan frequent stent exchange (the suggested indwelling time by the producer is one month indeed), due to the ease of stent encrustation.

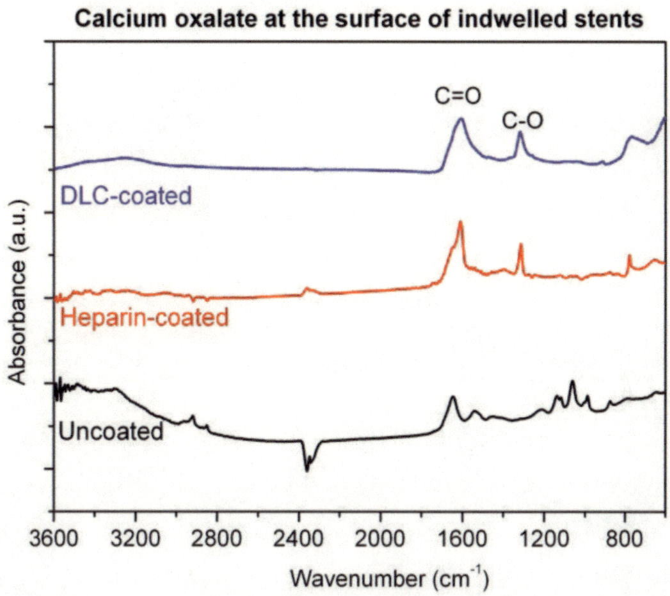

Figure 7. IR spectra of the three stents surfaces, also depicted in Figure 6, clearly showing the vibration modes of calcium oxalate.

2.4.2. Stents unilaterally indwelled longer than 3 months

In this paragraph we present the results obtained from the surface characterization of both heparin-coated and unmodified PU stents, indwelled longer than three months. This indwelling time actually exceed the recommendation of the stent producer, and the results are therefore quite interesting. All these patients refused the stent substitution after 3 months, thus the stents were substituted or definitely removed once the consent of patient was given. Again, the stent study has to be divided according to the patient's pathology, that is, stone-former patient or not.

The first example shows the comparison between PU and heparin-coated stents indwelled unilaterally into non-stone former patients after calculus removal by lithotripsy.

Despite the long indwelling time and the producer recommendation, both stents were quite free from encrustation (Figure 8 shows FESEM characterization). The encrustation thickness was measured 2.5 μm at the PU stent, whereas it was not enough compact to form a layer on the heparin-coated surfaces. The results obtained by EDS and IR spectroscopy confirmed the absence of bacterial biofilm on both stent surfaces. In addition, the PU surface showed a higher percentage of sodium chloride and silica with respect to the heparin-coated one. From these findings, one can conclude that the encrustation levels of the surface-treated stent were lower than the PU one.

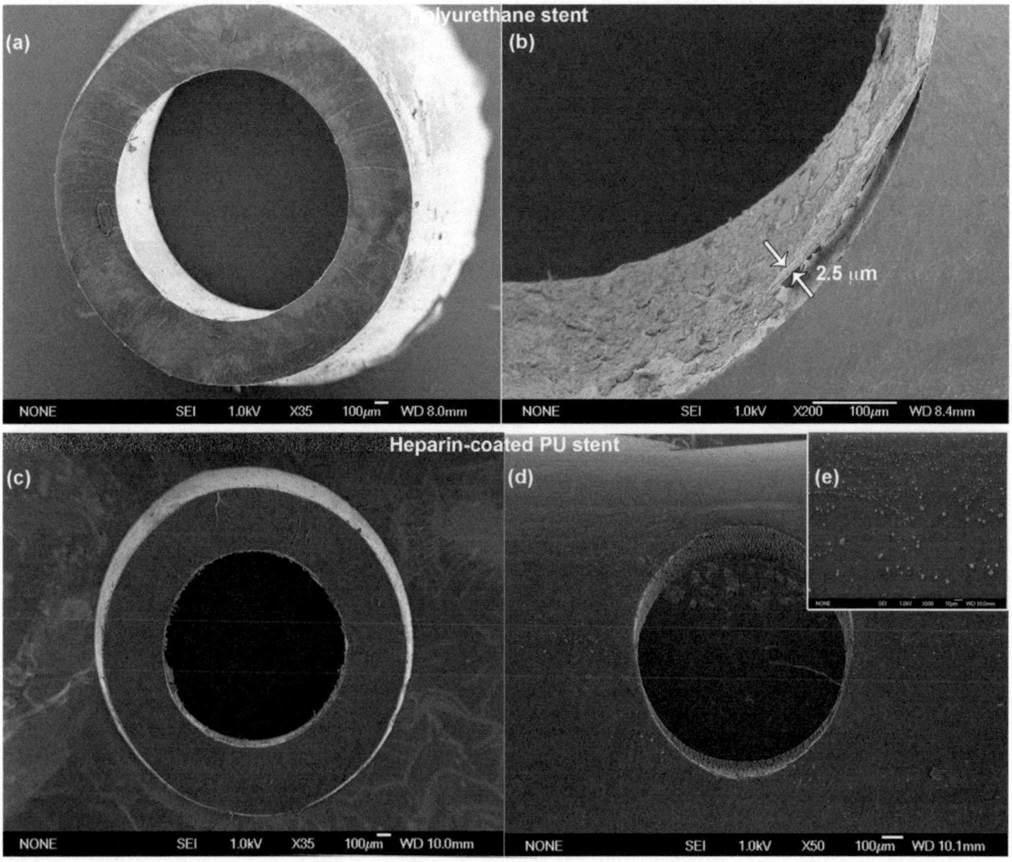

Figure 8. Surface morphology of PU and heparin-coated stents inserted unilaterally for more than 3 months.

The second example referred to the stents indwelled into stone-former patients for long-term periods (more than 3 months).

In contrast to the previous results, the heparin-coated stent showed higher degree of encrustations at both inner and outer surfaces with respect to the PU stent (Figure 9.a, b, c). In particular, needle-like crystals were observed a higher magnification at the internal stent surface (Figure 9.e and f). This morphology corresponded to the calcium oxalate crystals, and was confirmed by both EDS spectroscopy (due to the presence of calcium in high percentages in Table 3), and IR spectroscopy (Figure 10.b). Indeed both spectra collected at the inner and outer surfaces indicated a thick layer of pure calcium oxalate.

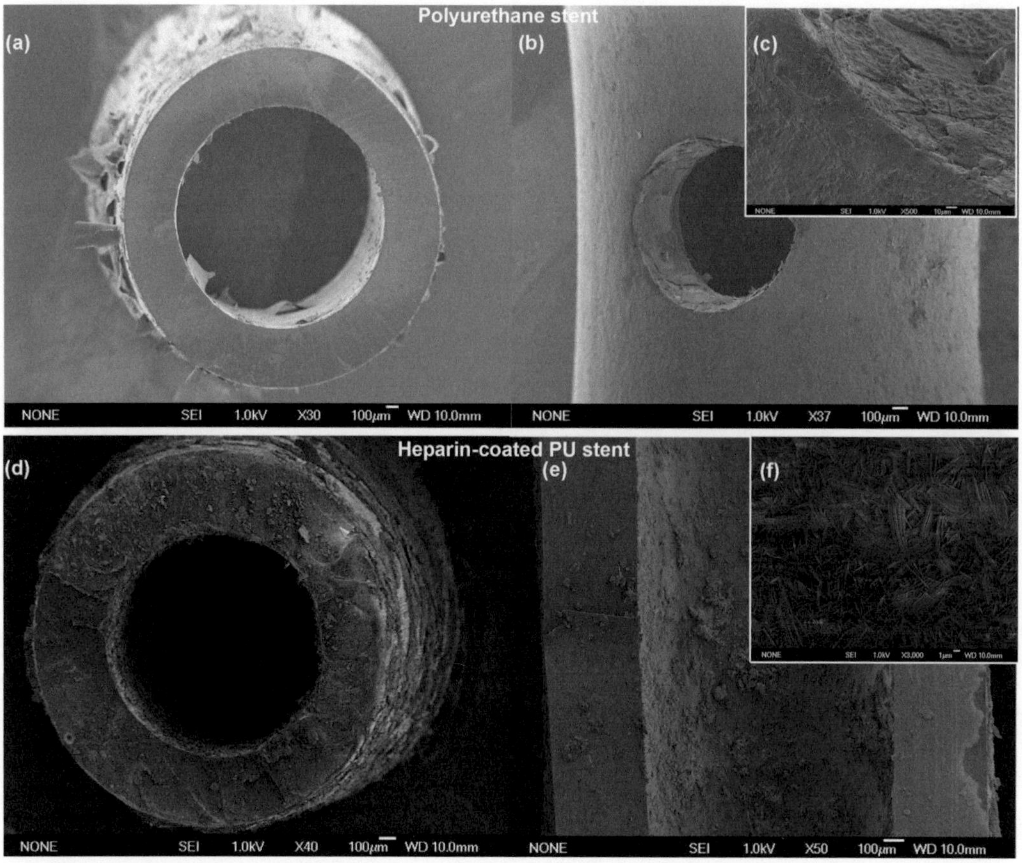

Figure 9. FESEM images of PU and heparin-coated stents indwelled unilaterally for more than 3 months in stone former patients.

In this example no general conclusion can be drawn concerning the comparison of the two stents, since the surface-modified stent and PU one were indwelled into two different patients, although both stone-formers.

PU stent				Heparin-coated stent			
Inner Surface		Outer Surface		Inner Surface		Outer Surface	
Element	Atom%	Element	Atom%	Element	Atom%	Element	Atom%
Na K	34.75	Na K	0.00	Na K	10.14	Na K	0.00
Mg K	0.00	Mg K	0.00	Mg K	0.00	Mg K	0.00
Si K	18.65	Si K	37.77	Si K	0.00	Si K	0.00
S K	4.39	S K	0.00	P K	0.00	P K	2.56
Cl K	32.53	Cl K	37.93	S K	24.46	S K	0.00
K K	0.00	K K	0.00	Cl K	3.42	Cl K	0.00
Ca K	0.00	Ca K	0.00	K K	0.00	K K	0.00
Br L	0.00	Br L	0.00	Ca K	43.82	Ca K	97.44
Bi M	9.69	Bi M	0.00	Ba L	18.17	Ba L	0.00

Table 3. Results of the EDS analysis carried out at both the inner and outer surfaces of PU and heparin-coated stents after prolonged unilateral indwelling into stone-former patients.

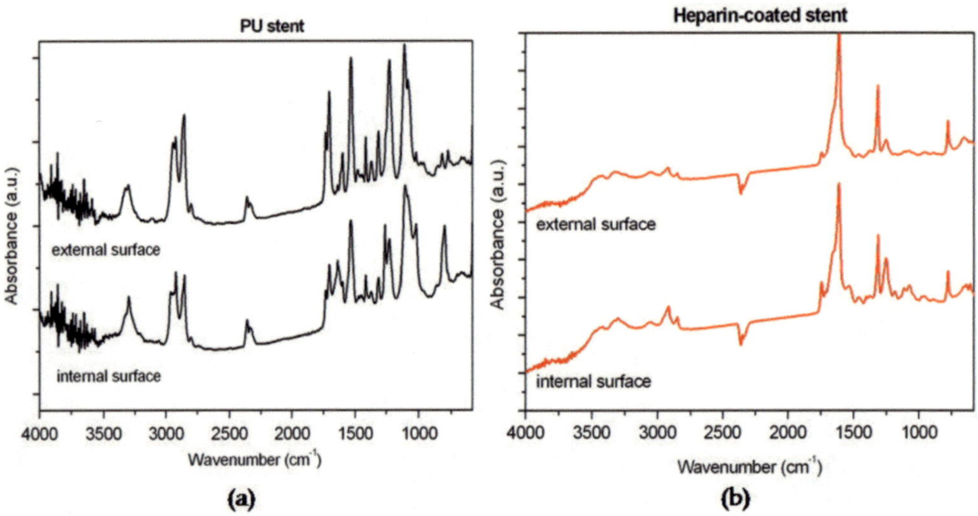

Figure 10. IR spectroscopy of the internal and external surfaces of (a) PU stent and (b) heparin-coated stent, unilaterally indwelled into stone-former patient for a period longer than 3 months.

2.4.3. Stents bilaterally indwelled for 1 month

The bilateral indwelling is the ideal condition to compare the effectiveness of the surface treatment in preventing encrustation, since both stents are exposed to the same patient's conditions.

Here we report on two examples of surface-modified and PU stents, bilaterally indwelled for one month into a non-stone former patient and, in a second case, into a stone-former one.

In the first example, shown in Figure 11 by FESEM, heparin-coated and PU untreated stents were indwelled bilaterally after calculus removal by lithotripsy with holmium laser. Both the treated and untreated surfaces showed low levels of encrustation. The EDS and IR spectroscopic characterizations (not shown here) revealed phosphorous, potassium and sodium chloride salts at both stents surfaces, with higher percentages of bacterial biofilm at the inner surface of the PU stent.

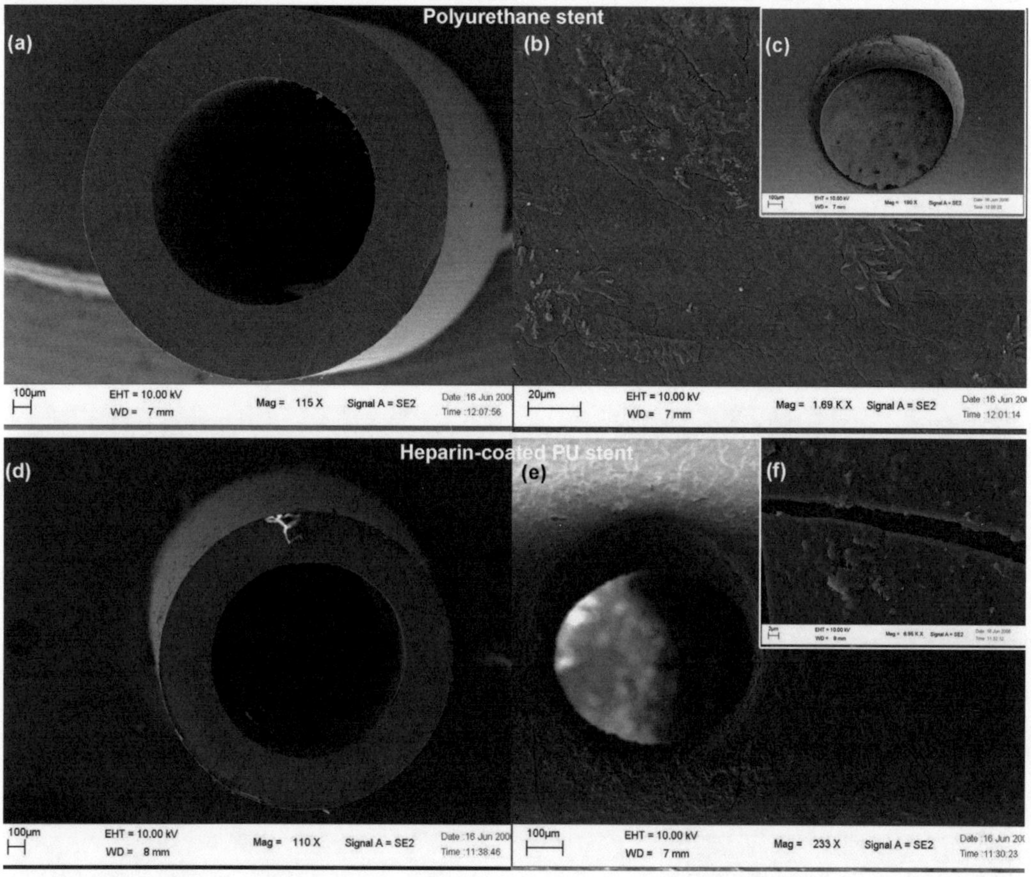

Figure 11. The surface morphology of PU and heparin-coated stents indwelled bilaterally into the same patient for one month.

In the second case, we examined the stents after bilateral indwelling into a patient presenting a recidivist calculosis into both her kidneys. In particular the patient had multiple lithiasis (3 calculus) at the lower calyx of the left kidney. Stenting into both the ureters was carried out after removal of stones by endoscopic lithotripsy with holmium laser; however residual lithiasis was still present. The DLC-treated stent was positioned into the left ureter, to verify its performance after calculus removal. The PU stent was indwelled at the same time into the right ureter. After one month both stents were removed, due to patient low tolerance. This patient underwent two repetitive stents indwelling, always having pain and very low tolerance towards every catheter.

Figure 12 shows the high degree of encrustation, in particular concerning the PU stent. In particular, the encrustation was thicker than 40 μm and was composed of bacterial biofilm and inorganic salts, such as brushite ($CaHPO_4{\cdot}2H_2O$) and sodium chloride.

In contrast, the encrustation level of the DLC-coated stent was not so compact and uniform as in the untreated PU stent. It was therefore assumed that the presence of the DLC surface modification prevented partially the encrustation deposition and the bacterial adhesion, despite the pathology of the patient. In particular the DLC-coating seemed to guarantee the stent lumen free for urine drainage (Figure 12.c).

From EDS analysis, no particular inorganic compounds were identified on the surface of the DLC-coated stent (data not shown).

For the overall 20 stents indwelled bilaterally for one month (see Table 1) it was concluded that both surface treatments effectively prevented or at least decreased the levels of encrustation by both bacterial biofilm and inorganic compounds, with respect to the pure PU catheter. It was found that the highest incidence of encrustations took place at the inner surface of the catheter.

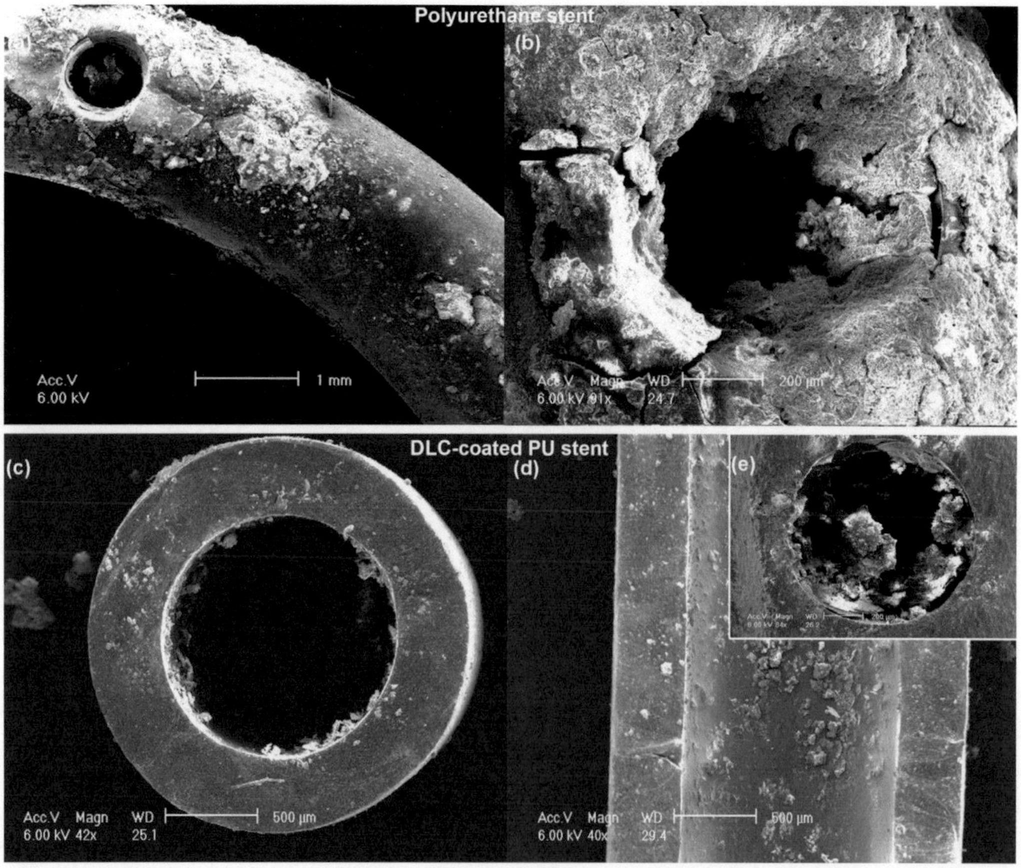

Figure 12. FESEM images of PU and DLC-coated ureteral stents indwelled bilaterally for one month into the same patient, suffering from recidivist calculosis.

2.5. Results summary

Summarizing the results obtained upon stent indwelling into 59 patients, some statistical considerations can be drawn. Considering only the surface-treated stents, the level of bacterial biofilm stabilized after three months of indwelling (incidence of the encrustation on the total amount of heparin-coated stents analyzed: 81.3%), remaining almost constant for longer indwelling time (78.6%). The incidence of sodium chloride, silica and salts of magnesium and potassium increased with the indwelling time or due to the recidivist calculosis of the patient. Interestingly, the highest levels of calcium oxalate and brushite were found at the surface-modified stents indwelled for 3 months into stone-former patients (75% for both compounds). For indwelling times longer than 3 months, the levels of calcium oxalate and brushite decreased (42.9 % for calcium oxalate and 57.1% for brushite).

Concerning the PU stents, high contents of sodium chloride and other salts were generally observed in high percentages (up to 100%), despite the indwelling time and the patient pathology. The highest percentages of both bacterial biofilm and calcium oxalate were observed after already 3 months of indwelling time (85% of biofilm , 50% of calcium oxalate, 43% of brushite).

3. Conclusion

In the present chapter we have summarized the results from the characterization of uncoated PU, heparin- and DLC-coated PU ureteral stents after indwelling into 59 patients. Field Emission Scanning Electron Microscopy, Energy Dispersive Spectroscopy, Infrared Spectroscopy were used to characterize both inner and outer catheter surfaces.

With these techniques two kinds of deposits were detected at the stent surfaces:

1. inorganic compounds, like calcium oxalate, sodium chloride, brushite and salts of potassium, magnesium and phosphorous;
2. bacterial biofilm;

In addition, the thickness of the encrustations was estimated at the stent cross sections.

We have divided the obtained data according to their indwelling time, the unilateral or bilateral indwelling and the patient's tendency to form calculus (stone-former or not).

Concerning the non-stone former patients, the encrustation levels were lower in the surface-treated stents with respect to the untreated PU surfaces. In particular, concerning the bilaterally indwelled stents, a direct comparison between the surface properties of the stent in preventing encrustation was clearly observed. It was indeed assessed that the formation of bacterial biofilm was lower at the surface-treated catheters, whereas the precipitation of inorganic compounds were not completely inhibited. We attributed reduction of the biofilm to the presence of the surface treatments (heparin- or DLC-coatings) on the polyurethane surface. No relevant differences were found between the two surface modifications in preventing the stent encrustation upon indwelling. It was also observed that both treated and untreated PU stents did not degrade in this kind of patients.

These considerations were no more valid when the patient was a stone-former. Indeed, the recidivist calculosis induced a continuous deposition of biofilm and salts at the stent surface, thus strongly reducing the effect of the surface modification in preventing encrustation. In addition, the stents were more stiff and brittle after already one month of insertion, thus inducing patient discomfort or pain.

It was noteworthy that both the formation of biofilm and inorganic encrustation and the success of the stent indwelling depended more significantly on the patient's pathology (i.e. stone former or not) than on the indwelling time. For these reasons, implanting a stent for a period of time longer than one month was feasible. A surface-treated polyurethane stent was also preferable with respect to the untreated PU one. However, frequent stent exchange, regardless of the surface treatment, is a general recommendation for patients suffering from a recidivist calculosis.

As a future outlook, new studies should expand in the direction of bio-degradable drug-eluting polymeric stents. The preparation of such highly engineered ureteral stents should require the following properties:

- Providing effective urine drainage without the formation of bacterial biofilm;
- Being biocompatible and preventing cytotoxicity;
- Being fully biodegradable to avoid the complications of the stent removal and the subsequent patient hospitalization;
- Effectively incorporating a therapeutic agent;
- Showing the capability to release the drug in a time-controlled manner;
- Releasing the required amount of drug, between the minimum effective level and the minimum toxic one;
- Preventing the patient discomfort and pain;
- Being deliverable and visible, with adequate radiopacity (or the presence of radiopaque markers) to enable precise positioning under X-ray fluoroscopic guidance.

We envision that such a commitment will require a strong interdisciplinary background, thus combining the fields of material science and technology to the clinical and endourological requirements.

Author details

Valentina Cauda
Center for Space Human Robotics CSHR@Polito, Italian Institute of Technology, Turin, Italy

Furio Cauda
Urology division, Koelliker Hospital,Turin, Italy

Acknowledgement

Cook Ireland Ltd. is gratefully acknowledged for the financial support and the stent supply.

4. References

Anne Thomson, L., et al. (1991). Biocompatibility of diamond-like carbon coating.Biomaterials Vol. 12, No 1,pp. 37-40.

Appelgren, P., et al. (1996). Surface heparinization of central venous catheters reduces microbial colonization in vitro and in vivo: Results from a prospective, randomized trial.Crit. Care Med. Vol. 24, pp. 1482-1489.

Arshad, M., et al. (2006). Applications and complications of polyurethane stenting in urology.J. Ayub. Med. Coll. Abbottabad Vol. 18, No 2,pp. 69-72.

Cadieux, P. A., et al. (2006). Triclosan loaded ureteral stents decrease proteus mirabilis 296 infection in a rabbit urinary tract infection model.J. Urol. Vol. 175, No 6,pp. 2331-2335.

Cauda, F., et al. (2009). Coated Ureteral Stent. Biomaterials and tissue engineering in urology. P. J. Denstedt and P. A. Atala. London, Woodhead Publishing Ltd.

Cauda, F., et al. (2008). Heparin Coating on Ureteral double J Stents Prevents encrustations: an In Vivo Case Study.J. Endourology Vol. 22, No 3,pp. 465-472.

Cho, Y. H., et al. (2001). Prophylactic efficacy of a new gentamicin-releasing urethral catheter in short-term catheterized rabbits.BJU International Vol. 87, No 1,pp. 104-109.

Cormio, L., et al. (2001). Bacterial adhesion to urethral catheters: Role of coating materials and immersion in antibiotic solution.Eur. Urol. Vol. 40, No 3, pp. 354-359.

Costerton, J. W. (2007). The Biofilm Primer. Springer, Hiedelberg. Pp 1- 200

Gorman, S. P. and A. D. Woolfson (2002). Novel biomimetic and bioactive silicones.Med. Device Technol. Vol. 13, No 7,pp. 14-15.

Grill, A. (1999). Diamond-like carbon: state of the art.Diamond and Related Materials Vol. 8, 428-434.

Hildebrandt, P., et al. (1999). Immobilisiertes heparin als inkrustierungresistente Beschichtung auf urologischen implantaten.Biomed. Techn. Vol. 42, pp. 123-124.

Jansen, B., et al. (1993). Bacterial adherence to hydrophilic polymer-coated polyurethane stents.Gastrointest. Endosc. Vol. 39, No 5,pp. 670-673.

John, T., et al. (2007). Antibiotic Pretreatment of Hydrogel Ureteral Stent.J. Endourology Vol. 21, No 10,pp. 1211-1216.

Laube, N., et al. (2006). Plasma-deposited carbon coating on urological indwelling catheters : Preventing formation of encrustations and consecutive complications.Urologe A. Vol. 45, No 9,pp. 1163-1169.

Laube, N., et al. (2007). Diamond-like carbon coatings on ureteral stents-a new strategy for decreasing the formation of crystalline bacterial biofilms?J. Urol. Vol. 177, No 5,pp. 1923-1927.

Leung, J. W., et al. (1992). Decreased bacterial adherence to silver-coated stent material: an in vitro study.Gastrointest. Endosc. Vol. 38, No 3,pp. 338-340.

Leung, J. W., et al. (2001). Effect of antibiotic-loaded hydrophilic stent in the prevention of bacterial adherence: A study of the charge, discharge, and recharge concept using ciprofloxacin.Gastrointest. Endosc. Vol. 53, No 4,pp. 431-437.

Lingeman, J. E., et al. (2003). Use of a temporary ureteral drainage stent (TUDS) after uncomplicated ureteroscopy: results from a phase II clinical trial.J. Urol. Vol. 169, No 5,pp. 1682-1688.
Mardis, H. K., et al. (1993). Comparative evaluation of materials used for internal ureteral stents.J. Endourol. Vol. 7, No 2,pp. 105-115.
Multanen, M., et al. (2000). Bacterial adherence to ofloxacin-blended poly-lactone-coated self-reinforced L-lactic acid polymer urological stents.BJU International Vol. 86, No,pp. 966-969.
Olweny, E. O., et al. (2002). Evaluation of the use of a biodegradable ureteric stent after retrograde endopyelotomy in a porcine model.J. Urol. Vol. 167, No 5,pp. 2198-2202.
Piper, J. (1946). The anticoagulant effect of heparin and synthetic polysaccharide-polysulphuric acid esters.Acta Pharmacol. Vol. 2, pp. 138-148.
Raad, I., et al. (1997). Central Venous Catheters Coated with Minocycline and Rifampin for the Prevention of Catheter-Related Colonization and Bloodstream Infections.Ann. Intern. Med. Vol. 127, No 4,pp. 267-274.
Reid, G. (2001). Oral fuoroquinolone theraphy results in drug adsorption on ureteral stents and prevention of biofilm formation.Int. J. Antimicr. Ag. Vol. 17, pp. 317-320.
Roy, R. K. and K.-R. Lee (2007). Biomedical applications of diamond-like carbon coatings: A review.J. Biomed. Mater. Res. Part B: Appl. Biomater. Vol. 83, No B,pp. 72-84.
Ruggieri, M. R., et al. (1987). Reduction of bacterial adherence to catheter surface with heparin.J. Urol. Vol. 138, pp. 423-426.
Tenke, P., et al. (2004). Bacterial biofilm formation on urologic devices and heparin coating as preventive strategy Int. J. Antimicr. Ag. Vol. 23, No S1,pp. 67-74.
Voevodin, A. A. and M. S. Donley (1996). Preparation of amorphous diamond-like carbon by pulsed laser deposition: a critical review.Surface and Coatings Technology Vol. 82, No 3,pp. 199-213.
Warren, J. W., et al. (1994). Long-term urethral catheterization increases risk of chronic pyelonephritis and renal inflammation.J. Am. Geriat. Soc. Vol. 42, No 12,pp. 1286-1290.
Zdrahala, R. J. and I. J. Zdrahala (1999). Biomedical application of polyurethanes: A review of past promises, present realities, and a vibrant future.J. Biomater. Appl. Vol. 14, pp. 67-90.

Chapter 8

Polyurethane as Carriers of Antituberculosis Drugs

Yerkesh Batyrbekov and Rinat Iskakov

Additional information is available at the end of the chapter

http://dx.doi.org/10.5772/35896

1. Introduction

Polyurethanes (PU) are an important class of polymers that have found many applications as biomaterials due to their excellent physical properties and relatively good biocompatibility. Basically, PU may be produced by two chemical processes: by polycondensation of a diamine with bischloroformates or by reaction between a diol and a diisocyanate. Many biomedical devices are made from segmented PU such as catheters, blood pumps, prosthetic heart valves and insulation for pacemakers (Lelah & Cooper, 1986, Lamba et al., 1997). A promising approach for the development of new controlled-releasing preparations is use of PU as the carriers in drug delivery systems.

Drug delivery systems have been progressively developed in the field of therapeutic administration owing to their advantages: providing drug concentration over a period of prolonged action, decreasing the total therapeutic dose and reducing the undesirable side effects, and, hence, improving the pharmaceutical efficiencies. These are achieved by the use of the controlled-release drug delivery systems (Hsien, 1988). Controlled release dosage forms are consist of the pharmacological agent and the polymer carrier that regulate its release. In general two types of drug delivery systems have been used: diffusion-controlled systems and dissolution-controlled systems. In the first cause the drug is usually dispersed or dissolved in the solid reservoir or membrane and the kinetics of drug release are generally controlled by diffusion through the polymer. In the second cause the drug are generally incorporated into a water-soluble or water-swellable polymer and the release of drug is controlled by swelling and dissolution of polymer. In both the causes polymer function is a principal component which controls the transport and the release rate of drug molecule. To be a useful drug carrier, a polymer needs to possess certain features. The polymeric carrier has to be non-toxic, non-immunogenic and biocompatible; the carrier must contain an effective dose of active agent; the material of system must be biodegradable and form biologically acceptable degradation products; the rate of drug release from the carrier must occur at an acceptable rate; the carrier must be able to be easily sterilized.

The design of the PU controlled-release forms for therapeutic drug administration is the subject of intense interest. Such systems are being used for sustained and controlled delivery of various pharmaceutical agents such as prednisolon (Sharma et al., 1988), morphine, caffeine (Graham et al., 1988), prostaglandin (Embrey et al., 1986) and theophylline (Reddy et al., 2006). The PU carrier is utilized to deliver iodine-containing drugs (Touitou & Friedman, 1984). Urethane-based hydrogels were prepared based on the reaction of diisocyanates with amphilic ethylene oxide and triol crosslinker to deliver propranolol hydrochloride, an antihypertensive drug (Van Bos & Schacht, 1987). Drug delivery systems on a PU base with various antitumorous drugs, such as cyclophosphane, thiophosphamide and vincristine, have been prepared (Iskakov et al., 1998, 2000). An in vitro technique was used to determine the release characteristics of the drugs into model biological media. It was shown the drug release occurs in accordance with first-order kinetics.

PU-based drug delivery systems have considerable potential for treatment of tuberculosis. Tuberculosis is widely spread disease in most developing countries. The main method of tuberculosis treatment is chemotherapy. Although current chemotherapeutic agents for tuberculosis treatment are therapeutically effective and well tolerated, a number of problems remain. The chemotherapy is burden some, extends over long periods and requires continuous and repeated administration of large drug doses. Thus, traditional drug chemotherapy has serious limitations because of increasing microbial drug resistance and toxico-allergic side effects. One of the ultimate problems in effective treatment of tuberculosis is patient compliance. These problems of increasing drug resistance, toxico-allergic side effects, patient compliance can be approached by the use of long-acting polymeric drug delivery systems (Sosnik et al., 2010). The design of implantable systems containing the antituberculosis drugs in combination with biocompatible polymers would make possible to achieve the significant progress in treatment of this global debilitating disease (Shegokar et al., 2011).

Biodegradable microsphere drug delivery systems have shown application for oral and parenteral administration. Administration of microparticles to the lungs (alveolar region) may provide the opportunity for the prolonged delivery active agent to tuberculosis infected macrophages. Microspheres can be produced to meet certain morphological requirements such as size, shape and porosity by varying the process parameters. However, the morphology of the lung is such that to achieve effective drug deposition it is necessary to control the particle size of microparticles.

The objective of the chapter is to develop an effective polymeric drug delivery systems based on PU for the treatment of tuberculosis. Polyurethane materials are investigated as carriers for sustained and controlled release of antituberculosis drugs. The synthesis and characterization of PU microcapsules are studied making use various molecular weight polyethylene glycol and tolylene-2,4-diisocyanate. Antituberculosis drug isoniazid (Is), rifampicin, ethionamide and florimicin were incorporated into the PU microcapsules and foams. The effects of nature and concentration of drugs and diols, molecular weight (Mw), morphology of polyurethanes on release behavior from polymeric systems were studied. The possibility of application of the polymeric drug delivery systems based on polyurethane for tuberculosis treatment was shown by some medical and biological tests.

2. Polymeric microparticles for tuberculosis treatment

Recent trends in polymeric controlled drug delivery have seen microencapsulation of pharmaceutical substances in biodegradable polymers as an emerging technology. Extensive progressive efforts have been made to develop various polymeric drug delivery systems to either target the site of tuberculosis infection or reduce the dosing frequency (Toit et al., 2006). Carriers as microspheres have been developed for the sustained delivery of antituberculosis drugs and have demonstrated better chemotherapeutic efficacy when investigated in animal models. Antituberculosis drugs have been successfully entrapped in microparticles of natural and synthetic polymers such as alginate (ALG), ALG-chitosan, poly-lactide-co-glycolide and poly-butyl cyanoacrylate (Gelperina et al., 2005, Pandey & Khuller, 2006).

ALG, a natural polymer, has attracted researchers owing to its ease of availability, compatibility with hydrophobic as well as hydrophilic molecules, biodegradability under physiological conditions, lack of toxicity and the ability to confer sustained release potential. The ability of ALG to co-encapsulate multiple antitubercular drugs and offer a controlled release profile is likely to have a major impact in enhancing patient compliance for better management of tuberculosis (Ahmad & Khuller, 2008).

Spherical microspheres able to prolong the release of Is were produced by a modified emulsification method, using sodium ALG as the hydrophilic carrier (Rastogi et al., 2007). The particles were heterogeneous with the maximum particles of an average size of 3.719 μm. Results indicated that the mean particle size of the microspheres increased with an increase in the concentration of polymer and the cross-linker as well as the cross-linking time. The entrapment efficiency was found to be in the range of 40-91%. Concentration of the cross-linker up to 7.5% caused increase in the entrapment efficiency and the extent of drug release. Optimized Is-ALG microspheres were found to possess good bioadhesion. The bioadhesive property of the particles resulted in prolonged retention in the small intestine. Microspheres could be observed in the intestinal lumen at 4h and were detectable in the intestine 24h post-oral administration. Increased drug loading (91%) was observed for the optimized formulation suggesting the efficiency of the method. Nearly 26% of Is was released in simulated gastric fluid pH 1.2 in 6h and 71.25% in simulated intestinal fluid pH 7.4 in 30h.

ALG microparticles were developed as oral sustained delivery carriers for antituberculosis drugs in order to improve patient compliance (Qurrat-ul-Ain et al., 2003). Pharmacokinetics and therapeutic effects of ALG microparticle encapsulated Is, rifampicin and pyrazinamide were examined in guinea pigs. ALG microparticles containing antituberculosis drugs were evaluated for in vitro and in vivo release profiles. These microparticles exhibited sustained release of Is, rifampicin and pyrazinamide for 3-5 days in plasma and up to 9 days in organs. Peak plasma concentration, elimination half-life and infinity of ALG drugs were significantly higher than those of free drugs. The encapsulation of drug in ALG microparticles resulted in up to a nine-fold increase in relative bioavailability compared with free drugs. Chemotherapeutic efficacy of ALG drug microspheres against experimental tuberculosis not detectable at 1:100 and 1:1000 dilutions of spleen and lung homogenates. Histopathological studies further substantiated these observations, thus suggesting that application of ALG-encapsulated drugs could be useful in the effective treatment of tuberculosis.

Pharmacokinetics and tissue distribution of free and ALG-encapsulated antituberculosis drugs were evaluated in mice at different doses (Ahmad et al., 2006). ALG nanoparticles encapsulating Is, rifampicin, pyrazinamide and ethambutol were prepared by controlled cation-induced gelification of ALG. The formulation was orally administered to mice at two dose levels. A comparison was made in mice receiving free drugs at equivalent doses. The relative bioavailabilities of all drugs encapsulated in ALG nanoparticles were significantly higher compared with free drugs. Drug levels were maintained at or above the minimum inhibitory concentration until 15 days in organs after administration of encapsulated drugs, whilst free drugs stayed at or above 1 day only irrespective of dose. The levels of drugs in various organs remained above the minimum inhibitory concentration at both doses for equal periods, demonstrating their equiefficiency.

Chemotherapeutic potential of ALG nanoparticle-encapsulated econazole and antituberculosis drugs were studied against murine tuberculosis (Ahmad et al., 2007). Econazole (free or encapsulated) could replace rifampicin and Is during chemotherapy. Eight doses of ALG nanoparticle-encapsulated econazole or 112 doses of free econazole reduced bacterial burden by more than 90% in the lungs and spleen of mice infected with Mycobacterium tuberculosis. ALG nanoparticles reduced the dosing frequency of azoles and antitubercular drugs by 15-fold.

Is was encapsulated into microspheres of ALG-chitosan by means of a complex coacervation method in an emulsion system (Lucinda-Silva & Evangelista, 2003). The particles were prepared in three steps: preparation of a emulsion phase and adsorption of the drug. The results showed that microspheres of ALG-chitosan obtained were of spherical shape. The emulsion used for microparticle formation allows the preparation of particles with a narrow size distribution. The adsorption observed is probably of chemical nature, i.e. there is an ionic interaction between the drug and the surface of the particles.

ALG-chitosan microspheres encapsulating rifampicin, Is and pyrazinamide, were formulated (Pandey & Khuller, 2004). A therapeutic dose and a half-therapeutic dose of the microsphere-encapsulated were orally administered to guinea pigs for pharmacokinetic and chemotherapeutic evaluations. The drug encapsulation efficiency ranged from 65% to 85% with a loading of 220-280 mg of drug per gram microspheres. Administration of a single oral dose of the microspheres to guinea pigs resulted in sustained drug levels in the plasma for 7 days and in the organs for 9 days. In Mycobacterium tuberculosis $H_{37}Rv$-infected guinea pigs, administration of a therapeutic dose of microspheres spaced 10 days apart produced a clearance of bacilli equivalent to conventional treatment for 6 weeks.

Polylactide-co-glycolide (PLG) polymers are biodegradable and biocompatible, they have been the most commonly used as carriers for microparticle formulations. Monodispersed poly-lactic-co-glycolic acid (PLGA) microspheres containing rifampicin have been prepared by solvent evaporation method (Makino et al., 2004, Yoshida et al., 2006). The microspheres were spherical and their average diameter was about 2 μm. The loading efficiency of rifampicin was dependent on the molecular weight of PLGA. The higher loading efficiency was obtained by the usage of PLGA with the lower Mw, which may be caused by the interaction of the amino groups of rifampicin with the terminal carboxyl groups of PLGA. PLGA with the monomer compositions of 50/50 and 75/25, of lactic acid/glycolic acid, were used in this study. From rifampicin-loaded PLGA microspheres formulated using PLGA with the Mw of 20,000, rifampicin was released with almost constant rate for 20 days after the lag phase was observed for the initial 7 days at pH 7.4. On the other hand, from rifampicin-loaded PLGA microspheres formulated using PLGA with the molecular weight of 5000 or 10,000, almost 90% of rifampicin-loaded in the microspheres was released in the initial 10 days. Highly effective delivery of rifampicin to alveolar macrophages was observed by the usage of rifampicin-loaded PLGA microspheres. Almost 19 times higher concentration of rifampicin was found to be incorporated in alveolar macrophages when rifampicin-loaded PLGA microspheres were added to the cell culture medium than when rifampicin solution was added.

Controlled release rifampicin-loaded microspheres were evaluated in nonhuman primates (Quenelle et al., 2004). These microsheres were prepared by using biocompatible polymeric excipients of lactide and glycolide copolymers. Animals received either 2.0 g of a large formulation (10–150 mcm, 23 wt% rifampicin) injected subcutaneously at Day 0 (118–139 mg rifampin/kg), 4.0 g of a small formulation (1–10 mcm, 5.8 wt% rifampicin) administered intravenously in 2.0 g doses on Day 0 and 7 (62.7–72.5 mg rifampicin/kg), or a combination of small and large microspheres (169–210 mg rifampin/kg). Extended rifampicin release was observed up to 48 days. Average rifampicin concentrations remaining in the liver, lung, and spleen at 30 days were 14.03, 4.09, and 1.98 μg/g tissue, respectively.

PLG nanoparticles encapsulating streptomycin were prepared by the multiple emulsion technique and administered orally to mice for biodistribution and chemotherapeutic studies (Pandey & Khuller, 2007). The mean particle size was 153.12 nm with 32.12±4.08% drug encapsulation and 14.28±2.83% drug loading. Streptomycin levels were maintained for 4 days in the plasma and for 7 days in the organs following a single oral administration of PLG nanoparticles. There was a 21-fold increase in the relative bioavailability of PLG-encapsulated streptomycin compared with intramuscular free drug. In *Mycobacterium tuberculosis (M.tuberculosis)* $H_{37}Rv$ infected mice, eight doses of the oral streptomycin formulation administered weekly were comparable to 24 intramuscular injections of free streptomycin.

PLG nanoparticle-encapsulated econazole and moxifloxacin have been evaluated against murine tuberculosis (drug susceptible) in order to develop a more potent regimen for tuberculosis (Ahmad et al., 2008). PLG nanoparticles were prepared by the multiple emulsion and solvent evaporation technique and were administered orally to mice. A single oral dose of PLG nanoparticles resulted in therapeutic drug concentrations in plasma for up to 5 days (econazole) or 4 days (moxifloxacin), whilst in the organs (lungs, liver and spleen) it was up to 6 days. In comparison, free drugs were cleared from the same organs within 12-24h. In *M. tuberculosis*-infected mice, eight oral doses of the formulation administered weekly were found to be equipotent to 56 doses (moxifloxacin administered daily) or 112 doses (econazole administered twice daily) of free drugs. Furthermore, the combination of moxifloxacin+econazole proved to be significantly efficacious compared with individual drugs. Addition of rifampicin to this combination resulted in total bacterial clearance from the organs of mice in 8 weeks. PLG nanoparticles appear to have the potential for intermittent therapy of tuberculosis, and combination of moxifloxacin, econazole and rifampicin is the most potent.

Antituberculosis drugs Is, rifampicin, streptomycin and moxifloxacin have been encapsulated in poly(butyl cyanoacrylate) nanoparticles (Anisimova et al., 2000, Kisich et al, 2007). Incorporation of drugs in polymeric nanoparticles not only increased the intracellular accumulation of these drugs in the cultivated human blood monocytes but also produced enhanced antimicrobial activity of these agents against intracellular *M. tuberculosis* compared with their activity in extracellular fluid. Encapsulated moxifloxacin accumulated in macrophages approximately three-fold times more efficiently than the free drug, and was detected in the cells for at least six times longer than free moxifloxacin at the same extracellular concentration.

This brief review suggested that micro- and nanoparticles based delivery systems have a considerable potential for treatment of tuberculosis. Their major advantages, such as improvement of drug bioavailability and reduction of the dosing frequency, may create a sound basis for better management of the disease, making directly observed treatment more practical and affordable.

3. Polyurethane microparticles as carriers of drug

PU microspheres can be prepared by interfacial polycondensation in emulsions. These techniques include polycondensation of two or more complimentary monomers at the interface of two-phase system, carefully emulsified for obtaining little drop-lets in emulsion phase. Usually, the interfacial polycondensation carried out by two steps: emulsification step (emulsion formation using a mechanical stirring during few minutes and one of the monomers is dissolved in the emulsion drops; polycondensation step (the second complementary monomer is added to the external phase of the emulsion and the polycondensation reaction takes place at the liquid-liquid emulsion interface). Interest in the PU microparticles in each day has being increased since products presents numerous advantages in biomedical, pharmaceutical and cosmetic applications.

PU microparticles could be interesting matrices for controlled drug delivery. Aliphatic PU Tecoflex was evaluated as microsphere matrix for the controlled release of theophylline (Subhaga et al., 1995). PU microspheres were prepared using solvent evaporation technique from a dichloromethane solution of the polymer containing the drug. A dilute solution of poly(vinyl alcohol) served as the dispersion medium. Microspheres of good spherical geometry having theophylline content of 35% could be prepared by the technique. The release of the drug from the microspheres was examined in simulated gastric and intestinal fluids. While a large burst effect was observed in gastric fluid, in the intestinal fluid a close to zero-order release was seen.

Microencapsulation of theophylline in PU was developed with 4, 4'-methylene-diphenylisocyanate, castor oil and ethylene diamine as chain extender (Rafienia et al., 2006). PU microspheres were prepared in two steps pre-polymer preparation and microspheres formation. Particle size investigation with optical microscopy revealed size distribution of 27–128 μm. Controlled release experiment of theophylline was performed in phosphate buffered saline at pH 7.4 with UV-spectrometer at 274 nm. Drug release profiles showed initial release of 2–40% and further release for more than 10 days.

The effect of chain-extending agent on the porosity and release behavior of biologically active agent diazinon from PU microspheres were studied (Jabbari & Khakpour, 2000). Microsphere was prepared using a two-step suspension polycondensation method with methylene diphenyl diisocyanate, polyethylene glycol 400 and 1,4-butanediol as the chain-extending agent. Chain-extending agent was used to increase the ratio of hard to soft segments of the PU network, and its effect on microsphere morphology was studied with SEM. According to the results, porosity was significantly affected by the amount of chain-extending agent. The pore size decreased as the concentration of chain-extending agent increased from zero to 50 mole%. With further increase of chain-extending agent to 60 and 67%, PU chains became stiffer and formation of pores was inhibited. Therefore, pore morphology was significantly affected by variations in the amount of chain-extending agent. The release behavior of microspheres was investigated with diazinon as the active agent. After an initial burst, corresponding to 3% of the incorporated amount of active agent, the release rate was zero order.

PU polymers and poly(ether urethane) copolymers were chosen as drug carriers for alpha-tocopherol (Bouchemal et al., 2004). This active ingredient is widely used as a strong antioxidant in many medical and cosmetic applications, but is rapidly degraded, because of its light, heat and oxygen sensitivity. PU and poly(ether urethane)-based nanocapsules were synthesized by interfacial reaction between two monomers. Interfacial polycondensation combined with spontaneous emulsification is a new technique for nanoparticles formation. Nanocapsules were characterized by studying particle size (150-500 nm), pH, yield of encapsulation and morphologies. Polyurethanes were obtained from the condensation of isophorone diisocyanate and 1,2-ethanediol, 1,4-butanediol , 1,6-hexanediol. Poly(ether urethane) copolymers were obtained by replacing diols by polyethylene glycol oligomers Mw 200, 300, 400 and 600. Mw of di- and polyols have a considerable influence on nanocapsules characteristics cited above. The increase of Mw of polyols tends to increase the mean size of nanocapsules from 232±3 nm using ethanediol to 615± 39 nm using PEG 600, and led to the agglomeration of particles. We also noted that the yield of encapsulation increases with the increase of polyol length. After 6 months of storage, polyurethanes nanocapsules possess good stability against aggregation at 4 and 25° C. Comparing results obtained using different monomers, it reveals that the PU based on hexanediol offers good protection of alpha-tocopherol against damaging caused by the temperature and UV irradiation (Bouchemal et al., 2006).

Ovalbumin (OVA)-containing PU microcapsules were successfully prepared by a reaction between toluene diisocyanate and different polyols such as glycerol, ethane diol, and propylene glycol (Hong & Park, 2000). The structural and thermal properties of the resultant microcapsules and the release profile of the OVA from the wall membranes were studied. In conclusion, the microcapsules from the glycerol showed the highest thermal stability, with the formation of many hydrogen bonds. From the data of release profiles, it was confirmed that the particle size distribution and morphologies of microcapsules determined the release profiles of the OVA from the wall membranes.

Bi-soft segmented poly(ester urethane urea) microparticles were prepared and characterized aiming biomedical application (Campos et al., 2011). Two different formulations were developed, using poly(propylene glycol), tolylene 2,4-diisocyanate terminated pre-polymer and poly(propylene oxide)-based tri-isocyanated terminated pre-polymer (TI). A second soft segment was included due to poly(ε-caprolactone) diol. Infrared spectroscopy, used to study the polymeric structure, namely its H-bonding properties, revealed a slightly higher degree of phase separation in TDI-microparticles. TI-microparticles presented slower rate of hydrolytic degradation, and, accordingly, fairly low toxic effect against macrophages. These new formulations are good candidates as non-biodegradable biomedical systems.

The synthesis of PU microsphere-gold nanoparticle "core-shell" structures and their use in the immobilization of the enzyme endoglucanase are described (Phadtare et al., 2004). Assembly of gold nanoparticles on the surface of polymer microspheres occurs through interaction of the nitrogens in the polymer with the nanoparticles, thereby precluding the need for modifying the polymer microspheres to enable such nanoparticle binding. Endoglucanase could thereafter be bound to the gold nanoparticles decorating the PU microspheres, leading to a highly stable biocatalyst with excellent reuse characteristics. The immobilized enzyme retains its biocatalytic activity and exhibits improved thermal stability relative to free enzyme in solution.

Microencapsulation of the water soluble pesticide monocrotophos (MCR), using PU as the carrier polymer, has been developed using two types of steric stabilizers polymethyllauril acrylate (PLMA) macrodiol and PLMA-g-PEO graft copolymer (Shukla et al., 2002). The microencapsulation process is carried out in non-aqueous medium and at a moderate temperature to avoid any chemical degradation of monocrotophos during the encapsulation process. Microcapsules were characterized by optical microscopy and SEM for particle size and morphology, respectively. The effects of loading of MCR, crosslinking density of PU, and nature of steric stabilizer on the release of MCR from PU microcapsules have been studied.

Poly(urea-urethane) microcapsules containing oil-soluble dye dioctyl phthalate as core material were prepared by the interfacial polymerization with using mixtures of tri- and di-isocyanate monomers as wall forming materials (Chang et al., 2003, 2005). The time course of the dye release in dispersing tetrahydrofuran was measured as a function of the weight fraction of tri-isocyanate monomer in the total monomer weight and the core/wall material-weight ratio. The dye release curves were well represented by an exponential function C=Ceq(1-e-t/tau), where C is the concentration of the dye in the dispersing medium, Ceq that at equilibrium state, t the elution time and tau is a time constant. tau increased linearly against weight at high contentration, suggesting controllability of the release rate of microcapsules by varying tri-isocyanate/di-isocyanate ratio

4. Polyurethane microparticle as carrier of antituberculosis drug

New polyurethane microcapsules incorporated with antituberculosis drug Is have been synthesized by interfacial polyaddition between toluene-2,4-diisocyanate (TDI) and various poly(ethylene glycol)s (PEG). Drug Is is hydrophilic water-soluble compound, and it is insoluble in toluene. Thus Is could be capsulated by interfacial polycondensation technique using water-in-toluene emulsion, which prevents transferring of Is to the external phase. And drug encapsulation is possible during the process of the polymer wall formation (Batyrbekov et al., 2009; Iskakov et al., 2004).

Isocyanate groups react with hydroxyl groups of PEG to form polyurethane chains according to the Scheme (Fig.1 a). TDI can also reacts with molecules of water at the border of reaction to form unstable NH-COOH group, which dissociates into amine and carbon dioxide (Fig.1 b). Chains with amine end-group reacts with the isocyanate groups of growing polymer with urea segments formation.

Polycondensation was carried out in a 1 L double-neck flask fitted with a stirrer. Polyethylene glycol with 4 various molecular weights 400, 600, 1000 and 1450 (Sigma, USA) - PEG 400, PEG 600, PEG 1000 and PEG 1450 respectively were used as diol monomers. TDI (Sigma, USA) was applied as a bifunctional monomer for the polycondensation. Three solutions were prepared separately. Solution A: 10 mg surfactant Tween 40 was dissolved in 100 ml of toluene; solution B: x mmol diol and Is were added in y ml of water; amount of Is was 10, 20, and 30 mol % from PEG; solution C: 2.5 x mmol TDI was dissolved in 10 ml of solution A. Water/oil ratio was 1:10 vol.%. Solution B was poured into the reaction flask, contaning 90 ml of solution A under the stirring at 1000 rpm during 15 minutes. After the formation of microemulsion, solution C was added dropwise. After 180 min the polymerization was stopped. Microparticles were filtered, carefully washed with distillated water and dried at ambient conditions. Yield of polymers was estimated from the total amount of introduced monomers compared to the weight of polycondensation products.

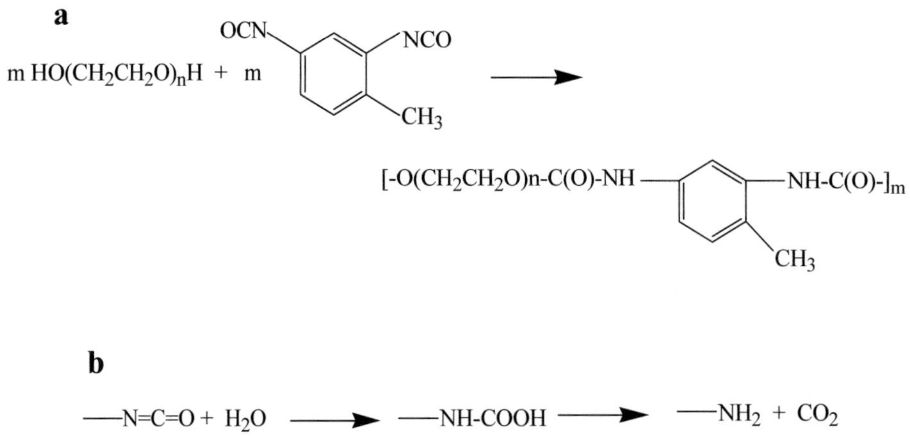

Figure 1. Scheme of reaction between PEG and TDI with polyurethane formation.

The completion of polycondensation process was estimated by IR-spectra from decreasing of the absorption band at 2270-2320 cm^{-1}, which correspond to -N=C=O isocyanate group. IR spectra were obtained by a Nicolet 5700 FT-IR (USA) infrared spectrophotometer in KBr.

In the IR-spectra of microparticles the N-H stretching vibration were observed at 3450–3300 cm^{-1}, absorption bands at 1740–1700 cm^{-1} for the C=O stretching of urethane and at 1690–1650 cm^{-1} for urethane-urea formation were also present (Fig.2). Absorption bands are present at 1100 cm^{-1} for C-O-C ether group and at 2850 -2950 cm^{-1} for C-H. In FT-IR spectra of microparticles containing Is, the new stretching vibrations band appeared at 1350 cm^{-1}, 1000 cm^{-1} and 690 cm^{-1}, which were also present in FT-IR spectra of pure Is that indicates the physical mechanism of Is capsulation.

In the process of interfacial polycondensation, two PU products of reaction were detected: the main product - microparticles, and the secondary product - linear precipitated polyurethane. The increase of PEG content in water phase resulted in increased amount of the secondary product, and as the PEG content in water phase reached 60 vol.%, maximum of the secondary product was observed (about 40%).

Decreasing PEG concentration in water phase leads to increased yield of polyurethane microparticles. Maximum of yield was reached at PEG concentration 22 - 27 vol.% and in that condition whole olygomer reacted at surface of emulsion drops with microparticles formation. Reduction of microparticle yield after the maximum is due mainly to increasing contribution of the hydrolysis process of isocyanate groups.

Appearance of the secondary product and increase of its yield, probably, can be attributed to the increase of PEG concentration and results in PEG partially transfer from the water phase to the internal phase of toluene and the process of polycondensation between PEG and TDI takes place with linear polyurethane formation. At the end of reaction rate of PEG diffusion to surface, namely at the reaction region, seems to be a limit stage of the process. Reducing of PEG concentration causes to decrease of system viscosity. Effectiveness of Is capsulation in PU microparticles ranged from 3.4 to 41.7 % and significantly depended on water/PEG ratio in the water phase of emulsion.

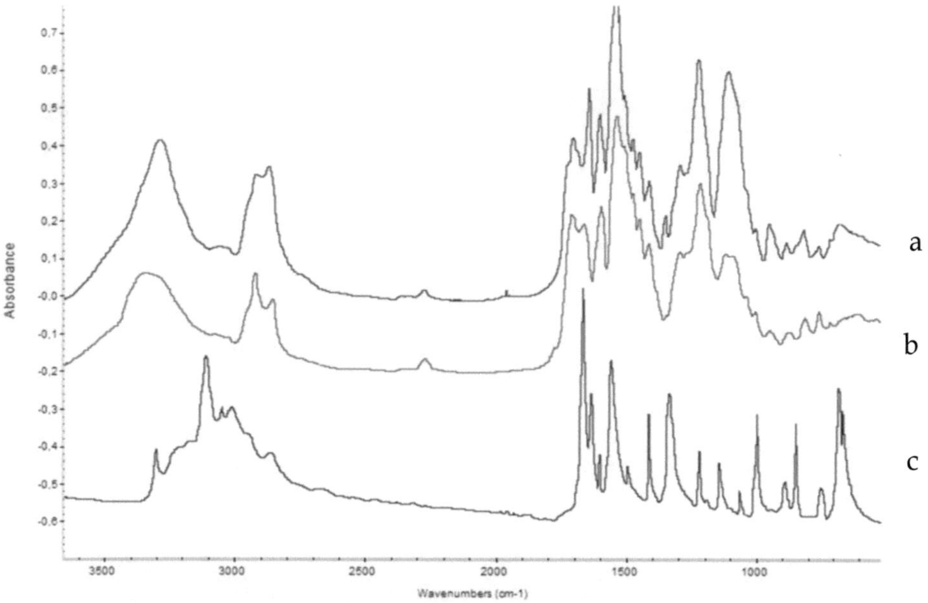

Figure 2. FT-IR-spectra of polyurethane microparticles containing isoniazid (a), polyurethane microparticles (b) and isoniazid (c).

Fig. 3 shows that composition of the water phase influences upon effectiveness of capsulation. Increase of PEG concentration results in decreasing effectiveness of capsulation and decrease of Is loading correspondingly. The high PEG concentration promotes miscibility of Is in the internal oil phase - toluene. Morphology of the surface of microparticles is very important factor, which affects release behavior of active agent. The wall structure depends on the conditions of interfacial polycondensation process, such as Mw and chemical structure of diol, the concentration of the monomers and other. The effect of water/PEG ratio in aqueous phase on morphology of microparticle was investigated. Microparticles prepared from PEG solutions of higher concentration have dense surface so that Is diffused much slower. At the high concentration of PEG reaction between PEG and TDI is significantly limited on the interface of the drops. Furthermore excessive PEG transfers to the external surface of microparticles and reacts with TDI and less penetrable wall was formed.

Fig.4 shows SEM photos of interfacial polycondensation products prepared by reaction between TDI and PEG 400 at water/PEG ratio 11,8 : 88,2 vol.% in water phase. According to Figs. 4a and 4b two products of polycondensation with different structure were formed. PU microparticles have spherical shape and size about 5 - 10 μm (Fig. 4a). The secondary product has fibril structure with diameter less then 500 nm (Fig 4b). The effect of water/PEG ratio on morphology of microparticle walls is shown in Figs. 4c and 4d. PU microparticles prepared at water/PEG ratio 82,4 :17,6 (Fig. 4c) have rough surface. On the contrary the surface of PU microparticles prepared at water/PEG ratio 11,8 : 88,2 (Fig 4d) were dense and smooth.

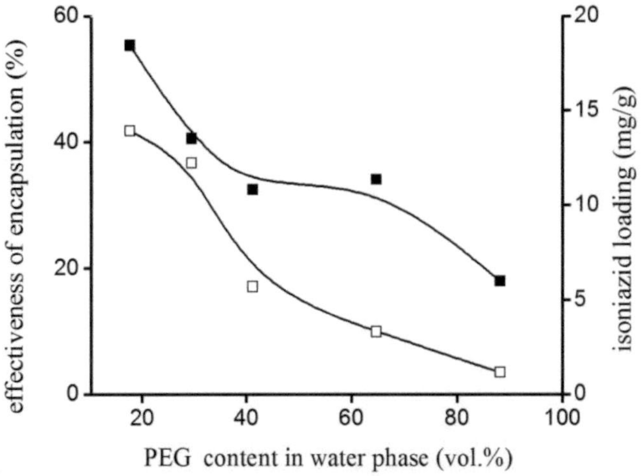

Figure 3. The effect of PEG content in water phase on effectiveness of encapsulation (□) and Is loading in PU microparticles (■).

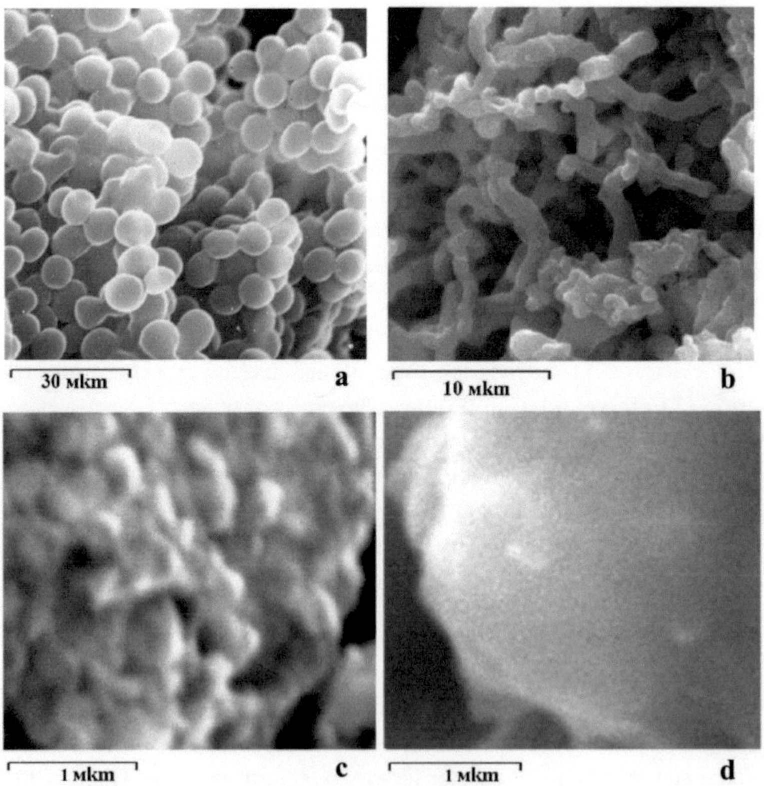

Figure 4. SEM photographs of products of interfacial polycondensation . between TDI and PEG 400 at 60°C. PU microparticles (a) and PU secondary product (b)synthesized at water/PEG ratio 11.8 : 88.2 in water phase. Surface of PU microparticles prepared at water/PEG ratio 82.4 :17.6 (c) and 11.8 : 88.2 (d) in water phase.

The release behavior of Is from PU microparticles was carried out and different conditions of synthesis such as water/PEG ratio, molecular weight of PEG and isoniazid concentration was investigated. The release behavior of microparticles loaded with isoniazid was studied with ultraviolet (UV) spectroscopy. For calibration, physiological solutions of isoniazid with concentration ranging from 0.004 to 0.05 mg/ml were prepared and their absorption was measured at 263.5 nm with Jasco UV/VIS 7850 spectrophotometer (Japan). 10 mg of isoniazid-loaded microparticles were dispersed in 10 ml of physiological solution under light stirring at constant temperature 37°C. After fixed time interval 2 ml of solution was taken out by the squirt equipped with the special filter. The efficiency of capsulation was calculated as ratio of introduced isoniazid to solution B compared with amount of delivered isoniazid into water during 3 weeks. Isoniazid loading was weight of isoniazid (mg) contained in 1 g of microparticles.

Fig. 5 shows the release behavior of Is from PU microparticles, synthesized at different water/PEG ratio. The most part of the drug delivered during the first three hours, then slow release of the residual Is was observed during the next two weeks. Microparticles prepared with less concentration of PEG in the water phase demonstrated faster diffusion of Is through walls of microparticles. The increased PEG content in water phase of reaction, results in decreasing Is diffusion, due to formation of PU microparticles with denser polymer wall. Microparticles prepared with PEG concentration 17.6, 29.4 and 41.2 vol.% showed the release 58 - 66 % of Is during 3 h. However, due to denser wall of microparticles prepared with PEG 64.7 и 88.2 vol. % demonstrated the release no more 35% of the drug within the same time.

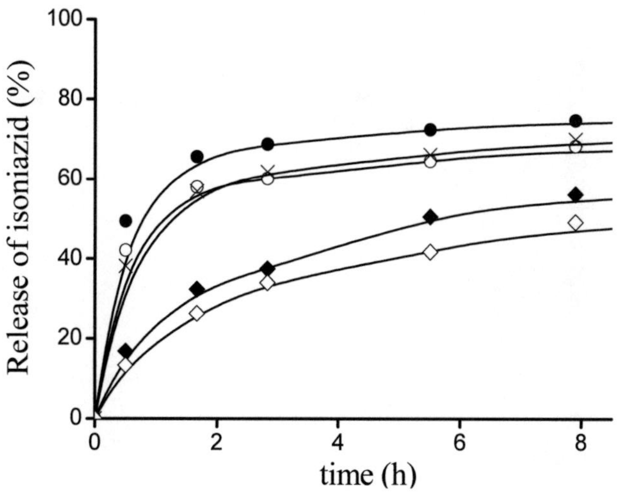

Figure 5. Release of Is from PU microparticles synthesized at various water/PEG ratio: • - 82.4:17.6, o - 70.6:29, ✕ - 58.8:41.2, ◆ - 35.3:64.7 and ◇ - 11.8:88.2.

The effect of molecular weight of PEG on release of Is from PU microparticles was investigated (Fig 6). Microparticles were prepared by using PEG with Mw 400, 600, 1000 and 1450. Increasing molecular weight of soft segments (PEG) results in the increase of diffusion rate of Is into solution. This phenomenon can be attributed to increasing Mw of PEG which leads to accelerating diffusion of water-soluble Is through hydrophilic PEG chains.

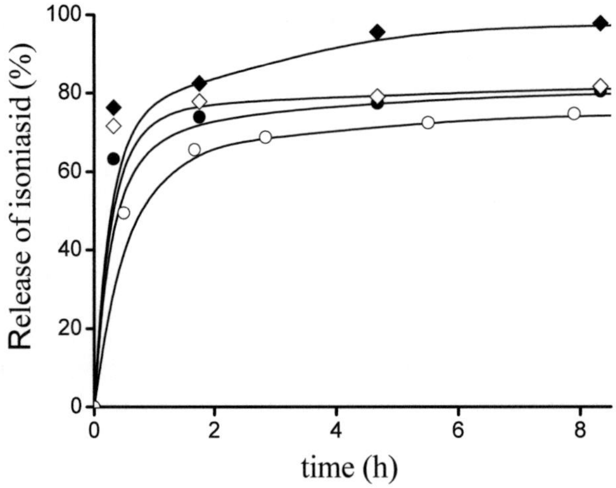

Figure 6. Release of Is from PU microparticles synthesized at various Mw of PEG. Mw = 400 (o), 600 (•), 1000 (◇) and 1450 (◆).

Microparticles with different Is loading, namely 18.4, 35.3 and 65.6 mg/g were produced. In Fig 7 release behavior of Is is shown.

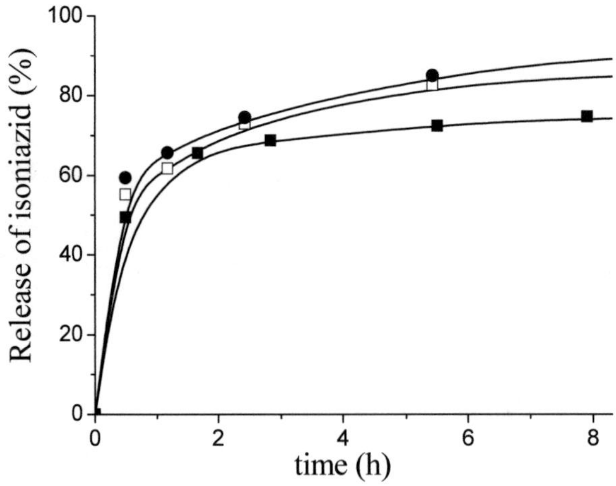

Figure 7. Release of Is from PU microparticles with different Is loading: 18.4 (■), 35.3 (□) and 65.6 mg/g (•).

Microparticles with higher Is loading demonstrate faster release rate of the drug due to increased gradient of concentrations between the external solution and core of microparticles.

PU microparticles were administrated subcutaneously to mice BL/6. Histologic analyses of the underskin tissue was carried out at a different period of microparticles administration in the mice by using electron microscope LEO F360, equipped with X-ray analyzer EDS Oxford ISI 300.

Fig 8 shows histologic analyses of tissue under skin. Within 5 days of the microparticles deposition, the thickening of the surrounding tissue due to primary macrophage reaction and fibrillar tissue formation were detected as shown in Fig. 8b. On day 21, some enzymatic lysis of polyurethane – C(O)–NH – group probably took place (Fig. 8c) and partial biodegradation of PU microparticles was observed. For all experimental animals no casting-off or necrosis of tissue were observed.

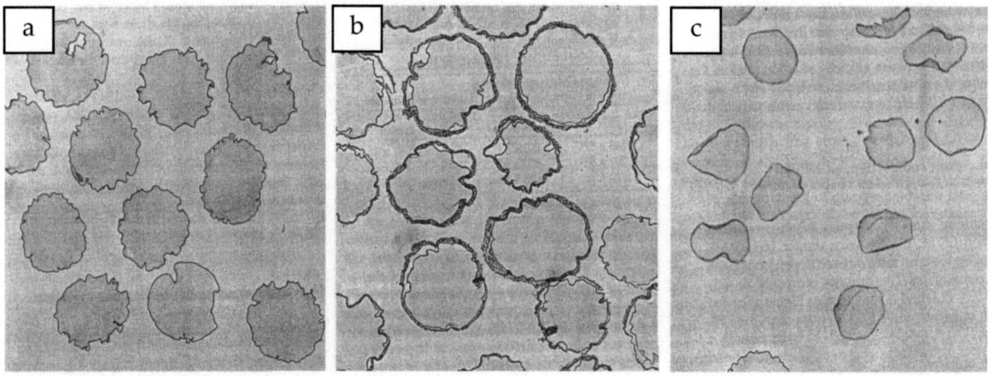

Figure 8. Histological slices of tissue under mice skin 1 (a), 5 (b), and 21 (c) days after deposition of is-containing PU microparticles to BL/6 mice provided by transparent electron microscope with 400x magnification.

Thus, the data obtained in the present work have demonstrated the possibility of using PU microparticles as a carrier for the controlled delivery of antituberculosis drug Is. PU microparticles were prepared by interfacial reaction between PEG and TDI in water in toluene emulsion. The effect of water/PEG ratio on the morphology of microparticles and release behavior was shown. The low PEG content in aqueous phase results in the formation of microparticles with rough surface, which demonstrate faster diffusion of Is in comparison to PU microparticles produced from more concentrated PEG solutions, they have smooth surface and less penetrable walls for Is. Increased Mw of PEG and Is loading leads to increased diffusion rate of isoniazid from polyurethane microparticles. For PU microparticles administered in mice BL/6 subcutaneously, biodegradation was observed due to enzymatic lysis of polyurethane group. Preliminary data indicate that PU microparticles could be perspective carriers for controlled delivery and respirable administration of antituberculosis drug Is.

5. Polyurethane foams as carriers of antituberculosis drugs

The use of soft PU foams as carriers of antituberculosis drugs is of considerable interest. In such systems pharmaceutical agents are dispersed or dissolved in the PU carrier and the kinetics of drug release are generally controlled by diffusion phenomena through the polymer. Such systems are being used for treatment of tuberculosis-infected cavities (wounds, pleural empyema, bronchial fistula). It is the purpose of this chapter to show the possibility of using polyurethane foams as carriers of some antituberculosis agents for tuberculosis treatment.

PU foams were synthesized by reaction of pre-polymer with isocyanate terminal groups with a small amount of branching agent and water. Other ingredients, such as catalyst and chain extenders, were not used in order to preserve medical purity. The scheme of synthesis is presented below in Fig.9.

- 2 O=C=N-R-N=C=O + HO-R′-OH → OCN-R-NH-COO-R′-OCO-NH-R-NCO

  diisocyanate macrodiol prepolymer (≈R*)

- ~ R*-NCO + H_2O → ~ R*-NH_2 + CO_2↑

- ~ R*-NH_2 + OCN-R*~ → ~R*- NH-CO-NH- R*~

Figure 9. Scheme of PU foams synthesis.

Antituberculosis drugs Is, ethionamide (Eth), florimicin (Fl) and rifampicin (Rfp) were incorporated as fine crystals in the polymeric matrix at the stage of PU synthesis. The PU contained 100-300 mg of heterogeneously dispersed antituberculosis drugs.

The release of drugs from PU was examined by immersing polymeric samples in a model biological medium (physiological solution, phosphate buffer pH 7.4 and Ringer-Locke solution) at 37°C. The amount of drug released was determined UV-spectrophotometrically by measuring the absorbance maximum characteristic for each drug.

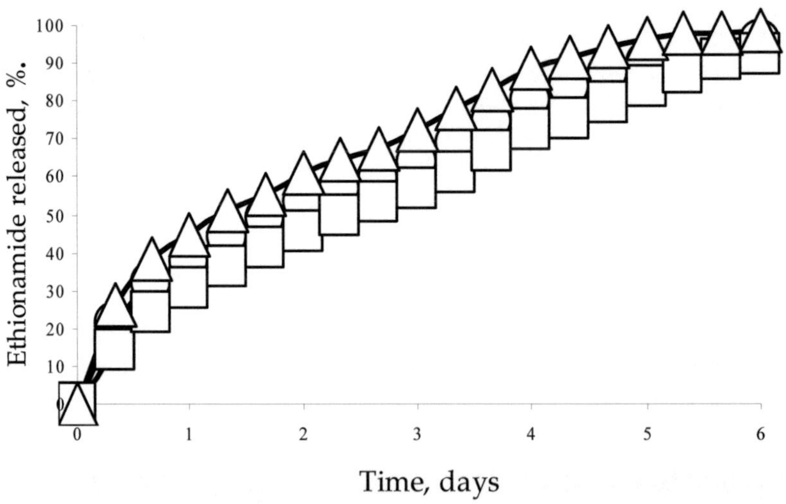

Figure 10. Release of Eth from polyurethane foam into Ringer-Locke solution at 37°C. Drug loadings (mg/g PU): 100(O), 200(Δ), 300(□).

All the release data show the typical pattern for a matrix-controlled mechanism. The cumulative amount of drugs released from the PU was linearly related to the square root of time and the release rate decreased with time. The process is controlled by the dissolution of the drug and by its diffusion through the polymer. The release is described by Fick's law and proceeds by first-order kinetics (Philip & Peppas, 1987). The structure of the drugs and their solubility influences the rate of release: the total amount of Is is released in 3-4 days, Eth in 5-6 days (Fig.10), Fl and Rfp in 14-16 days (Fig.11). The release time for 50% of Is is 22-26 h, for Eth 28-30 h, for Fl and Rfp 72-76 h, respectively. The rapid release of Is and Eth in comparison with Fl and Rfp is due to the higher solubility of these drugs in the dissolution medium. Increasing the drug loading from 100 to 300 mg/g resulted in an increase in the release rate.

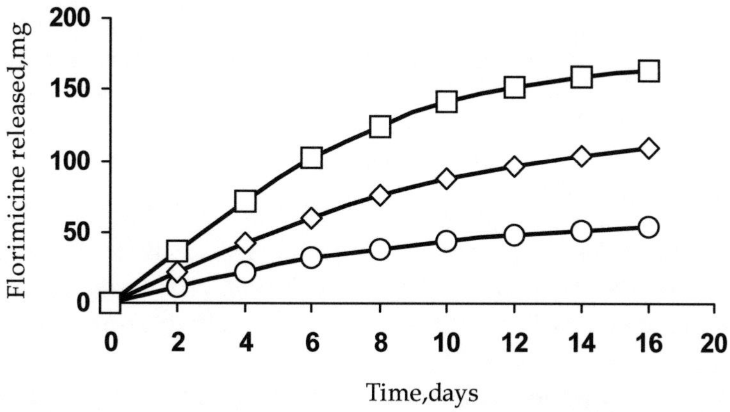

Figure 11. Release of Fl from PU foam into Ringer-Locke solution at 37°C. Drug loadings (mg/g PU): 100(O), 200(Δ), 300(□).

Table 1 presents the values of the diffusion coefficients for drug release into different media, calculated for the initial release stage by a modified Higuchi equation (Higuchi, 1963). With increase of drug loading, the diffusion coefficient is not significantly decreased. This is connected with the plasticizing action of the drug, resulting in the deterioration of the mechanical properties of the polymeric matrix. The medium into which the drugs are released has no significant effect upon the diffusion coefficient.

The release results show that the use of PU as a carrier of antituberculosis drugs provides a controlled release of drugs suitable for use in practical medicine, i.e. it allows prolonged action of drugs over some days.

The tuberculostatic activity of drugs released from the PU was determined by diffusion into dense Levenshtein-Iensen nutrient medium compared with a museum strain of *M. tuberculosis*.

It has been shown that drugs introduced into a polymeric matrix have tuberculostatic activity on the level of free drugs. Is formed a microorganism growth delay zone of 41 mm, Eth 35 mm and Fl 29 mm.

Drug	Loading (mg/g PU)	10^7 x D (cm^2 s^{-1})		
		Physiological solution	Ringer-Locke solution	Phosphate Buffer
Is	100	7,482	7,926	7,346
	200	7,026	7,150	7,158
	300	6,845	6,890	6,804
Eth	100	6,433	6,228	6,248
	200	6,237	5,928	6,142
	300	5,972	5,768	5,636
Fl	100	2,124	2,430	2,315
	200	1,980	2,068	2,112
	300	1,642	1,786	1,720
Rfp	100	1,116	1,224	1,226
	200	0,984	1,082	1,068
	300	0,922	0,944	0,896

Table 1. Diffusion coefficient (D) values for drug release from polyurethane foams into different media at 37°C for initial stage of release.

The efficiency of tuberculosis treatment by PU containing drugs was studied in experiments on guinea pigs (Batyrbekov et al., 1998). Several groups of animals, consisting of 20-25 guinea pigs, were infected with a 6-week culture of a laboratory strain of *M. tuberculosis*. Treatment was started 2 weeks after infection. Animals were treated by weekly administration of PU containing 5-day doses of the drugs (PU-Is, PU-Eth or PU-Fl), or by daily administration of a day's dose of Is, Eth or Fl. Animals of the control group were not treated (C). The weights of the guinea pigs and the dimensions of ulcers at the site of infection were periodically determined during the experiment. All untreated animals died 1.5-2 months after infection. The animals of the other groups were killed with 2.5 months after the beginning of the treatment. Guinea pigs were dissected and damage to lungs, livers, spleens and lymphatic ganglions was determined. The efficiency of the applied therapy is presented in Table 2.

Group	Index of damage, %				
	Lung	Liver	Spleen	Lymphatic ganglion	Summary
C	36.6	25,8	22,0	6,0	90,4
Is	5,0	12,2	12,2	2,5	31,9
PU-Is	4,0	11,4	12,0	2,5	29,9
Eth	8,0	13,6	14,6	3,5	39,7
PU-Eth	7,2	13,0	14,2	3,5	37,9
Fl	20,2	14,4	16,8	4,8	56,2
PU-Fl	18,8	13,0	16,6	4,4	52,0

Table 2. Macroscopic evaluation of damage to inner organs of guinea pigs.

The experimental observations show that the treatment of tuberculosis in the animals by the polymeric systems gave the same therapeutic effect as daily treatment with single doses of the drugs. The most effective action was displayed by PU containing Is. This is related to its greater tuberculostatic activity in comparison with Eth and Fl. The animals of the PU-Is and Is groups had the dissemination nidi in their inner organs practically cured: guinea pigs lost weight slightly (4·6% and 1·2%, cf. untreated 30·6%) and had small ulcers in the place of infection (3·0 mm and 3·2 mm in diameter, cf. untreated 11·2 mm) (Batyrbekov et al., 1997)

The values of weight loss and ulcer dimensions in the place of infection in animals of the another groups are following: 8·6% and 4·4 mm (PU-Eth); 9.0% and 4·8 mm (Eth); 11·4% and 5·2 mm (PU-F1); 11.0% and 5·0 mm (Fl). The treatment of experimental tuberculosis by the polymeric systems was analogous to daily treatment with free drugs. The use of a PU carrier provides a stable bacteriostatic concentration of chemotherapeutic agents for 5-7 days. Clinical observations have shown the efficiency of PU drug delivery systems for treatment of tuberculosis-infected cavities (wounds,pleural empyema, bronchial fistula).

The results obtained in the present work have shown the possibility of using PU foams as a matrix for drug delivery systems for prolonging the action of chemotherapeutic agents in tuberculosis treatment.

6. Conclusion

PU microparticles containing antituberculosis drugs were prepared by interfacial reaction between PEG and TDI in water in toluene emulsion. Two products of polycondensation were detected: the main product is spherical microparticles with size about 5-10 μm and the second product is fibrils of linear PU, which precipitate in toluene. The increase of PEG content in water phase results in increased amount of the secondary product, and as the PEG content in water phase reaches 60 vol.%, maximum of the secondary product was observed (about 40%). Decreasing PEG concentration in water phase leads to increased yield of PU microparticles. Maximum of yield was reached at PEG concentration 22 - 27 vol.% and in that conditions whole oligomer reacted at surface of emulsion drops with microparticles formation.

The release behavior of drugs from microparticles was carried out and different conditions of synthesis such as water/PEG ratio, molecular weight of PEG and drug concentration was investigated. The increase PEG content in water phase of reaction, results in decreasing drug diffusion, due to formation of PU microparticles with densere polymer wall. Increasing molecular weight of soft segments (PEG) results in the increase of diffusion rate of drug into solution. This phenomenon can be attributed to increasing molecular weight of PEG which leads to accelerating diffusion of water-soluble drug through hydrophilic PEG chains. It was shown that microparticles with higher drug loading demonstrate faster release rate of the drug due to increased gradient of concentrations between the external solution and core of microparticles.

The tuberculostatic activity of drugs released from the PU show that drugs introduced into PU have antimicrobial activity identical of low molecular drugs. The efficiency of the tuberculosis treatment by polyurethane drug delivery systems was shown in experiments on animals. The use of PU carrier provides a stable bacteriostatic concentration of the chemotherapeutical agents for 5-7 days. The treatment of animals infected with tuberculosis by PU systems was more effective than the treatment by free drugs. It was shown that is released from PU systems was 1.5-2 times less toxic in comparison with the low molecular drug. Minimal toxic action of PU on the native organism tissue was established hystologically. Medical-biological tests show that PU ensures sustained release of antituberculosis drugs and maintains effective drug concentration for long time.

The results obtained in the present chapter have shown the possibility and outlook of PU as carriers of antituberculosis drugs for the delivery systems for prolonging the action of chemotherapeutical agents in tuberculosis treatment.

Author details

Yerkesh Batyrbekov and Rinat Iskakov
Institute of Chemical Sciences,
Kazakh-British Technical University, Kazakhstan

Acknowledgement

This research was financially supported by a grant from the Ministry for Education and Science of Republic of Kazakhstan. The authors thank Dr. Mariya Kim for her assistance in carrying experimental studies.

7. References

Ahmad, Z.; Pandey, R.; Sharma, S. & Khuller, G.K. (2006). Pharmacokinetic and pharmacodynamic behaviour of antitubercular drugs encapsulated in alginate nanoparticles at two doses. *International Journal of Antimicrobial agents,* Vol.27, No.5, (May 2006), pp. 409-416

Ahmad, Z.; Sharma, S. & Khuller, G.K. (2007). Chemotherapeutic evaluation of alginate nanoparticle-encapsulated azole antifungal and antitubercular drugs against murine tuberculosis. *Nanomedicine,* Vol.3, No.3, (March 2007), pp. 239-243

Ahmad, Z.; Pandey, R.; Sharma, S. & Khuller, G.K. (2008). Novel chemotherapy for tuberculosis: chemotherapeutic potential of econazole- and moxifloxacin-loaded PLG nanoparticles. *International Journal of Antimicrobial agents*, Vol.31, No.2, (February 2008), pp.142-146

Ahmad, Z. & Khuller, G.K. (2008).. Alginate-based sustained release drug delivery systems for tuberculosis. *Expert Opinion on Drug Delivery*, Vol.5, No.12, (December 2008), pp. 1323-1334

Anisimova, Y.V.; Gelperina, S.E.; Peloquin, C.A. & Heifets, L.B. (2000). Nanoparticles as antituberculosis drugs carriers: effect on activity against *M. tuberculosis* in human monocyte-derived macrophages. *Journal of Nanoparticle Research,* Vol.2, No.1, (January 2000), pp. 165–171

Batyrbekov, E.O.; Rukhina, L.B.; Zhubanov, B.A.; Bekmukhamedova, N.F. & Smailova, G.A. (1997). Drug delivery systems for tuberculosis treatment. *Polymer International,* Vol.43, No.4, (April 1997), pp. 317-320

Batyrbekov, E.O.; Iskakov, R. & Zhubanov, B.A. (1998). Synthetic and natural polymers as drug carriers for tuberculosis treatment. *Macromolecular Symposia,* Vol.127, pp. 251-255

Batyrbekov, E.; Iskakov, R. & Zhubanov, B. (2009). Microcaparticles on the basis of Segmented Polyurethanes for the Treatment of Tuberculosis. *Life Sciences, Medicine, Diagnostics, Bio Materials and Composites. Proceedings of 2009 NSTI Nanotechnology Conference and Trade Show.* Vol.2, pp. 96-99, Houston, Texas, USA, May 3-7, 2009

Bouchemal, K.; Briançon, S.; Perrier, E.; Fessi, H.; Bonnet, I. & Zydowicz, N. (2004). Synthesis and characterization of polyurethane and poly(ether urethane) nanocapsules using a new technique of interfacial polycondensation combined to spontaneous emulsification. *International Journal of Pharmaceutics,* Vol.269, No.1, (January 2004), pp. 89-100

Bouchemal, K.; Briançon, S.; Couenne, F.; Fessi, H. & Tayakout, M. (2006). Stability studies on colloidal suspensions of polyurethane nanocapsules. *Journal of Nanoscience and Nanotechnology,* Vol.6, No.9-10, (October 2006), pp. 3187-3192

Campos, E.; Cordeiro, R.; Santos, A.C.; Matos, C. & Gil, M.H. (2011). Design and characterization of bi-soft segmented polyurethane microparticles for biomedical application. *Colloids and Surfaces. B:Biointerfaces,* Vol.88, No.1, (November 2011), pp. 477-482

Chang, C.P.; Yamamoto, T.; Kimura, M.; Sato, T.; Ichikawa, K. & Dobashi, T. (2003). Release characteristics of an azo dye from poly(ureaurethane) microcapsules. *Journal of Controlled Release,* Vol.86, No.2-3, (January 2003), pp. 207-211

Chang, C.P.; Chang, J.C.; Ichikawa, K. & Dobashi, T. (2005). Permeability of dye through poly(urea-urethane) microcapsule membrane prepared from mixtures of di- and tri-isocyanate. *Colloids and Surfaces. B:Biointerfaces,* Vol.44, No.4, (September 2005), pp. 187-190

Embrey, H.P.; Graham, N.B.; McNeill, M.E. & Hiller, K. (1986). Release characteristics and long term stability of polyethylene oxide hydrogels vaginal pessaries containing prostaglandin. *Journal of Controlled Release,* Vol.3. No.1, (January 1986), pp. 39-45

Gelperina, S.; Kisich, K.; Iseman, M.D. & Heifets, L. (2005). The Potential Advantages of Nanoparticle Drug Delivery Systems in Chemotherapy of Tuberculosis. *American Journal of Respiratory and Critical Care Medicine,* Vol.172, No. 12, (December 2005), pp. 1487-1490

Graham, N.B.; Zulfiqar, M.; McDonald, B.B. & McNeill, M.E. (1988). Caffeine release from fully swollen polyethylene oxide hydrogel. *Journal of Controlled Release,* Vol.5, No.2, (February 1988), pp. 243-252

Higuchi, T (1963). Mechanism of sustained-action medication. Theoretical analysis of rate of release of solid drugs dispersed in solid matrices. *Journal of Pharmaceutical Sciences,* Vol.52, No.12, (December 1963), pp.1145-1149

Hong, K. & Park, S. (2000). Characterization of ovalbumin-containing polyurethane microcapsules with different structures. *Polymer Testing,* Vol.19, No.8, (September 2000), pp. 975-984

Hsien, D.S. (1998). *Controlled Release Systems: Fabrication Technology,* CRC Press, Boca Raton, Florida, USA

Iskakov, R.; Batyrbekov, E.; Zhubanov, B. & Volkova, M. (1998). Polyurethanes as Carriers of Antitumorous Drugs. *Polymers for Advanced Technologies,* Vol.9, No.2. (February1988), pp. 266-270

Iskakov, R.M.; Batyrbekov, E.O.; Leonova, M.B. & Zhubanov, B.A. (2000). Preparation and release profiles of cyclophoshamide from segmented polyurethanes. *Journal of Applied Polymer Sciences,* Vol.75, No.1, (January 2000), pp. 35-43

Iskakov, R.; Batyrbekov, E.O.; Zhubanov, B.A. & Mooney, D.J. (2004). Microparticles on the basis of segmented polyurethanes for drug respiratory administration. *Eurasian ChemTechnology Journal,* Vol.6, No.1, (January 2004), pp. 51-55

Jabbari, E. & Khakpour, M. (2000). Morphology of and release behavior from porous polyurethane microspheres. *Biomaterials,* Vol.21, No.20, (October 2000), pp.2073-2079

Kisich, K.O.; Gelperina, S.; Higgins, M.P.; Wilson, S.; Shipulo, E.; Oganesyan, E. & Heifets. L. (2007). Encapsulation of moxifloxacin within poly(butyl cyanoacrylate) nanoparticles enhances efficacy against intracellular Mycobacterium tuberculosis. *International Journal of Pharmaceutics,* Vol.345, No.1-2, (December 2007), pp. 154-162

Lamba, N.M.K.; Woodhouse, K.A.; Stuart, L. & Cooper, S.L. (1997). *Polyurethanes in Medical Application,* CRC Press, Boca Raton, Florida, USA

Lelah, M.D. & Cooper, S.L. (1986). *Polyurethanes in Medicine,* CRC Press, Boca Raton, Florida, USA

Lucinda-Silva, R.M. & Evangelista, R.C. (2003). Microspheres of alginate–chitosan containing isoniazid. *Journal of Microencapsulation,* Vol.20, No.2, (Februry 2003), pp. 145–152

Makino, K.; Nakajima, T.; Shikamura, M.; Ito, F.; Ando, S.; Kochi, C.; Inagawa, H.; Soma. G. & Terada, H. (2004). Efficient intracellular delivery of rifampicin to alveolar macrophages using rifampicin-loaded PLGA microspheres: effects of molecular weight and composition of PLGA on release of rifampicin. *Colloids and Surfaces. B:Biointerfaces,* Vol.36, No.1, (July 2004), pp. 35-42

Pandey, R. & Khuller, G.K. (2004). Chemotherapeutic potential of alginate-chitosan microspheres as anti-tubercular drug carriers. *Journal of Antimicrobiology and Chemotherapy,* Vol.53, No.4, (April 2004), pp. 635-640

Pandey, R. & Khuller, G.K. (2006). Nanotechnology based drug delivery systems for the management of tuberculosis. *Indian Journal of Experimental Biology,* Vol.44, No.5, (May 2006), pp. 357-66

Pandey, R. & Khuller, G.K. (2007). Nanoparticle-Based Oral Drug Delivery System for an Injectable Antibiotic Streptomycin - Evaluation in a Murine Tuberculosis Model. *Chemotherapy* (*International Journal of Experimental and Clinical Chemotherapy*), Vol.53, No.6, (July 2007), pp. 437-441

Phadtare, S.; Vyas, S.; Palaskar, D.V.; Lachke, A.; Shukla, P.G.; Sivaram, S. & Sastry, M. (2004). Enhancing the reusability of endoglucanase-gold nanoparticle bioconjugates by tethering to polyurethane microspheres. *Biotechnology Progress,* Vol.20, No.6, (November-December 2004), pp. 1840-1846

Philip, L. & Peppas N.A. (1987). A simple equation for description of solute release II. Fickian and anomalous release from swellable devices. *Journal of Controlled Release,* Vol.5, No.1, (June 1978), pp. 37-42

Quenelle, D.C.; Winchester, G.A.; Staas, J.K.; Hoskins, D.E.; Barrow, E.W. & Barrow, W.W. (2004). Sustained Release Characteristics of Rifampin-Loaded Microsphere Formulations in Nonhuman Primates. *Drug Delivery,* Vol.11, No.4, (April 2004), pp. 239-246

Qurrat-ul-Ain; Sharma, S.; Khuller, G.K. & Garg, S.K. (2003). Alginate-based oral drug delivery system for tuberculosis: pharmacokinetics and therapeutic effects. *Journal of Antimicrobiology and Chemotherapy,* Vol.51, No.4, (April 2003), pp. 931-938

Rafienia, M.; Orang, F. & Emami, S.H. (2006). Preparation and Characterization of Polyurethane Microspheres Containing Theophylline. *Journal of Bioactive and Compatible Polymers,* Vol.21, No.4, (July 2006), pp. 341-349

Rastogi, R.; Sultana, Y.; Aqil, M.; Ali, A.; Kumar S.; Chuttani, K. & Mishra, A.K. (2007). Alginate microspheres of isoniazid for oral sustained drug delivery. *International Journal of Pharmaceutics,* Vol.334, No.1-2 (February 2007), pp. 71-77

Reddy, T.T.; Hadano, M. & Takahara, A. (2006). Controlled Release of Model Drug from Biodegradable Segmented Polyurethane Ureas: Morphological and Structural Features. *Macromolecular Symposia,* Vol.242, No.1, (October 2006), pp. 241–249

Sharma, K.; Knutson, K. & Kirn, S.W. (1988). Prednisolon release from copolyurethane monolithic devuces. Journal of Controlled Release, 1988. Vol.7, No.2, (February 1988), pp. 197-205

Shegokar, R.; Shaal, L.A. & Mitri, K. (2011). Present status of nanoparticle research for treatment of Tuberculosis. *Journal of Pharmacy & Pharmaceutical Sciences,* Vol.14, No.1, (January 2011), pp. 100-116

Shukla, P.G.; Kalidhass, B.; Shah, A. & Palaskar, D.V. (2002). Preparation and characterization of microcapsules of water-soluble pesticide monocrotophos using polyurethane as carrier material. *Journal of Microencapsulation,* Vol.19, No.3, (May 2002), pp. 293-304

Sosnik, A.; Carcaboso, A.M.; Glisoni, R.I.; Moretton, M.A. & Chiappetta, D.A. (2010). New old challenges in tuberculosis: Potentially effective nanotechnologies in drug delivery. *Advanced Drug Delivery Reviews.* Vol.62, No.4-5, (March 2010), pp. 547-559

Subhaga, C.S.; Ravi, K.G.; Sunny, M.C. & Jayakrishnan, A. (1995). Evaluation of an aliphatic polyurethane as a microsphere matrix for sustained theophylline delivery. *Journal of Microencapsulation,* Vol.12, No.6, (December 1995), pp. 617-625

Toit, L.K. ; Pillay, V. & Danckwerts, M.P. (2006). Tuberculosis chemotherapy: current drug delivery approaches. *Respiratory Research,* Vol.7 , No.1, (January 2006), pp. 118-132

Touitou, E. & Friedman, D. (1984). The release mechanism of drug from poly-urethane transdermal delivery system. *International Journal of Pharmaceutics,* Vol.19, No.3, (March 1984), pp. 323-332

Van Bos, M. & Schacht, E. (1987). Hydrophilic Polyurethanes for the Preparations of Drug Release Systems. *Acta Pharmaceutical Technology,* Vol.33, No.3, (March 1987), pp. 120-125

Yoshida, A.; Matumoto, M.; Hshizume, H.; Oba, Y.; Tomishige, T.; Inagawa, H.; Kohchi, C.; Makino, K.; Hori, H. & Soma, G. (2006). Selective delivery of rifampicin incorporated into poly(DL-lactic-co-glycolic) acid microspheres after phagocytotic uptake by alveolar macrophages, and the killing effect against intracellular Mycobacterium bovis Calmette-Guérin. *Microbes Infection,* Vol.8, No. 9-10, (August 2006), pp. 2484-2491

Use of Polyurethane Foam in Orthopaedic Biomechanical Experimentation and Simulation

V. Shim, J. Boheme, C. Josten and I. Anderson

Additional information is available at the end of the chapter

http://dx.doi.org/10.5772/47953

1. Introduction

Biomechanical experimentation and computer simulation have been the major tool for orthopaedic biomechanics research community for the past few decades. In validation experimentations of computer models as well as *in vitro* experimentations for joint biomechanics and implant testing, human cadaver bones have been the material of choice due to their close resemblance to the *in vivo* characteristics of bones. However, the challenges in using cadaveric bones such as availability, storage requirements, high cost and possibility of infection have made synthetic bone analogs an attractive alternative.

There are a variety of synthetic bone materials available but polyurethane foam has been used more extensively in orthopaedic experiments, especially in fracture fixation testing. The foams are produced by a polymerization reaction with a simultaneous generation of carbon dioxide by the reaction of water and isocyanate. The resultant product is a closed cell structure, which is different from the open porosity of cancellous bone. However the uniformity and consistency in their material properties make rigid polyurethane ideal for comparative testing of various medical devices and implants.

Therefore we have extensively used synthetic bones made of polyurethane foam in various orthopaedic biomechanical researches from optimization of bone graft harvester design to acetabular fractures and the stability of osteosynthesis. We identified important design parameters in developing bone graft harvester by performing orthogonal cutting experiment with polyurethane foam materials. We also validated the fracture prediction capability of our finite element (FE) model of the pelvis with a validation experiment with polyurethane foam pelvis. We also performed in vitro experimentation to compare the stability of different types of osteosynthesis in acetabular fractures and used this result again to validate our fracture fixed pelvis model. These results as well as reports from others that highlight the use of polyurethane in orthopaedic biomechanical experiment will be included in this chapter.

Specifically, there will be three sections in this chapter. The first section will describe the basic material properties of rigid polyurethane foam. We will especially highlight the similarities and differences between the foam and human bone. The second section will then present the review of the literature focusing on the use of polyurethane foam in biomechanical experimentations. We will conclude the chapter with our use of polyurethane foam in bone grafting harvester design, fracture predictions and stability testing of osteosynthesis.

2. Basic material properties of rigid polyurethane foam

Polyurethanes are characterized the urethane linkage (-NH-C(=O)-O-) which is formed by the reaction of organic isocyanate groups with hydroxyl groups as shown below

$$R\text{-}NCO + R'\text{-}OH = R\text{-}NH\text{-}C(=O)\text{-}O\text{-}R'$$

Polyurethanes can be turned into foam by means of blowing agents such as water. The cells created during the mixing process are filled and expanded with carbon dioxide gas, which is generated when water reacts with isocyanate group. The result is a closed foam structure, which is a cellular solid structure made up of interconnected network of solid struts or plates which form the edges and faces of cells. Thanks to its desirable material properties that give versatility and durability to the material, polyurethane has become one of the most adaptable materials that it is found everywhere such as carpet, sofa, beds, cars to name a few.

One unlikely place, however, is inside human body, that is human cancellous or spongy bone. The macroscopic structure of cancellous bone consists of a network of interconnecting rods and plates that forms complex struts and columns. Although this structure has strictly speaking open porosity structure, the overall macroscopic structure shows close resemblance to the closed foam structure of polyurethane foam (Figure 1)

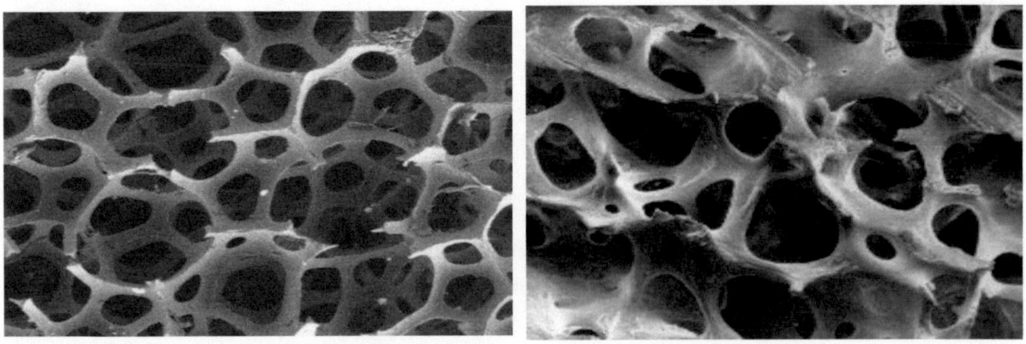

Polyurethane foam microscopic structure

Cancellous bone microscopic structure

Figure 1. Microstructures of cancellous bone and polyurethane foam.

The stress-strain curve of polyurethane foam exhibits similar pattern as cancellous bone. Figure 2 shows a schematic compressive stress-strain curve for polyurethane foams which shows three regions; firstly they show linear elasticity at low stresses followed by a long plateau, truncated by a regime of densification at which the stress rises steeply (Gibson and Ashby, 1988). Linear elasticity is controlled by cell wall bending while the plateau is associated with collapse of the cells by either elastic buckling or brittle crushing. When the cells have almost completely collapsed opposing cell walls touch and further strain compresses the solid itself, giving the final region of rapidly increasing stress.

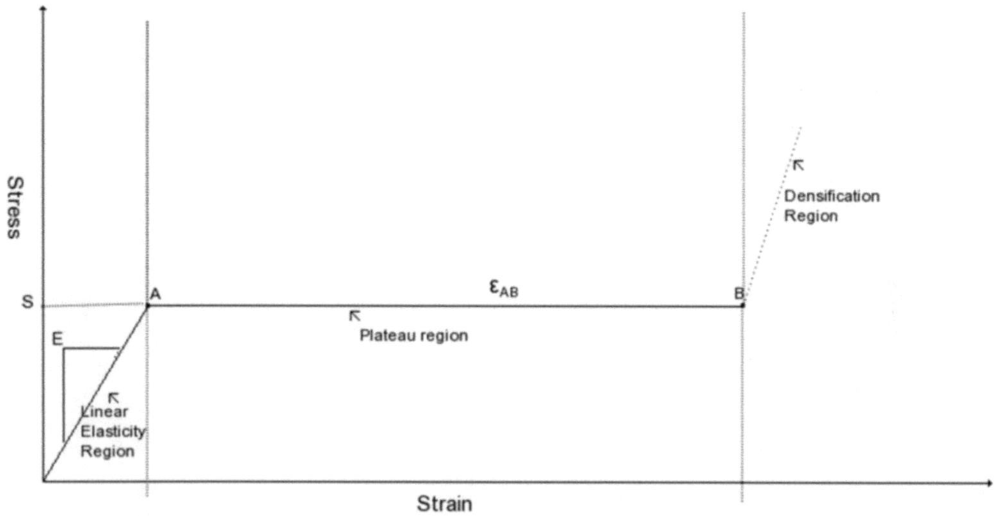

Figure 2. Typical compressive stress-strain curve of polyurethane foam

The compressive stress-strain curve of cancellous bone has the similar three regimes of behaviour (Figure 3). Firstly the small strain, linear elastic region appears which is mainly from the elastic bending of the cell walls. Then the linear-elastic region ends when the cells begin to collapse and progressive compressive collapse gives the long horizontal plateau of the stress-strain curve which continues until opposing cell walls meat and touch, causing the stress rise steeply.

Such similarities have made polyurethane (PU) foams as popular testing substitutes for human cancellous bones and many researchers have quantitatively characterized material properties of polyurethane foam to investigate the suitability and usefulness of PU foams as bone analog. Szivek, Thomas and Benjamin (Szivek et al., 1993)conducted the first study on mechanical properties of PU foams with different microstructures. Compression testing was done to identify elastic modulus and compressive strength. The same group conducted further studies with three compositions of PU foams and evaluated their properties as well(Szivek et al., 1995). Thompson and co-workers(Thompson et al., 2003) analyzed compressive and shear properties of commercially available PU foams. They tested samples of four grades of rigid cellular foam materials and found out that elastic behaviour was similar to cancellous bones and an appropriate density of PU foams can be determined for a particular modulus value. However the shear response showed some discrepancy and concluded that caution is required when simulating other behaviours than elastic behaviour with these foams. Calvert and coworkers (Calvert et al.) evaluated cyclic compressive properties of PU foams and examined the mechanical properties in terms of microstructural features. They found that microstructural properties such as cell size and volume were uniform and increased with decreasing density. And their cyclic testing revealed hysteresis in the low density foams but consistent modulus up to 10 cycles.

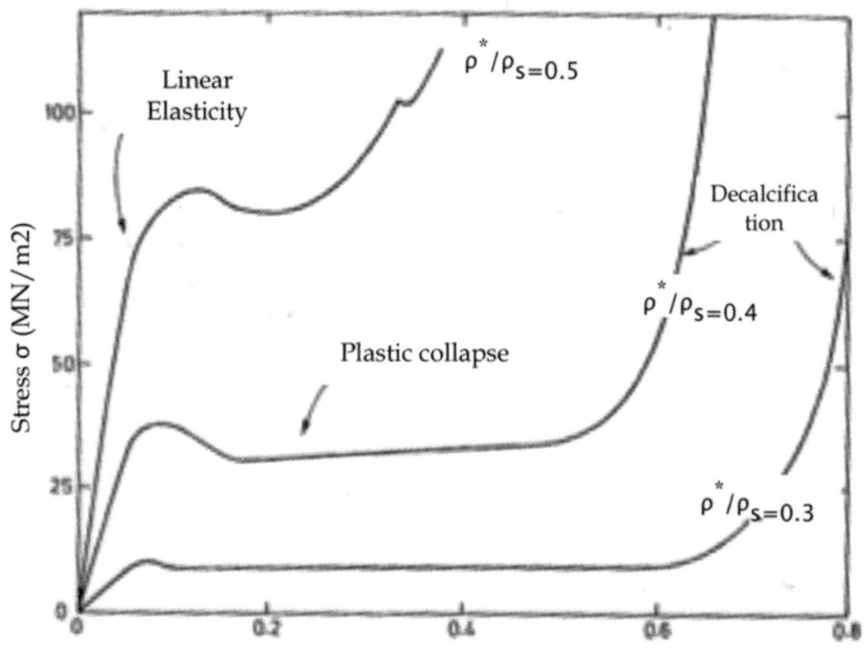

Figure 3. Compressive stress-strain curves for several relative densities (ϱ^*/ϱ_s)of wet cancellous bone (modified from Figure 11-5 in Gibson and Ashyby, 1997)

As the use of PU foams in orthopaedic implant testing and their use as bone analogs increased, the American Society for Testing and Materials (ASTM) developed ASTM F1839-97, "Rigid polyurethane foam for use as a standard material for testing orthopaedic devices and instruments." The aim of this standard is to provide a method for classifying foams as graded or ungraded based on the physical and mechanical behaviour with a given density. This standard has been revised twice since 1997 when it was originally introduced in order to include a wide range of properties and nominal densities(American Society for Testing and Materials, 2008a). As such the number of studies that used PU-foam in testing implant materials and function has increased dramatically after the introduction of the standard. The next section will give review of those studies.

3. The use of PU foams in orthopaedic implant testing

The number and variety of implants for osteosynthesis and joint replacement has increased dramatically over the past few decades along with the use of biomechanical testing of these implants to evaluate their performance. The most obvious material of choice will be fresh or embalmed cadaveric human or animal bones as they have the unique viscoelastic properties and internal structures of real bone. However such studies are often beset with a number of other problems such as issues in handling biological samples and huge variety in size, shape and material properties even in matched pairs to name a few. If reproducibility of experiments is important and comparable not absolute results are required, synthetic bones made from PU-foam can provide a great alternative to the real bones (Figure 4).

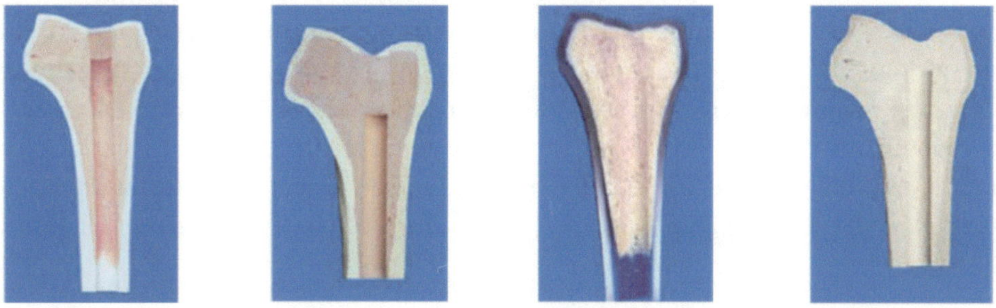

(a) plastic cortical shell with cellular PU-foam inside

(b) rigid PU-foam cortical shell with cellular PU-foam inside

(c) transparaent plastic cortical shell with cellular PU-foam inside

(d) Solid PU-foam throughout

Figure 4. Various synthetic bone material combinations with different types of PU foams and plastics (from www.sawbones.com)

The major use of PU foam blocks is comparatives studies for quantitatively measuring some important functional parameters of orthopaedic implants such as pull out strength, stability and stiffness. Bredbenner et al. (Bredbenner and Haug, 2000) investigate the suitableness of synthetic bone made of PU-foam in testing rigidity of fracture fixations by comparing pull out strength from cadaveric bones, epoxy red oak and PU foams. They found out that PU-foam bone substitutes generated comparable results to cadaveric bones, concluding that PU-foams can be used in mechanical investigation of human bones.

Indeed many researchers have used PU-foam in comparative studies measuring pullout strength of fixation screws. Calgar et al.(Caglar et al., 2005) performed biomechanical comparative studies of different types of screws and cables using Sawbone models and found out that the load to failure of screws was significantly greater than that of the cables. Farshad et al. (Farshad et al., 2011)used PU foam blocks to test bone tunnels drilled during anterior cruciate ligament reconstruction. They found that screw embossed grafts achieved higher pull out strengths. Krenn et al. (Krenn et al., 2008)investigated the influence of thread design on screw fixation using PU-foam blocks with different densities.

PU-foams are also extensively used in biomechanical studies for finding optimal surgical parameters in orthopaedic surgeries. For example, osteotomy is a surgical procedure where a bone is cut to shorten or lengthen to rectify abnormal alignment. One variation of that technique is Weil osteotomy where the knuckle bone in the foot is cut to realign the bones and relieve pain. There are two separate independent studies on Weil osteotomy involving PU-foams and cadaver bones. Melamed et al. (Melamed et al., 2002) used 40 PU-foams to find the optimal angle for the osteotomy and found that an angle of 25° to the metatarsal shaft give the best result. This result was confirmed a year later by Trnka et al. (Trnka et al., 2001)who performed the similar study on fresh frozen cadaver feet. They also found that the range between 25°-35° give the optimal results, confirming that the use of PU-foams in such studies. Nyska et al. (Nyska et al., 2002) analyzed osteotomy for Bunion deformity using 30 PU-foams and found that displacement osteotomies provided good correction for middle and intermediate deformity. Acevedo et al. (Acevedo et al., 2002) compared five different first metatarsal shaft osteotomies by analyzing the relative fatigue endurance. They used 74 polyurethane foam synthetic bones to determine the two strongest of the five osteotomy techniques and they found that Chevron and Ludloff osteotomies showed superior endurance than the other techniques.

Nasson and coworkers (Nasson et al., 2001)used eight foam specimens for tibia and talus to evaluate the stiffness and rigidity of two different arthrodesis techniques where artificial joint ossification is induced between two bones either with bone graft or synthetic bone substitutes. They performed arthrodesis on these artificial bones made up of PU-foams and tested rotation and bending strength and recommended that the use of crossed screws for the strength, simplicity , speed and minimal tissue dissection.

Another interesting development in the use of PU foam in orthopaedic biomechanics is the development of so called composite bones. Since PU foam closely resembles cancellous bone structure and properties, a composite material made up of epoxy resin with fibre glass along with PU foam was used to create a synthetic bone where cortical and cancellous bone materials are simulated with epoxy resin and PU foam respectively (Figure 5). Zdero et al (Zdero et al., 2007, Zdero et al., 2008) tested the performance of these composite bones by measuring bone screw pullout forces in such composite bones and comparing them with cadaver data from previous literature. They found out that composite bones provide a satisfactory biomechanical analog to human bone at the screw-bone interface.

When such composite bone material is shaped according to the actual bony shapes of human bones such as femur and tibia, they can be a great alternative for cadaver bones in research and experiment. Since they closely mimic both geometry and material properties of actual bones and yet have consistency that is lacking in cadaver bones, they can lower variability significantly, offering a more reliable testing bed. Composite replicates of femur and tibia were first introduced in 1987 and then have undergone a number of design changes over the years. The currently available composite bones are fourth-generation composite bones where a solid rigid PU-foam is used as cancellous core material while a mixture of glass fibres and epoxy resin was pressure injected around the foam to mimic cortical bone. Chong et al. (Chong et al., 2007a, Chong et al., 2007b) performed extensive mechanical testing with these synthetic bones and found out that the fourth-generation material has better fatigue behavior and modulus, strength and toughness behaviours a lot closer to literature values for fresh-frozen human bones than previous composite bones. Heiner (Heiner, 2008) tested stiffness of the composite femurs and tibias under bending, axial and torsional loading as well as measuring longitudinal strain distribution along the proximal-medial diaphysis of the femur. She found out that the fourth-generation composite bones average stiffness and strains that were close to values for natural bones (Table 1). Papini et al.(Papini et al., 2007) performed an interesting study where they compared the biomechanics of human cadaveric femurs, synthetic composite femurs and FE femur models by measuring axial and torsional stiffness. They found that composite femurs represents mechanical behaviours of healthy rather than diseased femur (e.g. osteoporosis), hence caution is required in interpreting the data from experiment with composite bones.

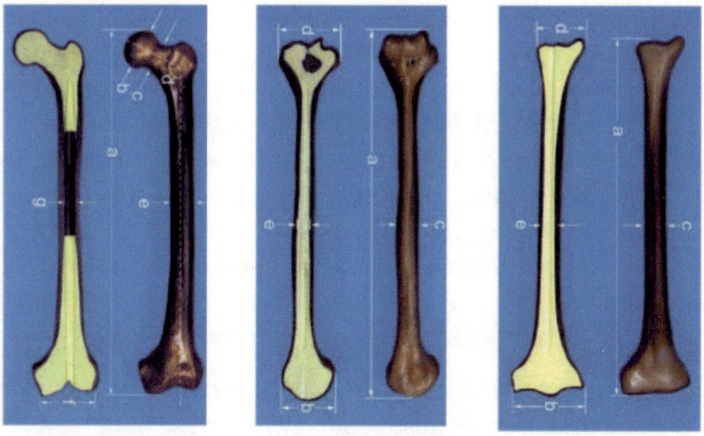

(a) Femur (b) Humerus (c) Tibia

Figure 5. Various composite bones made up of PU-foam core covered with a cortical shell of short fiber filled epoxy (from www.sawbones.com)

Property	Bone Type	Value
Anterior flexural rigidity (N m^2)	Natural	317
	Composite	241
Lateral flexural rigidity (N m^2)	Natural	290
	Composite	273
Axial stiffness (N/μm)	Natural	2.48
	Composite	1.86
Torsional rigidity (N m^2/deg)	Natural	4.41
	Composite	3.21

Table 1. Structural properties of natural human and 4th generation femurs (modified from Table 2 of (Heiner, 2008))

The advert of such biomechanically compatible bone analog greatly widened the use of PU-foam in orthopaedic biomechanics experiments as more mechanically meaningful parameters such as strength and surface strains were possible to be introduced in the design of the experiment.

Agneskirchner et al. (Agneskirchner et al., 2006) investigated primary stability of four different implants for high tibial osteotomy with composite bones and found that the length and thickness as well as the rigidity of the material strongly influence the load to failure of tibial osteotomy. Gulsen et al. (Gulsen et al.) used composite bone in testing biomechanical function of different fixation methods for periprosthetic femur fractures and compared the yield points of these techniques. Cristofolini et al. (Cristofolini et al., 2003) performed in vitro mechanical testing with composite femurs to investigate difference between good design and bad design in total hip replacement femoral stems. They placed two different implants (one good design and the other bad design) to synthetic femurs and applied one million stair climbing loading cycles and measured interface shear between the stem and cement mantle to see the result of long term performances. Their set-up involving composite bones was sensitive enough to detect the result of design difference and was able to predict long term effects of different implant designs. Simoes et al. (Simoes et al., 2000) investigated the influence of muscle action on the strain distribution on the femur. They measured strain distributions for three loading conditions that involve no muscle force, abductor muscle force only and then 3 major muscle forces in the hip. They placed 20 strain gauges on the composite femur and applied muscle and joint forces accordingly. They found out that strain levels were lower when muscle forces were applied than when only joint reaction force was used, indicating that the need to constrain the femoral head to reproduce physiological loading conditions with joint reaction force only.

As discussed up till now, the use of PU-foam based material is almost limitless and the list discussed here is by no means an exhaustive survey of the use of PU-foam based materials in orthopaedic experiment. However, our group has also been working with the PU-foam materials in our orthopaedic biomechanics extensively. The following chapters give summary of these works.

4. Use of PU-foams in device and implant testing for bone grafting and acetabular fractures

4.1. Identification of optimal design parameters in bone grafting tools with PU-foams

Bone grafting is a reconstructive orthopaedic procedure in which a bone substitute is used to fuse broken bones and to repair skeletal defects (Arrington et al., 1996, Lewandrowski et al., 2000). Bone grafting is performed worldwide around 2.2 million times per year, with approximately 450 000 procedures in the United States alone (Russell and Block, 2000). Most popular method is autograft where the graft material is extracted from the patient itself. The graft can be harvested from the patient's femur, tibia, ribs and the iliac crest of the pelvis(Betz, 2002). Autograft has the advantages of being histocompatible and non-immunogenic, it eliminates the risk of transferring infectious diseases and has osteoinductive and osteoconductive properties (Arrington et al., 1996). However, the harvesting procedure often requires a second incision to extract the graft from the donor site, which can extend operation time by up to 20 min (Russell and Block, 2000). The ensuing donor site morbidity is regarded as "a serious postoperative concern for both patient and surgeon" (Silber et al., 2003). Ross et al. (Ross et al., 2000) reported an overall complication rate of 3.4–49%, of which 28% suffered persistent pain, which can last as long as 2 years and often exceeds the pain from the primary operation.

The reason for donor site pain remains unclear, however, it might be proportional to the amount of dissection needed to obtain the graft (Kurz et al., 1989). Conventional bone grafting tools usually require great exposure of the donor site with accompanied trauma to nerves and muscles. Damage to nerves and muscles may be reduced by using minimally invasive bone grafting techniques (Russell and Block, 2000), which shorten the incision length by approximately 60%, reduce the amount of dissection and are roughly two times faster than conventional methods (Burstein et al., 2000). Minimally invasive tools are usually rotational cutting tools, which include trephines, bone grinders (Burstein et al., 2000) and bone graft harvesters as depicted in Figure 6 A.

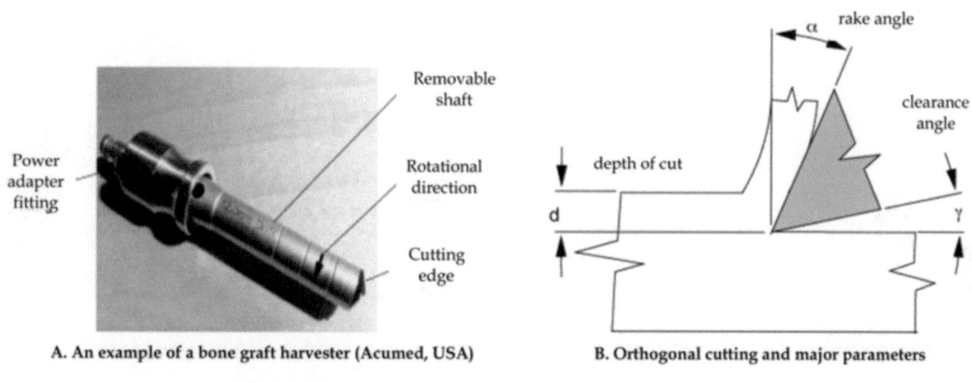

Figure 6. Bone graft harvester and its major parameters

The harvester collects the graft, i.e. bone chips consisting of cancellous bone fragments and bone marrow, in its barrel as it turns and penetrates deeper into the bone. Despite the advantages of using minimally invasive tools such as the bone graft harvester, cell morbidity is yet unavoidable, because both fracturing of the bone architecture and heat generation accompany every bone cutting process. However, mechanical and thermal damage could be reduced by improving tool geometry and by applying appropriate cutting parameters.

Many researchers have studied various cutting operations, such as orthogonal cutting (Jacobs et al., 1974), drilling (Saha et al., 1982, Natali et al., 1996) milling (Shin and Yoon, 2006) and sawing (Krause et al., 1982), in order to identify some of the critical parameters that influence heat generation and to gain an overall understanding of bone cutting mechanisms. For a bone graft harvester, these parameters are the rake and point angles of tool, the rotational speed and the feed rate (Figure 6 B).

The influence of such parameters can be measured by characterizing chip types formed when cutting the bone. Smaller chips imply more fracturing per volume of bone material collected and due to the linkage between fracturing of the bone architecture and cell morbidity, and larger chips are believed to act as "life rafts" for the bone cells and will increase the rate of survival for embedded living cells. We have conducted two-part study where we identified various chip types during orthogonal cutting process(Malak and Anderson, 2008, Malak and Anderson, 2005).

In Part I (Malak and Anderson, 2005), we used polyurethane foams of various densities and cell sizes to investigate chip formation and surface finish. An optical arrangement made up of dynamometer, microscope and camera system (Figure 7) was used to visually record the cutting process, while horizontal and vertical cutting forces were measured. A total of 239 measurement were performed using rake angles of 23°, 45° and 60° with depths of cut from 0.1 to 3 mm (increments of 0.1 and 0.2 mm). Cutting events were observed on the video and then linked to simultaneous force events by merging both sets of data into a combined video stream, generating force plot images.

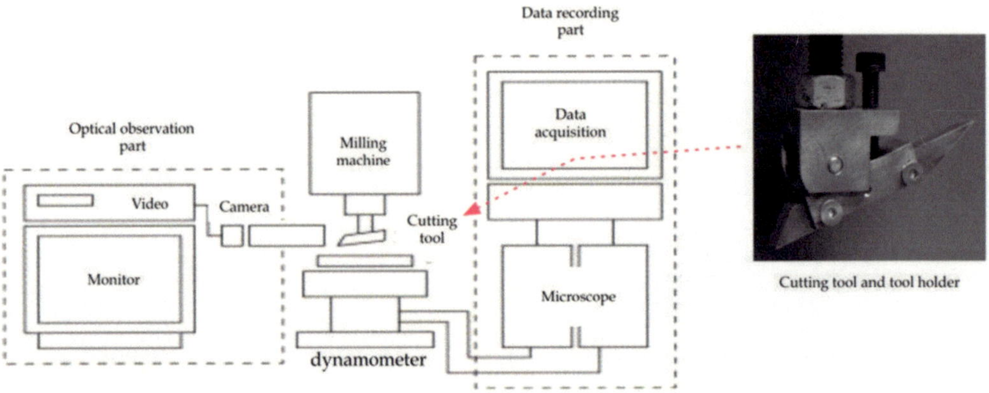

Figure 7. Experimental set-up for measuring chip formation during orthogonal cutting procedure

Three types of cutting response were identified and categorized as 1) surface fragmentation; 2) continuous chip formation; 3) discontinuous chip formation depending on tool rake angle, depth of cut, foam density and cell size (Figure 8). Surface fragmentation was associated with cutting depth less than the PU-foam cell size. By cutting an order of magnitude of the cell size deeper, continuous chips were produced, which is a desirable feature whenever good surface finish after cutting is desired as in the case of bone harvester. A large rake angle (60°) was also inductive of continuous chip formation. Discontinuous chip formation was associated with 1) foam compaction followed by chevron shaped chip; 2) crack propagation in front of the tool. Compaction of the foam could be minimized by using a tool with a large rake angle and normal cut depth.

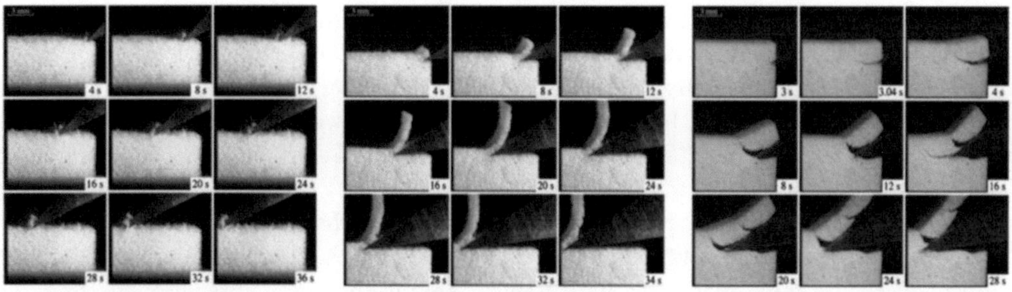

a) Surface fragmenation when PU-foam density 160 kg/m3, depth of cut 0.2mm & rake angle 45°

b) Continuous chip formation when PU-foam density 160 kg/m3, depth of cut 0.7mm & rake angle 60°

c) discontinous chip formation when PU-foam density 320 kg/m3, depth of cut 2mm & rake angle 60°

Figure 8. Various chip types formed during orthogonal cutting of polyurethane foams of various densities

The same experiment was repeated with bovine fresh cancellous bone(Malak and Anderson, 2008) from the patella, the femur and the iliac crest. Similar orthogonal cutting experiments were conducted using the same device used in Part I to identify major parameters that influence the formation of chips after cutting. Three groups of experiments were done where the effect of the depth of the cut, rake angle and cutting speed. Similar chip types as the experiment with PU-foams in Part I were observed which were found to be dependent on rake angle and depth of cut (Figure 9).

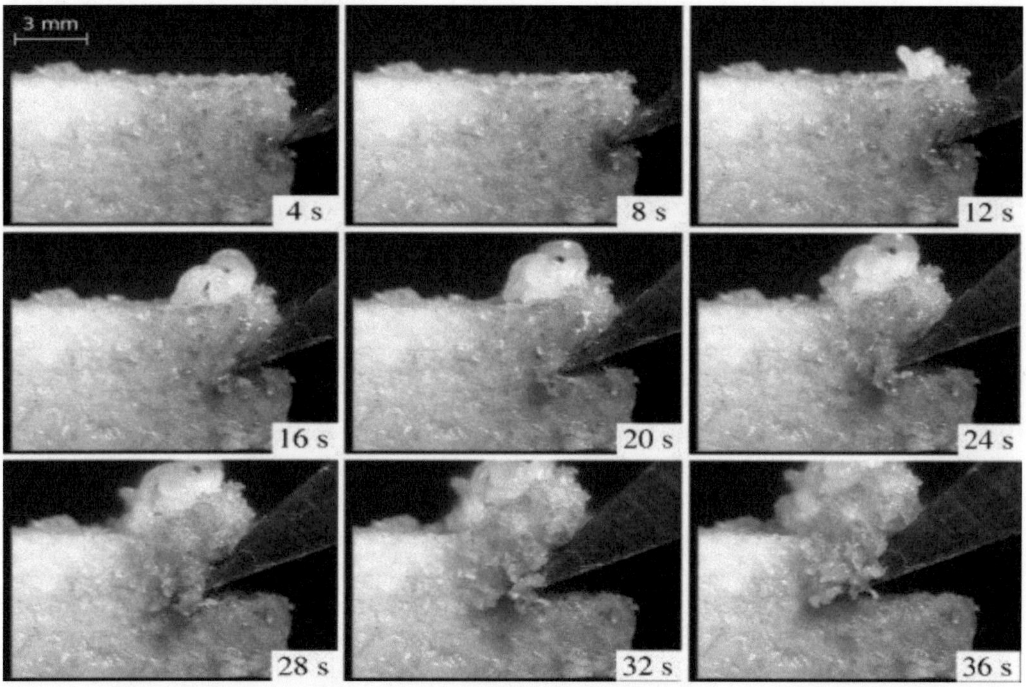

Figure 9. Chip foramtion during orthogonal cut of cancellous bone.

When the results from cancellous bone were compared with those from PU-foams, both showed surface fragmentation, continuous and discontinuous chip formations. During polyurethane foam cutting, chip types were influenced by rake angle and depth of cut. Bone cutting showed similar trend, however, resulting chip type were mainly influenced by the tool rake angle. We also identified the depth of cut that marked the transition from surface fragmentation to continuous or discontinuous chip types. In PU-foams, such a transition is indicated by a change in cutting forces. A similar trend was observed in bone; a depth of cut of 0.5 mm ~ 0.8 mm led to continuous or discontinuous chips. These values approximately correspond to the trabecular separation values. This is in accordance with our findings during the cutting of polyurethane foam, where depths of cut had to reach values of the foam cell size diameter in order to be either continuous or discontinuous.

Therefore polyurethane foam was successfully used in identifying optimal parameters for designing minimally invasive bone graft harvester. The next section will describe how PU-foam based synthetic bone was used in FE modeling of hip fracture.

4.2. Development and validation of finite element fracture predictions with PU-foam based synthetic bones

Acetabular fractures are one of the big challenges that trauma surgeons face today. Despite the great stride made in treating this fracture in the past few decades, one medical text book states that "fractures of the acetabulum remains an enigma to the orthopaedic surgeon(Tile et al., 2003)." The main reason for this difficulty lies on the complexity of acetabular fractures. Acetabular fractures are usually a result of indirect trauma where the major impact is transmitted via the femur after a blow to the greater trochanter, to the flexed knee or to the foot with the knee extended (Ruedi et al., 2007). Moreover acetabular fractures are dependent on a number of variables such as the type of force that caused the fracture, the direction of displacement, the damage to the articular surface as well as the anatomical types of the fracture (i.e. the shapes of the fragments). The relative rareness of acetabular fractures makes matters worse since general orthopaedic surgeons may not gain wide experience with them.

The past researches on acetabular fracture can be broadly divided into two categories. The first is experimental studies where the stability of different acetabular fracture fixation techniques was investigated with in-vitro mechanical experiments (Goulet et al., 1994, Konrath et al., 1998a, Konrath et al., 1998b, Olson et al., 2007). There are also clinical studies that examined the effectiveness and longer-term results of different fracture fixation techniques (Borrelli et al., 2005, Cole and Bolhofner, 1994, Giannoudis et al., 2005). Finite element (FE) models can enhance greatly the body of knowledge obtained from such experimental and clinical studies. FE models can overcome the limitations of in-vitro experimental studies because they can be used to simulate the behaviour of the fractured acetabulum under physiological loading conditions that include muscle forces. FE models can also elevate the results from clinical studies into a new dimension as they can be used to predict the outcome of particular fixation techniques after the surgery. If the model performance in fracture prediction is validated, it can be used to evaluate various fixation techniques depending on their fracture types. The problem is how to validate the model. If cadaver bone is to be used, the issue of sample variability and the requirement of ethics approval, special storage and high cost need to be resolved first. Therefore the aim of this study is to develop a finite element model of the pelvis that can accurately predict the fracture load and locations of acetabular fractures and validate its performance with synthetic PU-foam based bones(Shim et al., 2010).

4.2.1. *Fracture experiment with PU-foam synthetic pelvic bones*

Ten synthetic male pelves made with polyurethane foam cortical shell and cellular rigid cancellous bone (Full Male Pelvis 1301-1, Sawbones, Pacific Research Laboratories, INC, Washington, WA, USA) were used for fracture experiment. A similar set-up as (Shim et al., 2008) was used where the pelvis was placed upside down in a mounting pane filled with acrylic cement. Two different loading conditions – seating (or dashboard) fracture and fall from standing fracture – were tested. When the angle (α) between the vertical line and the line formed by joining the pubic tubercle and the anterior superior point of the sacrum (Figure 10) was 30 °, the whole set-up mimicked the standing position, hence simulating standing fracture. When the angle α was raised to 45°, the set-up mimicked the position of the pelvis when seated, hence simulating seating fracture. Force was exerted from the femoral head attached to the crosshead of the Instron machine (Instron 5800 series, Norwood, MA, USA). The femoral head was also from a matching Sawbone femur (Large Left femur 1130, Sawbones, Pacific Research Laboratories, INC, Washington, WA, USA) to the pelvis used. The femur was first chopped at the neck region and then attached to a custom made holding device connected to the crosshead of the Instron machine (Figure 10). The femoral head was dipped into liquid latex to ensure a complete and stable seating of the femoral head to the acetabulum. The force was applied from the femoral head to the acetabulum at a constant speed of 40N/s until failure. Total ten pelves were used for testing. Five were tested for standing fracture and the rest was tested for seating fracture. The fracture loads and patterns were recorded for comparison with finite element simulations

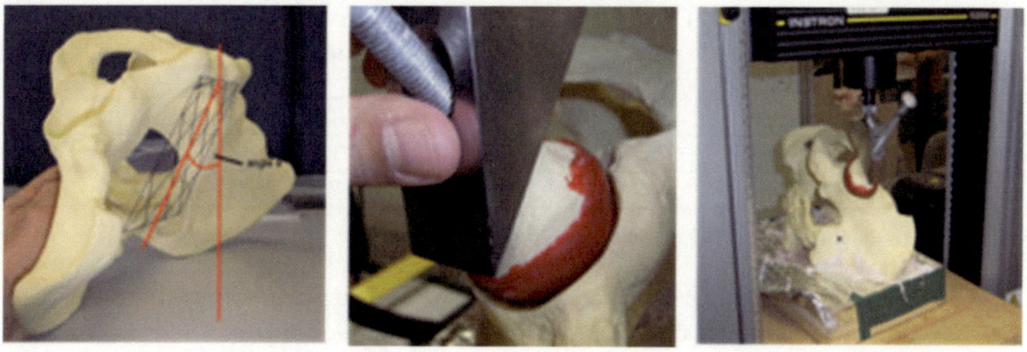

Figure 10. Photos of the experiment: The photo on the left shows the angle alpha that determined seating or standing positions; the center photo shows the close-up of the chopped femoral head attached to the holding device that goes into the crosshead of the Instron machine; the photo on the right shows the overall set-up.

4.2.2. Finite element analysis of PU-foam based synthetic pelvis

The Sawbone pelvis used in the experiment was CT scanned (Philips Brilliance 64, Philips). Each CT image was manually segmented and a finite element model of the hemi pelvis was generated using the previously validated procedure (Shim et al., 2007, Shim et al., 2008)(Figure 11). Our model is both geometrically and materially non-linear. Geometric non-linearity was achieved by using finite elasticity governing equations rather than linear elasticity approximations (Shim et al., 2008). Material non-linearity was incorporated in a similar manner as in (Keyak, 2001). In our model, the material behaviour of the synthetic PU-foam based pelvis was divided into two regions – 1) an elastic region with a modulus E; 2) perfectly plastic regions with a plastic strain εAB (Figure 11) according to the material behaviour of polyurethane materials(Thompson et al., 2003). The value for εAB was obtained from the specifications provided by the manufacturer (Pacific Research Laboratories, INC, Washington, WA, USA).

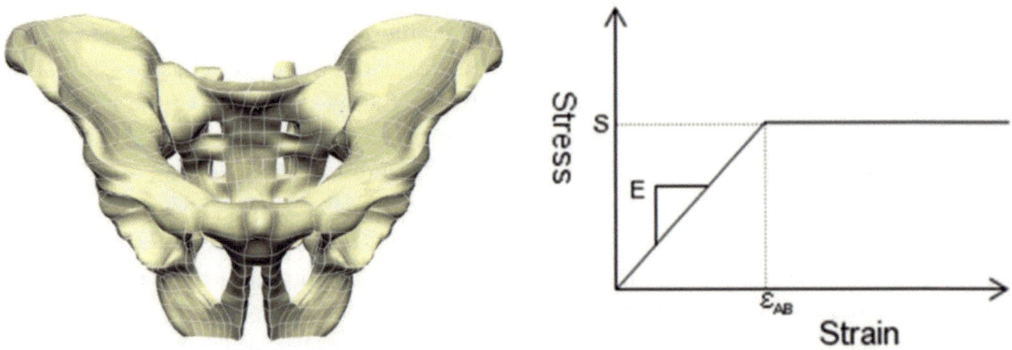

Figure 11. FE model and it nonlinear material behaviour. The polyurethane foam material behavior was represented by an elastic region with modulus E until stress S, followed by a perfectly plastic region with plastic strain εAB.

Two materials were incorporated in our model – 1) solid polyurethane foam that mimicked the cortical bone property; 2) cellular rigid polyurethane foam for cancellous bone. The material properties for the two polyurethane foams are given in Table 2.

	Density (g/cc)	Strength (MPa)	Modulus (MPa)
Solid polyurethane foam	0.32	8.8	260
Cellular polyurethane foam	0.16	2.3	23

Table 2. Material properties of solid and cellular polyurethane foam

The previously developed algorithm that determines cortical thickness was used to distinguish between solid polyurethane and cellular polyurethane foam regions from CT scans. We used Gauss points inside the mesh to assign material properties and those points placed in the solid region was given the solid polyurethane foam material properties while those in the cellular region was assigned with the cellular polyurethane foam material property(Shim et al., 2008). This allowed our model to have location dependent cortical thickness.

The contact between the femoral head and the acetabulum was modeled as frictional contact ($\mu = 0.3$) and the boundary conditions used in FE simulation were the same as the experiment. The nodes on the superior region of the iliac crest were fixed and two different loading conditions used in the experiment were used as the boundary condition. As in the experiment, the standing and seating positions were differentiated by varying the angle α defined in Figure 10. A vertically directed force was exerted on the femoral head mesh until failure.

The failure behaviour was characterized using the distortion energy (DE) theory of failure (Keyak et al., 1997). The DE theory is a simplified form of the Hoffman failure theory which was proposed for brittle fracture of orthotropic materials (Hoffman, 1967). Assuming isotropy, the fracture condition is reduced to the following (Lotz et al., 1991)

$$C_1\left[\sigma_2-\sigma_3\right]^2+C_2\left[\sigma_3-\sigma_1\right]^2+C_3\left[\sigma_1-\sigma_2\right]^2+C_4\sigma_1+C_5\sigma_2+C_6\sigma_3=1 \quad (1)$$

where

$$C_1=C_2=C_3=\frac{1}{2S_tS_c}$$

$$C_4=C_5=C_6=\frac{1}{S_t}-\frac{1}{S_c}$$

In this equation, σ_i is the principal stresses and S_t is the tensile strength and Sc is the compressive strength. If S_t and S_c are equal, Equation (1) becomes the distortion energy (DE) theory of failure, which was used in our study as a failure criterion. Since we used Gauss points in assigning material properties, we calculated a factor of safety (FOS) for every Gauss point (Equation 2) and if the FOS value was predicted to be less than one the Gauss point was regarded as in failure (Keyak et al., 1997). The fracture load and location were recorded for each loading condition and compared with the experimental results.

$$FOS=\frac{\text{Gauss point strength (from CT and material property)}}{\text{Gauss point von Mises stress (from FE simulation)}} \quad (2)$$

4.2.3. Finite element model sensitivity analysis

Sensitivity analysis was performed to find out which material parameters affect the strength of the pelvic bone most. The parameters of interest for sensitivity analysis were: 1) cortical thickness; 2) cortical modulus; 3) trabecular modulus. Since we used synthetic bones made of polyurethane foam, the three corresponding material parameters for the Sawbone FE model were 1) solid polyurethane foam thickness; 2) solid polyurethane modulus; 3) cellular polyurethane modulus. The values for the parameters were varied to see their effects on the predicted fracture load. The following equation (Equation 3) was used to measure the sensitivity of the chosen parameters.

$$S=\frac{\%\ \text{Change in predicted fracture load}}{\%\ \text{Change in input parameter}} \quad (3)$$

The range of simulated variation in the input parameters is given in Table 3. Since we used Gauss points in assigning material properties, the number of Gauss points in the transverse direction was varied from 4 to 6, which had the equivalent effect of varying the solid polyurethane thickness by -40% to +50% (Shim et al., 2008). As for the modulus values, the values were varied by ±25%, which was the next available value in the manufacturer's specification for material properties (Table 3). Multiple FE simulations were run with these values and the change in the predicted fracture load was recorded.

	Simulated of variation
Solid polyurethane foam thickness	-40% and +50% from the original thickness obtained from CT scans
Solid polyurethane foam modulus	±25 % from the original density value of 0.32g/cc
Cellular polyurethane foam modulus	±25% from the original density 0.16 g/cc

Table 3. Sensitivity analysis of polyurethane foam thickness and modulus

4.2.4. Fracture experiment with PU-foam pelvis and corresponding FE model predictions

The fracture behaviour of Sawbone pelves was linear elastic fracture of brittle material (Figure 12 (a)), which coincides with other studies involving fractures of polyurethane (Mcintyre and Anderston, 1979)

The fracture loads from the mechanical experiments are given in Figure 12 (b). The standing case has a slightly higher fracture load (mean 3400N) than the seating case (mean 2600N). Our FE model predicted the fracture load for both cases with a good accuracy as the predicted values are within the standard deviation of the experimental values for both case (Figure 12). The predicted fracture loads from the FE model are 3200N and 2300N for standing and seating cases respectively.

The predicted and actual fracture locations were consistent for both experiments and FE simulations and the fractures occurred mainly in the posterior region of the acetabulum. Different fracture patterns were obtained from two different loading conditions. For the fall from standing experiment, the fracture pattern resembled posterior column fracture according to the Letournel's classification (Letournel, 1980) while the dashboard experiment produced posterior wall fractures. Since the main cause of posterior wall fracture is car accidents (Spagnolo et al., 2009), our experimental set-up was able to capture the main features present in this fracture. The fracture locations predicted by the FE model were similar to the actual fracture patterns from the experiment. For the fall from heights fracture case, the failed Gauss points were concentrated at the region that extends from the dome of the acetabulum to the posterior superior region and then to the posterior column of the pelvis. This resembled the posterior column fracture that was observed from the experiment. For the seating case, on the other hand, the failed Gauss points were more or less limited in the posterior wall region of the acetabular rim, resembling the posterior wall fracture (Figure 12 (a)).

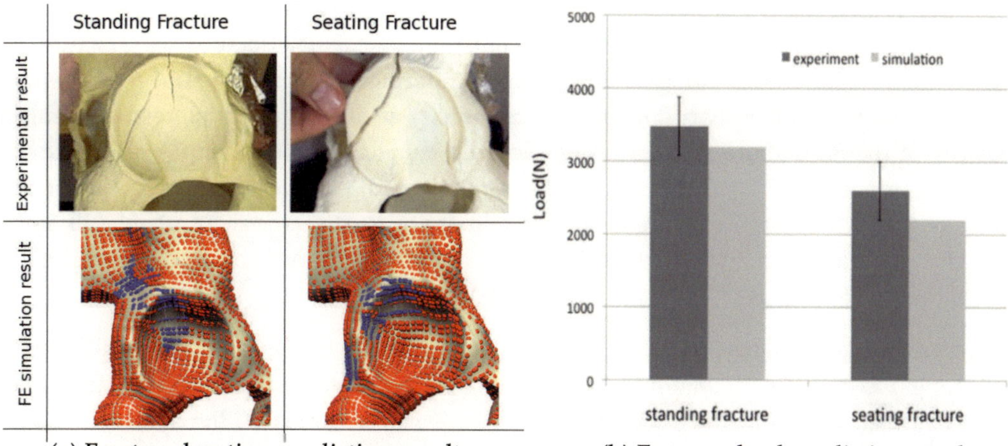

(a) Fracture location prediction results (b) Fracture load prediction results

Figure 12. Fracture location (a) and load (b) predictions from FE models

Once the fracture prediction was done, the sensitivity analysis was performed. Three input parameters (solid polyurethane thickness and modulus, cellular polyurethane modulus) were varied to examine the effect of their variation on the predicted fracture load. Among the three input parameters solid polyurethane foam modulus had the greatest impact on the resulting fracture load. Solid polyurethane foam thickness also had some effect on the predicted fracture load, but the sensitivity of this parameter was not as high as the modulus. Cellular polyurethane foam modulus, on the other hand, did not have any significant impact on the predicted fracture load as can be seen in Table 4. Since the pelvis has a sandwich structure where the outer cortical shell bears most of the load, our results indicate that this structural characteristic is also preserved even when the pelvis undergoes fracture.

Type of input parameters	Amount of variations in input parameters	Predicted fracture load	Sensitivity
Solid polyurethane foam thickness	0.6	3200	0.395
	1.4	4500	0.461
Solid polyurethane foam Modulus	153	2200	0.858
	400	5000	0.874
Cellular polyurethane foam Modulus	12.4	3200	0.128
	47.5	3500	0.028

Table 4. Results of sensitivity analysis

4.2.5. Feasibility of the use of synthetic PU-based bone in validating FE fracture predictions

FE models have been extensively used in predicting fracture load. Our approach to fracture mechanics was based on the work by Keyak and co-workers (Keyak et al., 1997, Korn et al.,

2001) which used the DE theory of fracture as well as material non-linearity. However, our approach differs from their work in that we simulated acetabular fractures not fractures of the proximal femur which are generally more complicated than fractures of the proximal femur. Moreover, rather than applying force directly to the bone of interest as done in majority of the FE fracture studies, we employed a contact mechanics approach where the force was applied to the acetabulum via the femoral head. Another novel approach of our study is that we incorporated geometric non-linearity to the model by using full finite elasticity governing equations, which has been found to enhance the fracture prediction capabilities of FE models (St^lken and Kinney, 2003).

However the most notable feature of our approach is the use of PU-based synthetic bone in validating FE fracture predictions. At present, it is not known whether our model can predict human bone fractures with the same degree of accuracy as the synthetic bones. Therefore caution is required when interpreting the data. However we are confident that our result will translate into human bones due to the following reasons. Firstly our experimental results with PU-foam pelves showed similar results as other human cadaver results as the fracture patterns generated in seating and standing cases correspond well with clinical results. Moreover the sensitivity analysis revealed that our model behaves in a similar manner as the cadaver bones despite the apparent difference in absolute magnitudes in modulus values between PU-foams and bones.

In fact, PU-foam based synthetic bone served our purpose of model validation very well due to their uniformity and consistency (Nabavi et al., 2009). As such, the ASTM standard states that it is "an ideal material for comparative testing" of various orthopaedic devices (American Society for Testing and Materials, 2008b). Although the fracture load is expected to be different from the fracture load of human pelvis, the material behaviour is expected to be comparable to human bones, both of which exhibit brittle fracture (Schileo et al., 2008, Thompson et al., 2003). Therefore the model's ability to predict fracture load and location of the synthetic bone can be regarded as a positive indication that it will also be applicable to human cases. Therefore we continued to use this approach in developing and validating FE model predictions for fracture stability with PU-foam based synthetic bones, which will be described in the next section(Shim et al., 2011).

4.3. Development and validation of finite element predictions of the stability of fracture fixation with PU-foam based synthetic bones

The posterior wall fracture is the most common fracture type of the acetabulum(Baumgaertner, 1999). Depending on the fragment size, open reduction and internal fixation (ORIF) is performed especially when the fracture involves more than 50% of the posterior wall. But ORIF requires considerable exposure that often leads to major blood loss and significant complications(Shuler et al., 1995). Percutaneous screw fixations, on the other hand, have become an attractive treatment option as they minimize exposure, blood loss and risk of infection. As such, they have been advocated by some authors

(Parker and Copeland, 1997) for the treatment of minimally displaced acetabular fractures without comminution or free fragment in the joint. However the biomechanical stability of percutaneous fixation has not been studied thoroughly, especially in terms of interfragmentary movement. In particular, the stability of percutaneous fixation in acetabular fractures has not been compared with the more conventional ORIF involving a plate with screws. There have been previous biomechanical studies that compared different types of stabilization in posterior wall fractures (Goulet et al., 1994, Zoys et al., 1999). But the main focus of such studies was to compare the strength of several types of osteosynthesis. However it is interfragmentary movement that exerts major influences on the primary stability and fracture healing (Klein et al., 2003, Wehner et al., 2010). As discussed in Section 3, PU-foam based synthetic bones have been used extensively in testing stability of various osteosynthesis techniques. Therefore we have further developed our FE model capable of prediction acetabular fractures to simulate stability in osteosynthesis. Specifically, we have developed a fast and efficient way of predicting the interfragmentary movement in percutaneous fixation of posterior wall fractures of the acetabulum and validated with a matching biomechanical experiment using PU-foam based synthetic pelves.

4.3.1. Mechanical experiment with PU-foam based synthetic pelvis

Seven synthetic pelves (Full Male Pelvis 1301-1, Pacific Research Laboratories Inc) were loaded until failure with the loading condition that resembled seating fracture[10], creating posterior wall fractures[11]. The fractures were then reduced and fixed with two fixation methods –with two screws (3.5mm Titan Screws, Synthes) and then with a 10-12 hole plates (3.5mm Titan Reconstruction- or LCDC-Plates, Synthes) by an experienced surgeon (JB) (Figure 13 A and B). The maximum remaining crack was 0.7 mm.

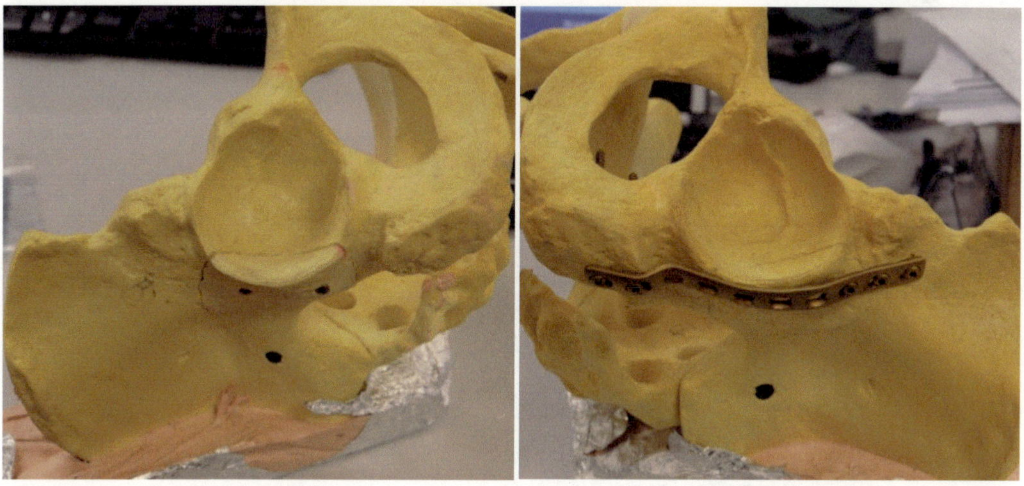

(a) Screw fixation (b) Plate fixation

Figure 13. Two fixation methods performed on the fractured acetabulum

The pelves were then loaded in the Instron Machine (Instron 5800 series) with a cyclic load that oscillated between 0N to 900N at 40N/s. The force was applied using a synthetic femoral head (Large Left Femur, 1129, Pacific Research Laboratories Inc) attached to the crosshead of the Instron Machine (Figure 14). At the multiple of 300N the loading was paused for 3 seconds to measure the displacement between the fragment and the bone by taking photographs of the crack opening (Figure 14 (a)). A digital SLR camera (Nikon D70) with a 50mm macro lens (NIKKOR dental lens) was used to accurately measure the amount of crack openings (resolution of 10 μm (Figure 14 (b))). A resolution of 10μm was achieved (Figure 15).

(a) Measurement set-up

(b) Optical set-up with macro

Figure 14. Interfragmentary movement measuring set-up with a digital single-lens reflex camera and an Instron machine

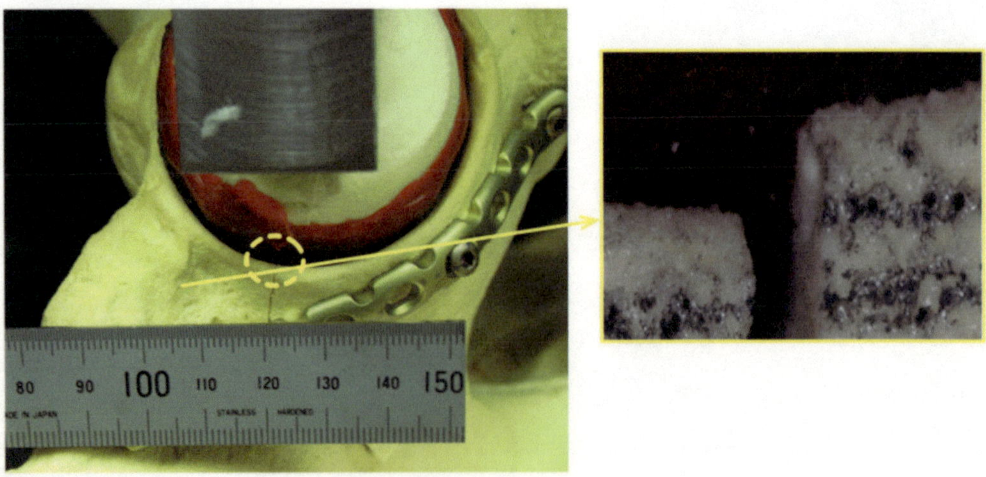

Figure 15. A photo taken with the macro lens. The magnified view shown in the box left has the resolution of 10μm

The displacement was measured in two different positions – front and side - to obtain the fragment movement in three directions – frontal, vertical and lateral directions (Figure 16). 10 photographs were taken at each angle and load and the mean value was taken.

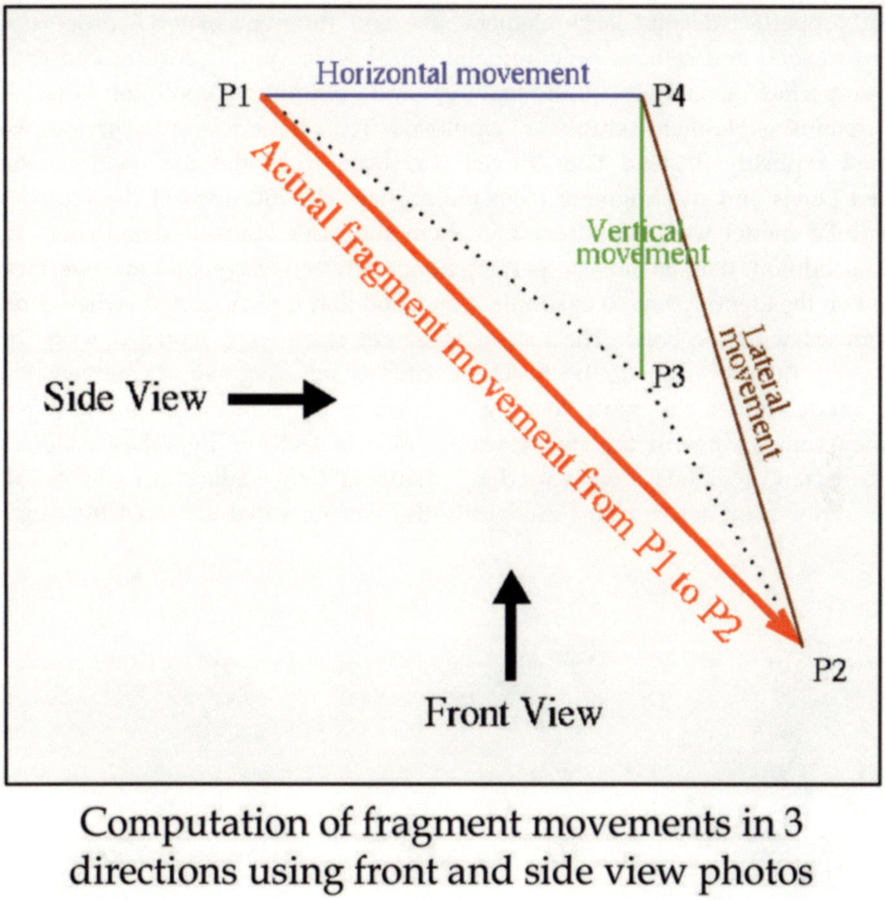

Figure 16. Getting fragment movements in three directions – horizontal, vertical and lateral – using photos taken at two different views. The actual movement of the fragment is from P1 to P2 from no load to full load conditions. The front view photo gives the triangle P1P4P3, allowing us to calculate horizontal and vertical movement. The side view photo gives the triangle P4P2P3 which allows us to calculate the lateral movement.

4.3.2. Finite element simulation

The stability of screw fixation was analyzed with finite element models. We developed the models of the fractured pelvis, fragment and femoral head in order to perform the mechanical testing that we did in silico. Firstly, one of the PU-foam based synthetic fractured pelves that had been fixed with two screws was dismantled. The resulting fractured pelvis and its fragment were scanned separately with a Faro Arm (Siler Series Faro Arm) and a laser scanner (Model Maker H40 Laser Scanner). Two sets of data point clouds, which accurately described the shapes of the fragment and the fractured pelvis, were obtained (Figure 17). The fractured pelvis model was developed from our previous FE model of the pelvis, which was generated from CT scans of the synthetic pelvis used in the experiment (Shim et al., 2010) . Our elements had inhomogeneous location dependent material properties despite large element size and different material properties were assigned to solid and cellular polyurethane foams which mimic cortical and cancellous bone properties separately. The loading and boundary condition that mimics the mechanical experiment setup were employed. The FE models of the screws were not generated explicitly. Instead tied contact was used to model the bond between the fractured pelvis and the fragment from the screws. The locations of the screws on the fragment FE model were identified first from the laser scanned data. Then, the tied contact condition that ensures a perfect bond between slave and master faces was imposed on the identified faces to simulate the bond that screws provide when connecting the fragment with the bone. The rest of fragment faces were modeled with frictional contact (μ=0.4)(Gordon et al., 1989). The predicted interfragmentary movements from the FE model under the same loading and boundary conditions as the experiment were then compared with the experimental value to test our hypothesis. Once tested, then, the screw positions were varied by changing tied contact faces to simulate all possible screw positions in order to identify the positions that achieved the most stable fixation.

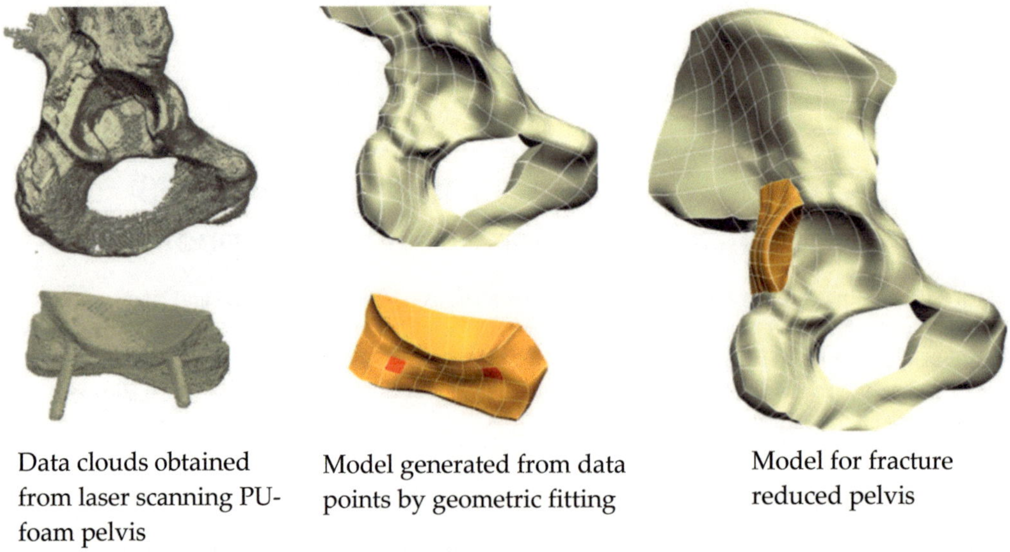

Figure 17. The far left column shows clouds of data points obtained from laser scanning. The center column shows the meshes for the fragment and fractured pelvis that were generated by geometric fitting to laser scanned data points. The red faces on the fragment mesh indicate where tied contact conditions were imposed in order to simulate the support provided by the screws. The final mesh is shown on the far right column.

4.3.3. Interfragmentary movement in acetabular fracture osteosynthesis measured with PU- foam based synthetic bones

The overall amount of displacement between the pelvis and the fragment was relatively small and the main direction of the fragment movement was in the lateral posterior direction (in body directions). The average displacement was around 0.4 – 0.9 mm for both screw and plate fixations. The plates gave higher stability especially in the horizontal and lateral directions (Figure 18). However, screw fixations also gave good stability of less than 1mm on average in all directions. Therefore, considering the fact that the maximum load of our experiment was higher than normally allowed weight bearing (around 20kg after the surgery for 3 months), the stability of screw fixation was sufficient for the cyclic loading condition used.

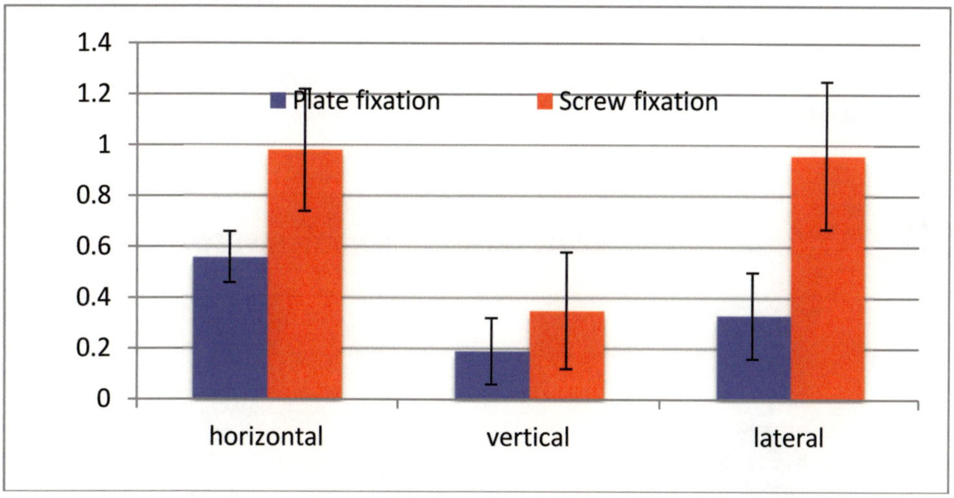

Figure 18. Comparison of interfragmentary movement between plate and screw fixation

4.3.4. Accuracy of FE model interfragmentary movement predictions validated with PU-foam based synthetic bones

The FE model predicted the movement of the fragment in screw fixations with a good accuracy. The values predicted by the FE simulation were within one standard deviation of the experimental measurements (Figure 19) in the horizontal and lateral directions. The predicted vertical direction movement was a little bit greater than the upper limit of the experimental measurement (mean + 1 SD) but still very close to it as the difference was 0.05 mm.

The optimized screw positions were found to be on the two diagonal corners of the fragment. When the virtual screws were placed in this manner, the stability of screw fixations improved dramatically to the level that is comparable to plate fixation (Figure 20). The fragment movements in the horizontal and lateral directions were smaller than the average movements in the plate fixation in these directions. Although the movement in the vertical direction was bigger than the upper limit of the experimental measurement, the difference was small 0.1mm.

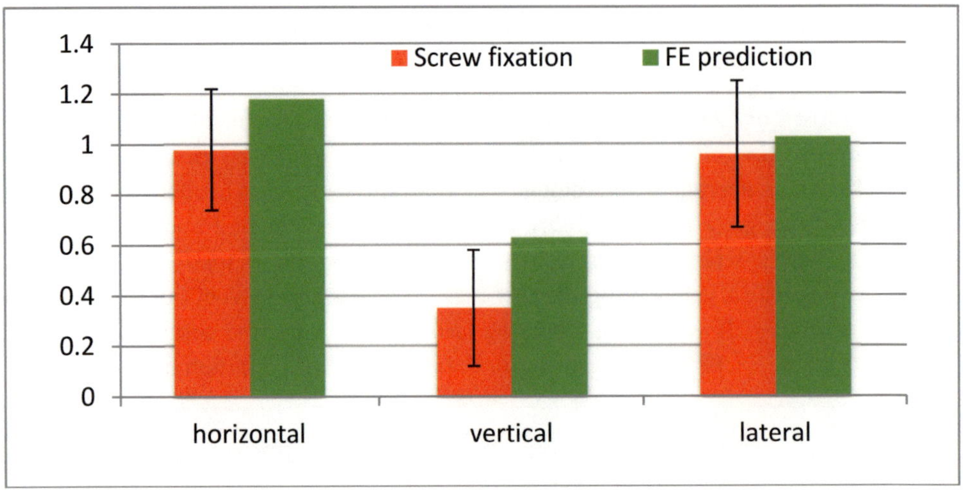

Figure 19. Comparison between FE prediction and experimental measurement

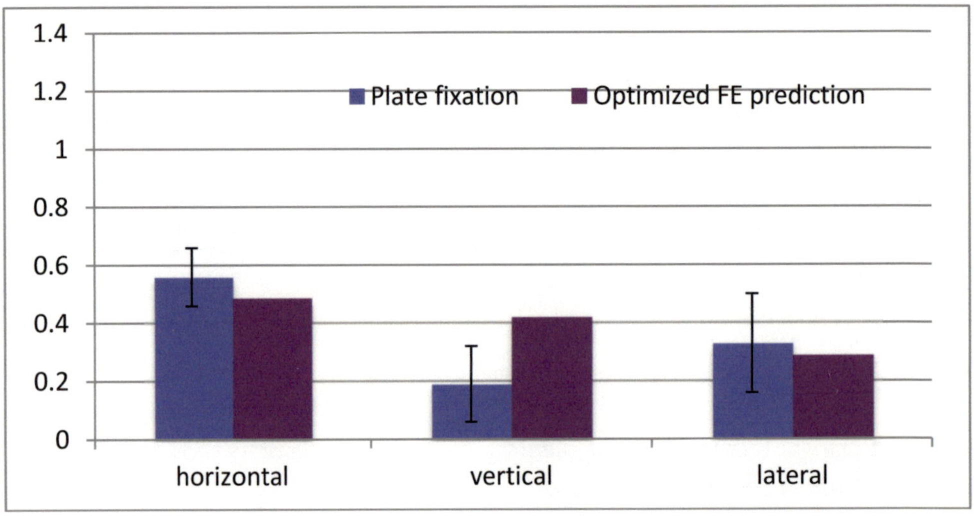

Figure 20. Comparison between plate fixation and optimized screw fixation predicted from FE model

4.3.5. Feasibility of the use of synthetic PU-foam based bone in validating FE predictions of fracture stability

We have developed a new and efficient way of simulating interfragmentary movement in acetabular fractures using a FE model. We validated our method with a biomechanical experiment involving PU-foam based synthetic bones. There are numerous studies that employed PU-foam based synthetic bones in measuring fracture stability as discussed in Section 3. However this data has not been used in validating FE model predictions of fracture stability. The use of FE models in fracture analysis is not new. However, the major focus has been to analyze the stress distribution on the implant or the overall stiffness of bone/implant composite after fracture fixation (Eberle et al., 2009, Stoffel et al., 2003).

Therefore we have performed a biomechanical experiment with PU-foam based synthetic bones to measure interfragmentary movements in 3D and used the result to validate our FE model.

The mechanical experiment with synthetic pelves showed that the displacement between fragment and bone was relatively small for both plate and screw fixations, indicating that screw fixations in single fragment fractures may be a good alternative to the current gold standard of plate fixation. In particular the excellent stability displayed by the screw FE model with the optimized screw positions indicate that screw fixations along with computer navigation should be an option considered by trauma surgeons if available.

The FE model showed a great potential for use in analyzing fracture fixation techniques. Our model was able to predict the movement of the fragment with a reasonable accuracy. Although we have not modeled screws explicitly, our modeling approach was able to accurately predict the fragment movement, which was the main aim of the model. Moreover the computational efficiency of the approach allowed us to perform a parametric study for optimization of screw positions.

Since we have used PU-foam based synthetic pelves in our study it is not known if our model predictions will be as accurate when cadaver bones are used. However the use of synthetic bones has some advantages due to their uniformity and consistency (Nabavi et al., 2009). Moreover the main aim was to make comparisons between different osteosynthesis techniques and between experimental and FE simulations and the ASTM standard states that it is "an ideal material for comparative testing (American Society for Testing and Materials, 2008a)." In this regard, the use of PU-foam based synthetic bones in comparative studies in orthopaedic biomechanics can provide useful data for FE model validation as well as testing hypothesis.

5. Concluding remarks

In this chapter we discussed the use of polyurethane in orthopaedic biomechanical experiments. Due to the similarity of polyurethane foam with cancellous bone in terms of microstructure and material properties, polyurethane has found a unique and important position in orthopaedic biomechanics studies. The main use of PU-foams was in experimental studies to find optimum values in various surgical procedures and to test stability of fracture or joint replacement implants. Due to the uniformity and consistency in material properties, PU-foam based synthetic bones are capable of generating reproducible results that are so difficult to obtain when using human cadaver bones. Therefore we used PU-foam materials in designing bone graft harvesters and obtaining validation data for FE model predictions in fracture load and stability. Although material properties of PU-foams are not identical to natural bone, they are able to generate comparable results that can provide important insight into surgical procedures or function of implants or devices. Moreover thanks to the advent of new composite bones made up of PU-foams and other relevant materials that mimic the geometry, structure and material properties of human bone, only the imagination of biomechanical engineers is the limit in ways that PU-foam based materials can be used in orthopaedic biomechanical studies in the future.

Author details

V. Shim and I. Anderson
Auckland Bioengineering Institute, University of Auckland, New Zealand

J. Boheme and C. Josten
University of Leipzig, Germany

Acknowledgement

This work was supported in part by Faculty Development Research Fund (FRDF) from the University of Auckland awarded to V. Shim and Federal Ministry of Education and Research (BMBF) grant awarded to J. Böhme. The authors would like to thank Mr. Sharif Malak for his work in orthogonal cutting of PU-foams.

6. References

Acevedo, J. I., Sammarco, V. J., Boucher, H. R., Parks, B. G., Schon, L. C. & Myerson, M. S. (2002) Mechanical comparison of cyclic loading in five different first metatarsal shaft osteotomies. *Foot Ankle Int*, 23, 711-6.

Agneskirchner, J. D., Freiling, D., Hurschler, C. & Lobenhoffer, P. (2006) Primary stability of four different implants for opening wedge high tibial osteotomy. *Knee Surg Sports Traumatol Arthrosc*, 14, 291-300.

American Society for Testing and Materials, A. (2008a) ASTM F1839 - 08 standard Specification for Rigid Polyurethane Foam for Use as a Standard Material for Testing Orthopedic Devices and Instruments.

American Society for Testing and Materials, A. (2008b) ASTM F1839 - 08 standard Specification for Rigid Polyurethane Foam for Use as a Standard Material for Testing Orthopedic Devices and Instruments.

Arrington, E. D., Smith, W. J., Chambers, H. G., Bucknell, A. L. & Davino, N. A. (1996) Complications of iliac crest bone graft harvesting. *Clin Orthop Relat Res*, 300-9.

Baumgaertner, M. R. (1999) Fractures of the posterior wall of the acetabulum. *J Am Acad Orthop Surg*, 7, 54-65.

Betz, R. R. (2002) Limitations of autograft and allograft: new synthetic solutions. *Orthopedics*, 25, s561-70.

Borrelli, J., JR., Ricci, W. M., Steger-May, K., Totty, W. G. & Goldfarb, C. (2005) Postoperative radiographic assessment of acetabular fractures: a comparison of plain radiographs and CT scans. *J Orthop Trauma*, 19, 299-304.

Bredbenner, T. L. & Haug, R. H. (2000) Substitutes for human cadaveric bone in maxillofacial rigid fixation research. *Oral Surg Oral Med Oral Pathol Oral Radiol Endod*, 90, 574-80.

Burstein, F. D., Simms, C., Cohen, S. R., Work, F. & Paschal, M. (2000) Iliac crest bone graft harvesting techniques: a comparison. *Plast Reconstr Surg,* 105, 34-9.

Caglar, Y. S., Torun, F., Pait, T. G., Hogue, W., Bozkurt, M. & Ozgen, S. (2005) Biomechanical comparison of inside-outside screws, cables, and regular screws, using a sawbone model. *Neurosurg Rev,* 28, 53-8.

Calvert, K. L., Trumble, K. P., Webster, T. J. & Kirkpatrick, L. A. Characterization of commercial rigid polyurethane foams used as bone analogs for implant testing. *J Mater Sci Mater Med,* 21, 1453-61.

Chong, A. C., Friis, E. A., Ballard, G. P., Czuwala, P. J. & Cooke, F. W. (2007a) Fatigue performance of composite analogue femur constructs under high activity loading. *Ann Biomed Eng,* 35, 1196-205.

Chong, A. C., Miller, F., Buxton, M. & Friis, E. A. (2007b) Fracture toughness and fatigue crack propagation rate of short fiber reinforced epoxy composites for analogue cortical bone. *J Biomech Eng,* 129, 487-93.

Cole, J. D. & Bolhofner, B. R. (1994) Acetabular fracture fixation via a modified Stoppa limited intrapelvic approach. Description of operative technique and preliminary treatment results. *Clin Orthop Relat Res,* 112-23.

Cristofolini, L., Teutonico, A. S., Monti, L., Cappello, A. & Toni, A. (2003) Comparative in vitro study on the long term performance of cemented hip stems: validation of a protocol to discriminate between "good" and "bad" designs. *J Biomech,* 36, 1603-15.

Eberle, S., Gerber, C., Von Oldenburg, G., Hungerer, S. & AUGAT, P. (2009) Type of hip fracture determines load share in intramedullary osteosynthesis. *Clin Orthop Relat Res,* 467, 1972-80.

Farshad, M., Weinert-Aplin, R. A., Stalder, M., Koch, P. P., Snedeker, J. G. & Meyer, D. C. (2011) Embossing of a screw thread and TCP granules enhances the fixation strength of compressed ACL grafts with interference screws. *Knee Surg Sports Traumatol Arthrosc.*

Giannoudis, P. V., Grotz, M. R., Papakostidis, C. & Dinopoulos, H. (2005) Operative treatment of displaced fractures of the acetabulum. A meta-analysis. *J Bone Joint Surg Br,* 87, 2-9.

Gibson, L. J. & Ashby, M. F. (1988) *Cellular Solids - Structure and properties,* Oxford, Pergamon Press.

Gordon, J., Kauzlarich, J. J. & Thacker, J. G. (1989) Tests of two new polyurethane foam wheelchair tires. *J Rehabil Res Dev,* 26, 33-46.

Goulet, J. A., Rouleau, J. P., Mason, D. J. & Goldstein, S. A. (1994) Comminuted fractures of the posterior wall of the acetabulum. A biomechanical evaluation of fixation methods. *J Bone Joint Surg Am,* 76, 1457-63.

Gulsen, M., Karatosun, V. & Uyulgan, B. The biomechanical assessment of fixation methods in periprosthetic femur fractures. *Acta Orthop Traumatol Turc,* 45, 266-9.

Heiner, A. D. (2008) Structural properties of fourth-generation composite femurs and tibias. *J Biomech,* 41, 3282-4.

Hoffman, O. (1967) The Brittle Strength of Orthotropic Materials. *J. Composite Materials,* 1, 200-206.

Jacobs, C. H., Pope, M. H., Berry, J. T. & Hoaglund, F. (1974) A study of the bone machining process-orthogonal cutting. *J Biomech*, 7, 131-6.

Keyak, J. H. (2001) Improved prediction of proximal femoral fracture load using nonlinear finite element models. *Medical Engineering and Physics*, 23, 165-173.

Keyak, J. H., Rossi, S. A., Jones, K. A. & Skinner, H. B. (1997) Prediction of femoral fracture load using automated finite element modeling. *Journal of Biomechanics*, 31, 125-133.

Klein, P., Schell, H., Streitparth, F., Heller, M., Kassi, J. P., Kandziora, F., Bragulla, H., Haas, N. P. & Duda, G. N. (2003) The initial phase of fracture healing is specifically sensitive to mechanical conditions. *J Orthop Res*, 21, 662-9.

Konrath, G. A., Hamel, A. J., Sharkey, N. A., Bay, B. & Olson, S. A. (1998a) Biomechanical evaluation of a low anterior wall fracture: correlation with the CT subchondral arc. *J Orthop Trauma*, 12, 152-8.

Konrath, G. A., Hamel, A. J., Sharkey, N. A., Bay, B. K. & Olson, S. A. (1998b) Biomechanical consequences of anterior column fracture of the acetabulum. *J Orthop Trauma*, 12, 547-52.

Korn, B., Weissman, S., Werner, T. & Gardiner, K. (2001) Report on the tenth international workshop on the identification of transcribed sequences 2000. Heidelberg, Germany, October 28-31, 2000. *Cytogenet Cell Genet*, 92, 49-58.

Krause, W. R., Bradbury, D. W., Kelly, J. E. & Lunceford, E. M. (1982) Temperature elevations in orthopaedic cutting operations. *J Biomech*, 15, 267-75.

Krenn, M. H., Piotrowski, W. P., Penzkofer, R. & Augat, P. (2008) Influence of thread design on pedicle screw fixation. Laboratory investigation. *J Neurosurg Spine*, 9, 90-5.

Kurz, L. T., Garfin, S. R. & Booth, R. E., JR. (1989) Harvesting autogenous iliac bone grafts. A review of complications and techniques. *Spine (Phila Pa 1976)*, 14, 1324-31.

Letournel, E. (1980) Acetabulum fractures: classicifaction and management. *Clin Orthop Relat Res*, 151, 81-106.

Lewandrowski, K. U., Gresser, J. D., Bondre, S., Silva, A. E., Wise, D. L. & Trantolo, D. J. (2000) Developing porosity of poly(propylene glycol-co-fumaric acid) bone graft substitutes and the effect on osteointegration: a preliminary histology study in rats. *J Biomater Sci Polym Ed*, 11, 879-89.

Lotz, J. C., Cheal, E. J. & Hayes, W. C. (1991) Fracture prediction for proximal femur using finite element models: Part II - Nonlinear analysis. *Journal of Biomechanical Engineering*, 113, 361-365.

Malak, S. F. & Anderson, I. A. (2008) Orthogonal cutting of cancellous bone with application to the harvesting of bone autograft. *Med Eng Phys*, 30, 717-24.

Malak, S. F. F. & Anderson, I. A. (2005) Orthogonal cutting of polyurethane foam. *International Journal of Mechanical Sciences*, 47, 867-883.

Mcintyre, A. & Anderston, E. (1979) Fracture properties of a rigid polyurethane foam over a range of densities. *Polymer*, 20, 247-253.

Melamed, E. A., Schon, L. C., Myerson, M. S. & Parks, B. G. (2002) Two modifications of the Weil osteotomy: analysis on sawbone models. *Foot Ankle Int*, 23, 400-5.

Nabavi, A., Yeoh, K. M., Shidiac, L., Appleyard, R., Gillies, R. M. & Turnbull, A. (2009) Effects of positioning and notching of resurfaced femurs on femoral neck strength: a biomechanical test. *Journal of orthopaedic surgery,* 17, 47-50.

Nasson, S., Shuff, C., Palmer, D., Owen, J., Wayne, J., Carr, J., Adelaar, R. & May, D. (2001) Biomechanical comparison of ankle arthrodesis techniques: crossed screws vs. blade plate. *Foot Ankle Int,* 22, 575-80.

Natali, C., Ingle, P. & Dowell, J. (1996) Orthopaedic bone drills-can they be improved? Temperature changes near the drilling face. *J Bone Joint Surg Br,* 78, 357-62.

Nyska, M., Trnka, H. J., Parks, B. G. & Myerson, M. S. (2002) Proximal metatarsal osteotomies: a comparative geometric analysis conducted on sawbone models. *Foot Ankle Int,* 23, 938-45.

Olson, S. A., Kadrmas, M. W., Hernandez, J. D., Glisson, R. R. & WEST, J. L. (2007) Augmentation of posterior wall acetabular fracture fixation using calcium-phosphate cement: a biomechanical analysis. *J Orthop Trauma,* 21, 608-16.

Papini, M., Zdero, R., Schemitsch, E. H. & Zalzal, P. (2007) The biomechanics of human femurs in axial and torsional loading: comparison of finite element analysis, human cadaveric femurs, and synthetic femurs. *J Biomech Eng,* 129, 12-9.

Parker, P. J. & Copeland, C. (1997) Percutaneous fluroscopic screw fixation of acetabular fractures. *Injury,* 28, 597-600.

Ross, N., Tacconi, L. & Miles, J. B. (2000) Heterotopic bone formation causing recurrent donor site pain following iliac crest bone harvesting. *Br J Neurosurg,* 14, 476-9.

Ruedi, T. P., Buckley, R. E. & Moran, C., G (2007) *AO Principles of fracture management* Davos, Switzerland, AO Publishing.

Russell, J. L. & Block, J. E. (2000) Surgical harvesting of bone graft from the ilium: point of view. *Med Hypotheses,* 55, 474-9.

Saha, S., Pal, S. & Albright, J. A. (1982) Surgical drilling: design and performance of an improved drill. *J Biomech Eng,* 104, 245-52.

Schileo, E., Taddei, F., Cristofolini, L. & Viceconti, M. (2008) Subject-specific finite element models implementing a maximum principal strain criterion are able to estimate failure risk and fracture location on human femurs tested in vitro. *J Biomech,* 41, 356-67.

Shim, V., Bohme, J., Vaitl, P., Klima, S., Josten, C. & Anderson, I. (2010) Finite element analysis of acetabular fractures--development and validation with a synthetic pelvis. *J Biomech,* 43, 1635-9.

Shim, V. B., Boshme, J., Vaitl, P., Josten, C. & Anderson, I. A. (2011) An efficient and accurate prediction of the stability of percutaneous fixation of acetabular fractures with finite element simulation. *J Biomech Eng,* 133, 094501.

Shim, V. B., Pitto, R. P., Streicher, R. M., Hunter, P. J. & Anderson, I. A. (2007) The use of sparse CT datasets for auto-generating accurate FE models of the femur and pelvis. *Journal of Biomechanics,* 40, 26-35.

Shim, V. B., Pitto, R. P., Streicher, R. M., Hunter, P. J. & Anderson, I. A. (2008) Development and validation of patient-specific finite element models of the hemipelvis generated from a sparse CT data set. *Journal of Biomechanical Engineering,* 130, 051010.

Shin, H. C. & Yoon, Y. S. (2006) Bone temperature estimation during orthopaedic round bur milling operations. *J Biomech,* 39, 33-9.

Shuler, T. E., Boone, D. C., Gruen, G. S. & Peitzman, A. B. (1995) Percutaneous iliosacral screw fixation: early treatment for unstable posterior pelvic ring disruptions. *J Trauma,* 38, 453-8.

Silber, J. S., Anderson, D. G., Daffner, S. D., Brislin, B. T., Leland, J. M., Hilibrand, A. S., Vaccaro, A. R. & Albert, T. J. (2003) Donor site morbidity after anterior iliac crest bone harvest for single-level anterior cervical discectomy and fusion. *Spine (Phila Pa 1976),* 28, 134-9.

Simoes, J. A., Vaz, M. A., Blatcher, S. & Taylor, M. (2000) Influence of head constraint and muscle forces on the strain distribution within the intact femur. *Med Eng Phys,* 22, 453-9.

Spagnolo, R., Bonalumi, M., Pace, F. & Capitani, D. (2009) Minimal-invasive posterior approach in the treatment of the posterior wall fractures of the acetabulum. *Musculoskeletal Surgery,* 93, 9-13.

St^lken, J. S. & Kinney, J. H. (2003) On the importance of geometric nonlinearity in finite-element simulations of trabecular bone failure. *Bone,* 33, 494-504.

Stoffel, K., Dieter, U., Stachowiak, G., Gachter, A. & Kuster, M. S. (2003) Biomechanical testing of the LCP--how can stability in locked internal fixators be controlled? *Injury,* 34 Suppl 2, B11-9.

Szivek, J. A., Thomas, M. & Benjamin, J. B. (1993) Characterization of a synthetic foam as a model for human cancellous bone. *J Appl Biomater,* 4, 269-72.

Szivek, J. A., Thompson, J. D. & Benjamin, J. B. (1995) Characterization of three formulations of a synthetic foam as models for a range of human cancellous bone types. *J Appl Biomater,* 6, 125-8.

Thompson, M. S., McCarthy, I. D., Lidgren, L. & Ryd, L. (2003) Compressive and shear properties of commercially available polyurethane foams. *J Biomech Eng,* 125, 732-4.

Tile, M., Helfet, D. L. & Kellam, J. F. (Eds.) (2003) *Fractures of the pelvis and acetabulum,* Philadelphia, PA USA, Lippincott Williams & Wilkins.

Trnka, H. J., Nyska, M., Parks, B. G. & Myerson, M. S. (2001) Dorsiflexion contracture after the Weil osteotomy: results of cadaver study and three-dimensional analysis. *Foot Ankle Int,* 22, 47-50.

Wehner, T., Penzkofer, R., Augat, P., Claes, L. & Simon, U. (2010) Improvement of the shear fixation stability of intramedullary nailing. *Clin Biomech (Bristol, Avon),* 26, 147-51.

Zdero, R., Olsen, M., Bougherara, H. & Schemitsch, E. H. (2008) Cancellous bone screw purchase: a comparison of synthetic femurs, human femurs, and finite element analysis. *Proc Inst Mech Eng H,* 222, 1175-83.

Zdero, R., Rose, S., Schemitsch, E. H. & Papini, M. (2007) Cortical screw pullout strength and effective shear stress in synthetic third generation composite femurs. *J Biomech Eng,* 129, 289-93.

Zoys, G. N., Mcganity, P. L., Lanctot, D. R., Athanasiou, K. A. & Heckman, J. D. (1999) Biomechanical evaluation of fixation of posterior acetabular wall fractures. *J South Orthop Assoc,* 8, 254-60; discussion 260.

Chapter 10

Biocompatibility and Biological Performance of the Improved Polyurethane Membranes for Medical Applications

Maria Butnaru, Ovidiu Bredetean, Doina Macocinschi,
Cristina Daniela Dimitriu, Laura Knieling and Valeria Harabagiu

Additional information is available at the end of the chapter

http://dx.doi.org/10.5772/34653

1. Introduction

Polyurethanes (PUs) are one of the most "pluripotent" synthetic polymer classes used in medical applications. Due to their structural versatility, they have been widely discussed as materials appropriate for biomedical applications (Abd El-Rehim & El-Amaouty, 2004; Guelcher et. al., 2007; Guelcher, 2008; Kavlock et. al, 2007; J.S. Lee et. al., 2001; Lelah & Cooper, 1987; Siepe et. al., 2007). Up to now, new PUs have been synthesized that possess good mechanical properties. Most of them are considered biocompatible on account of *in vitro* cytotoxicity evaluation.

However, it is well known that structural and mechanical adaptability of PUs is not always accompanied by cell and tissue biocompatibility. Therefore, numerous data in the literature are focused on biocompatibilization or functionalization of PUs (Yao, 2008; Sartori, 2008, Huang & Xu, 2010). Some promising methods for the improvement of biological response of PUs are conjugation, blending or coating with natural polymers. Thus, polysaccharides as chitosan, cellulose and their derivatives (Raschip, 2009; Zia, 2009; Zuo, 2009), proteins and glycoproteins as collagen, fibrin, fibronectin (R. Chen et. al., 2010; Sartori et. al., 2008), proteoglycans and glycosaminoglycans (Gong et. al., 2010) and other molecules (Hwang & Meyerhoff, 2008; Hsu et. al., 2004; Makala et. al., 2006; Song et. al., 2005; Verma & Marsden, 2005) are employed successfully for PUs modification. Owing its specific properties, hydroxypropylcellulose (HPC) is already used as binder, thickener, lubricating material (artificial tears) and emulsion stabilizer in pharmaceutical and food industry. Moreover, HPC may provide interactions through its hydroxyl radicals, being an excellent compound for copolymerization in scaffolds for tissue engineering and in drug delivery systems (Berthier et. al., 2011; D. Chen & Sun, 2000; Gutowska et. al., 2001; Raschip et. al., 2009;

Valenta & Auner, 2004). In previous studies we found that when added to PU structure, HPC improves hydrophilicity and mechanical properties of PUs by increasing the elasticity of the resulted materials (Macocinschi et. al., 2009).

Considering the reviewed concept of biocompatibility as "the ability to exist in contact with tissues of the human body without causing an unacceptable degree of harm to the body" (Williams, 2008), our interdisciplinary work was focused on the synthesis of PU-based materials with improved ability to long-time functional integration. PU/HPC membranes were prepared by blending method. HPC was chosen due to its physical-chemical properties, its demonstrated biocompatibility and accessibility. The aim of the chapter is to highlight the most important criteria, able to predict the behaviour of material-tissue interfaces and the long-term material-tissue integration, in order to select most suitable compositions and morphologies for specific medical application. Thus, surface zeta (ζ) potential, wettability (as contact angle measurement and water uptake), pH modification after long time hydration and autoclaving, protein adsorption at protein physiological concentration and some relevant elements of bulk and surface morphology are treated as screening criteria for suitable membrane choice in the first part of the chapter. Biological performance evaluation, such as oxidative stress action, thrombogenicity and *in vivo* behaviour of PU/HPC membranes are further discussed.

2. Materials and methods

2.1. Preparation of polymer samples

Preparation of PU/HPC samples was performed according to Fig. 1 as previously reported (Macocinschi et. al. 2009; Vlad et. al, 2010).

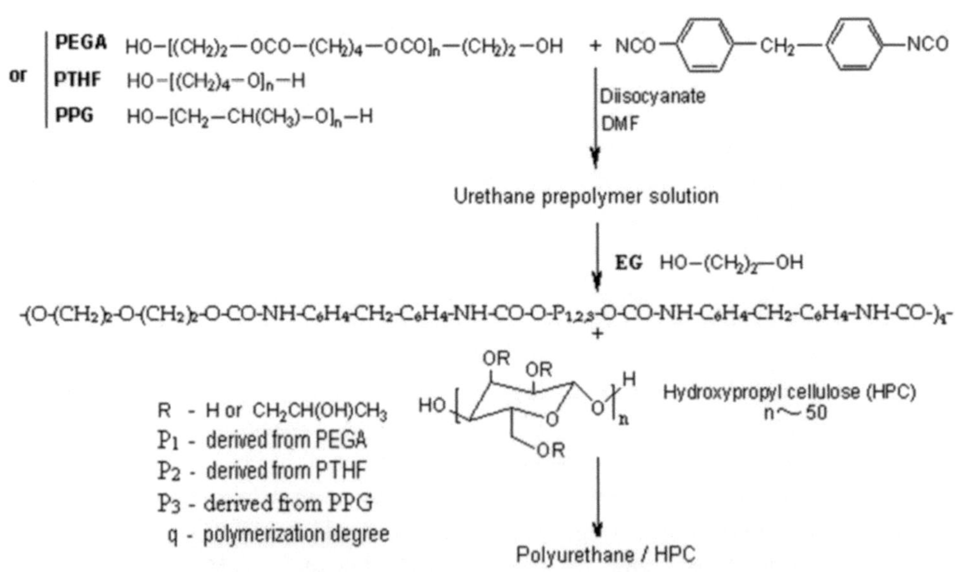

Figure 1. Scheme of chemical structure and synthesis way of PUs/HPC

Briefly, isocyanate terminated urethane prepolymers were first synthesized by the polyaddition reactions between 4',4'-diphenylmethane diisocyanate (MDI) and macrodiols in N,N-dimethylformamide (DMF) as solvent. Poly(ethylene adipate)diol (PEGA, Mn = 2000 g/mol), polytetrahydrofuran (PTHF, Mn = 2000 g/mol) or poly(propylene)glycol (PPG, Mn = 2000 g/mol) were used as macrodiols. The urethane prepolymers were treated in a subsequent step with ethylene glycol (EG) as chain extender. Finally, HPC (average weight molecular weight Mw = 95 000 g/mol) was added to PU solutions to obtain the following compositions for all PU/HPC samples: macrodiol/MDI/EG/HPC = 52.24 /36.57/7.27/3.92 (weight ratios). As the molar ratio between isocyanate groups in MDI and the sum of hydroxylic groups in macrodiol and EG was 1.02, the excess of isocyanate groups linked to PU prepolymers were available to bind a part of HPC chains. Membranes with about 1 mm thickness were prepared by pouring PU/HPC DMF solutions in distilled water, at 40 °C. The formed films were then dried under vacuum for several days and kept in distilled water for solvent removing.

To half of PUs with PEGA macrodiol in the soft segment no HPC was added to obtain PU-PEGA reference sample. HPC containing samples based on PEGA, PTHF and PPG macrodiols were codified as PU-PEGA/HPC; PU-PTHF/HPC and PU-PPG/HPC, respectively.

2.2. ζ potential determination

ζ potential of the PU membranes was measured by streaming potential method using a commercial electrokinetic analyzer SurPASS, (Anton Paar GmbH, Graz, Austria). For each sample, ζ potential has been measured in 0.1 M NaCl solution at physiological 7.4 pH value, a 300 mbar electrolyte pressure and a 80 ml/min flow rate. For statistical reasons, four streaming potentials were measured. The mean value of these data was used for potential calculation by Fairbrother–Mastin equation, considering also the effect of surface conductivity (Luxbacher, 2006)

2.3. Wettability

Wettability of the PU membranes was determined by measuring the surface contact angle and water uptake. For surface contact angle, uniform drops of the tested liquid (double-distilled water) with a volume of 2 μl were deposited on the film surface and the contact angles were measured after 30 s, using a video-based optical contact angle measuring device equipped with a Hamilton syringe in a temperature-controlled environmental chamber. All measurements were performed at room temperature of 25 °C. Repeated measurements of a given contact angle were all within the range of ± 3 degrees. *Water uptake* was calculated as the ratio between fully hydrated and dried sample weights.

2.4. Material extraction in a simulated biological microenvironment

Material extraction in a simulated biological microenvironment was done for long period of time (over 2 months) in Hank's Balanced Salt Solution (HBSS) without Ca^{2+} and Mg^{2+}, with glucose, and phenol red as pH indicator. For extraction experiments, 0.2 g of each membrane, cut in very small pieces (see Fig. 2), were incubated in 2 ml of HBSS solution at 37°C. pH variation was monitored daily, based on phenol red indicator colour and measured after 1, 2, 3, 30 and 60 days of incubation using Mettler Toledo SevenGo SG2ELK pH-meter.

2.5. Scanning Electron Microscopy (SEM)

SEM analysis of PU/HPC membrane cross-sections was performed using a VEGA TESCAN microscope, in high vacuum mode, at an acceleration voltage of 30 kV.

2.6. Protein adsorption

Amount of protein adsorption on membrane surfaces was measured in three different conditions: (a) on individual protein solutions of fibrinogen (FB) at 3 mg/ml (95% clotable from Sigma-Aldrich) and serum albumin (SA) at 45 mg/ml (bovine SA (BSA) from Sigma-Aldrich); (b) FB and BSA mixed solutions of physiological concentrations (3 mg/ml for BSA and 45 mg/ml for FB); (c) complex protein conditions (platelet poor blood plasma (PPP)). Prior adsorption experiment, the PU/HPC films were brought to equilibrium with phosphate buffer saline (PBS) up to reaching maximum hydration, for about 72 h. Briefly, PU/HPC hydrated membranes with 0.5 cm x 0.5 cm surface area were covered with 0.25 ml of one of the protein solutions or with blood plasma and kept at 37 °C for 30 min. FB and BSA concentration in incubated medium was determined before and after incubation. A turbidimetric method based on the formation of an insoluble complex with Na_2SO_4 was used for FB determination. The method based on antigen–antibody reaction was performed for SA measuring, using a Dialab kit, Austria. FB and SA reaction products were assessed on a Piccos 05 UV–VIS spectrophotometer at λ = 530 nm for FB and λ = 340 nm for SA. The adsorbed amount of proteins was calculated with the following relation:

$$\text{Adsorbed proetin}\,(\text{mg/cm}^2) = \frac{(Co - Ce)\cdot V}{S} \quad (1)$$

where Co and Ce are the initial and post-incubation concentrations of protein solution (mg/ml), V is the incubated volume of the protein solution (ml) and S is the surface of the incubated PU/HPC sample

2.7. Total Antioxidant Status (TAS)

TAS was measured in blood plasma obtained by human blood centrifugation at 1000 G for 20 min. PU samples were incubated in blood plasma for 1, 2 and 3 days at 37 °C and mild orbital shacking. The TAS measurement was made by standard protocol provided by Randox TAS kit. Thus, 2,2'-azino-di-[3-ethylbenzthiazoline sulphonate] (ABTS)® was incubated with a peroxidase (metmyoglobine) and H_2O_2 to produce the ABTS®+ radical cations having a stable blue-green colour that was measured at 600 nm on a spectrophotometer mentioned in the previous section. By adding blood plasma containing antioxidants a suppression of this colour to a degree which is proportional to their concentration is observed. Control serum ("standard" provided by the determination kit) was used for data validation. TAS values were calculated based on the measured absorbance in the standard, blood plasma sample and blank (buffer provided by the kit) before and after H_2O_2 adding. The absorbance differences (ΔA) between measurement before and after H_2O_2 adding for standard, sample or blank solutions were used for calculation of TAS concentration according to relations 2 and 3:

$$\text{Factor} = \frac{\text{concentration of standard}}{\Delta\text{A blank} - \Delta\text{A standard}} \tag{2}$$

$$\text{TASmMol/L} = \text{Factor} \cdot \Delta\text{A blank} - \Delta\text{A sample} \tag{3}$$

2.8. Haemocompatibility testing

Haemocompatibility of membrane surface was evaluated by haemolysis and coagulation tests. All tests were performed on well swollen PU samples in PBS. *Haemolysis* was determined using 0.25 ml of blood (human blood from healthy voluntary donors, collected on 3.8 % sodium citrate solution as anticoagulant in 9:1 v/v ratio) that was incubated with 1 cm^2 surface area PU samples for 30 min at 37 ^oC. Haemoglobin released from lysed erythrocytes was measured by spectrophotometric method at $\lambda = 545$. *Prothrombin time* was measured after 1 hour incubation of polymer sample in blood plasma. Standard laboratory method was applied using PT kit (Biodevice, Italy) and ACL 100 coagulometer. Blood plasma was obtained by blood centrifugation at 1000 G for 10 min.

Platelet adhesion on material surface was determined based on number of platelet counted in 0.1 ml platelet rich blood plasma (PRP), before and after membrane (0.5 cm x 0.5 cm) incubation for 1 hour at 37 ^oC. PRP was obtained by blood centrifugation at 400 G for 20 min. Improved Neubauer haemocytometer was used for platelet counting. *Clot weight test* was performed by adding 0.2 ml of human blood upon well swollen samples with 1cm^2 surface area. The thrombus formation was started by adding 0.05 ml $CaCl_2$ solution (0.025 mol/l). Each formed thrombus was weighed and compared with control. Collagen film was used as positive pro-coagulant control and normal blood plasma without polymer sample as negative control.

2.9. *In vivo* biocompatibility

Subcutaneous implantation experiment was performed on Wistar 200 g weight male rats. Testing protocol was designed according to ISO 10993-2 (Animal Welfare Requirements) and the guidelines of Council for International Organizations of Medical Sciences (CIOMS). The pieces of autoclaved purified or unpurified membranes (0.5 x 0.5 cm size) were implanted under both sites (right and left) of dorso-lateral skin. Material purification was performed by immersion in sterile distilled water for 1 week and equilibration in physiological salted sterile solution for 24 hours before subcutaneous implantation. All surgical procedures were done under thiopental anaesthesia, using a dosage of 35 mg/kg body weight. Lots of six animals for each material were taken in each experiment. The period of 10 or 30 days was chosen for material examination. Explanted samples together with surrounding tissue were fixed in 10% formaldehyde solution embedded in paraffin wax, sliced in 15 μm pieces and stained using Hematoxylin – Eosin (HE) method for cell examination and Masson's trichrome for collagen fibres.

3. Results and discussions

3.1. Primary screening criteria for the appropriate selection of PU/HPC membranes for medical usage

Biocompatibility of PUs, seen in terms of specific application, is a result of a "bio-appropriate" expression of surface and bulk properties achieved by synthesis and scaffold fabrication methods. Thus, surface ζ potential and surface wettability are important characteristics responsible for specific tissue-material interaction mechanisms, starting with protein adsorption that can be influenced in turn by specific physiological/pathological tissue environment.

3.1.1. Surface ζ potential and wettability

Surface charge plays an important and active role in tissue-material interaction and must be considered in accordance with the targeted application. The importance of surface charge on cell adhesion, biofilm formation or thrombogenesis was demonstrated (Cai et. al., 2006; Colman & Schmaier, 1997; Kang et. al., 2006; Khorasani et. al., 2006). These phenomena are a consequence of adsorptive behavior of proteins on charged surface rather than the effect of electrostatic interactions with cells (Keselowsky et. al., 2003; Wilson et. al., 2005). Many data refer to the effect of surface charge on biological phenomena (Jelinek et. al., 2010; Kang et. al., 2006). However, there are not many data reporting surface charge and its clear relevance for biocompatibility of PU-based membranes. Moreover, it is difficult to estimate the electro-kinetic properties of such surfaces, mainly due to the complexity of the chemical composition but also due to membrane variable porosity and swelling behavior that can influence surface charge values (Yaroshchuk & Luxbacher, 2010).). Surface ζ potential of material is a property that reflects surface charge. Some reported data have shown that poly(ether-urethane)s exhibit a very negative (– 25 mV) ζ potential, while poly(ester-urethane)s are less negative (- 12 mV). Contradictory data were published on the beneficial effect of positively (Khorasani et. al., 2006) or negatively charged surfaces (Sanders et. al., 2005) on cells attachment and proliferation.

Thus, this section is aimed to predict the influence of surface potential and wettability on the biocompatibility and biological performances of PU-based samples. Table 1 shows hydrophilic/hydrophobic properties and ζ potential of examined PU-based samples.

Material samples	Contact angle						WU (%)	ζ (mV)
	First immersion			*Second immersion*				
	θ_{adv}(°)	θ_{rec}(°)	H(%)	θ_{adv}(°)	θ_{rec}(°)	H(%)		
PU-PEGA	85.3±1.1	54.3±0.6	36.3	51.0±0.5	54.1±0.6	5.6	141±10	- 4.31
PU-PEGA/HPC	84.8±1.1	44.2±0.5	47.9	52.6±0.5	43.7±0.5	16.9	140±4	+ 3.14
PU-PTHF/HPC	77.4±1.1	42.9±0.5	44.5	31.6±0.4	42.3±0.4	25.2	167±3	+ 0.78
PU-PPG/HPC	85.6±1.1	44.8±0.5	47.7	60.3±0.6	44.1±0.5	27.0	92±6	+ 4.85

Table 1. Dynamic contact angle values (θ) in contact with water, hysteresis (H) resulted from advanced (*adv*) and receded (*rec*) contact angles, water uptake (WU) (Macoconschi et. al., 2009) and ζ potential of the PU samples

As one can see from Table 1, PU-PEGA has a slightly negative ζ potential, probably due to the presence of carboxylic groups resulted by the hydrolysis of residual isocyanate groups during membrane precipitation in water. After blending with HPC, the residual isocyanated groups linked to PU prepolymer are reacted with the hydroxyl groups of HPC and all PU/HPC membranes showed a slightly positive surface. The most hydrophilic sample (PU-PTHF/HPC) exhibited the most neutral ζ potential. This observation is in accordance to other data that report dependence of surface charge on water swelling capacity (Aranberri-Askargorta et. al., 2003).

3.1.2. Extraction microenvironment

The material biocompatibility can be appreciated through its effects on the physico-chemical properties of the physiological environment, especially on the pH. Thus, pH modification of HBSS buffer solutions after unsterilized and sterilized membranes incubation was measured. The results are shown in Figs 2 and 3 (1, PU-PEGA; 2- PU-PEGA/HPC; 3, PU-PTHF/HPC; 4, PU-PPG/HPC).

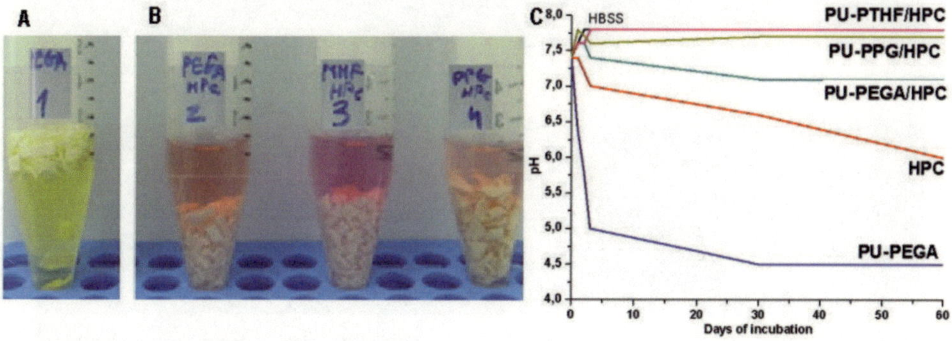

Figure 2. pH variation of HBSS buffer in which unsterilized membranes were incubated: A, PU-PEGA; B, PUs/HPC; C, pH variation curves

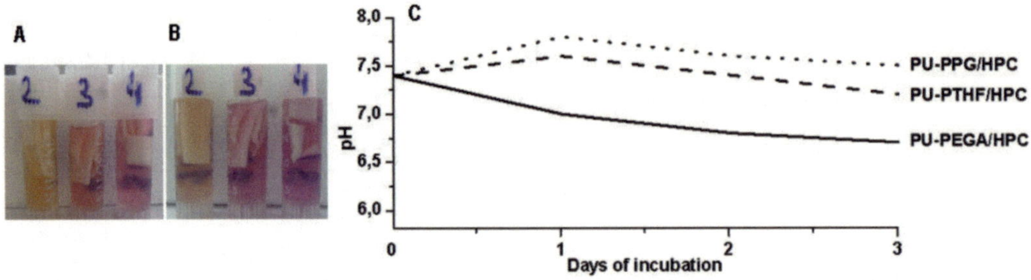

Figure 3. pH variation of HBSS buffer in which autoclaved PU/HPC membranes were incubated: A, 24 h of incubation; B, 72 h of incubation; C, pH variation curves

As one can see from Figs. 2 and 3, a long-period of incubation of unsterilized and sterilized (by autoclaving at 121 °C and 1 atm) PU/HPC membranes in simulated biological fluid did not meaningfully modify the physiological range pH value of the incubation environment, while a pronounced decrease of the environment pH was observed for pure PU-PEGA sample (Fig. 2 A). Thus, one can say that HPC gives an important contribution to hydrolytic stability of urethane and ester bonds of PU chains.

For autoclaved samples, the variation of pH values of the environment for poly(ether-urethane)s remains in the range of the physiological value, while PU-PEGA/HPC membrane induced a higher decrease of pH (Fig. 3), a normal result owing the higher thermal degradability of poly(ester-urethane)s (Guelcher, 2008).

Another property that was changed by modifying PU membranes with HPC was the floatability (see Fig. 2B). As the surface wettability and water uptake for PU-PEGA sample is similar to HPC modified one (see Table 1), the reason of these different behavior could reside in different morphologies, as seen from SEM images of membrane cross-sections (Fig. 4).

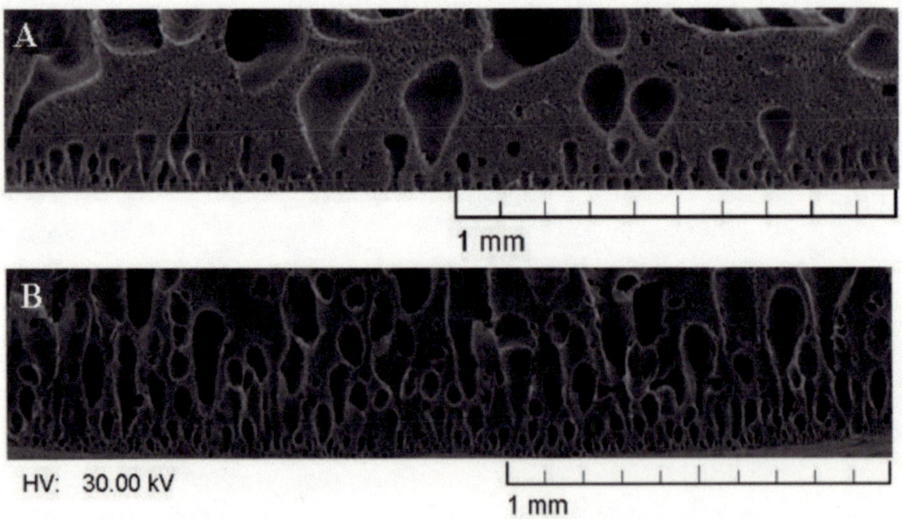

Figure 4. SEM image of PU-PEGA (A) and PU-PEGA/HPC (B) membrane cross-sections

PU-PEGA sample showed important bulk microporosity, with isolated pores, while PU-PEGA/HPC presented smaller but interconnected pores allowing water diffusion and the decrease of the floatability.

Thus one can conclude that PU/HPC membranes are slightly positively charged and they possess interconnected porous morphology influencing the wettability and floatability. They also showed a less pronounced influence on the biological media as compared to the pure PU membrane.

3.2. Protein adsorption

There are many data concerning mechanisms of protein adsorption on different surfaces (Gray, 2004; Scott & Elbert, 2007; Van Tassel, 2006; Wilson et. al., 2005). It was clearly demonstrated that proteins have amphoteric properties, being able to adsorb on both negatively and positively charged surfaces (Michelsen et. al., 2000; Van Tassel, 2006). The amount of adsorbed proteins is depending on their isoelectric points as well as on surface chemistry and hydrophilicity (Keselowsky et. al., 2003; Wertz & Santore, 2001). Hydrophobic surfaces manly interact with hydrophobic protein core that leads to the modification of the protein physiological conformation and its functionality. Opposite to hydrophobicity, superficial water maintains native protein conformation and specific functionality (Keselowsky et. al., 2003; Noinville & Revault, 2006).

Many authors have reported protein adsorption behaviour on different surfaces using simulated solutions and highly sensible methods in which very low protein concentrations are detected. Thus, the adsorption of albumin solutions of different concentrations on pure silica or on silica modified with NH_2 and CH_3 terminated self-assembled monolayers (SAMs) (Noinville & Revault, 2006) and on silica-titanium surfaces (Kurrat et. al. 1997) was studied. Other authors reported the competitive adsorption of fibrinogen on mica (Gettens et. al., 2005; Tsapikouni & Missirlis, 2007). Surface adsorption of SA, FB, fibronectin (FN), immunoglobulins (IGs) and lysozyme were investigated to evaluate the surface biocompatibility (Bernsmann et. al., 2010; Pompe et. al., 2006; Rezwan et. al., 2005), each class of these proteins providing specific surface properties for targeted application. Thus, FN adsorption is relevant for the prediction of cell adhesion, lysozyme – for enzymatic degradability predisposition, IGs - for immune-specific interactions, while SA and FB adsorption have haemocompatibility predictive value.

In order to estimate protein adsorption (retention) capacity of materials at blood or tissues contact, simulated physiological environment, close to normal blood conditions is required. For example, Bajpai, 2005 followed SA adsorption capacity of biomaterials at SA bulk concentration from 1 to 6 mg/ml, while Alves et. al., 2010 used mix protein conditions, considering physiological value for each protein. The mix protein adsorption conditions are considered to better reflect the complex interactions that occur between different proteins (Latour, 2008).

PU/HPC membranes were previously demonstrated to possess good mechanical properties (elongation at break for dried/hydrated PU-PEGA/HPC = 71/84; for PU-PTHF/HPC = 72/159 and for PU-PPG/HPC = 53/55), appropriate for cardio-vascular applications (Macocinschi et. al., 2009). The physisorption of SA and FB is further highlighted as screening criteria for biocompatibility and, more specifically, haemocompatibility. Very short characteristics of SA and FB, important for protein-material interaction are given below.

SA is a protein belonging to the so called "soft" class of proteins, with a molar mass of about 65 kD for BSA and 67 kD for human SA (HSA). This protein represents about 60% of the blood proteins. Normal blood concentration of HSA is 35 – 50 mg/ml. This protein is involved in many physiological phenomena as carrier protein for fatty acids, metals, cholesterol, bile pigment, hormones and drugs. SA is also characterised by antioxidant properties (Bourdon et. al. 1999; Kouoh et. al., 1999; H. Lee et. al., 2000) that is higher in alkaline pH, up to 8 (H. Lee et. al., 2000). SA is preponderantly negatively charged, its isoelectric point being close to 4.8 (Carter & Ho 1994; Noinville & Revault, 2006). Approximately 67% of the secondary SA structure is represented by the α-helix. It was demonstrated that the stability of SA secondary structure strictly depends on pH (Freeman, 2006) that influence the protein conformation. Thus, at pH = 5, SA takes almost spherical, native, unfolded shape that forms a thick layer on the adsorptive surfaces. At pH = 7 (close to physiological pH), due to molecular spreading, SA forms an extended contact area with adsorptive surfaces. This behavior can be influenced by surface charge, surface functionality and functionality distribution, surface morphology or wettability conditions (Wilson et. al., 2005). The role of adsorbed SA on biomaterial biocompatibility is still ambiguously described in the literature. While some authors have demonstrated biocompatibility improvement of material with increased adsorption of SA (Eberhart et.al. 1987; Marconi et. al., 1996; Randrasana et. al., 1994), others demonstrated a better biocompatibility of SA-resistant surfaces (Ostuni et. al., 2001; Wan et. al., 2006).

FB is a high molecular weight (340 kD) complex glycoprotein that has 2 molecular domains, each of them consisting of three polypeptide chains called Aα, Bβ and γ. Molecular updated analysis of FB can be found in recent reports (Cardinali et.al, 2010). FB is an important factor of haemostasis. Through fibrin network formation as first cell scaffold, FB is involved in wound healing and tissue regeneration. Its normal blood concentration varies from 2 to 4 mg/ml. In inflammations or in other pathological statuses - as cardiovascular diseases - FB can reach up to 7 mg/ml, therefore adsorption properties of biomaterials for this protein should be carefully analysed, especially for those targeted for blood contact applications.

The results obtained in adsorption experiments of SA and FB from both individual and mixed solutions on PU/HPC membranes are presented in Fig. 5.

No significant differences between adsorption behavior of both proteins in their pure and mixed solutions were registered, except a small tendency to decrease adsorbed BSA from mixed solution as compared to individual solution, especially on PU-PEGA and PU-PEGA/HPC membranes, where FB, with a higher molecular weight, showed a higher affinity.

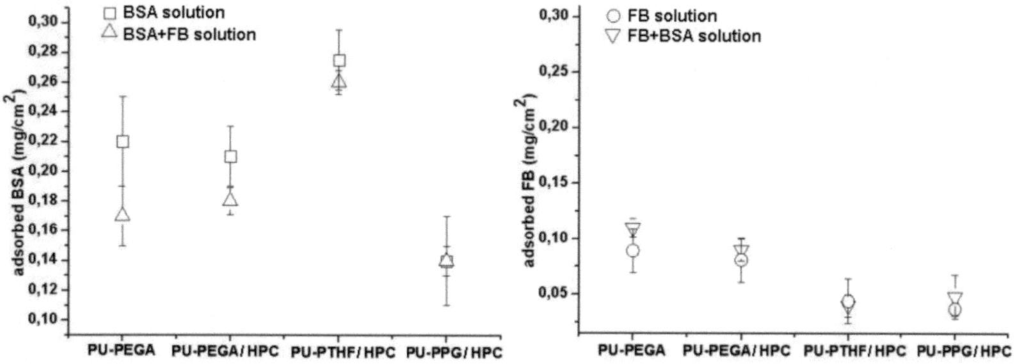

Figure 5. Amount of adsorbed BSA (left) and FB (right) from individual protein solutions and in co-adsorptive environment (mixed protein solution) of physiological concentrations, i.e., 3 mg/ml for FB and 45 mg/ml for BSA

Figure 6 shows the ratios of adsorbed BSA and FB from mixed protein solutions and from blood plasma. In both studied conditions and for all membranes, the amount of adsorbed SA is higher than that of adsorbed FB, a normal result considering the lower concentration of FB in solutions. The total amount of the adsorbed SA and FB proteins from blood plasma is lower as compared to that adsorbed from mixed solutions due to the competitive adsorption of some other blood plasma proteins. Moreover, the ratio between adsorbed FB and SA is lower in blood plasma than in mixed solutions.

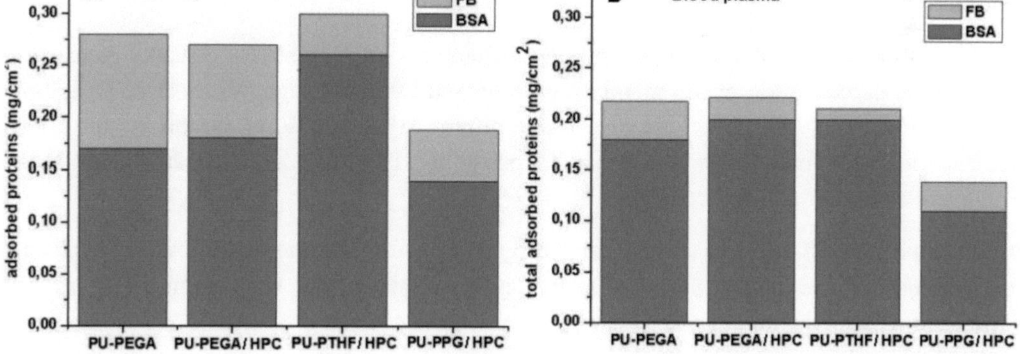

Figure 6. Total amount of adsorbed FB and BSA from: A – mixed protein solution at 3 mg/ml FB and 45 mg/ml BS physiological concentration; B - human blood plasma with 2.98 mg/ml initial FB concentration and 45.3 mg/ml initial SA concentration

As conclusion, comparing PU-PEGA and PU-PEGA/HPC membranes one can observe that small amount of polysaccharide rich in functional substituents can bio-stabilize PU structures and improve their resistance for autoclaving procedures as important step in ready to use biomaterials preparation. From all the data presented in this section, one can say that the more hydrophilic PU-PTHF/HPC membrane could be the most appropriate for biomedical applications.

3.3. *In vitro* and *in vivo* performances of PU/HPC membranes

The biocompatibility of PUs are widely discussed and questioned, mostly in the past. In the last two decades new generation of PUs that combine mechanical advantages with the biological performances emerged (Gisselfa et. al., 2002; Jordan & Chaikof et. al., 2007; Jun et. al., 2005; Kavlock et. al., 2007; Parveen et. al., 2008). For many years it has been considered that PUs biocompatibility is spotless due to their products of degradation, e.g., aromatic polyamines. As it is well known for the most part of biocompatible materials, the life time of their *in vitro* functionality is quite short. This is a consequence of their intrinsic physico-chemical properties, on one hand, and of the tissue action on the material, on the other hand (Anderson, 2001; Guelcher, 2008; Shen & Horbett, 2001).

3.3.1. Oxidative in vitro behavior

Oxidative degradation of PUs caused by hydrolytic or enzymatic mechanism was intensively discussed (Christenson et. al., 2004; Guelcher, 2008; Gary & Howard, 2002; Sutherland et. al., 1993). First of all, PUs designed for tissue-contact devices undergo hydrolytic degradation as a result of watering with physiological solutions. This process has an impact especially on poly(ester-urethane)s that can generate hydroxy-acids, being susceptible to induce reactive oxygen species (ROS) production following the material-tissues interaction. By means of this mechanism, PUs can be implied in the sustained oxidative degradation and a wide range of pathological states.

As it is well known, ROS can trigger subtle mechanisms responsible for diseases generation through the peroxidation of cell membrane lipids and DNA damage (Marnett, 2002; Tribble et. al., 1987; Yagi, 1987). The most susceptible organs to oxidative aggression are the heart, vessels, lung, gut, liver, brain and nerves (Ames et. al., 1993; Förstermann, 2008; Paradis et. al., 1997; Rahman et. al., 2002; Sayre et. al., 1997).

In a normal body state, ROS appear constantly as a result of some biological errors or as a consequence of some short living reactive intermediate products generated by the cell aerobic metabolism. Endogenous enzymatic and nonenzymatic pathways are responsible for the formation of free radicals. These pathways are balanced by two endogenous antioxidant pathways, which form the TAS (see fig. 7).

While some harmful material characteristics can be marked as cytotoxic or proinflammatory by standard testing, others, such as oxidative stress (that causes long-time material failure), are undetectable by using short period testing. Thus, well known biocompatible materials were found to display surface alteration or cracking after long-time implantation. Adding antioxidant compounds to materials can improve their resistance against tissue degradation (Oral et. al., 2006; Stachelek et. al., 2006; Wattamwar et. al., 2010).

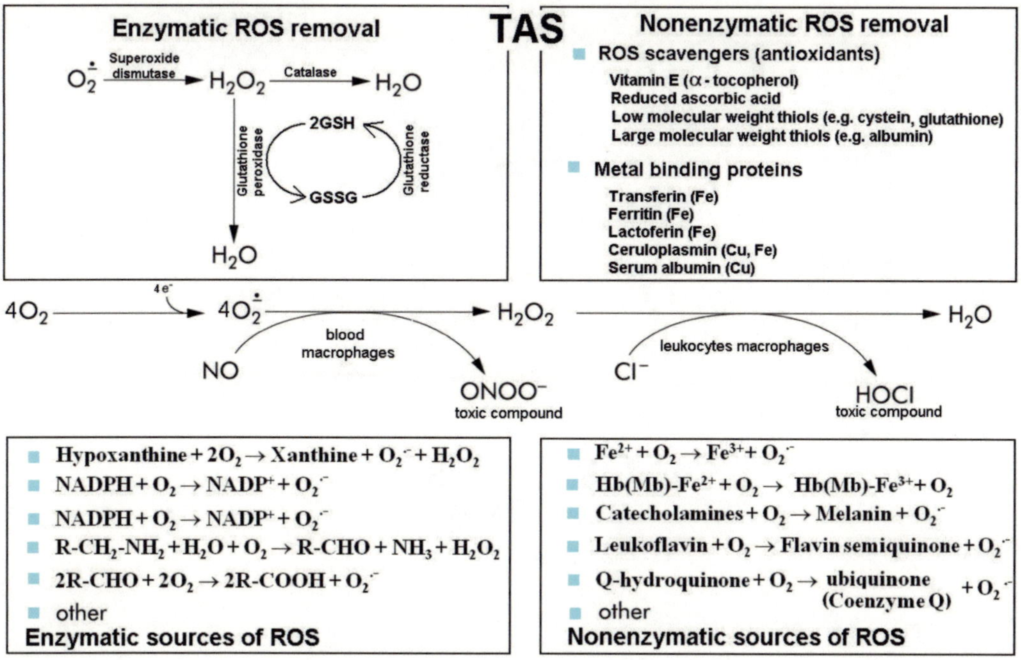

Figure 7. Schematic representation of the oxidative/antioxidative balance with enzymatic and nonenzymatic tissue pathways

Antioxidant defensive systems are present in both cells and extracellular environment. SA molecules are the most important antioxidants in blood. Due to their high concentration and polyvalent possibilities to fit with oxygen free radicals, SA molecules are considered to be the main plasmatic components of defence that assure neutralisation of more than 70% of ROS (Bourdon & Blache, 2001).

Assigning to SA molecules the main role in protective effect, we analysed the interaction of PU/HPC membranes with blood plasma, following the plasma antioxidant status. To define the importance of SA adsorption on material surface, the membranes were incubated at 37 °C in blood plasma and TAS was measured periodically. The results are shown in Fig. 8.

Two PU samples (PU-PEGA and the more hydrophobic PU-PPG/HPC) had significant tendency to quickly decrease TAS activity in the first 48 hours. Due to the complexity of TAS, it is difficult to speculate on the mechanism by which the decreasing phenomenon arises and certainly more examinations are needed. However, one can suppose that PU-PEGA alter the TAS activity as a result of plasma pH modification that leads to sustained free radical generation in the presence of the material. The mechanisms by which TAS activity is lowered after PU-PPG/HPC incubation could not be related directly to SA antioxidant activity, but to some other oxidant pathways that need further investigations.

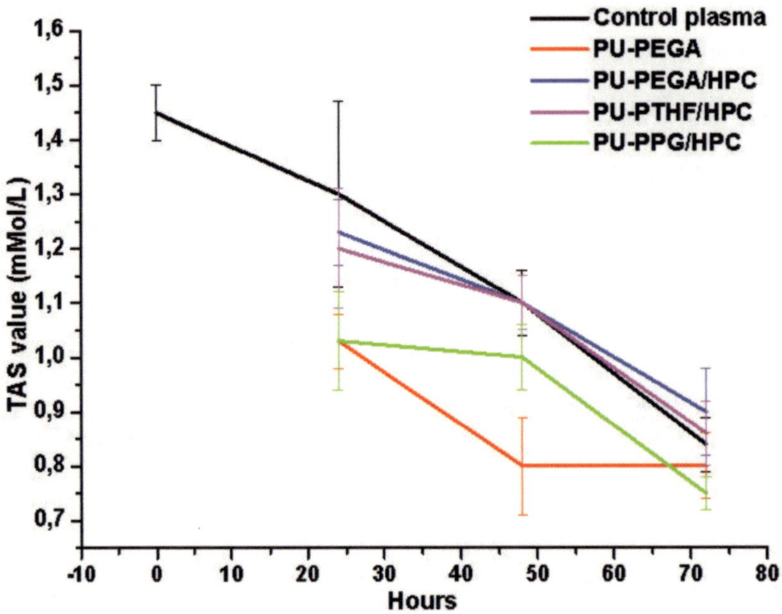

Figure 8. TAS evolution after PU/HPC incubation in blood plasma at 37 °C

3.3.2. *In vitro haemocompatibility*

Haemocompatibility involves compatibility with blood cells and blood plasma in other words, nonhaemolytic and nonthrombogenic behavior. Haemolysis is a mechanism by which erythrocytes (red blood cells) are destroyed through cell membrane lyses. Erythrocyte membrane lyses may occur as a result of environment pH modification or by cytotoxic action on erythrocyte membrane. Thus, both lipid (by lipid peroxidation) and protein (by protein modification) compounds can be affected.

Thrombogenesis is a complex phenomenon by which thrombus is formed by blood clotting. As a physiological event, haemostasis implies the activation of the enzymatic cascades in which three main factors are involved – vascular, cellular and plasmatic (Edmunds, 1998).

A synthetic material can induce haemostasis activation by surface charge, hydrophobicity and/or released products of degradation. It is widely recognised that both positively and negatively charged as well as hydrophobic surfaces can induce thrombus formation. This can be explained by involvement of several mechanisms as presented in Fig. 9.

A positive charge can be favourable for FB adsorption, followed by its conformational modification and adhesion of platelets and leukocytes (monocytes). Adherent cells are activated and they release numerous molecules that lead finally to FB cleavage with fibrin network formation (clot). Among platelet secreted factors are platelet thromboplastin, fibrin stabilizing factor, serotonin, anti-heparin factor, and others. Adherent (activated) monocyte releases thrombogen tissue factor (TF). Mechanism triggered by positive and hydrophobic surfaces is mainly related to extrinsic coagulation pathway (B. Furie & B. C. Furie, 2008).

A negative charge acts as an activator of plasmatic factor XII (Hageman factor) that involves contact system and intrinsic coagulation pathway (Zhuo et.al. 2006). This mechanism also involves high molecular weight and positively charged kininogen (HMW) and plasma thromboplastin (factor XI). Contact mechanism is tightly related to inflammatory events because some intrinsic pathways factors are direct activators of neutrophils. Whole mechanism of contact blood coagulation is still unclear. Some authors hypothesized that it can also be induced by adsorbed FB (Colman & Schmaier, 1997) and hydrophobic surfaces (Zhuo et. al., 2006). As for intrinsic coagulation mechanism of thrombus formation, this can also be activated by negatively charged low density lipoproteins (LDL), the molecules that adhere to the vessel walls in some pathologic conditions associated with cardiovascular risks (Krieter et. al., 2005). This possible mechanism should be taken into account because almost all pathological situations in which blood-assisted devices are used are accompanied by high level of cardiovascular risk factors (high level of LDL, cholesterol, triglycerides and modified blood pressure).

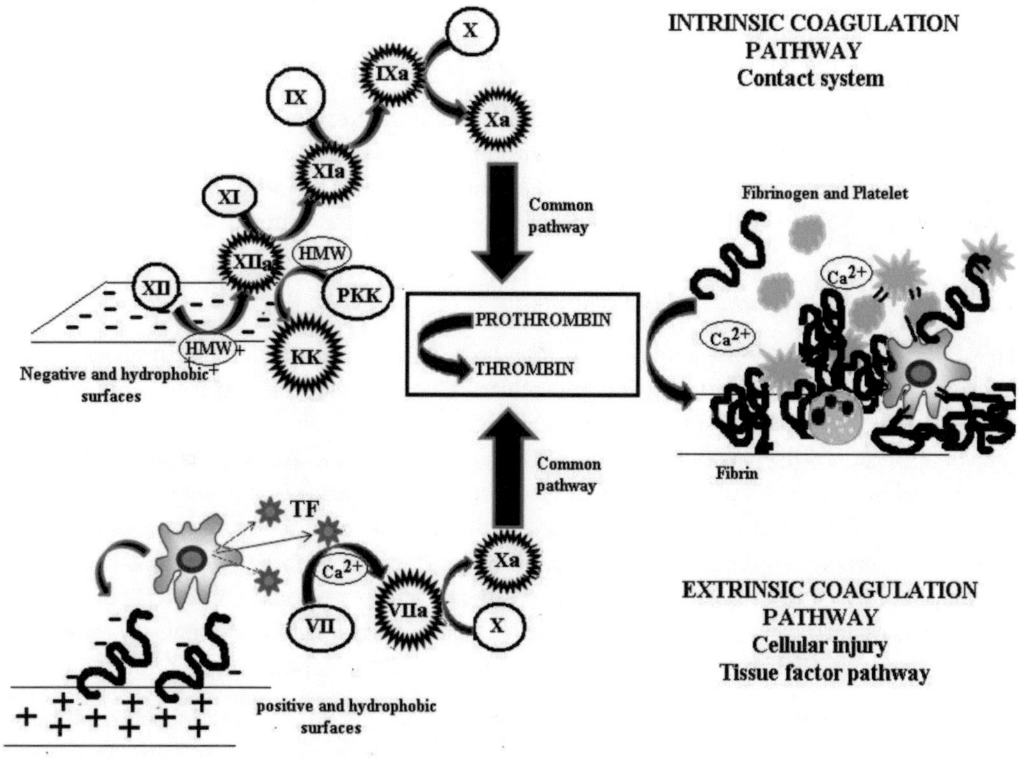

Figure 9. Contact coagulation and extrinsic (tissue factor) coagulation blood pathways

PUs are promising materials for implantable and non-implantable blood-interacting devices. They combine an increased elasticity with good mechanical resistance. For haemocompatibility evaluation of PU/HPC materials discussed above, the haemolytic and thrombotic potentials were determined by standard and adapted methods (see section 2). The obtained results are summarised in Table 2.

All studied membranes showed a low haemolytic activity, lower for PU/HPC than for pure PU-PEGA sample.

Material samples	Haemolytic potential	Thrombotic potential			
	Released Hb (%)[1]	FB (mg/ml)[2]	PT (s)[3]	Adhered platelet (cells x 10^5/mm^2)	% blood clot amount[4]
PU-PEGA	6,7±0,2	2,79±0,04	11,06±0,4	1,40±0,08	40%
PU-PEGA/HPC	5,2±0,1	2,87±0,04	10,9±0,09	0,82±0,05	29%
PU-PTHF/HPC	4,2±0,2	2,90±0,01	10,9±0,09	0,86±0,05	15%
PU-PPG/HPC	5,5±0,1	2,77±0,07	10,9±0,07	1,25±0,09	89%

Table 2. Haemolytic and thrombotic potential of the PU/HPC samples: Hb-haemoglobin; FB-fibrinogen; PT- prothrombin time

As for thrombotic action, a correlation between adsorbed FB, platelet adhesion and amount of formed clot was registered, while no significant variation was recorded for PT. This latter parameter was kept within the normal limits (see footnote 3).

The judgement strictly based on the haemocompatibility results permits to state that all examined materials have an acceptable thrombotic potential (referring to physiological requirements). Considering clot amount and all the other characteristics discussed above, it is obvious that PU-PEGA and PU-PPG/HPC are not suitable for long-time functional integration.

3.3.3. *In vivo biocompatibility and performance*

The technological progress achieved in the last decades in apparently unrelated areas (biomaterials, biotechnology, cell and molecular biology, tissue engineering, and polymer science) has generated a boost in the development and use of devices for medical and/or other type of applications (e.g. artificial organs, biosensors, catheters, heart valves) (Shastri, 2003). In spite of real improvement of this sort of devices there are still some important problems to face since implanted medical devices usually reveal different degree of loss of functionality over time after insertion (Göpferich, 1996). Tissue or blood-device interface interactions or a lack of biocompatibility resulting from the normal homeostatic response of the body to the implantation injury, determining an inadequate *in vivo* functionality and longevity, remains a serious concern (Callahan & Natale, 2008; Fujimoto et. al., 2007; Morais et. al., 2010).

In order to protect the body from the foreign object, under normal physiological conditions, the body reacts by several nonspecific mechanisms (immune and inflammatory cells recruitment), usually termed foreign body reaction (FBR) (Anderson, 2001). There is an imperative call for knowing the degree to which the pathophysiological conditions are

[1] Percentage of released Hb over negative control

[2] FB concentration remained in blood plasma after incubation. FB control was 2,98 ± 0,04 mg/ml

[3] Physiological normal value according to related laboratory are between 8,3 s and 11,3 s

[4] Percentage of blood clotting over negative control (blood without incubated material)

created, the homeostatic mechanisms are disturbed, and the resolution of the inflammatory response (simple put, the measure of the host reaction). All of these will finally establish the effective compatibility of a specific device. In the same time, understanding these reactions (the implant versus the host and the host versus the implanted device) will reduce health problems to the beneficiary of the device and device malfunction. Usually, for practical reasons, the homeostatic mechanisms are separately assessed even if it is well known that they are profoundly interrelated (Sieminski & Gooch, 2000).

The first event after a device/material insertion is that the body generates quickly a sort of "interface" *via* nonspecific adsorption of plasma/tissue soluble proteins on the implant surface (Shen & Horbett, 2001). There are some well identified elements that determine the FBR strength: device material composition, surface chemistry, size and shape, porosity, degradation, velocity as well as the place of device insertion (Ratner & Bryant, 2004)

As presented shortly below, tissue injury associated with device implantation, initiates a complex set of events (nonspecific inflammatory reaction and wound healing responses) that will bring about a FBR (Wahl et. al., 1989). The stages of inflammatory responses are well studied and can be separate in acute and chronic inflammatory periods.

The initial phase, acute stage, starts quickly in matter of hours, lasts for several days (up to 14 days) and is underlined by rapid device interface generation and typical for this phase, different degree of neutrophil leucocytes responses (Jiang et. al., 2007). The main result of this stage is the building of temporary interface material-tissue, the cleaning-up of the injury place and the vasodilation that bring more blood in the affected area.

The acute inflammatory reaction typically decline in maximum 14 days with a "biocompatible" material. Some local conditions (extent of surgical injury, body reactivity) or properties of the implanted device can trigger a chronic inflammatory evolution (Kirkpatrick et al., 1998).

Numerous blood and tissue proteins such as cytokines (e.g. tumor necrosis factor (TNF), interleukins (IL-6, IL-8), matrix metalloproteases (MMP-1, MMP-3), granulocyte-macrophage growth factors (GM-CSF)) are released, and leukocytes adhere to the endothelium of the blood vessels and infiltrate the injury site. These proteins are strong calling factors for monocytes, cells which will migrate to the site of inflammation where they will differentiate into macrophage. If inflammatory stimuli persist, the conditions that can lead to chronic inflammation are created. Cell population of this stage of inflammatory reaction is usually characterized by the presence of monocytes, macrophages, and lymphocytes (Bhardwaj et al., 2010). Also, in this step it can be noticed that the proliferation of blood vessels (angiogenesis), and connective tissue occurs that participate in remodelling of the affected area. The formation of blood vessels is crucial for wound healing, supplying necessary factors for tissues reconstruction. In the end, the granulomatous tissue is replaced by an extracellular matrix (ECM) that acts not only as a physical scaffold but also as an essential modulator of the biological processes, including differentiation, development, regeneration, repair, as well as tumour progression. The end phase of the FBR draws in wrapping the implant by a collagenic fibrous capsule that limits the implant and therefore prevents it from interacting with the surrounding tissue. The main tissue events of the material-tissue interaction and wound healing are schematically presented in Fig. 10.

Morphologic aspects (light microscopy) of the acute tissue reaction to subcutaneous implanted polyurethane (PU-PTHF/HPC) at 10 days of implantation and chronic inflammation at 30 days of implantation are shown in Fig. 11. The study was conducted on Wistar male rats using the protocol described in section 2.

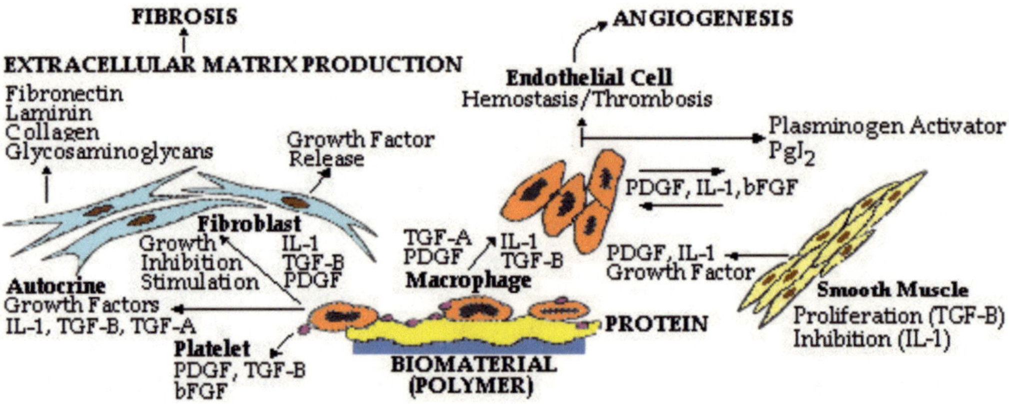

Figure 10. Fibrosis and fibrous encapsulation. End stage healing response to biomaterials. GF – growth factor (PD – platelet derived, T – transforming, bF – basic fibroblastic); IL – interleukin; PGL – prostaglandin.

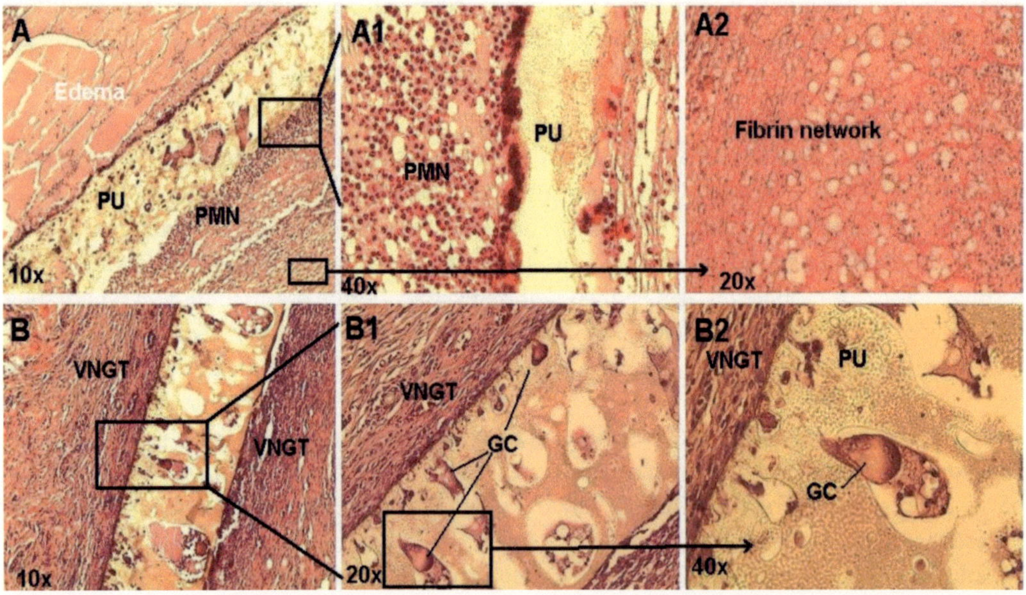

Figure 11. Light microscopy images of tissue response to implantation of subcutaneous non-washed PU-PTHF/HPC (PU): A - A2, 10 days of implantation; B -B2, 30 days of implantation. All images – HE staining. Objective magnifications are indicated in the left bottom corner

Unwashed (unpurified) material was implanted first, to highlight the importance of the properly prepared biomaterial for medical usage. From Fig. 11, A-A2 images, it can be easily seen as an intense acute inflammation reaction with numerous neutrophil polymorphonucleate leucocytes (PMN), edema and early fibrin network formation away from implantation site. These results suggest that an inappropriately prepared material at some stage in manufacture and/or manipulation can delay wound healing. As we expected, at 30 days of implantation (Fig.11, B-B2 images), inflammatory chronic reaction was really strong for related material, with the characters of neovascularised granulomatous tissue (VNGT) and giant cells (GC).

In the end of this chapter, comparative study concerning long-time potential functionality based on evolution of chronic inflammation of PUs/HPC discussed above was done. The histological images of 30 days implanted, properly purified PUs/HPC are shown in Fig. 12.

There were found chronic inflammations with VNGT and FBR with GC for PU-PEGA/HPC (A-A2 images) and PU-PPG/HPC samples (C-C2 images). Moreover, granuloma formation (G) as result of macrophage material degradation was present at material-tissue interface of PU-PPG/HPC (C and C1 images in Fig. 12).

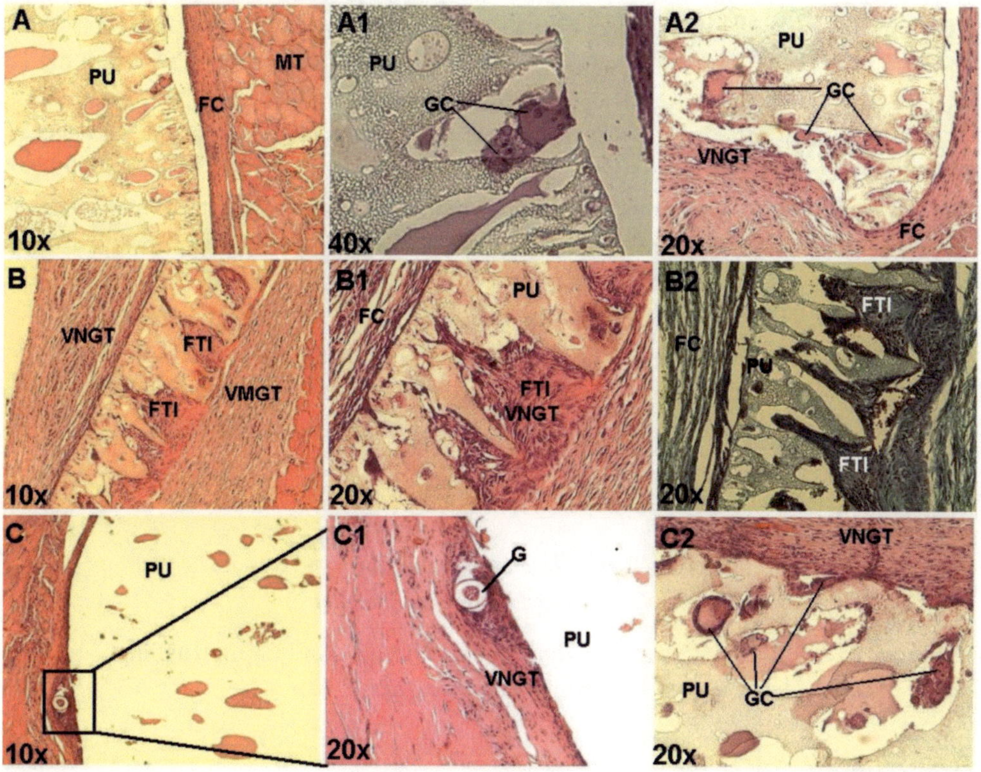

Figure 12. Light microscopy images of tissue response following 30 days subcutaneous implantation of washed PUs/HPC. A–A2, PU-PEGA/HPC; B-B2, PU-PTHF/HPC and C-C2, PU-PPG/HPC. B2, Masson's trchrome staining; all other images - HE staining. Objective magnifications are indicated in left bottom corner

The absence of GC and rich granulomatous tissue ingrowth through large material pores was observed for PU-PTHF/HPC sample (Fig.12, B-B2 images). Morphological aspect for PU-PTHF/HPC implant suggests a material-tissue integration and regenerative remodelling. Moreover, the fibroblast-rich tissue ingrowth only from one side of the membrane highlights the bifacial behavior of the implanted sample, with potentially tubular or cavity-like device performances. Thus, considering PU-PTHF/HPC increased haemocompatibility, oxidative and other biocompatibility advantages discussed above, we presume a cardiovascular-device performance for this PU sample.

4. Conclusions and further perspectives

Polyester and polyether urethane structures with improved bulk and surface characteristics by blending with small amount of biocompatible cellulose derivative, HPC, are screened for long-time functional integration. The stability of the pH value of biological media and the ratio of adsorbed albumin and fibrinogen from blood plasma were found to be the most valuable screening criteria to evaluate the blood-interface functionality, but not only. These criteria could provide information on material capacity to keep stability of the main body balances (oxidant/antioxidant, haemostasis/haemolysis) that are responsible for material acceptance in the early phase, followed by structural and functional integration in the later stages. These characteristics together with other important material properties as surface neutral charge and desired porous structure are keys points for good results expectance as was demonstrated. Another PU characteristic highlighted in our study was washability for potentially proinflammatory compounds removal. Due to interconnected mechanisms of thrombosis and inflammation, even haemocompatible PU, but with chronic prolonged inflammatory capacity (through itself or some released compound) will certainly get to fail its haemocompatibility *in vivo*. From this point of view we demonstrate an acceptable stability of some PU membrane by autoclaving and long-time watering in biological buffers. Further studies are necessary on extended classes of polyurethanes in the aim to prepare and keep ready to use pre-equilibrated and safe PUs for medical applications.

Author details

Maria Butnaru, Doina Macocinsch and Valeria Harabagiu
"Petru Poni" Institute of Macromolecular Chemistry, Iasi, Romania

Maria Butnaru, Ovidiu Bredetean, Cristina Daniela Dimitriu and Laura Knieling
"Grigore T. Popa " University of Medicine and Pharmacy, Iasi, Romania

Acknowledgment

The financial support of European Social Fund – "Cristofor I. Simionescu" Posdoctoral Fellowship Programme (ID POSTDRU/89/1.5/S/55216), Sectoral Operational Programme Human Resources Development 2007 – 2013 is acknowledged.

5. References

Abd El-Rehim, H.A.A., & El-Arnaouty, M.B. (2004). Properties and Biocompatibility of Polypropylene Graft Copolymer Films. *Journal of Biomedical Materials Research Part B: Applied Biomaterials*, Vol. 68B, No.2, (February 2004), pp.209-215, ISSN 1552-4973

Alves, C.M., Reis, R.L., & Hunt, J.A. (2010). The Competitive Adsorption of Human Proteins Onto Natural-Based Biomaterials. *Journal of the Royal Society, Interface / The Royal Society*, Vol. 50, No. 7, (February 2010), pp. 1367-1377, ISSN 1742-5689

Ames, B.N., Shigenaga, M.K., & Hagen, T.M. (1993). Oxidants, Antioxidants and The Degenerative Diseases of Aging. *Proceedings of the National Academy of Sciences of the United States of America*, Vol. 90, No. 17, (September 1993), pp. 7915-7922, ISSN 0027-8424

Anderson, J.M. (2001). Biological Responses to Materials. *Annual Review of Materials Research*, Vol. 31, No. 1, (May 2001), pp. 81–110, ISSN 1531-7331

Aranberri-Askargorta, I., Lampke, T., & Bismarck, A. (2003). Wetting Behavior of Flax Fibers as Reinforcement for Polypropylene. *Journal of Colloid and Interface Science*, Vol.263, No.2, (July 2003), pp.580–589, ISSN 0021-9797

Bajpai, A.K. (2005). Blood Protein Adsorption Onto a Polymeric Biomaterial of Polyethylene Glycol and Poly[(2-Hydroxyethyl Methacrylate)-Coacrylonitrile] and Evaluation of In Vitro Blood Compatibility. *Polymers International*, Vol. 54, No. 2, (February 2005), pp. 304-315, ISSN 0959-8103

Bernsmann, F., Frisch, B., Ringwald, C., & Ball, V. (2010). Protein Adsorption on Dopamine–Melanin Films: Role of Electrostatic Interactions Inferred from f-Potential Measurements Versus Chemisorptions. *Journal of Colloid and Interface Science*, Vol. 344, No. 1, (April 2010), pp. 54–60, ISSN 0021-9797

Berthier, D.L., Herrmann, A., & Ouali, L. (2011). Synthesis of Hydroxypropyl Cellulose Derivatives Modified with Amphiphilic Diblock Copolymer Side-Chains for The Slow Release of Volatile Molecules. *Polymer Chemistry*, Vol. 2, No. 9, (May 2011), pp. 2093-2101, ISSN 1759-9954

Bhardwaj, U., Radhacirsshana, S., Papadimitrakopoulos, F., Burgess, D.J. (2010). PLGA/PVA Hydrogel Composites for Long-Term Inflammation Control Following Subcutaneous Implantation. *International journal of pharmaceutics*, Vol. 484, No. 1-2, (January 2010), pp. 78–86, ISSN 0378-5173

Bourdon, E., & Blache, D. (2001). The Importance of Proteins in Defense Against Oxidation. *Antioxidants & Redox Signaling*, Vol. 3, No. 2, (April 2001), pp. 293-311, ISSN 1523-0864

Bourdon, E., Loreau, N., & Blanche, D. (1999). Glucose and Free Radicals Impair the Antioxidant Properties of Serum Albumin, *The FASEB Journal*, Vol. 13, No. 2, (February 1999), pp. 233-234, ISSN 0892-6638

Cai, K., Frant, M., Bossert, J., Hildebrand, G., Liefeith, K., & Jandt, K.D. (2006). Surface Functionalized Titanium Thin Films: Zeta-potential, Protein Adsorption and Cell Proliferation. *Colloids and Surfaces B: Biointerfaces*, Vol. 50, No.1, (June 2006), pp. 1–8, ISSN 0927-7765

Callahan, T.D. 4th, & Natale, A. (2008). Catheter Ablation of Atrial Fibrillation. *The Medical Clinics of North America,* Vol. 92, No.1, (Janury 2008), pp.179–201, ISSN 0025-7125

Cardinali, B., Profumo, A., Aprile, A., Byron, O., Morris, G., Harding, S.E., Stafford, W.F., & Rocco, M. (2010). Hydrodynamic and Mass Spectrometry Analysis of Nearly-Intact Human Fibrinogen, Chicken Fibrinogen, and of a Substantially Monodisperse Human Fibrinogen Fragment X. *Archives of Biochemistry and Biophysics,* Vol. 493, No. 2, (January 2010), pp. 157–168, ISSN 0003-9861

Carter, D.C. & Ho, J.X. (1994). Structure of Serum Albumin. *Advances in Protein Chemistry* Vol. 45, (1994), pp.155–203, ISSN 0065-3233

Chen, D., & Sun, B. (2000). New Tissue Engineering Material Copolymers of Derivatives of Cellulose and Lactide: Their Synthesis and Characterization. *Materials Science and Engineering: C,* Vol. 11, No. 1, (June 2000), pp. 57-60, ISSN 0928-4931

Chen, R., Huang, C., Ke, Q., He, C., Wang, H., & Mo, X. (2010). Preparation and Characterization of Coaxial Electrospun Thermoplastic Polyurethane/Collagen Compound Nanofibers for Tissue Engineering Applications. *Colloids and Surfaces B: Biointerfaces,* Vol. 79, No. 2, (September 2010), pp. 315–325, ISSN 0927-7765

Christenson, E.M., Anderson, J.M., & Hiltner A. (2004). Oxidative Mechanisms of Poly(carbonate urethane) and Poly(ether urethane) Biodegradation: In Vivo and In Vitro Correlations. *Journal of Biomedical Materials Research,* Vol. 70A, No. 2, (August 2004), pp. 245–255, ISSN 0021-9304

Colman, R.W., & Schmaier, A.H. (1997). Contact System: A Vascular Biology Modulator With Anticoagulant, Profibrinolytic, Antiadhesive, and Proinflammatory Attributes. *Blood,* Vol. 90, No. 10, (November 1997), pp 3819-3843, ISSN 0006-4971

Eberhart, R.C., Munro, M.S., Wiliams, G.B., Kulkarni, P.V., Shannon, W.A., Brink, B.E., & Fry, W.J. (1987). Albumin Adsorption and Retention on C18-alkyl-derivatized Polyurethane Vascular Grafts. *Artificial Organs,* Vol. 11, No. 5, (October 1987), pp. 375-382, ISSN 0160-564X

Edmunds, L.H.Jr . (1998). Inflammatory Response to Cardiopulmonary Bypass. *The Annals of Thoracic Surgery,* Vol. 66, No. 5, (November 1998), pp.12-16, ISSN 0003-4975

Förstermann, I. (2008). Oxidative Stress in Vascular Disease: Causes, Defense Mechanisms and Potential Therapies. *Nature Clinical Practice. Cardiovascular Medicine,* Vol. 5, No. 6, (June 2008), pp. 338-349, ISSN 1743-4297

Freeman, N. (2006). Analysis of the Structure at the Interface, In: *Proteins at Liquid – Solid Interfaces (Principle and practice),* P. Dejardin, (Ed.), pp. 75-104, Springer, ISBN-10 3-540-32657-X, Berlin, Germany

Fujimoto, K.L., Guan, J., Oshima, H., Sakai, T., & Wagner, W.R. (2007). In Vivo Evaluation of a Porous, Elastic, Biodegradable Patch for Reconstructive Cardiac Procedures. *The Annals of Thoracic Surgery,* Vol. 83, No. 2, (February 2007), pp. 648 –54, ISSN 0003-4975

Furie, B., & Furie, B.C. (2008). Mechanisms of Thrombus Formation. *The New England Journal of Medicine,* Vol. 359, (August 2008), pp. 938-949, ISSN 0028-4793

Gary, T. & Howard, G.T. (2002). Biodegradation of Polyurethane: A Review. *International Biodeterioration & Biodegradation,* Vol. 49, No. 4 (June 2002), pp. 245 – 252, ISSN 0964-8305.

Gettens, R.T., Bai, Z., & Gilbert, J.L. (2005). Quantification of The Kinetics and Thermodynamics of Protein Adsorption Using Atomic Force Microscopy. *Journal of Biomedical Materials Research. Part A,* Vol. 72, No.3, (March 2005), pp. 246-257, ISSN 1549-3296

Gisselfa, K., Edberg, B., & Flodin, P. (2002). Synthesis and Properties of Degradable Poly(urethane urea)s To Be Used for Ligament Reconstructions. *Biomacromolecules,* Vol. 3, No. 5, (September-October 2002), pp. 951-958, ISSN 1525-7797

Gong, F., Lu, Y., Guo, H., Cheng, S., & Gao, Y. (2010). Hyaluronan Immobilized Polyurethane as a Blood Contacting Material. *International Journal of Polymer Science,* Vol. 2010, Article ID 807935, (March 2010), pp.1-8, ISSN 1687-9422

Göpferich, A. (1996). Polymer Degradation and Erosion: Mechanisms and Applications. *European Journal of Pharmaceutics and Biopharmaceutics,* Vol. 42, No. 1, (1996), pp. 1–11, ISSN 0939-6411

Gray, J.J. (2004). The Interaction of Proteins With Solid Surfaces. *Current Opinion in Structural Biology,* Vol. 14, No.1, (February 2004), pp. 110–115, ISSN 0959-440X

Guelcher, S., Srinivasan, A., Hafeman, A., Gallagher, K., Doctor, J., Khetan, S., Mcbride, S., & Hollinger, J. (2007). Synthesis, In Vitro Degradation, and Mechanical Properties of Two-Component Poly(Ester Urethane)Urea Scaffolds: Effects of Water and Polyol Composition. *Tissue Engineering,* Vol.13, No. 9, (September 2007), pp. 2321-2333, ISSN 1076-3279

Guelcher, S.A. (2008). Biodegradable Poliurethanes: Syntesis and Applications in Regenerative Medicine. *Tissue Engineering,* Vol. 14, No.1, (March 2008), pp. 3-17, ISSN 2152-4947

Gutowska, A., Jeong, B., & Jasionowski, M. (2001). Injectable Gels for Tissue Engineering. *The Anatomical Record,* Vol. 263, No.4, (August 2001), pp. 342–349, ISSN 0003-276X

Hsu, H.S., Kao, Y.C., & Lin, Y.C. (2004). Enhanced Biocompatibility in Biostable Poly(carbonate)urethane. *Macromolecular Bioscience,* Vol.4, No. 4, (April 2004), pp. 464–470, ISSN 1616-5187

Huang, J., & Xu, W. (2010). Zwitterionic Monomer Graft Copolymerization Onto Polyurethane Surface Through a PEG Spacer. *Applied Surface Science,* Vol. 256, No. 12, (April 2010), pp. 3921–3927, ISSN 0169-4332

Hwang, S., & Meyerhoff, M.E. (2008). Polyurethane With Tethered Copper(II)ecyclen Complex: Preparation, Characterization and Catalytic Generation of Nitric Oxide from S-nitrosothiols. *Biomaterials,* Vol. 29, No. 16, (June 2008), pp. 2443-2452, ISSN 1552-4973

Jelinek, M., Kocourek, T., Remsa, J., Mikšovský, J., Zemek, J., Smetana, K. Jr., Dvořánková, B., & Luxbacher, T. (2010). Diamond/Graphite Content and Biocompatibility of DLC Films Fabricated by PLD. *Applied Physics A, Materials Science & Processing,* Vol. 110, No.4, (June 2010), pp.579-583, ISSN 0947-8396

Jiang, W.W., Su, S.H., Eberhart, R.C., & Tang, L. (2007). Phagocyte Responses to Degradable Polymers. *Journal of Biomedical Materials Research. Part A,* Vol. 82, No. 2, (August 2007), pp. 492–497, ISSN 1549-3296

Jordan, S.W. & Chaikof, E.L. (2007). Novel Thromboresistant Materials. *Journal of Vascular Surgery,* Vol. 45, Suppl A, (June 2007), pp. 104A-115A, ISSN 0741-5214

Jun, H.W., Taite, L.J., & West, J.L. (2005). Nitric Oxide-Producing Polyurethanes. *Biomacromolecules*, Vol. 6, No. 2, (March-April 2005), pp. 838-844, ISSN 1525-7797

Kang, S., Hoek, E.M.V., Choi, H., & Shin, H. (2006). Effect of Membrane Surface Properties During the Fast Evaluation of Cell Attachment. *Separation Science and Technology*, Vol. 41, No. 7, (2006), pp. 1475–1487, ISSN 0149-6395

Kavlock, K.D., Pechar, T.W., Hollinger, J.O., Guelcher, S.A., & Goldstein, A.S. (2007). Synthesis and Characterization of Segmented Poly(esterurethane urea) Elastomers for Bone Tissue Engineering. *Acta Biomaterialia,* Vol. 3, No.4, (July 2007), pp. 475–484, ISSN 1742-7061

Keselowsky, B.G., Collard, D.M., & García, A.J. (2003). Surface Chemistry Modulates Fibronectin Conformation and Directs Integrin Binding and Specificity to Control Cell Adhesion. *Journal of Biomedical Materials Research Part A*, Vol. 66A, No. 2, (August 2003), pp. 247–259, ISSN 1549-3296

Khorasani, M.T., MoemenBellah, S., Mirzadeh, H., & Sadatnia, B. (2006). Effect of Surface Charge and Hydrophobicity of Polyurethanes and Silicone Rubbers on L929 Cells Response. *Colloids and Surfaces B: Biointerfaces,* Vol. 51, No. 2, (August 2006), pp.112–119, ISSN 0927-7765

Kirkpatrick, C.J., Bittinger, F., Wagner, M., Köhler, H., van Kooten, T.G., Klein, C.L., & Otto, M. (1998). Current trends in Biocompatibility Testing. *Proceedings of The Institution of Mechanical Engineers. Part H, Journal of Engineering in Medicine*, Vol. 212, No. 2, (1998), pp. 75–84, ISSN 0954-4119

Kouoh, F., Gressier, B., Luyckx, M., Brunet, C., Dine T., Cazin M., & Cazin, J.C. (1999). Antioxidant Properties of Albumin: Effect on Oxidative Metabolism of Human Neutrophil Granulocytes. *II Farmaco,* Vol. 54, No. 10, (October 1999), pp. 695-699, ISSN 0014-827X

Krieter, D.H., Steinke, J., Kerkhoff, M., Fink, E., Lemke, H.D., Zingler, C., Müller, G.A., & Schuff-Werner, P. (2005). Contact Activation in Low-Density Lipoprotein Apheresis Systems. *Artificial Organs*, Vol. 29, No.1, (January 2005), pp. 47-52, ISSN 0160-564X

Kurrat, R., Prenosil, J. E., & Ramsden, J. J. (1997). Kinetics of Human and Bovine Serum Albumin Adsorption at Silica–Titania Surfaces. *Journal of Colloid and Interface science,* Vol. 185, No. 1 (January 1997), pp. 1-8, ISSN 0021-9797

Latour, R.A.Jr. (2008) Biomaterials: Protein–Surface Interactions. In: Encyclopedia of Biomaterials and Biomedical Engineering, Vol. 1, Wnek G.E., Bowlin G.L. (Ed.), pp. 270-285, Informa Healthcare, ISBN-10 1-4200-7953-0, New York, USA.

Lee, H., Cha, M.K., Kim, I.H. (2000). Activation of Thiol-Dependent Antioxidant Activity of Human Serum Albumin by Alkaline pH is Due to the B-like Conformational Change. *Archives of Biochemistry and Biophysics,* Vol. 380, No. 2, (August 2000), pp. 309-318, ISSN 0570-6963

Lee, J.S., Cho, Y.S, Lee, J.W, Kim, H.J., Pyun, D.J., Park, M.H., Yoon, T.R., Lee, H.J. & Kuroyanagy, Y., (2001). Preparation of Wound Dressing Using Hydrogel Polyurethane Foam. *Trends in Biomaterials & Artificial Organs,* Vol. 15, No. 1, (July 2001), pp. 4-6, ISSN 0971-1198

Lelah, M.D., & Cooper, J.L. (1987) *Polyurethanes in Medicine*, CRC Press, ISBN 0849363071, Boca Raton, Florida, U.S.A.

Luxbacher, T. (2006). Electrokinetic Characterization of Flat Sheet Membranes by Streaming Current Measurement. *Desalination*, Vol. 199, (March 2006), pp. 376–377, ISSN 0011-9164

Macocinschi, D., Filip, D., Vlad, S., Cristea, M., & Butnaru, M. (2009). Segmented Biopolyurethanes for Medical Applications. *Journal of Materials Science: Materials in Medicine*, Vol. 20, No. 8, (August 2009), pp. 1659–1668, ISSN 0957-4530

Makala, U., Wood, L., Ohmanb, D.E., & Wynnea, K.J. (2006). Polyurethane Biocidal Polymeric Surface Modifiers. *Biomaterials*, Vol. 27, No. 8, (March 2006), pp. 1316–1326, ISSN 1552-4973

Marconi, W., Galloppa, A., Martellini, A., & Piozzi, A. (1996). New Polyurethane Compositions Able to Bond High Amounts of Both Albumin and Heparin. II: Copolymers and Polymer Blends. *Biomaterials*, Vol. 17, No. 18, (September 1996), pp.1795-1802, ISSN 0142-9612

Marnett, L.J. (2002). Oxy radicals, Lipid Peroxidation and DNA Damage. *Toxicology*, Vol. 181-182, (December 2002), pp. 219-222, ISSN 0300-483X

Michelsen, A.E., Santi, C., Holme, R., Lord, S.T., Simpson-Haidaris, P.J., Solum, N.O., Pedersen, T.M., & Brosstad, F. (2000). The Charge-Heterogeneity of Human Fibrinogen as Investigated by 2D Electrophoresis. *Thrombosis Research*, Vol. 100, No.6, (December 2000), pp. 529-535, ISSN 0049-3848

Morais, J.M., Papadimitrakopoulos, F., & Burgess, D.J. (2010). Biomaterials/Tissue Interactions: Possible Solutions to Overcome Foreign Body Response, *The AAPS Journal*, June, Vol. 12, No. 2, (June 2010), pp. 188-196, ISSN 1550-7416

Noinville, S., & Revault, M. (2006). Conformations of Proteins Adsorbed at Liquid–Solid Interfaces. In *Proteins at Liquid – Solid Interfaces (Principle and Practice),* P. Dejardin, (Ed.), pp. 119-150, Springer, ISBN-10 3-540-32657-X, Berlin, Germany

Oral, E., Rowella, S.L., & Muratoglu, O.K. (2006). The effect of α-Tocopherol on The Oxidation and Free Radical Decay in Irradiated UHMWPE. *Biomaterials*, Vol. 27, No. 32, (November 2006), pp. 5580–5587, ISSN 0142-9612

Ostuni, E., Chapman, R.G., Holmlin, R.E., Takayama, S., & Whitesides, G.M. (2001). A Survey of Structure –Property Relationships of Surfaces that Resist the Adsorption of Protein. *Langmuir: the ACS Journal of Surfaces and Colloids,* Vol. 17, No. 18, (September 2001), pp. 5605–5620, ISSN 0743-7463

Paradis, V., Kollinger, M., Fabre, M., Holstege, A., Poynard, T., & Bedossa, P. (1997). In Situ Detection of Lipid Peroxidation by-Products in Chronic Liver Diseases. *Hepatology*, Vol. 26, No. 1, (July 1997), pp. 135–142, ISSN 0270-9139

Parveen, N., Khan, A.A., Baskar, S., Habeeb, M.A., Babu, R., Abraham, S., Yoshioka, H., Mori, Y., & Mohammed, H.C. (2008). Intraperitoneal Transplantation of Hepatocytes Embedded in Thermoreversible Gelation Polymer (Mebiol Gel) in Acute Liver Failure Rat Model. *Hepatitis Monthly*, Vol. 8, No.4, (2008), pp. 275-280, ISSN 1735-143X

Pompe, T., Renner, L., & Werner, C. (2006). Fibronectin at Polymer Surfaces with Graduated Characteristics. In *Proteins at Liquid – Solid Interfaces (Principle and Practice),* P. Dejardin, (Ed), pp. 175-198, Springer, ISBN-10 3-540-32657-X, Berlin, Germany

Rahman, I., van Schadewijk, A.A.M., Crowther, A.J., Hiemstra, P.S., Stolk, J., MacNee, W., & De Boer, W.I. (2002). 4-Hydroxy-2-Nonenal, a Specific Lipid Peroxidation Product, Is Elevated in Lungs of Patients with Chronic Obstructive Pulmonary Disease. American Journal of Respiratory and Critical Care Medicine, Vol. 166, No. 4, (August 2002), pp. 490-495, ISSN 1073-449X

Randrasana, S., Baquey, C.H., Delmond, B., Daudé, G., & Filliatre, C. (1994). Polyurethanes Grafted by Pendent Groups With Different Sizes and Functionality. *Clinical Materials,* Vol. 15, No. 4, (1994), pp. 287-292, ISSN 0267-6605

Raschip, I.E., Vasile, C., & Macocinschi, D. (2009). Compatibility and Biocompatibility Study of New HPC/PU Blends. *Polymers International,* Vol. 58, No.1, (January 2009), pp. 4–16, ISSN 0959-8103

Ratner, B.D., & Bryant, S.J. (2004). Biomaterials: Where We Have Been and Where We Are Going. *AnnualReview of Biomedical Engineering,* Vol. 6, (2004), pp. 41–75, ISSN 1523-9829

Rezwan, K., Meiera, L.P., & Gauckler, L.J. (2005). Lysozyme and Bovine Serum Albumin Adsorption on Uncoated Silica and AlOOH-Coated Silica Particles: The Influence of Positively and Negatively Charged Oxide Surface Coatings. *Biomaterials,* Vol. 26, No. 21 (July 2005), pp. 4351–4357, ISSN 0142-9612

Sanders, J.E., Lamont, S.E., Karchin, A., Golledge, S.L., & Ratner, B.D. (2005). Fibro-Porous Meshes Made From Polyurethane Micro-Fibers: Effects of Surface Charge on Tissue Response. *Biomaterials,* Vol. 26, No. 7, (March 2005), pp. 813–818, ISSN 0142-9612

Sartori, S., Rechichi, A., Vozzi, G., D'Acunto, M., Heine, E., Giusti, P., & Ciardelli, G. (2008). Surface Modification of A Synthetic Polyurethane By Plasma Glow Discharge: Preparation and Characterization of Bioactive Monolayers. *Reactive & Functional Polymers,* Vol. 68, No. 3 (March 2008), pp. 809–821, ISSN 1381-5148

Sayre, L.M., Zelasko, D.A., Harris, P.L., Perry, G., Alomon, R.G., & Smith, M.A. (1997). 4-Hydroxynonenal-Derived Advanced Lipid Peroxidation End Products Are Increased in Alzheimer's Disease. *Journal of Neurochemistry,* Vol. 68, No. 5, (May 1997), pp. 2092-2097, ISSN 0022-3042

Scott, E.A., & Elbert, D.L. (2007). Mass Spectrometric Mapping of Fibrinogen Conformations at Poly(Ethylene Terephthalate) Interfaces. *Biomaterials,* Vol. 28, No. 27, (September 2007), pp. 3904–3917, ISSN 0142-9612

Shastri, V.P. (2003). Non-Degradable Biocompatible Polymers In Medicine: Past, Present and Future. *Current Pharmaceutical Biotechnology,* Vol. 4, No. 5, (October 2003), pp. 331–337, ISSN 1389-2010

Shen, M., & Horbett, T.A. (2001). The Effects of Surface Chemistry and Adsorbed Proteins on Monocyte /Macrophage Adhesion to Chemically Modified Polystyrene Surfaces. *Journal of Biomedical Materials Research,* Vol. 57, No. 3, (December 2001), pp. 336–345, ISSN 0021-9304

Sieminski, A.L., & Gooch, K.J. (2000). Biomaterial–Microvasculature Interactions. *Biomaterials,* Vol. 21, No. 22, (November 2000), pp. 2232–2241, ISSN 0142-9612

Siepe, M., Giraud, M.N., Liljensten, E., Nydegger, U., Menasche, P., Carrel, T., & Tevaearai, HT. (2007). Construction of Skeletal Myoblast-Based Polyurethane Scaffolds for

Myocardial Repair. *Artificial Organs,* Vol. 31, No.6, (June 2007), pp. 425–433, ISSN 0160-564X

Song, M., Xia, H.S., Yao, K.J., & Hourston, D.J. (2005). A Study on Phase Morphology and Surface Properties of Polyurethane/Organoclay Nanocomposite. *European Polymer Journal,* Vol. 41, No. 2, (April 2005), pp. 259–266, ISSN 0014-3057

Stachelek, S.J., Alferiev, I., Choi, H., Chan, C.W., Zubiate, B., Sacks, Composto, M.R., Chen, I.W., & Levy, R.J. (2006). Prevention of Oxidative Degradation of Polyurethane by Covalent Attachment of Di-Tert-Butylphenol Residues. *Journal of Biomedical Materials Research. Part A,* Vol. 78, No. 4, (September 2006), pp. 653-661, ISSN 1549-3296

Sutherland, K., Mahoney, J.R., Coury, A.J, & Eatonil, J.W. (1993). Degradation of Biomaterials by Phagocyte-Derived Oxidants. *The Journal of Clinical Investigation,* Vol. 92, No.5, (November 1993), pp. 2360-2367, ISSN 0021-9738

Tribble, D.L., Aw, T.Y., & Jones, D.P. (1987). The Pathophysiological Significance of Lipid Peroxidation in Oxidative Cell Injury. *Hepatology,* Vol. 7, No. 2, (March-April 1987), pp. 377–386, ISSN 0270-9139

Tsapikouni, T.S., & Missirlis, Y.F. (2007). pH and Ionic Strength Effect on Single Fibrinogen Molecule Adsorption on Mica Studied With AFM. *Colloids and Surfaces B: Biointerfaces,* Vol. 57, No. 1, (May 2007), pp. 89–96, ISSN 0927-7765

Valenta, C., Auner, B.G. (2004). The Use of Polymers for Dermal and Transdermal Delivery. *European Journal of Pharmaceutics and Biopharmaceutics,*Vol. 58, No. 2, (September 2004), pp. 279-289, ISSN 0939-6411

Van Tassel, P.R. (2006). Protein Adsorption Kinetics: Influence of Substrate Electric Potential. In *Proteins at Liquid – Solid Interfaces (Principle and Practice),* P. Dejardin, (Ed.), pp. 1-22 , Springer, ISBN-10 3-540-32657-X, Berlin, Germany

Verma, S., & Marsden, P.A. (2005). Nitric Oxide-Eluting Polyurethanes -- Vascular Grafts of the Future? *The New England Journal of Medicine,* Vol. 353, No. 7, (August 2005), pp. 730-731, ISSN 0028-4793

Vlad, S., Butnaru, M., Filip, D., Macocinschi, D., Nistor, A., Gradinaru L.M., & Ciobanu, C. (2010). Polyetherurethane Membranes Modified with Renewable Resource as a Potential Candidate for Biomedical Applications. *Digest Journal of Nanomaterials and Biostructures,* Vol. 5, No. 4, (October-December 2010), pp. 1089-1100, ISSN 1842 - 3582

Wahl, S.M., Wong, H., & McCartney-Francis, N. (1989). Role of Growth Factors in Inflammation and Repair. *Journal of Cellular Biochemistry,* Vol. 40, No. 2, (June 1989), pp. 193–199, ISSN 0730-2312

Wan, L.S., Xu, Z.K., & Huang, X.J. (2006). Approaches to Protein Resistance on The Polyacrylonitrile-Based Membrane Surface: An Overview. In: *Proteins at Liquid – Solid Interfaces (Principle and Practice),* P. Dejardin, (Ed.), pp. 245-270, Springer, ISBN-10 3-540-32657-X, Berlin, Germany

Wattamwar, P.P., Mo, Y., Wan, W., Palli, R., Zhang, Q., & Dziubla, T.D. (2010). Antioxidant Activity of Degradable Polymer Poly(trolox ester) to Suppress Oxidative Stress Injury in The Cells. *Advanced Functional Materials,* Vol. 20, No.1, (January 2010), pp. 147–154, ISSN 1616-301X

Wertz, C.F., & Santore, M.M. (2001). Effect of Surface Hydrophobicity on Adsorption and Relaxation Kinetics of Albumin and Fibrinogen: Single-Species and Competitive Behavior. *Langmuir: The ACS Journal of Surfaces and Colloids,* Vol. 17, No. 10, (2001), pp. 3006-3016, ISSN 0743-7463

Williams, D.F. (2008). On the Mechanisms of Biocompatibility. *Biomaterials,* Vol. 29, No. 20 (July 2008), pp. 2941–2953, ISSN 0142-9612

Wilson, C.J., Clegg, R.E., Leavesley, D.I., & Pearcy, M.J. (2005). Mediation of Biomaterial–Cell Interactions by Adsorbed. Proteins: A Review. *Tissue Engineering,*Vol. 11, No. 1/2, (January-February 2005), pp.1-18, ISSN 1076-3279

Yagi, K. (1987). Lipid Peroxides and Human Diseases. *Chemistry and Physics of Lipids,* Vol. 45, No. 2-4, (November-December 1987), pp. 337-351, ISSN 0009-3084

Yao, C., Li, X., Neohb, K.G, Shib, Z., & Kangb, E.T. (2008). Surface Modification and Antibacterial Activity of Electrospun Polyurethane Fibrous Membranes with Quaternary Ammonium Moieties. *Journal of Membrane Science,* Vol. 320, No. 1-2, (July 2008), pp. 259–267, ISSN 0376-7388

Yaroshchuk, A., & Luxbacher, T. (2010). Interpretation of Electrokinetic Measurements with Porous Films: Role of Electric Conductance and Streaming Current within Porous Structure. *Langmuir: the ACS Journal of Surfaces and Colloids,* Vol. 26, No. 13, (July 2010), pp. 10882–10889, ISSN 0743-7463

Zhuo, R, Siedlecki, C.A., & Vogler, E.A. (2006). Autoactivation of Blood Factor XII at Hydrophilic and Hydrophobic surfaces. *Biomaterials,* Vol. 27, No. 24, (April 2006), pp. 4325-4332, ISSN 0142-9612

Zia, K.M., Barikanib, M., Zuberc, M., Bhattia, I.A., & Barmarb, M. (2009). Surface Characteristics of Polyurethane Elastomers Based on Chitin/1,4-Butane Diol Blends. *International Journal of Biological Macromolecules,* Vol. 44, No.2, (March 2009), pp. 182–185, ISSN 0141-8130

Zuo, D.Y., Tao, Y.Z., Chen, Y.B., & Xu, W.L. (2009). Preparation and Characterization of Blend Membranes of Polyurethane and Superfine Chitosan Powder. *Polymer Bulletin,* Vol. 62, No.5, (January 2009), pp. 713–725, ISSN 0170-0839

Chapter 11

HTPB-Polyurethane: A Versatile Fuel Binder for Composite Solid Propellant

Abhay K. Mahanta and Devendra D. Pathak

Additional information is available at the end of the chapter

http://dx.doi.org/10.5772/47995

1. Introduction

One of the most promising applications of polyurethane (PU) polymers is as fuel-*cum*-binder material in composite solid propellant. Since the last two decades, PU filled with oxidizer and metallic fuel is being widely used for rockets propulsion. Ariane boosters, shuttles Apogee motors, Peacekeeper (also called the MX-Missile Experimental) missile, Indian Augmented Satellite Launch Vehicle(ASLV) and Polar Satellite Launch Vehicle (PSLV) boosters are some of the motors that are fuelled by PU propellant. PU composite propellant (PCP) is a heterogeneous mixture of polymeric binder, inorganic oxidizer and metallic fuel as the major ingredients. It can be classified as a highly filled PU system in which the three dimensional elastomeric matrix binds the oxidizer and metallic fuel to form a rubbery material. It imparts necessary mechanical properties to the propellant grain to maintain its structural integrity. A PU propellant grain should have sufficient tensile strength and elongation to withstand various types of stresses experienced during handling and transportation, thermal cycling, sudden pressurization on ignition, and acceleration load during flight of the rocket motor. A tensile strength of approximately 7-8 kgf/cm^2, an elongation of 40-50 % and initial modulus of 40-50 kgf/cm^2 are reasonable for a typical case bonded rocket motor (Manjari et al., 1993). The PU binder accounts to 10-15 % of the composite propellant, and usually consists of three components: (1) a prepolymer (polyol), (2) an isocyanate curator, and (3) a chain extender (butan-1,4-diol) and cross-linking agent (trimethylol propane). The most commonly used polyol in recent time is the Hydroxyl Terminated Polybutadiene (HTPB). This liquid prepolymer has excellent physical properties such as low glass transition temperature, high tensile and tear strength, and good chemical resistance (Eroglu, 1998). The hydrocarbon nature of HTPB (98.6%) along with low viscosity (5000 mPas at 30 °C) and low specific gravity (0.90 g/cm^3), makes it a promising fuel binder for PU propellant. It is capable of taking solid loading up to 86-88% without sacrificing the ease of processibility (Muthiah et al., 1992). In addition, it is also a major reducing agent and gas producing fuel. It is physically and chemically compatible with the conventional oxidizers and other ingredients at normal storage conditions. As it contains mostly carbon and hydrogen, during combustion, it is decomposed to give large volume of stable molecules like carbon monoxide, carbon dioxide, and water vapours increasing the specific impulse of the rocket motor. Additionally, PU obtained from HTPB offers many advantages over conventional polyether and polyester based urethane systems. Properties exhibited by polyurethanes(PUs) prepared from HTPB include (a) excellent hydrolytic stability, (b) low water absorption, (c) excellent low temperature flexibility, (d) high compatibility with fillers and extenders, and (e) formulation flexibility (Sadeghi et al., 2006). Of late, there is a growing demand of segmented HTPB PUs as these PUs have a unique combination of toughness, durability and flexibility, biocompatibility and biostability that makes them suitable materials for use in a diverse range of biomedical applications (Poussard et al., 2004). HTPB based pervaporation membrane technology is the current wave of innovation. It has introduced a new dimension to PU elastomeric technology.

The polymer chemo-rheology and thermo-oxidative degradation are the two relevant key areas of interest, where in-depth knowledge is essential for the effective performance assessment of PU propellants. Chemo-rhelogy is related with the PU processiblity, whereas thermo-oxidation is related to the stability and combustion performance. The information of change on viscosity during the curing process is critical in modelling the PU flow behaviour. Though extensive works have been carried out on this topic in the last decade (Muthiah et al., 1992, Lakshmi & Athithan, 1999, Singh et al., 2002 & Mahanta et al., 2007), it is still a fascinating research area at present. The thermal decomposition of HTPB has been studied exclusively in inert atmosphere (Panicker & Ninan, 1997). However, thermo-oxidative degradation in air, which is the most relevant in view of combustion of the polymer, has not been studied thoroughly. Additionally, the HTPB prepolymer being the decisive component in HTPB PUs, characterization of this polymer (HTPB) at macro as well as micro levels has been of paramount importance in last decade. Two types of HTPB prepolymer are currently in use: i.e., free radical HTPB and anionic HTPB. The free radical grade HTPB is widely used in composite PU propellants because of its low cost and wide availability. The current chapter is focused on prepolymer characterization, rheology, and oxidative degradation of the polymer and the PU systems.

2. Experimental

2.1. Analytical equipments

NMR measurements: The NMR spectra were recorded on a Bruker 800 MHz NMR spectrometer. The HTPB samples (10 % (w/v) for the ^{1}H NMR and 30 % (w/v) for the $^{13}C\{^{1}H\}$ NMR analysis) were recorded in $CDCl_3$ at room temperature. The ^{1}H NMR acquisition parameters were: spectral width = 16 ppm, acquisition time =2 s, relaxation delay = 1 s, pulse width = 90 °, and number of scans =1000. Similarly, $^{13}C\{^{1}H\}$ NMR spectra were recorded using spectral width = 220 ppm, acquisition time = 2 s, relaxation delay = 10 s, pulse width =

90 °, and number of scans = 300. HMQC spectra of HTPB samples (30 % (w/v) in $CDCl_3$ at room temperature) were recorded on a Bruker 500 MHz with a 5 mm inverse Z-gradient probe. Spectral widths: F2 (^1H)=8000 Hz, F1(^{13}C)=27500 Hz. Time domains : (^1H)=1024 and (^{13}C) =515, acquisition time (^1H) =0.23s, delay (^1H) =2s. In processing, the FID was zero-filled to 32 K data points and the resulting 32 K time domain was Fourier transformed. Additionally, Gaussian apodization was also applied in both ^1H and ^{13}C domains. **Viscosity measurements:** A Brookfield HADV-II+ programmable rotational type viscometer equipped with a motorized stand (helipath stand) was used to perform isothermal viscosity measurement at different temperatures. The temperature was controlled by a thermostatic temperature control bath (Brookfield). The temperature control accuracy was ± 0.5 °C. Polymer samples were sheared at different shear rate (rpm). The spindle used for binder slurry was AB-4, whereas for propellant slurry, T-E was used. For each experiment, data was collected after one complete revolution. For each successive revolution, total 10 readings, each at an interval of one second were recorded at the set rotational speed by using Wingather Software. The average viscosity value was calculated and used for data analysis and modelling. **DSC experiments**: Mettler FP-900 thermal analysis system equipped with FP-85 standard cell and FP90 central processor was used for DSC measurement. The heat flow and temperature calibration of DSC were carried out using pure indium metal as per the procedure recommended by the manufacturer (ΔH = 26.7 J/g, MP = 158.9 °C). All experiments were carried out in an air atmosphere at different heating rates, ranging from 2-15 °C/min. Aluminum sample pans (40 μL) were used for the DSC experiments. Almost constant sample mass of 5 ± 1 mg was used. **Tensile properties**: The tensile stress-strain measurements were performed at room temperature, using samples previously kept at 23±2 °C and relative humidity of 50± 5% for 48 hrs, according to ASTM D 618. Elastomeric test specimens were punched from the cured slab using a die prepared in accordance with ASTM D 412-68. Tensile testing was performed in an Instron Universal Testing Machine (UTM) using dumb-bell shaped specimens of cured PUs as well as propellants. A 100 kg load was applied at a crosshead speed of 50 mm/min. Hardness was measured by a Shore A Durometer as per the standard procedure.

2.2. Synthesis of PUs

2.2.1. Unfilled PUs: PU-I and PU-II

The basic compositions that were studied in the present work are shown in Table 1. The binder system studied consists of PU formed by reacting mixture of alcohols [(HTPB, OH value = 42 mg KOH/g), Butanediol (BDO, OH value = 1232 mg KOH/g) as chain extender and trimethylol propane (TMP, OH value = 1227 mg KOH/g) as cross linking agent] with toluene diisocyanate (TDI, purity > 99 percent and a mixture of 2, 4 and 2, 6-isomers in 80:20 ratio). The BDO and TMP were mixed in a fixed ratio (2:1) and dried under vacuum to reduce the moisture content (< than 0.25%) of the mixture. The mixture thus obtained had the hydroxyl value of 1242 mg of KOH/g.

Polyurethane system	Binder component	Fillers (%)	Hard segment content (% w/w)*
PU-I	HTPB/TDI	---	4.34
PU-II	HTPB/TDI/ (BDO +TMP)	---	7.25/7.34/7.43/7.52/7.61
PU-IIp	HTPB/TDI/ (BDO +TMP)	AP- 68, Al -18,	7.25/7.34/7.43/7.52/7.61

*Hard segment content= $\{[w_{TDI} + w_{BDO+TMP}]/w_{total}\} \times 100, w = weight.$

Table 1. The basic composition of the one step PUs.

The PUs were prepared in bulk by one step procedure. Mixing was carried out in a pilot mixer with facility for circulation of hot/cold water around the mixer jacket. The HTPB and BDO-TMP mixture were taken in the pilot mixer and stirred for 10 minutes. The calculated amount of TDI was added to the mixer, and the contents were stirred for 20 minutes at 40 ± 1 °C. The binder slurry was cast in to a Teflon coated mould and cured at 60 °C for 3 days.

2.2.2. Filled PUs (propellant): PU-IIp

The basic propellant composition that uses 68% ammonium perchlorate (AP) and 18% aluminum (Al) powder was taken up for study. AP (with purity > 99%) was used in bimodal distribution (3:1) having average particle size 280 μm and 49 μm, respectively. Particle size of AP and Al powder (mean diameter = 33.51 μm) were measured by a CILAS Particle Size Analyzer-1180 model. Dioctyl adipate (saponification value = 300 mg KOH/g) was used as a plasticizer. The mixing was carried out in two phases. In the first phase, all the ingredients, except the curing agent, were premixed thoroughly for about 3 h at 38 ± 2 °C. Hot water was circulated through the jacket of the mixer bowl to keep a constant temperature throughout the mixing cycle. A homogeneous test of the slurry was carried out after completion of the premix to confirm the uniform dispersion of AP and Al powder. In the second phase of mixing, a calculated amount of curing agent, *i.e.* toluene diisocyanate (TDI) was added to the premixed slurry, and further mixed for 40 minutes at 40 ±1 °C. The propellant slurry was cast in to the Teflon coated mould and cured at 60 °C for 5 days.

3. Results and discussion

3.1. Prepolymer characterization by high field NMR.

The substrate polymer (HTPB) is the key component that affects the elastomeric properties of PUs. Knowledge on the polymer structure and composition is essential for synthesis of PUs with required properties and understanding the various advantages, the polymer can offer. We have used the high field 1D and 2D NMR techniques for characterization of HTPB prepolymers. Analysis of microstructure and sequence distribution of monomer units can be discerned from the analysis of quantitative $^{1}H/^{13}C$

NMR spectra. Although ^{13}C NMR spectroscopy is good in terms of a wider range of chemical shifts and thus offering less possibility of overlapping peaks, problems associated with questionable assignments occasionally arise from steric-sensitive environments in the carbon skeleton. Additionally, the Nuclear Overhauser Enhancement (NOE) of different types of carbon is usually not equal and the wide spin-lattice relaxation time (t_1) range makes quantitative measurements of carbon signals difficult. A combination of NMR techniques such as ^1H, $^{13}C\{^1H\}$, $^{13}C\{^1H\}$-DEPT (Distortionless Enhancement By Polarization Transfer) and $^1H/^{13}C$-HMQC (Hetero-nuclear Multiple Quantum Coherence) permit assignments of all ^1H and ^{13}C resonance peaks. To our knowledge, hitherto the actual physical characteristics of the HTPB are not precisely known, particularly its absolute number-average molecular weight ($\bar{M}_n$). In the current work, we have examined the ^1H and ^{13}C NMR spectra of HTPB in order to precisely determine its number-average degree of polymerization ($\overline{DP}_n$), and thus, $\bar{M}_n$ of the polymer. A typical $^{13}C\{^1H\}$ NMR spectrum (200MHz, $CDCl_3$) of free radical HTPB prepolymer is shown in Fig.1. For convenience, resonances in the spectrum can be divided in to three distinct regions, *i.e.* (a) an olefinic region: δ 113-144, (b) a carbon bearing hydroxyl end group region: δ 56 – 65, and (c) an aliphatic region: δ 24-44. However, due to complex nature of the prepolymer, a complete assignment of all signals was not possible. The, methine and methylene carbons were distinguished by using the DEPT technique. The $^{13}C\{^1H\}$-DEPT spectrum of the polymer, recorded in $CDCl_3$, is depicted in Fig.2.

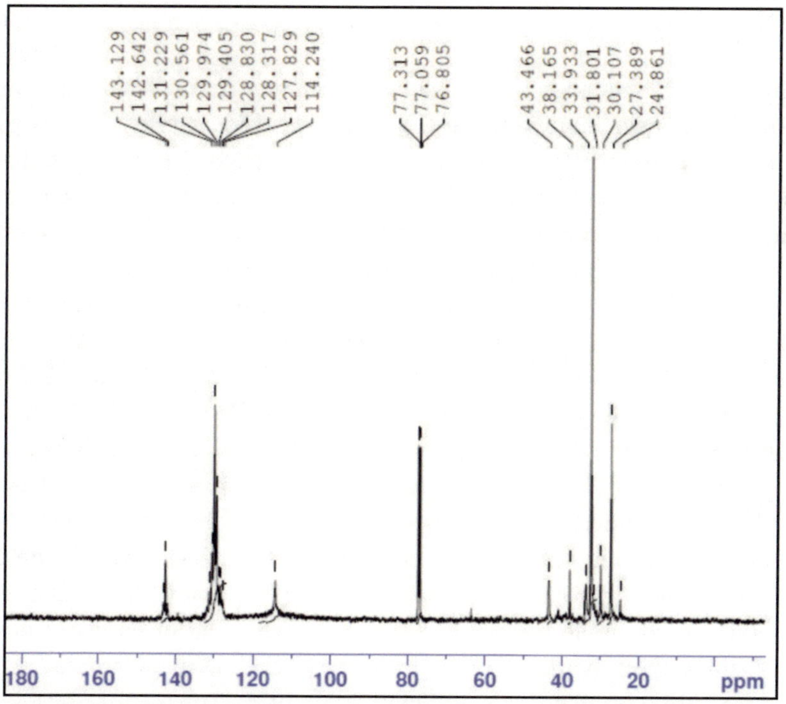

Figure 1. $^{13}C\{^1H\}$ NMR ($CDCl_3$, 200 MHz) spectrum of free radical HTPB prepolymer.

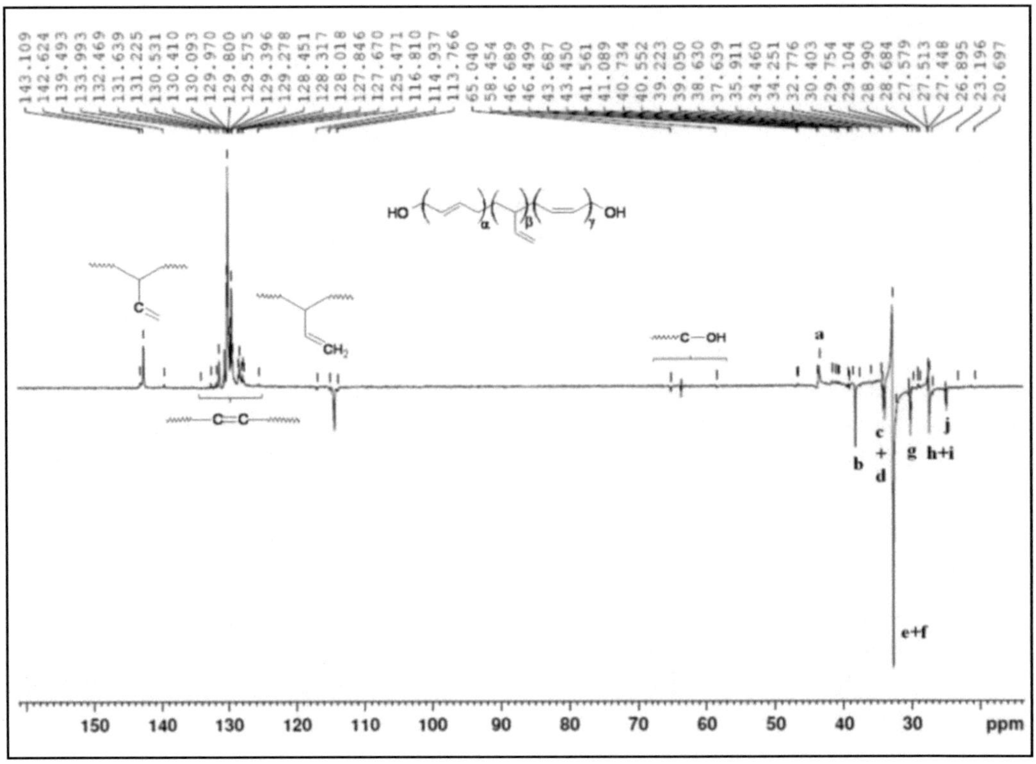

Figure 2. $^{13}C\{^{1}H\}$DEPT-135 spectrum of free radical HTPB prepolymer.

The delay in the DEPT sequence was chosen in such a way that methine carbons appeared as positive peak, whereas both methyl and methylene carbons appeared as negative peak. In the olefinic region, the DEPT spectrum showed a set of positive signals in the range of δ 142-144, that corresponds to methine (-CH=) carbons, whereas a set of negative signal at δ 113-115, corresponds to the methylenic ($=CH_2$) carbons of *vinyl*-1,2- unit. The fine splitting of the signals is due to the tacticity of the monomer units. A set of positive signals in the range of δ 125 – 134 was ascribed to the compositional splitting of the two olefinic carbons (-CH=CH-) in central *cis*-1,4- or *trans*-1,4- unit, present in different combination of homotriads, heterotriads, and symmetric and non-symmetric isolated triads (Frankland et al., 1991). A total of thirteen signals were observed in the olefinic double bond region, *i.e.*δ 127-132. (Fig.3). Each of the resonance line has been assigned to the methine carbon of 1,4-unit in the possible set of three consecutive monomer units (*cis*-1,4-; *trans*-1,4-; and *vinyl*-1,2-unit). When surrounded by 1,2-units, the methine carbon of 1,4-unit would have different chemical shift due to their different distance from the *vinyl*-1,2-side group. The chemical shifts of methine carbon signals in various possible triad sequences were calculated by a known method and then, compared with that of observed one to assign the signals. Besides, the assignment of the triad resonances was made based on the values reported in literature for polybutadiene (Elgert et al., 1975). The results, thus, obtained are summarized in Table 2.

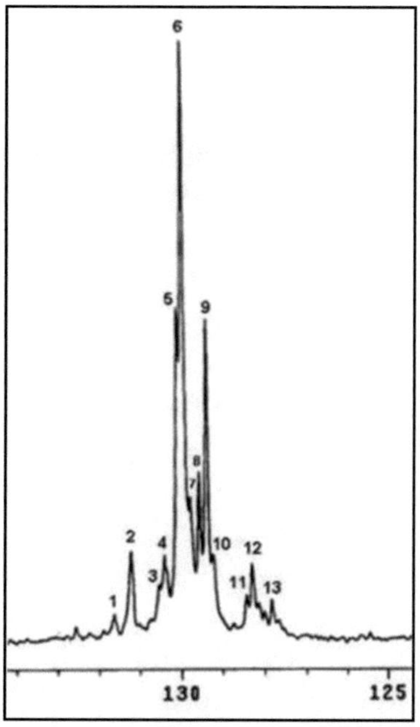

Figure 3. Expanded $^{13}C\{^{1}H\}$ NMR spectrum of δ 127-132 region of free radical HTPB prepolymer.

Signal	Sequence#	Chemical shift (δ values)	
		calculated	observed
1	v-T*-v	131.01	131.6
2	c-T*-v , t-T*-v	130.55	131.2
3	v-C*-v	129.87	130.5
3	v-T*-c ,v-T*-t	129.76	130.4
4	c-T*-c, c-T*-t	129.30	130.1
5	t-T*-c, t-T*-t	129.30	129.9
6	t-*T-v, c-*T-v	129.11	129.8
7	c-*C-c, t-*C-c	128.91	129.4
8	c-*C-t, t-*C-t	128.91	129.3
9	c-*C-v, t-*C-v	128.60	128.8
10	v-*T-t	127.64	128.4
11	v-*T-v	127.45	128.3
12	v-*C-c	127.31	128.0
13	v-*C-t	127.31	127.8

c = *cis*-1, 4-unit; t = *trans*-1, 4-unit; v = *vinyl*-1, 2-unit; v-T*-v = *vinyl*-1, 2-CH_2-CH=CH*-CH_2- *vinyl*-1, 2-unit; and v-*C-t = *vinyl*-1, 2-CH_2-*CH=CH-CH_2- *trans*-1, 4-unit.

Table 2. $^{13}C\{^{1}H\}$ Assignment of triad sequence of free radical HTPB prepolymer (δ 127-132 region).

In the aliphatic region (δ 24-44), the DEPT spectrum showed six sharp negative resonances at δ 38.6, 34.4, 32.8, 30.4, 27.4, and 24.9. A positive signal at δ 43.4, was assigned to the methine carbon of *vinyl*-1,2- unit. The chemical shift of each aliphatic carbon atom in HTPB polymer can be calculated by using empirical equation for branched and linear alkanes. According to Furukawa, the equation for calculating chemical shift of aliphatic carbon atom is given as $\delta_c(K) = A + \sum_l B_l N_{kl} + C_k$, where $\delta_c(K)$ is the chemical shifts of K carbon, A is a constant, B_l are the parameters away from various positions of K carbon, N_{kl} is the number of carbon away from various positions of K carbon, C_k is the parameter of characteristic structure for K carbon itself. The numerical values of all these parameters were taken from literature (Zheyen et al., 1983). The chemical shifts of the aliphatic carbon atoms in various sequence distribution were calculated and then, compared with the observed one to assign the signals. Besides, the assignment of the diad/triad resonances was made based on the values reported by Sato et al., (1987). The results, thus, obtained are given in Table 3. In the carbon bearing hydroxyl end group region (δ 56-65), the DEPT spectrum showed only the negative resonances (-CH_2-). Therefore, all the resonance signals belong to the adjacent methylene carbon to hydroxyl end group of HTPB prepolymer. Fig.4 shows the expanded $^{13}C\{^1H\}$ NMR spectrum of δ 56-65 region along with the assignment of carbon signals. The assignment of various resonances in this region was based on the report by Haas, (1985). The resonance at δ 58.50 is assigned to methylene carbon of *cis*-1,4-hydroxyl structure while other resonances at δ 63.67 and 65.06 are assigned to the methylene carbon of *trans*-1,4-hydroxyl and *vinyl*-1,2-hydroxyl structure, respectively. Further, the resonance line at δ 56.66 is attributed to the *cis*-1,4-epoxide carbon, while the resonance line at δ 58.26 is assigned to the *trans*-1,4-epoxide carbon.

Signal	Sequence*	Chemical shift (δ values)	
		calculated	observed
a	(1,4)-V-(1,4)	43.10	43.4
b	(1,4)-v-T	35.80	38.6
c	(1,4)-V-v (m)	35.70	34.5
d	(1,4)-V-(1,4)	34.80	34.2
e	T-(1,4) +(1,4)-v-C	33.30-33.40	32.8
f	v-v-C (m)	34.60	32.1
g	T-v/v-V-v	31.0/31.4	30.4
h	(1,4)-C	28.10	27.5
i	C-(1,4)	28.10	27.4
j	C-v	26.40	24.9

*C: *cis*-1, 4-unit; T: *trans*-1, 4-unit; V: *vinyl*-1, 2-unit; (1, 4): C+T; and m: meso.

Table 3. Assignment of $^{13}C\{^1H\}$ NMR resonances of Diad and Triad sequences of free radical HTPB prepolymer (δ 24-44 region).

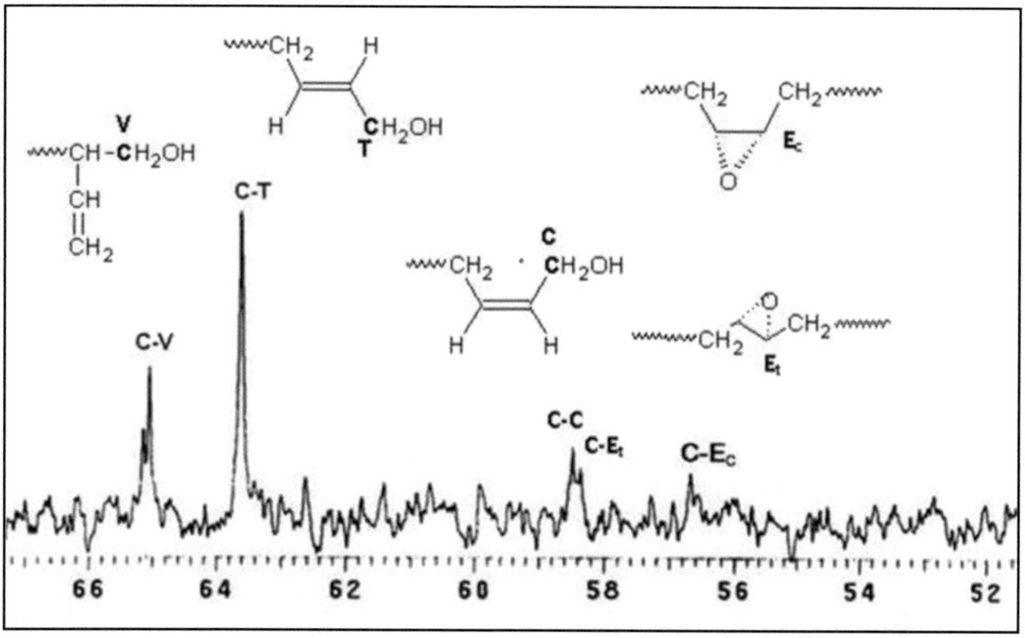

Figure 4. Expanded $^{13}C\{^{1}H\}$ NMR spectrum of δ 56-65 region of HTPB prepolymer along with the assignment of carbon signals.

The assignment of the various methylene and methine carbons from $^{1}H/^{13}C$-HMQC helped to assign the corresponding protons in the ^{1}H NMR spectrum.Figs.5, 6, and 7 show the $^{1}H/^{13}C$-HMQC spectrum of HTPB prepolymer in olefinic region, carbon bearing hydroxyl end group region, and aliphatic region respectively. In the olefinic region, the ^{13}C resonances at δ 113-144, showed three contours in the 2D HMQC spectrum (Fig.5) which corresponded to δ 4.9-5.7, in the ^{1}H NMR spectrum. Further, the fine splitting may be attributed to compositional sequences and tactic reasons. Thus, resonances observed in the HMQC spectrum (Fig.5) at δ 142.8-142.04, 132-127, and 114.9-114.2 corresponded to the protons in the ^{1}H NMR spectrum at δ 5.7-5.4, 5.44-5.41, and 5.0-4.9, respectively. Further, three signals seen in the HMQC spectrum at δ 65.06, 63.67, and 58.5 correspond to the protons at δ 3.4-3.7, 4.1-4.0, and 4.2, respectively (Fig.6). Similarly, in the aliphatic region (Fig.7), the ^{13}C resonance at δ 43.4 and 41.8 is correlated to protons at δ 2.12. The remaining resonances at δ 32.1, 27.4 and 24.9 correspond to the protons at δ 2.10, while the signals at δ 38.6, 32.8 and 30.4 belong to carbons associated with proton signals at δ 2.06. The signals at δ 30.02 and 29.0 are correlated to the protons at δ 1.48 and 1.23 respectively. Based on the above assignments, chemical shifts of various protons observed in the ^{1}H NMR spectrum of the polymer are summarized in Table 4. The proton resonance at 1.23 is assigned to the methyl group of isopropyl ether end group of the polymer. This isopropyl ether end group could be formed as isopropyl alcohol used as solvent in the synthesis of HTPB prepolymer also takes part in the free radical reactions. In presence of hydroxyl radical, isopropoxy radical is formed that leads to the formation of ether terminated polymer (Poletto & Pham, 1994).

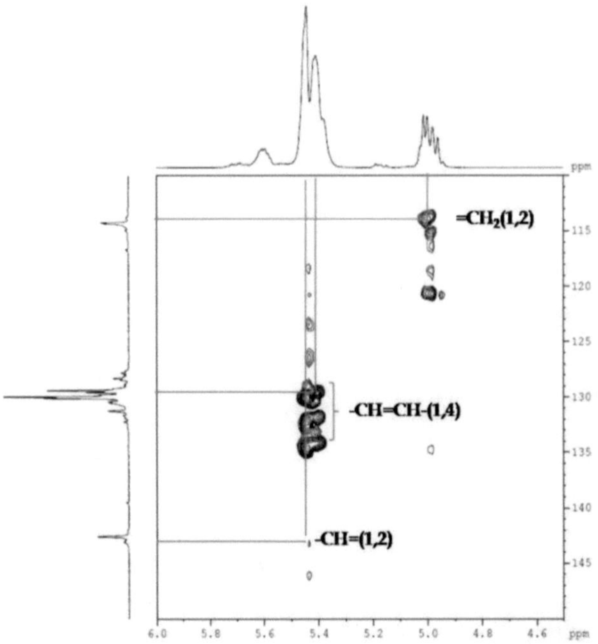

Figure 5. $^1H/^{13}C$ HMQC spectra of free radical HTPB prepolymer: olefinic region.

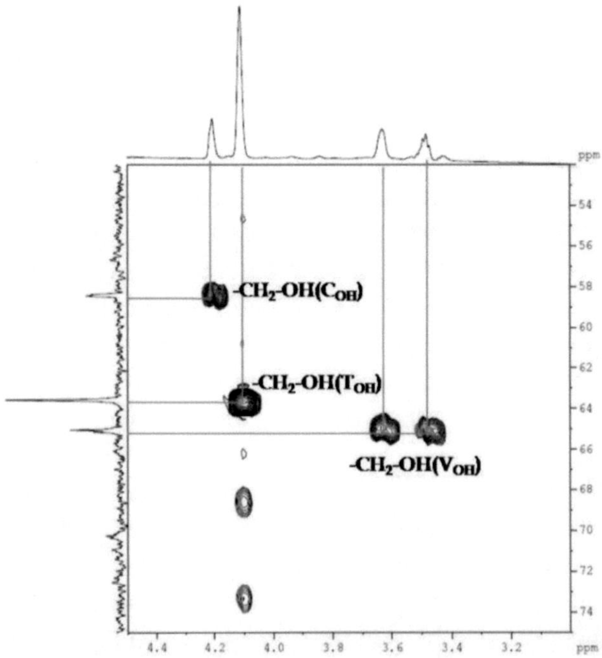

Figure 6. $^1H/^{13}C$ HMQC spectra of free radical HTPB prepolymer: carbon bearing hydroxyl end group region.

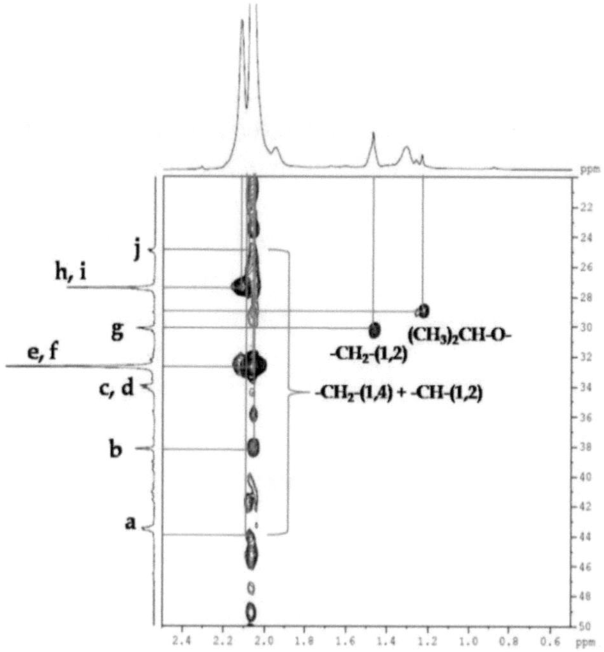

Figure 7. $^1H/^{13}C$ HMQC spectra of free radical HTPB prepolymer: aliphatic region.

Hydrogen	Chemical Shifts (δ)
-CH=(1,2) + -CH=CH-(1,4)	5.7-5.4
-CH=CH-(1,4)	5.44-5.41
=CH_2-(1,2)	5.0-4.9
-CH_2-OH(C_{OH})	4.20-4.16
-CH_2-OH(T_{OH})	4.1-4.0
-CH_2-OH (V_{OH})	3.7-3.4
-CH_2-(1,4) + -CH-(1,2)	2.12-1.90
-CH_2-(*cis*-1, 4-unit) + -CH-(1,2)	2.1
-CH_2-(*trans*-1, 4-unit)	1.90-2.06
-CH_2-(1,2)	1.6-1.3
$(CH_3)_2$CH-O-	1.23

Table 4. Assignments of Chemical shifts (δ) in the ^1H NMR spectrum of free radical HTPB prepolymer.

3.1.1. Chain microstructure and relative distribution

The integration of a resonance in NMR is directly proportional to the number of equivalent nuclei contributing to the particular resonance, under suitable experimental conditions. In polymer molecule, these nuclei are part of the chemical structure of a particular repeating unit. Therefore, quantitative result may be obtained by determining the ratio of resonance areas that corresponds to different structural units of the polymer. In Fig.8, the peak areas corresponding to olefinic ($a_1$and a_2) and aliphatic protons ($a_3, a_4$and a_5) can be measured separately. Therefore, the mole % of total olefinic protons in HTPB prepolymer can be determined as $olefinic\ protons\ (\%) = (a_1 + a_2)100/(a_1 + a_2 + a_3 + a_4 + a_5)$, where a_1, a_2, a_3, a_4 and a_5 are the integrated areas of peak clusters, as shown the Fig.8. The integrated peak area of a resonance due to the analyte nuclei is directly proportional to its molar concentration and to the number of nuclei that give rise to that resonance. So, we have $a_1 = K_s[2(y + z) + x]$, $a_2 = K_s[2x]$, $a_3 = K_s[4y + x]$, and $a_4 = K_s[4z]$, where K_s is the constant of proportionality. The x, y, and z are the mole fractions of *vinyl*-1,2-; *cis*-1,4-; and *trans*-1,4-content of HTPB, respectively, and $x + y + z = 1$. It can be seen in Fig.8 that each peak cluster is separated from the adjoining one by a sufficient amount of baseline to allow precise measurements. Thus $x, y,$ and z could be calculated by Eqs. (1), (2), and (3), respectively.

$$x = 2a_2/(2a_1 + a_2) \tag{1}$$

$$y = [(2a_3 - a_2)(2a_1 - a_2)]/[(2a_3 - a_2 + 2a_4)(2a_1 + a_2)] \tag{2}$$

$$z = [2a_4(2a_1 - a_2)]/[(2a_3 - a_2 + 2a_4)(2a_1 + a_2)] \tag{3}$$

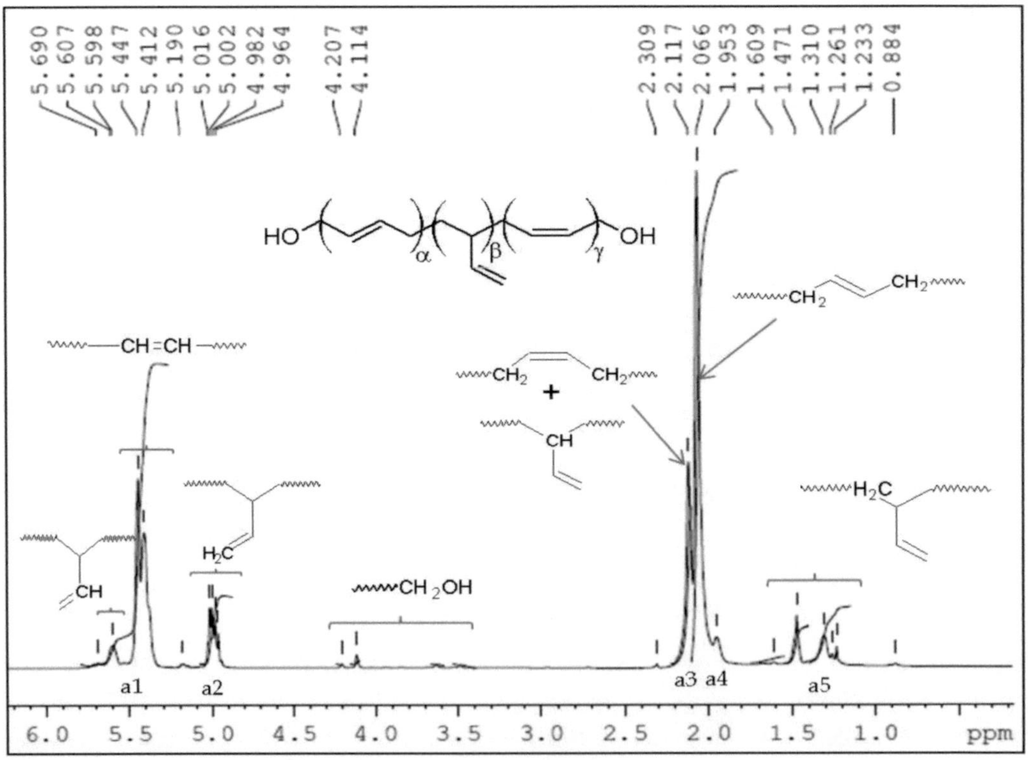

Figure 8. The ^{1}H NMR ($CDCl_3$, 800 MHz) spectrum of free radical HTPB with the assignments of proton signals.

The known microstructures of two commercially available anionic HTPB, *i.e.* Krasol LBH-2000 and LBH-3000 were also examined and the results are given against its standard value for comparison (Table 5). The values given in the parenthesis are the standard value. It clearly shows that the calculated values of the microstructures are very close to that of actual values. This shows the validity of the quantitative FT-NMR (FT-qNMR) method. Following the same procedure, the free radical HTPB prepolymer was analyzed by FT-qNMR method. Table 5 presents the ^{1}H NMR analysis results obtained on backbone microstructure content of the polymer. The microstructures obtained are typical of HTPB prepolymer synthesized by free radical method. Further, HTPB prepolymer has hydroxyl groups attached to the carbon which are in *cis-*, *trans-* or *vinyl-* configuration. Fig.9 shows the expanded ^{1}H NMR spectrum of δ 3.0-4.2 region. This region indicates the resonances of adjacent methylene protons to hydroxyl group of HTPB. Their assignments are shown in Fig.9.

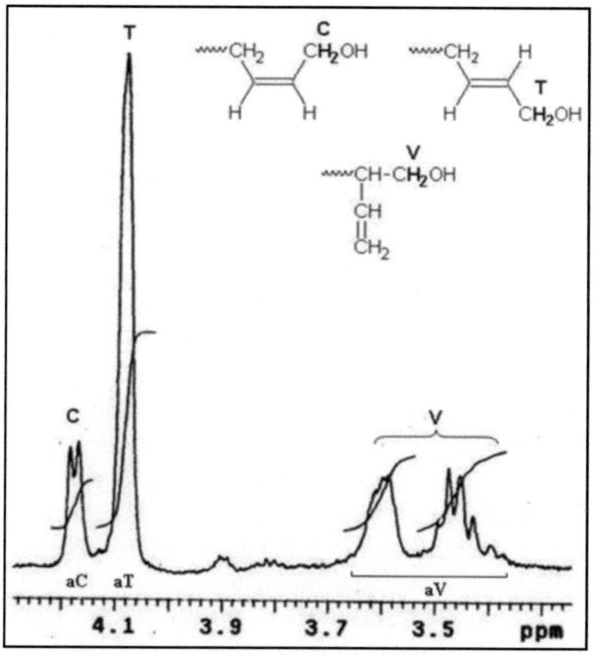

Figure 9. Expanded ^{1}H NMR spectrum of δ 3.0-4.2 region of free radical HTPB prepolymer.

The doublet shown by *cis*-1,4-unit is attributed to the difference in the nature of 1,2- or 1,4-butadiene unit adjacent to it. The complex feature of methylene resonance between δ 3.40 to 3.65 is due to H_A and H_B protons being non-equivalent because of steric hindrance (Fig.10).

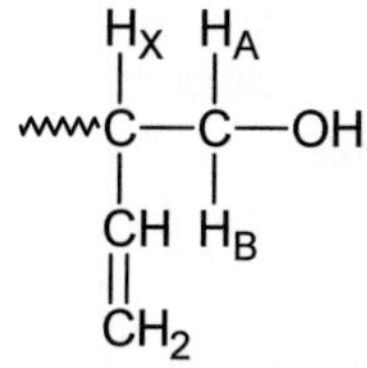

Figure 10. Chemical structure of vinyl-1,2- unit of HTPB prepolymer.

The chemical structure in Fig.10 reveals that all the three protons, *i.e.* H_A, H_B and H_X are magnetically non-equivalent, and therefore, have three different chemical shifts. Further, each of the three signal would split into four peaks (*i.e.* the signal for H_A is split into two by H_B and again into two by H_X proton). The coupling constants between any two of the protons would be different. In fact, two pairs of doublet are observed at δ 3.4-3.5 (H_A) and at δ 3.55 - 3.65 (H_B). The vinylic methine proton (H_X) resonates at δ 2.07. The coupling constant (^3J) between H_A and H_X protons is calculated to be ~8 Hz, where as between H_B and H_X protons is ~6 Hz. The ^2J between H_A and H_B protons is found to be ~18 Hz. The calculated values of coupling constant are in agreement with that of an ABX spin-spin system (Kalsi, 1995). The mole % of *cis*-1,4-; *trans*-1,4-; and *vinyl*-1,2-hydroxyl units were obtained by integrating the corresponding resonances as shown in Fig.9, *i.e.* adding together the integrated amounts and dividing the total into the integrated peak area obtained for each configuration and listed in Table 5. However, in case of Krasol LBH-2000 and LBH-3000 only single broad peak was obtained at δ 3.8.This peak is assigned to the secondary proton of the hydroxyl end group.

3.1.2. Determination of degree of polymerization and molecular weight

The molecular weight of the HTPB prepolymer has significant impact on the end-use properties of PUs. Thus, it needs to be estimated with a high degree of accuracy. More often, the absolute molecular weights of the prepolymer are required for higher precision in performance evaluation. Conventional measurement techniques, such as Vapor Pressure Osmometry (VPO) and Gel Permeation Chromatography (GPC) used for determination of molecular weights of the polymer, being the relative methods do have limitations to produce authentic results. The number average-molecular weight of the polymer could be estimated by ^1H qNMR end-group analysis. The area of an absorption peak in the ^1H qNMR spectrum is proportional to the number of equivalent nuclei and these nuclei are part of the chemical structure of a particular repeating unit. Therefore, in case of HTPB prepolymer, the number-average degree of polymerization ($\overline{DP}_n$) would be the ratio of the sum of olefinic protons integrals to that of hydroxylated methylene protons. The chemical structures of HTPB, synthesized by free radical and anionic polymerization method, are depicted in Fig.11, where α, β and γ are the number of *trans*-1,4- ; *vinyl*-1,2-; and *cis*-1,4- micro structural units respectively.

The number-average degree of polymerization ($\overline{DP}_n$) of HTPB prepolymer would be $\overline{DP}_n = \bar{\alpha}_n + \bar{\beta}_n + \bar{\gamma}_n$, and can be determined by Eq. (4).

$$\overline{DP}_n = \bar{\alpha}_n + \bar{\beta}_n + \bar{\gamma}_n = [(a_2 + 2a_1) \times \bar{F}_n(OH)]/[2(a_C + a_T + a_V)] \quad (4)$$

where a_1 , a_2 , a_C, a_T, and a_V are the integrated peak area of the peak clusters as shown in Figs.8 & 9, and $\bar{F}_n(OH)$ is the average functionality of the prepolymer. Thus, number-average molecular weight ($\bar{M}_n$) can be calculated as $\bar{M}_n(NMR) = (\overline{DP}_n \times 54) + (\bar{F}_n(OH) \times 17)$. Similarly, for Krasol LBH-2000 and LBH-3000, it will be $\bar{M}_n(NMR) = (\overline{DP}_n \times 54) + (\bar{F}_n(OH) \times 59)$. Table 5 presents the results obtained on the three polymers under investigation. The $\bar{M}_n(NMR)/\bar{M}_n(GPC)$ average ratio for free radical HTPB is found to be

0.69, which is well compared with the literature value of 0.67 (Kebir et al., 2005), where as for, Krasol LBH-2000 and LBH-3000, it is 1.03 and 1.15, respectively. This deviation may be due to the narrow distribution of Krasol LBH-2000 and LBH-3000 prepolymer. The polydispersity index (PI) obtained by GPC for Krasol LBH-2000 and LBH-3000 are 1.8 and 1.6, respectively.

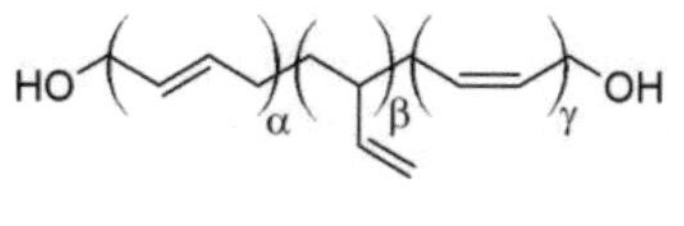

(a)

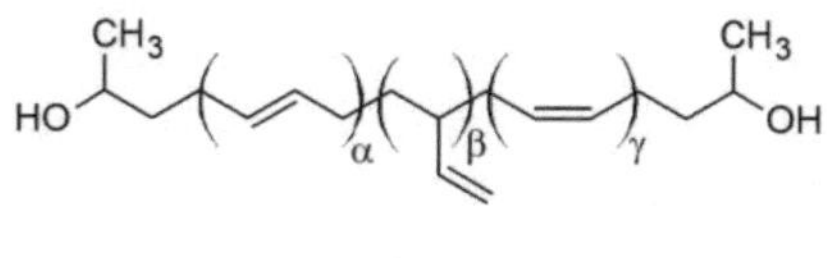

(b)

Figure 11. Molecular structure of (a) free radical HTPB, and (b) anionic HTPB prepolymer (Krasol LBH-3000)

HTPB types	Olefinic protons (%)	Microstructure	Hydroxyl types	$\bar{M}_n$		
		C/T/V (%)	C/T/V (%)	NMR	GPC/PI	VPO
LBH-2000	39.96	9.6/21.9/68.4 (12.5/22.5/65.0)	Secondary OH	4208	4068/1.8	2440 (2100)
LBH-3000	40.36	11.4/23.1/65.5 (12.5/22.5/65.0)	Secondary OH	7029	6094/1.6	2630 (3121)
Free radical HTPB	34.50	19.4/59.6/21.0	14.4/57.6/28.0 (Primary OH)	4087	5891/2.1	2590

C: *Cis*-1,4 ; T: *Trans*-1,4 ; and V: *Vinyl*-1,2

Table 5. Results of backbone microstructure, types of hydroxyl end group, and $\bar{M}_n$ of HTPB determined by ^{1}H FT-qNMR method.

3.2. Rheology of HTPB prepolymer: Temperature modelling

The viscosity (μ) of a polymer liquid depends on several variables such as shear rate (γ), molecular weight (M_w), time (t), and temperature (T), *i.e.* $\mu = f(M_w, \gamma, T, t)$. Isothermal viscosity of the prepolymer (HTPB) was obtained at the temperatures of 40 °C, 50 °C, 60 °C, and 70 °C by the Brookfield Viscometer with AB-4 spindle. To check whether shear thinning was occurring, viscosity of the polymer samples was measured at different shear rates (rpm) ranging from 5 to 100 rpm. The samples were also sheared for 10 minutes at a constant shear rate to check the thixotropy nature of the prepolymers. We observed that the viscosity remained more or less same with respect to shear rates indicating a Newtonian characteristic of the prepolymer. Also no effect was observed with time of shearing. Fig.12

shows the effect of temperature on viscosity of the prepolymers (HTPB). The viscosity versus temperature data for Krasol LBH-2000 and LBH-3000 are also included for comparison. It is evident from Fig. 12 that the viscosity decreases with increase in temperature. The temperature dependence of viscosity followed the Arrhenius exponential relation as $\mu(T) = \mu_0 e^{E_{vf}/RT}$, where μ_0=1.32 x10^{-3}, 7.7 x 10^{-7} and 2.11 x10^{-6} mPas and activation energy of viscous flow of the prepolymers E_{vf} = 38.7, 59.23 and 56.54 kJmol^{-1} for free radical HTPB, Krasol LBH-2000 and LBH-3000, respectively (with μ in mPas and T in Kelvin). The viscosity dependence on temperature can also be fitted with a Power Law model of the form $\mu(T) = BT^n$. The Power Law index is the characteristic parameter of the prepolymer. It shows the sensitivity of viscosity to temperature changes ($d\mu/dT$) of the prepolymer. The n values were determined from the log-log plot of viscosity versus temperature (°C) and found to be -2.09, -3.16 and -3.07 for free radical HTPB, Krasol LBH-2000, and LBH-3000, respectively. This indicates that the anionic HTPB prepolymers are more sensitive to the temperature change as compared to the free radical one. Both the Arrhenius and Power Law model satisfactorily described the viscosity dependence on temperature of the polymers as the correlation coefficients were > 0.98.

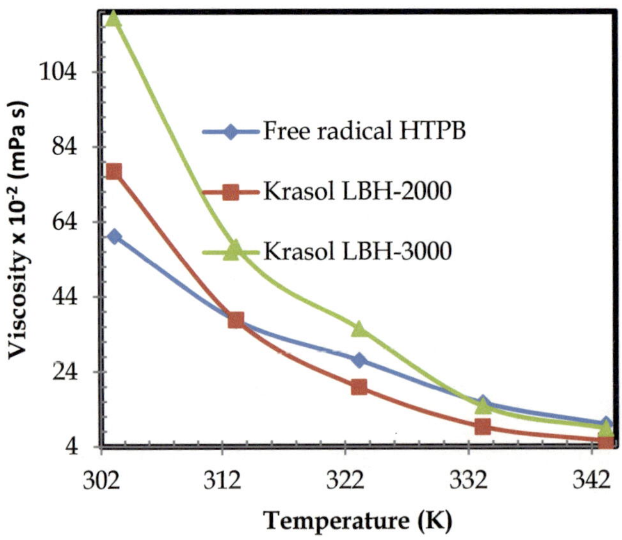

Figure 12. Plots of viscosity(μ) vs. temperature for prepolymers (HTPB).

3.3. Chemo-rheology of PU-I, PU-II and PU-IIp: Temperature and time modelling

Chemo-rheology is the study of chemo-viscosity which is the variation of viscosity caused by chemical reactions. Although the exact reaction mechanism of PU formation is more complex, the kinetics of reaction of diisocyanate with dihydroxyl compound is often expressed successfully by a second order rate equation, *i.e.* $-d[NCO]/dt = k_\mu[NCO][OH]$, where k_μ is the kinetic rate constant. The $[NCO]$and $[OH]$ are the concentration of isocyanate and hydroxyl groups, respectively. The viscosity of the curing PU system is determined by two factors: (a) the degree of cure, and (b) the temperature. We have carried out the chemo-rheological experiments at different temperatures and different shear rates. As the cure proceeds, the molecular size increases and so does the cross linking density, which in turn, decreases the mobility of the molecules. On the other hand, the temperature exerts direct effects on the dynamics of the reacting molecules and so, on the viscosity. To check whether shear thinning was occurring, viscosity of all the PU samples was measured at different shear rates (rpm) ranging from 5 to 100 rpm (AB-4 spindle) for PU-I and PU-II, whereas for PU-IIp, the shear rate was ranged from 0.5 to 10 rpm (T-E spindle). The samples were also sheared for 10 minutes at a constant shear rate to check the time dependent effect of the PUs. For PU-I and PU-II, we observed that the viscosity remained more or less same with respect to shear rates, which revealed the Newtonian characteristic of the binder resin. Also, no significant effect was observed with time of shearing. However, the PU-IIp (propellant slurry) is found to be shear sensitive. Fig. 13 depicts a typical viscosity build up plots with cure time at various temperatures for PU-I. We observed that viscosity decreased with an increase in temperature. In the initial period of the reaction, when the polymer molecules were small in size, viscosity varied considerably with temperature, higher temperatures resulted in lower viscosities. As the reaction proceeds and molecular size goes up, viscosity rises sharply with respect to time and temperature. This is because the effect of curing reaction overtakes the effect of temperature on viscosity (Reji et al., 1991).

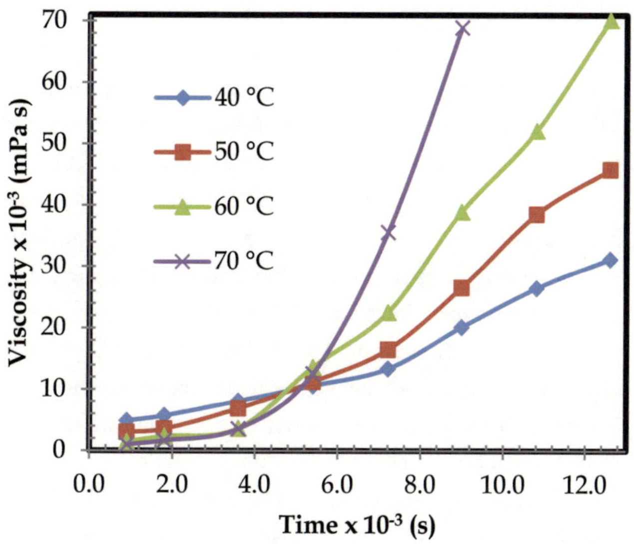

Figure 13. Plots of viscosity (μ) vs. time at various temperatures for PU-1.

The fact that the temperature changes the viscosity of the slurry means that special consideration must be given to kinetic and thermodynamic factors. In kinetics, the emphasis is on the reaction rate. Navarchian et al. (2005) used exponential function to model the viscosity versus time data and found that the semi-logarithmic plots were of good linearity.

The model representing the change of viscosity (μ) with reaction time (t) has the following form:

$$\mu(t) = \mu_0 e^{k_\mu t} \tag{5}$$

where μ_o is the viscosity at t = 0 and k_μ is the rate constant for viscosity build up. This exponential model was applied to the experimental data. The initial viscosity and the rate constants at each temperature were calculated from the intercept and slope of the straight line of $\ln \mu\ vs. t$ plots before the gel point and their values at each isothermal temperature are listed in Table 6.

Temp. (°C)	Unfilled polyurethane				Filled polyurethane	
	PU-I		PU-II		PU-IIp	
	μ_o(mPas)	k_μ(min-1)	μ_o(mPas)	k_μ(min-1)	μ_o (mPas)x10-2	k_μ(min-1)
40	4349	9.76 x 10-3	6320	12.08 x10-3	10869	1.80 x10-3
50	2647	14.62 x10-3	4142	17.70 x10-3	5226	2.80 x10-3
60	1352	20.70 x10-3	2590	27.40 x10-3	3150	4.11 x10-3
70	606	32.49 x10-3	2008	34.80 x10-3	---	---

Table 6. Values of viscosity (μ_o) and rate constants(k_μ) for various PU systems.

The results indicated that the rate constants increased with increase in temperature from 40 to 70 °C, while the μ_o decreased. However, the filled PU (PU-IIp) has shown a very slow build up as it is evident from the very low reaction rate constants. This could be due to the effect of various fillers molecules, which restrict the mobility of the reacting molecules, hence slow down the reaction rate. Further, the relationship of rate constant and viscosity (μ_o) with temperature followed the Arrhenius exponential relationship, *i.e.* $k_\mu(T) = A_\mu \exp(-E_\mu/RT)$, and $\mu_o(T) = A_o \exp(E_o/RT)$, where A_μ and A_o are the apparent rate constant, and initial viscosity at $T = \infty$, E_μ and E_o are the kinetic activation energy, and the viscous flow activation energy, respectively. The values of A_μ, E_μ, A_0, and E_0 of the PU reaction on different systems were determined from the Arrhenius plots and listed in Table 7. Further, unlike unfilled PUs (PU-I and PU-II), the filled PU (PU-IIp) showed the shear thinning behaviour. The effect of shear rate on viscosity is shown in Fig.14. For non-Newtonial material, if the viscosity decreases with shear, the rate of decrease is the measure of pseudoplasticity of the material. The flow of highly loaded propellant slurry (86 % solid loading) can be more closely approximated by the Power Law fluid model (Mahanta et al., 2007). The pseudoplasticity index (PI) and viscosity index were calculated from the curve by fitting to a Power Law equation i.e. $\mu(\gamma) = K\gamma^m$, where μ is the apparent viscosity, γ is the shear rate in rpm, m is the pseudoplasticity index, and K is the viscosity index. Newtonian fluid are the special case of Power Law fluid, when m = 0, viscosity is independent of shear rate. For dilatent fluid m is positive, while for pseudoplastics m varies from 0 and -1. In the current work, for the purpose of characterizing the PU-IIp, the minus sign of the m was excluded and reported in percentage.

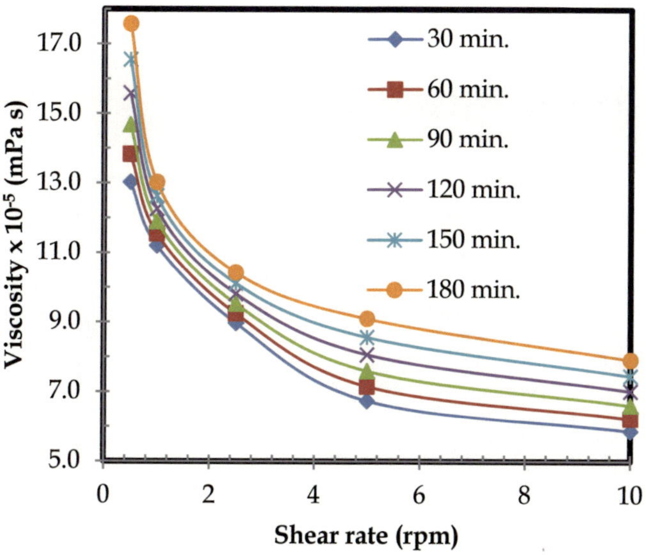

Figure 14. Viscosity (at various intervals) versus shear rate of PU-IIp at 40 °C.

Parameters	Unfilled polyurethane		Filled polyurethane
	PU-I	PU-II	PU-IIp
E_{μ}(kJ mol^{-1})	35.3	32.3	35.8
E_0(kJ mol^{-1})	58.6	35.0	53.8
A_{μ}(min^{-1})	7423	3020	1707
A_0(mPas)	7.99 x 10^{-7}	9.03 x 10^{-3}	1.13x10^{-5}
$\Delta S_{\mu}^{\#}$ (J mol^{-1} K^{-1})	-180	-187	-192
$\Delta H_{\mu}^{\#}$ (kJ mol^{-1})	33.1	29.6	33.1
$\Delta G_{\mu}^{\#}$ (kJ mol^{-1}) at 40 °C	89.5	88.3	93.3

Table 7. Kinetic and thermodynamic parameters for different PU-systems.

The pseudoplasticity indexes calculated from the Power Law equation are plotted as a function of cure time (Fig.15).

It is observed that the PI is higher at higher temperature. This indicates that at higher temperature the PU-IIp becomes more non-Newtonian. Interestingly, the PI decreases at 40°C and 60 °C with cure time, whereas at 50 °C, it is almost consistent within the pot life of 3 hours, usually required for casting of the propellant slurry into the rocket case. However, the viscosity index decreased initially with temperature, and afterwards, it increased with the cure time. This is attributed to the increase in cross linking, caused by PU reaction. The flow behaviour of HTPB propellant slurry assumes to have great importance as this is the cause of many grain defects in large scale motor. To make a logical decision regarding propellant mixing and casting, not only the effect of temperature and time on viscosity of the propellant slurry should be thoroughly studied, but pseudoplasticity of the slurry should also be equally emphasised. This study has indicated that at 50 °C, the PI remains consistent within the required pot life, so it is assumed that propellant mixing and casting at this temperature may result in a better quality grain.

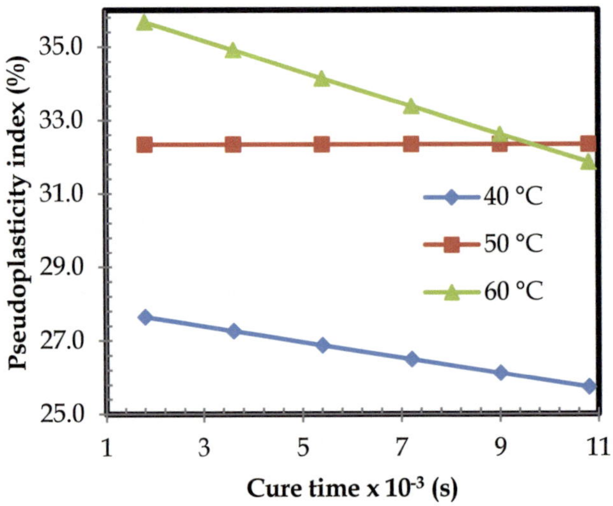

Figure 15. Pseodplasticity index (PI) of PU-IIp as a function of cure time at different temperatures.

A quantitative study of thermodynamic parameters ($\Delta H^{\#}, \Delta S^{\#}$, and $\Delta G^{\#}$) helps in understanding the reaction mechanism. It is also used to optimise the cure cycle of the PU reaction, both in terms of time and energy. Wynne-Jones-Eyring–Evans theory (Arlas et al., 2007) presents the temperature dependent pre-exponential factor, and the kinetic constant is given as:

$$k_{\mu} = \frac{k_B T^n}{h} e^{[N+(\Delta S^{\#}/R)]} e^{[-(\Delta H^{\#}+NRT)/RT]} \tag{6}$$

where T is the temperature (K), R = 8.314 Jmol $^{-1}$K^{-1} is the universal gas constant, k_{μ} is the kinetic rate constant, N is called molecularity, $h = 6.62 \times 10^{-34}$ Js is the Planck's constant, k_B is the Boltzmann constant, $\Delta H^{\#}$ is the activation enthalpy, and $\Delta S^{\neq}$ is the activation entropy. The classical Arrhenius constant have $N = 0$ and N equals to 1 for reactions occurring in liquid state. Thus, assuming $N = 1$, plotting $\ln(k_{\mu}/T)$ $vs. 1/T$, the values of $\Delta H^{\#}$ & $\Delta S^{\#}$ were calculated from the slope and the intercept of the straight line obtained. Also, the $\Delta G^{\#}$ value can be calculated from the fundamental thermodynamic relation, *i.e.* $\Delta G^{\#} = \Delta H^{\#} - T\Delta S^{\#}$. The results thus obtained are listed in Table 7.

It is observed that the activation entropy is negative and quite low. This suggests that the polymerization path is more ordered, that makes the reaction thermodynamically disfavoured. Negative values for activation entropy also indicate the association of reactants prior to chemical reaction.

3.4. Thermo-oxidative degradation of prepolymers (HTPB) and PU-II

The HTPB polymers are vulnerable to oxidative degradation due to it reactive carbon-carbon double bonds and hydroxyl functionality. These prepolymers are exposed to air, humidity, increased temperature and a lot of shear, during processing for PUs manufacturing. Oxygen and water can ingress into the system by several ways during storage, handling as well as processing, leading to oxidative degradation of the polymer. Oxidative degradation is due to reaction with oxygen from air, which can lead to deterioration of the polymer properties. As discussed earlier, the olefinic groups of HTPB may be present in three configurations namely, *cis*-1,4-; *trans*-1,4-; and *vinyl*-1,2-units. The content of these units varies from polymer to polymer. Generally, these olefinic groups are of different reactivity in the oxidation reaction (Duh et al., 2010). As a result, the per centage of *cis*-1,4-; *trans*-1,4-; and *vinyl*-1,2- olefinic groups in the HTPB may have great effect on oxidation rates and product composition. The typical DSC curves obtained for free radical HTPB as well as PU-II are shown in Fig.16. The DSC thermogram of Krasol LBH-3000 is also given for comparison. It is seen that the HTPB prepolymer degrades in two distinct stages, i.e. (1) 170-260 °C, and (2) 290-400 °C when heated up to 400 °C. The first exotherm is attributed to the thermal oxidation reaction of HTPB prepolymer. Upon heating in air atmosphere, HTPB and oxygen are involved in a variety of free-radical reactions as shown in Fig.17. The oxidation reactions, as indicated by the first stage exothermic peak in the DSC thermogram, are attributed to oxygen uptake *via* (a) peroxidation, (b) hydroperoxidation, and (c) crosslinking by peroxide linkage. In the first exotherm, the DSC thermogram of free radical HTPB depicted two peaks, one at 205.0 °C and the other at 244.3 °C, which clearly established that two different oxidative paths ((*i.e.*, peroxidation, and hydroperoxidation) were involved in the oxidation process. In contrary to this, a single peak was observed for Krasol LBH-3000 at 234.5 °C. The plausible explanation for this anomaly could be that the per centage of *vinyl*-1, 2-units in the sample of LBH-3000 was higher as compared to free radical HTPB. Owing to the higher reactivity of *vinyl*-1,2 content, the reaction rate escalates initially resulting in the disappearance of the peak.

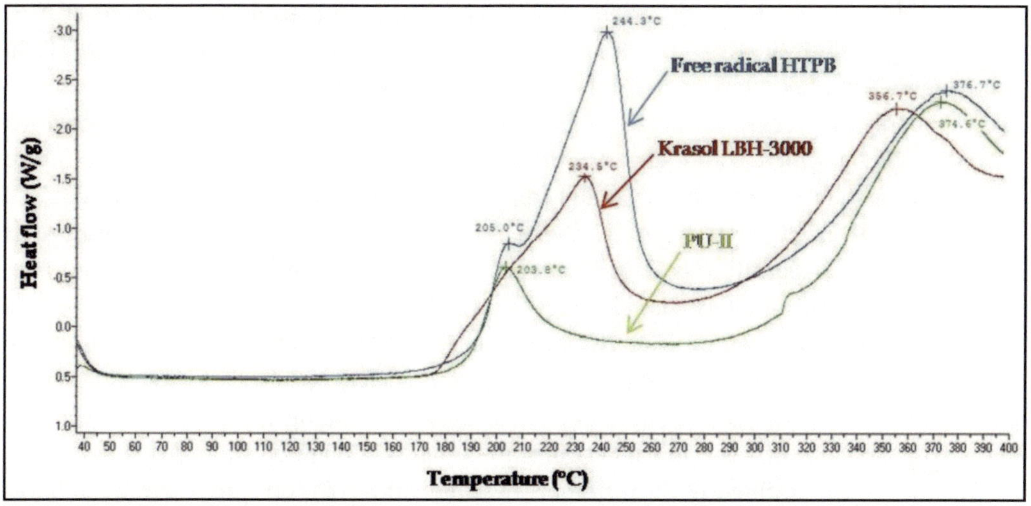

Figure 16. Dynamic DSC scans of HTPB prepolymers and PU-II at the heating rate of 10 °C/min.

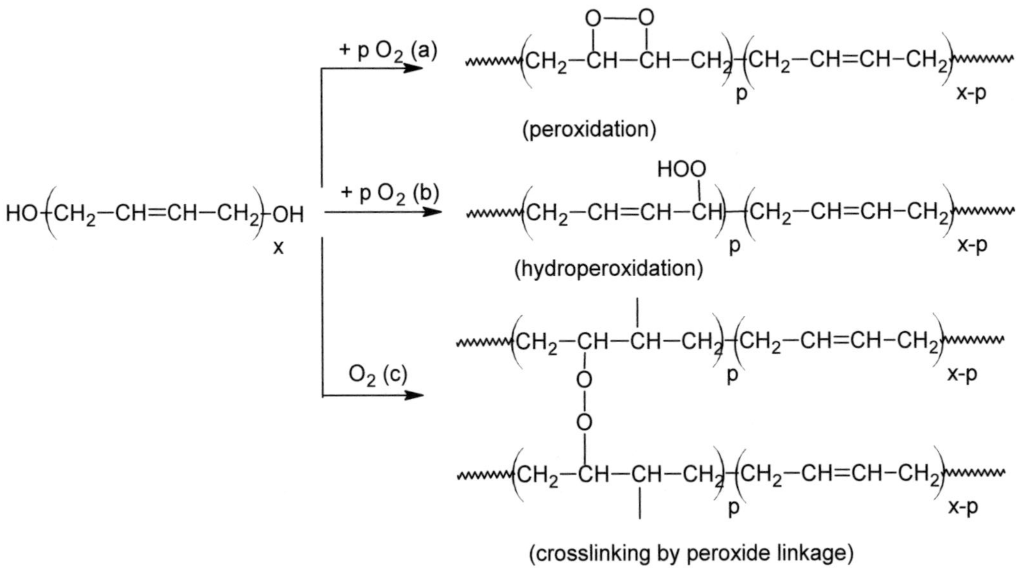

Figure 17. Thermo-oxidative reactions of HTPB prepolymers: (a) peroxidation, (b) hydroperoxidation, and (c) cross-linking by peroxide linkage.

The second exothermic peak occurred at 290-400 °C. The broad exothermic peak is attributed to the major oxidative degradations of HTPB prepolymer involving chain unzipping. It results from the endothermic depolymerisation, exothermic cyclization, and oxidative cross-linking processes of the HTPB prepolymer. The exothermicity is due to the energy released in the formation of new bonds during cross-linking and cyclization, which is greater than the absorbed energy for bond scission during depolymerisation. For PU-II, it is seen that the thermo-oxidative profile has a pattern very similar to that of HTPB prepolymer. This is expected as the PU-II constitutes HTPB more than 92 % of its weight. However, the most important difference is that the thermo-oxidative peak (first exotherm) is slightly less pronounced and occurs somewhat earlier than HTPB prepoymer. The peak temperature of PU-II is 203.8 °C which is 40 °C less as compared to HTPB prepolymer. In second stage i.e. between 290-400 °C, a small elevation is observed around 315 °C, which could be attributed to the cleavage of urethane linkages and subsequent loss of toluene diisocyanate, followed by depolymerization, cyclization, and crosslinking of HTPB prepolymer giving a broad exotherm with peak temperature of 374.6 °C, which is slightly less than its prepolymer peak temperature. This finding is in well agreement with the fact that cleavage of urethane linkages in HTPB PUs is the first step during thermal decomposition (Chen &Brill, 1991). As our objective was to study the thermo-oxidative behaviour of the polymer, we restricted only to the first exothermic peak of the DSC thermogram. The influence of different heating rates (β) on the thermo-oxidative behaviour of free radical HTPB prepolymer is illustrated in Fig.18.The insert in Fig.18 shows the influence of different heating rates (β) on the thermo-oxidative behaviour of PU-II.

We observed from Fig.18 that in both the cases, the thermograms shifted towards higher temperatures as the heating rate (β) increased. This shift of thermograms to higher temperature with increasing heating rate is anticipated since a shorter time is required for the samples to reach a given temperature at a faster heating rate. However, the shapes of the exothermic curves at all heating rates are similar. It indicates that similar reaction mechanisms are involved in oxidative degradation, irrespective of heating rates. The measured values of the onset temperature (T_i), peak temperature (T_p), final temperature (T_f), and oxidation enthalpy (ΔH_{ox}) for HTPB prepolymers and PU-II are listed in Table 8 and 9, respectively.

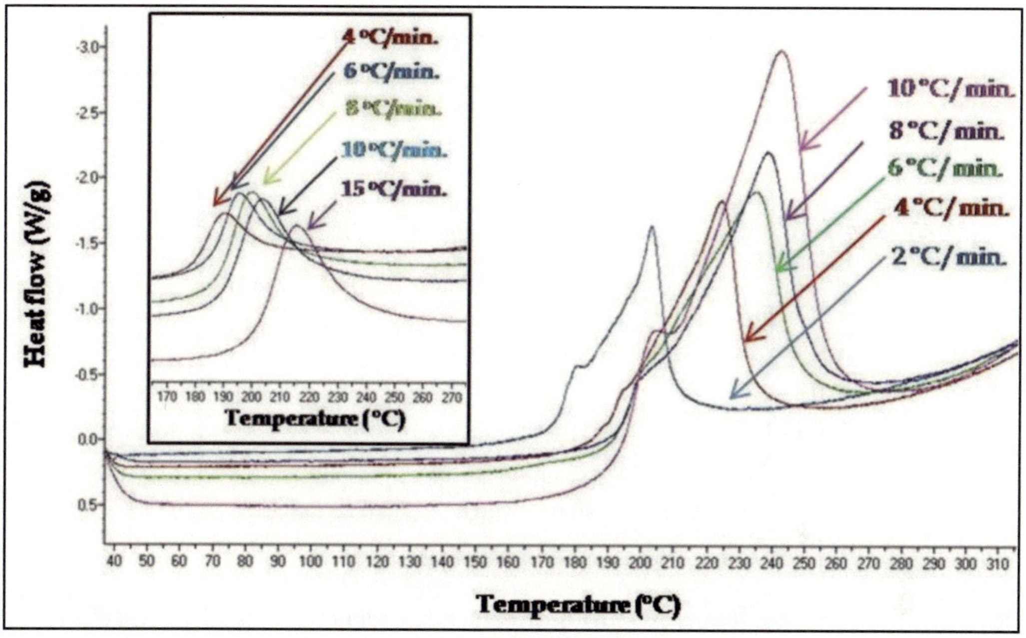

Figure 18. DSC thermogrames for decomposition of free radical HTPB at various heating rates (the insert Fig. is for PU-II).

β (°C min^{-1})	Thermo-oxidative properties of substrate polymers							
	Free radical HTPB				Krasol LBH-3000			
	T_i (°C)	T_p (°C)	T_f (°C)	ΔH_{ox} (Jg^{-1})	T_i (°C)	T_p (°C)	T_f (°C)	ΔH_{ox} (Jg^{-1})
2	159.6	204.1	233.4	896	147.0	188.9	212.4	510
4	177.0	225.4	252.4	653	160.7	206.8	236.1	454
6	186.5	235.7	270.4	595	166.0	216.0	238.9	426
8	187.0	239.1	271.0	649	170.1	222.8	247.1	443
10	187.7	244.3	279.0	628	173.9	234.5	264.4	369

Table 8. Thermo-oxidative properties of HTPB prepolymer at various heating rates (β).

β (°C min^{-1})	Thermo-oxidative properties of PU-II			
	T_i (°C)	T_p (°C)	T_f (°C)	ΔH_{ox} (Jg^{-1})
4	171.4	190.1	216.1	188
6	173.4	195.6	219.3	198
8	175.2	200.4	230.4	134
10	178.7	203.8	233.9	124
15	188.1	215.2	247.7	105

Table 9. Thermo-oxidative properties of PU-II at various heating rates (β).

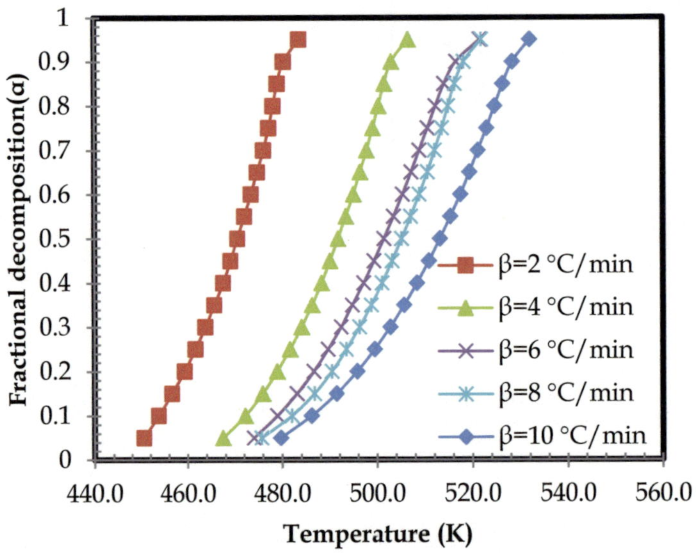

Figure 19. Plots of α versus temperature at different heating rates for free radical HTPB prepolymer.

The fractional decomposition (α) is experimentally determined from the measurement such as heat evolution or mass loss, depending upon the type of experiment performed. In DSC, it is calculated as $\alpha = \Delta H / \Delta H_0$, where ΔH and ΔH_0 are the released heat at certain degree of decomposition and the total heat of decomposition, respectively. Fig.19 shows the variation of fractional conversion as a function of temperature at various heating rates for free radical HTPB. It was seen that the temperature at same conversion increased with the increase of heating rate. A similar trend of conversion change versus temperature was found for PU-II and Krasol LBH-3000 also, under the same range of heating rates.

3.4.1. Kinetics of thermo-oxidation reaction (Model-Free Method)

For a complex reaction like thermo-oxidation reaction with an uncertain reaction mechanism, activation energy is not constant. Therefore, the isoconversional method is the method of choice for studying the kinetics. The isoconversional methods evaluate the effective activation energy as a function of the extent of conversion. It is assumed that the rate of conversion is proportional to the concentration of reacting molecules. The basic equation used in all kinetics studies is generally described as:

$$\frac{d\alpha}{dt} = k(T)f(\alpha) \tag{7}$$

where, α is the fractional decomposition, $f(\alpha)$ is the single reaction model function, T is the absolute temperature (K), and $k(T)$ is the Arrhenius rate constant. The temperature dependence of the rate constant $k(T)$ is described by Arrhenius equation $k(T) = A\,exp(-E_a/RT)$, where A, R and E_a are the pre-exponential factor, the universal gas constant, and the apparent activation energy, respectively. In non-isothermal conditions, the temperature varies linearly with time. Thus a constant heating rate (β) is defined as $\beta = dT/dt$. Upon introducing the heating rate, $\beta = dT/dt$, Eq.(7) can be modified to

$$\frac{d\alpha}{f(\alpha)} = \frac{\mathrm{A}}{\beta}\exp\left(-\frac{E_a}{RT}\right)dT \tag{8}$$

Therefore, Eq.(8) is the fundamental expression to determine kinetic parameters on the basis of DSC data. In the current work, we have used three different isoconversional methods i.e. (1) Kissinger, (2) Flynn-Wall-Ozawa (FWO), and (3) Kissinger-Akahira-Sunose (KAS) to evaluate the kinetic parameters for thermo-oxidative reaction of the prepolymers (HTPB) and PU-II.

3.4.1.1. Kissinger method

Kissinger (Kissinger, 1956) developed a model-free non isothermal method to evaluate kinetic parameters. In this method, the activation energy is obtained from a plot of $\ln(\beta/T_p^2)$ against $1/T_p$ for a series of experiments at different heating rates, where T_p is the peak temperature on the DSC curve.

$$\ln\left(\frac{\beta}{T_p^2}\right) = \ln\left(\frac{AR}{E_a}\right) - \frac{E_a}{RT_p} \tag{9}$$

The activation energy and pre-exponential factor can be calculated from the slope and intercept of the straight line plots of $\ln(\beta/T_p^2)$ versus $1/T_p$.

3.4.1.2. Flynn-Wall-Ozawa method (Flynn &Wall, 1966 and Ozawa,1965)

The integral form of Eq.(8) can be written as

$$g(\alpha) = \frac{\mathrm{A}}{\beta}\int_0^{\mathrm{T}}\exp\left(-\frac{E_a}{RT}\right)dT = \frac{\mathrm{AE_a}}{\beta\mathrm{R}}\mathrm{p(x)} \tag{10}$$

where $x = \frac{E_a}{RT}$ and $p(x) = -\int_{\infty}^{\frac{E_a}{Rx}}\frac{\exp(-x)}{x^2}dx$. p(x) is the so-called temperature or exponential integral which cannot be exactly calculated. To describe the thermal degradation kinetics, Ozawa assumed $\ln p(x) \approx -5.330 - 1.052x$ for $20 < x < 60$ for the non-plateau region of the curves, thus Eq. (10) can be written as :

$$\ln g(\alpha) = \ln\frac{AE_a}{\beta R} - 5.330 - 1.052\frac{E_a}{RT} \quad (11)$$

As A and R are constants, and for a particular conversion, g (α) is constant. Then Eq.(11) becomes

$$\ln\beta = C - 1.052\frac{E_a}{RT}, \text{where } C = \ln\frac{AE_a}{g(\alpha)R} - 5.330 \quad (12)$$

It is inferred from Eq. (12) that for a constant conversion, a plot of $ln\beta$ versus $1/T$ at different heating rates, should lead to a straight line whose slope provides E_a values. This method is known as Flynn-Wall-Ozawa method (FWO).

3.4.1.3. Kissinger-Akahira-Sunose (KAS) method (Arlas et al., 2007)

In KAS method, the expression p(x) is expressed using the Coats-Redfern approximation. It is $p(x) \cong \frac{\exp[-x]}{x^2}$, substituting this into Eq.(10) and taking logarithms, we get

$$\ln\left(\frac{\beta}{T^2}\right) \cong \ln\left(\frac{AR}{g(\alpha)E_a}\right) - \frac{E_a}{RT} \quad (13)$$

A plot of $ln\,(\beta/T^2)$ versus $1/T$ for a constant conversion gives the E_a at that conversion. We have evaluated the activation energy of prepolymers and PU-II by Kissinger, FWO and KAS methods. The activation energy and pre-exponential factor were calculated from Eq.(9), where T_p is the peak temperature in the DSC curve. The results obtained from Kissinger method are E_a= 68.1, 63.4, 90.6 kJmol^{-1}, $\ln A$ = 14.5, 13.9 and 22.0 min^{-1} for free radical HTPB, LBH-3000 and PU-II respectively. The fact that Kissinger method gives a single value of the E_a and $\ln A$ for the whole process, so it does not reveal the complexity of the reaction. On the other hand, FWO and KAS methods allow evaluating the activation energy at different degree of conversion. For illustration, a typical FWO plots of $ln\beta_i$ versus $1/T_{\alpha i}$ for different values of conversion for free radical HTPB prepolymer are shown in Fig.20. Fig.21 shows the corresponding KAS plots of $ln(\beta_i/T_{\alpha i}^2)$ versus $1/T_{\alpha i}$ at different values of conversion. Similar plots were obtained for Krasol LBH-3000 and PU-II also, and the $E_a$values were calculated from the slope of the regression lines and are listed in Table 10 and 11. As can be seen that in all the cases, the E_a values obtained from the Kissinger method are well within the range of activation energies ($\alpha = 0.1 - 0.9$) obtained by FWO and KAS methods. We observed that the E_a obtained by FWO method agreed reasonably well to that obtained by KAS method. Moreover, the linear correlation coefficients are all very close to unity. So the results are credible. Additionally, the $E_a$values obtained from FWO method were somewhat higher than the values from the KAS method. This could be due to the approximation techniques used in the integration of the former method.

Further, we observed that E_a decreased with the increase of conversion in both the prepolymer as well as the PU-II. Moreover, the E_a varied with the conversion in a systematic trend, which followed a Power Law function ($E_a = k\alpha^n, r^2 \geq 0.90$). The variation of activation energy with degree of conversion indicates the self-accelerating phenomenon.

In the first step of degradation, the reaction is accelerated once the decomposition starts owing to the decrease of the activation energy at higher conversion. The related hydroperoxidation and peroxidation reaction products are formed with simultaneous loss of un-saturation. The variation of E_a with conversion revealed the existence of a complex multistep mechanism. Moreover, initially the apparent activation energy of the anionic HTPB was marginally higher up to 60% conversion, after that it was same /lower as compared to its free radical counterpart. This reveals that the initiation requires approximately the same activation energy but as the reaction proceeds, the rate of thermo-oxidation is higher for anionic HTPB (Krasol LBH-3000) as compared to free radical HTPB, because of self accelerating effect as anionic HTPB contains the higher per centage of *vinyl*-1,2-units. Also, the activation energy for the PU-II is higher than its prepolymer, which indicates that the PUs are more thermally stable and are less susceptible to oxidation than the substrate polymer. Although, the FWO and KAS methods have advantages in terms of evaluating the activation energy as a function of conversion, the major flaw in the approach is that they do not provide a direct way of evaluating either the pre-exponential factor or the reaction model. On the other hand, model-fitting methods help in fitting different models to α-temperature curves and simultaneously determining the activation energy and pre-exponential factor. There are several non-isothermal model-fitting methods, and the most widely used one is the Coats - Redfern method (Reza et al., 2007).

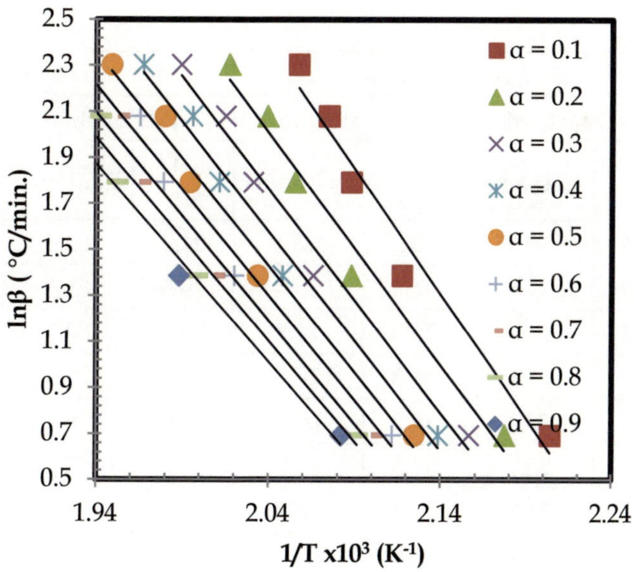

Figure 20. Iso-conversional plots of $ln\beta$ versus $1/T$ (FWO method) for prepolymer (free radical HTPB).

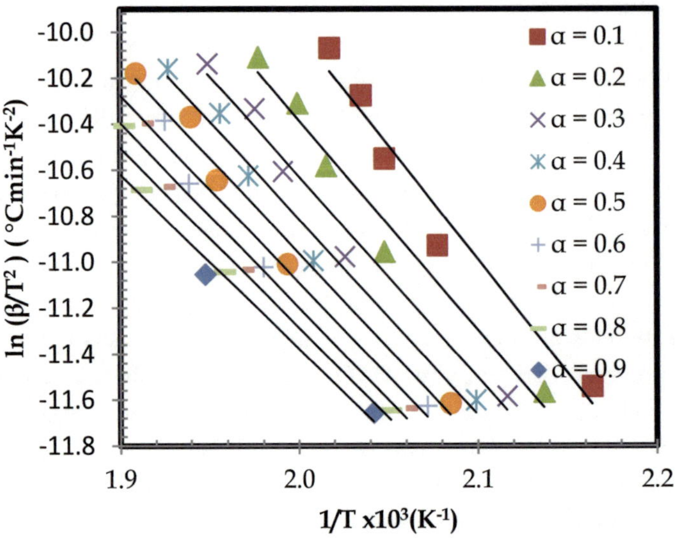

Figure 21. Iso conversional plots of $ln\,(\beta/T^2)$ versus $1/T$ (KAS method) for prepolymer (free radical HTPB).

Conv.	Free radical HTPB				Krasol LBH-3000			
	FWO method		KAS method		FWO method		KAS method	
	ln A * min^{-1}	E_a kJmol^{-1}	ln A * min^{-1}	E_a kJmol^{-1}	ln A * min^{-1}	E_a kJmol^{-1}	ln A * min^{-1}	E_a kJmol^{-1}
0.1	19.4	85.5	18.2	82.2	21.3	88.9	20.2	86.0
0.2	18.0	79.6	16.6	75.8	20.5	86.1	19.4	82.9
0.3	17.2	76.7	15.8	72.7	19.6	82.7	18.4	79.3
0.4	16.8	75.1	15.3	70.9	18.7	79.3	17.3	75.6
0.5	16.4	73.4	14.8	69.0	17.9	76.2	16.4	72.3
0.6	16.1	72.1	14.5	67.7	17.2	73.7	15.7	69.6
0.7	16.0	71.3	14.3	66.7	16.6	71.0	14.9	66.7
0.8	15.6	69.7	13.9	65.0	16.0	68.8	14.3	64.4
0.9	15.3	67.8	13.4	63.0	15.8	67.5	14.0	62.9

Table 10. Kinetic parameters for thermo-oxidative reaction of the prepolymers (HTPB) *(ln A values are calculated assuming the $g(\alpha) = [-\ln(1-\alpha)]^{1/2}$).

Conversion	Polyurethane: PU-II (Free radical HTPB)			
	FWO method		KAS method	
	$\ln A$ *(min^{-1})	E_a (kJmol^{-1})	$\ln A$ *(min^{-1})	E_a (kJmol^{-1})
0.1	24.2	99.0	23.3	96.3
0.2	23.4	96.0	22.5	93.2
0.3	23.0	94.2	22.0	91.3
0.4	22.6	92.6	21.5	89.5
0.5	22.3	91.5	21.2	88.3
0.6	21.9	89.8	20.7	86.5
0.7	21.5	88.4	20.3	85.0
0.8	20.9	86.2	19.7	82.6
0.9	20.2	83.6	18.8	79.8

Table 11. Kinetic parameters for thermo-oxidative reaction of the PU-II *($\ln A$ values are calculated assuming the $g(\alpha) = [-\ln(1-\alpha)]^{1/3}$).

3.4.2. Modelling of thermo-oxidation reaction of prepolymers (HTPB) and PU-II

The activation energies obtained from above three model-free methods (Kissinger, FWO and KAS) could be used to study the possible thermal degradation mechanism of prepolymer and its PU. We have used the Coats-Redfern method (CR) i.e. Eq. (14) to investigate the thermal degradation mechanism of the prepolymers and PU-II.

$$\ln\frac{g(\alpha)}{T^2} = \ln\left[\frac{AR}{\beta E_a}\left(1-\frac{2R\bar{T}}{E_a}\right)\right] - \frac{E_a}{RT} \qquad (14)$$

where $\bar{T}$ is the average value of the experimental temperatures. According to CR equation, if a correct model is selected for the thermal decomposition, the plot of $\ln[g(\alpha)/T^2]$ versus $1/T$ will be linear with high correlation coefficient giving the same kinetic parameters as obtained experimentally. First, the probable reaction model was selected and then the parameters were optimised by linear regression to obtain a precise model, which accurately fits the kinetic data. We found that model equation with $g(\alpha) = [-\ln(1-\alpha)]^{1/n}$ (Avrami-Erofeev equation) reasonably fit the kinetic data derived from FWO and KAS method. For prepolymer, $n = 2$, whereas, for PU-II $n = 3$. Table 12 lists the kinetic parameters along with the correlation co-efficient calculated by Coats -Redfern method taking $g(\alpha) = [-\ln(1-\alpha)]^{1/n}$ at different heating rates.

On comparison of values of E_a and $\ln A$ calculated by the model equations with those obtained by Kissinger, FWO, and KAS methods, we observed that they were reasonably in good agreement with each other. So, we concluded that the most probable kinetic model function of the thermo-oxidative degradation of prepolymers and PU-II could be described by Avrami-Erofeev equation with $f(\alpha) = 2(1-\alpha)[-\ln(1-\alpha)]^{1/2}$ and $f(\alpha) = 3(1-\alpha)[-\ln(1-\alpha)]^{2/3}$, respectively.

β °Cmin^{-1}	Free radical HTPB			Krasol LBH-3000			Polyurethane: PU-II (Free radical HTPB)		
	ln A min^{-1}	E_a kJmol^{-1}	r^2	ln A min^{-1}	E_a kJmol^{-1}	r^2	ln A min^{-1}	E_a kJmol^{-1}	r^2
2	21.1	91.8	0.994	19.3	82.5	0.997	---	---	---
4	19.4	86.6	0.995	16.3	71.7	0.997	21.0	87.7	0.957
6	16.1	73.7	0.998	15.7	69.0	0.996	22.0	91.0	0.966
8	17.1	77.2	0.995	15.0	66.3	0.996	19.0	79.4	0.955
10	14.4	66.7	0.995	12.4	56.3	0.994	17.9	75.1	0.957

Table 12. Kinetic parameters for no-isothermal oxidation by Coats -Redfern equation.

3.5. HTPB polyurethanes: Stress-strain properties

PU elastomers exhibit good elasticity in a wide range of hard segment contents. This is due to the change of soft or hard segments in different proportion and structure. PUs are composed of short alternating hard and soft segments. The hard segment of PUs usually consists of diisocyanate linked to a low molecular weight chain extender such as butanediol. Meanwhile, the thermodynamic incompatibility between hard and soft segments can lead to the micro-phase separation and hence make a significant contribution to elastomeric properties. Basically, soft segments provide the elasticity, while hard segments play a role in reinforcing the filler and physical cross-linking. In a condensed structure, hard segments usually exist in glassy state or crystalline state. Because of the strong hydrogen bonds of hard segments, their domains can be formed and distributed in the soft segments. The PU elastomeric properties obtained for different systems are reported in Table 13. As a generic trend, it was observed that increase in hard segment content corresponded to higher values of hardness, tensile strength and modulus. The increase in mechanical properties with hard segment content was attributed to the progressive effect of hydrogen bonds within the hard domains of the cross-linked PUs.

Parameters	Unfilled polyurethanes		Filled polyurethanes
	PU-I	PU-II	PU-IIp
Hard segment (% w/w)	4.34	7.25/7.34/7.43/7.52/7.61	7.25/7.34/7.43/7.52/7.61
Elastomeric properties: TS (kgf/cm2)	2.4	3.5/3.6/4.0/4.2/4.4	7.3/8.6/8.9/10.8/11.8
Elong. (%)	350	759/631/627/520/437	44/42/39/35/33
Mod.(kgf/cm2)	---	---	45/52/59/78/83
Hardness (Shore-A)	10	10/14/15/18/20	65/79/80/83/85

Table 13. Elastomric properties of different PU systems.

4. Conclusion

The chapter provides an insight into the microstructure and sequence distribution of the substrate polymer obtained from analysis of 1D and 2D ^{13}C and ^{1}H NMR techniques. The absolute molecular weight of the prepolymer has been determined by high field NMR method. This study pointed out that the HTPB prepolymer was a Newtonian fluid and viscosity decreased exponentially with temperature. The activation energy for viscous flow for free radical HTPB was less than that of anonic prepolymer. The chemo-rheological analysis concludes that the shear rate has no significant effect on the viscosity of the PU reaction within the cure time. The viscosity of various PU systems rises exponentially with cure time. The rate of viscosity build up for filled PU (propellant) is quite low as compared to the unfilled PU systems. Unlike the unfilled PUs, the filled PU slurry showed pseudoplastic behavior, *i.e.* the shear rate had significant effect on viscosity of the propellant slurry. For a typical composition with 86% solid loading, the pseudoplasticity index was found to be higher at higher temperature. It shows that at higher temperature, it becomes more non-Newtonian. Additionally, it also revealed that the pseudoplasticity index remained unchanged within the cure time studied (*i.e.*, 3 h), when maintained at 50 °C, which is desirable in view of propellant flow during casting of the propellant slurry. Further, the filled PU (propellant) gave excellent elastomeric properties, which were apt for solid rocket motor requirement. Additionally, the desired properties can be easily accentuated by simply tailoring the hard segment content of the PU composition. Thermo-oxidative behavior, as studied by DSC of the substrate polymer and the PU elastomers, confirms that PU elastomers are more resistant to thermo-oxidation as compared to the substrate polymer. The thermo-oxidative degradation could be modeled well by an empirical equation given by Avrami-Erofeev. Endowed with so many advantages, HTPB PUs is undoubtedly a versatile and ubiquitous fuel binder for solid rocket motors. However, in order to gain an in depth insight into the multi-step reaction mechanism, further analysis of the DSC data is warranted. Future studies aim at the simulation of the thermo-oxidative profile of HTPB PUs by using a suitable Computer Software in order to understand its complexity.

Author details

Abhay K. Mahanta
Defence Research & Development Organization, SF Complex, Jagdalpur, India

Devendra D. Pathak
Department of Applied Chemistry, Indian School of Mines, Dhanbad, India

Acknowledgement

Authors are thankful to the General Manager SF Complex, Jagdalpur for his kind permission to publish the article.

5. References

Arlas, B.F.; Rueda, L.; Stefani, P.M.; Caba, K.; Mondragon, I. & Eceiza A. (2007). Kinetic and Thermodynamic Studies of the Formation of a Polyurethane Based on 1,6-Hexamethylene Diisocyanate and Poly(carbonate-*co*-ester) Diol. *Thermochimica Acta*, Vol. 459, pp. 94-103

Duh, Y. S.; Ho T.C.; Chen, J.R. & Kao, C. S. (2010). Study on Exothermic Oxidation of Acrylonitrile-butadiene-styrene (ABS) Resin Powder with Application to ABS Processing Safety.*Polymers*, Vol.2, pp. 174-187

Chen, J. K. & Brill, T.B. (1991).Chemistry and Kinetics of Hydroxyl-terminated Polybutadiene(HTPB) and Diisocyanate-HTPB Polymers during Slow Decomposition and Combustion-like Conditions. *Combustion and Flame*, Vol. 87, pp. 217-232

Eroglu, M. S. (1998). Characterization of Network Structure of Hydroxyl Terminated Poly (butadiene) Elastomers Prepared by Different Reactive Systems. *Journal of Applied Polymer Science*, Vol.70, pp. 1129-1135

Elgert, K. F.; Quack, G. & Stutzel, B. (1975). On the Structure of Polybutadiene: 4.^{13}C n. m. r. Spectrum of Polybutadienes with cis-1, 4-, trans-1,4- and 1,2-units. *Polymer*, Vol.16, pp. 154-156

Frankland, J. A.; Edwards, H. G. M.; Johnson, A.F.; Lewis, I.R. & Poshyachinda, S. (1991). Critical Assessment of Vibrational and NMR Spectroscopic Techniques for the Microstructure Determination of Polybutadienes. *Spectrochimica Acta.*, Vol.47A, No. 11, pp.1511-1524

Flynn, J.H. & Wall L. A. (1966).A Quick, Direct Method for the Determination of Activation Energy from Thermogravimetric Data. *Journal of Polymer Science Part B: Polymer Letters*, Vol.4, No. 5, pp. 323-328

Haas, L. W. (1985). Selecting Hydroxy-terminated Polybutadiene for High Strain Propellants. US Patent Number 4536236, pp. 1-8

Kalsi, P. S. (1995). Proton Nuclear Magnetic Resonance Spectroscopy (PMR), *Spectroscopy of Organic Compounds*.2nd ed., pp.165-296, Wiley Eastern Limited, New Delhi

Kebir, N.; Campistron, I.; Laguerre, A.; Pilard, J. F.;Bunel, C.; Couvercelle, J. P. & Gondard, C. (2005). Use of Hydroxytelechelic *cis*-1,4-polyisoprene (HTPI) in the Synthesis of Polyurethanes (PUs). Part 1. Influence of Molecular Weight and Chemical Modification of HTPI on the Mechanical and Thermal Properties of PUs. *Polymer*, Vol.46, pp. 6869-6877

Kissinger, H. E. (1956).Variation of Peak Temperature with Heating Rate in Diffrential Thermal Analysis. *J. of Research of the National Bureau of Standards*,Vol. 57,No.4,pp. 217-221

Lakshmi,R.; & Athithan, S.K. (1999). An Empirical Model for the Viscosity Buildup of Hydroxy Terminated Polybutadiene Based Solid Propellant Slurry. *Polymer Composites* Vol. 20, No.3, pp. 346-356

Manjari, R.; Somasundaran, U. I.; Joseph, V. C. & Sriram, T. (1993). Structure-Property Relationship of HTPB-Based Propellants II. Formulation Tailoring for Better Mechanical Properties. *Journal of Applied Polymer Science*, Vol. 48, pp.279-289

Muthiah, R. M.; Krishnamurthy, V.N. & Gupta B.R. (1992). Rheology of HTPB Propellant.1.Effect of Solid Loading, Oxidizer Particle Size, and Aluminum Content. *Journal of Applied Polymer Science*, Vol. 44, pp. 2043-2052

Mahanta, A. K.; Dharmsaktu, I. & Pattnayak, P.K. (2007). Rheological Behaviour of HTPB-based Composite Propellant: Effect of Temperature and Pot Life on Casting Rate. *Defence Science Journal*, Vol.57, No.4, pp. 435-442

Navarchian, A. H.; Picchioni, F. & Janssen, L. P. B. M. (2005). Rheokinetics and Effect of Shear Rate on the Kinetics of Linear Polyurethane Formation. *Polymer Engineering and Science*, pp. 279-287

Ozawa, T. (1965). A New Method of Analyzing Thermogravimetric Data. *Bulletin of the Chemical Society of Japan*, Vol.38, pp. 1881-1886

Panicker, S. S. & Ninan, K.N. (1997). Influence of Molecular Weight on the Thermal Decomposition of Hydroxyl Terminated Polybutadiene. *Thermochimica Acta* Vol.290, pp. 191-197

Poussard, L; Burel, F.; Couvercelle, J. P.; Merhi, Y.; Tabrizian, M. & Bunel, C. (2004). Hemocompatibility of New Ionic Polyurethanes: Influence of Carboxylic Group Insertion Modes. *Biomaterials*, Vol. 25, pp. 3473-3483

Poletto, S. & Pham, Q. T. (1994).Hydroxytelechelic Polybutadiene, 13[a)] Microstructure, Hydroxyl Functionality and Mechanisms of the radical polymerization of Butadiene by H_2O_2.*Macromol. Chem.Phys.* Vol.195, pp.3901-3913

Reji, J.; Ravindran, P.; Neelakantan, N.R. & Subramanian, N. (1991). Viscometry of Isothermal Urethane Polymerization.*Bull.Chem.Soc.Jpn*.Vol. 64, pp.3153-3155

Reza, E.K.; Hasan A.M. & Ali, S. (2007). Model-Fitting Approach to Kinetic Analysis of Non-Isothermal Oxidation of Molybdenite.*Iran.J.Chem.Chem.Eng*.,Vol.26, No.2, pp.119-123

Sadeghi, G. M. M.; Morshedian J. & Barikani, M. (2006). The Effect of Solvent on the Microstructure, Nature of Hydroxyl End Groups and Kinetics of Polymerization Reaction in Synthesis of Hydroxyl Terminated Polybutadiene. *Reactive & Functional Polymers*, Vol. 66, pp. 255-266

Singh, M.; Kanungo, B.K. & Bansal, T.K. (2002). Kinetic Studies on Curing of Hydroxy-Terminated Polybutadiene Prepolymer-Based Polyurethane Networks. *Journal of Applied Polymer Science*, Vol. 85, pp. 842-846

Sato, H.; Takebayashi, K. & Tanaka, Y. (1987). Analysis of ^{13}C NMR of Polybutadiene by Means of Low Molecular Weight Model Compounds. *Macromolecules*, Vol.20, pp. 2418-2423

Zheyen, Z.; Zinan, Z. & Huimin, M. (1983). ^{13}C-NMR Study on the equibinary (cis-1,4;1,2) Polybutadiene Polymerized with Iron Catalyst. *Polm. Comm.*, No. 1, pp. 92-100

Chapter 12

Synthesis of a New Sorbent Based on Grafted PUF for the Application in the Solid Phase Extraction of Cadmium and Lead

Rafael Vasconcelos Oliveira and Valfredo Azevedo Lemos

Additional information is available at the end of the chapter

http://dx.doi.org/10.5772/36501

1. Introduction

The ability to determine trace elements in various types of samples is important in many areas of science including environmental, food, geochemical, forensic, and pharmaceutical. The amount of certain elements can indicate the level of contamination in a region, the nutritional value of a food and the quality of a manufactured product, among other things. However, the matrix of the samples can be chemically complex due to the large number of substances that are present. Some of these substances can hinder the determination of trace elements due to incompatibility with some detectors, especially those based on spectrometry. For example, substances can influence the viscosity of a solution that is introduced into the flame atomic absorption spectrometer (FAAS), interfering with the nebulization process (Teixeira et al., 2005). Additionally, some substances may interfere with the pyrolysis of a sample that is introduced into the graphite tube of an electrothermal atomization atomic absorption spectrometer (ETAAS) (Zambrzycka et al., 2011; Serafimovska, et al., 2011). Another common problem occurs when the content of the element is measured at a very low level in some matrices. In this case, the technique does not provide a detection limit sufficient to determine the element in the sample. These difficulties can be resolved or reduced by improving the selectivity and sensitivity of the analytical method or by including separation and preconcentration steps in the procedure. Separation is the removal of measurable constituents or interfering substances from the sample matrix. Preconcentration is a procedure based on the separation of a measured quantity of constituents for a medium volume that is smaller than the sample matrix.

Many procedures used for separation and preconcentration that involve different techniques are found in the literature, such as liquid-liquid extraction, coprecipitation, cloud-point extraction and solid phase extraction (Zeeb & Sadeghi, 2011; Tuzen et al.,2008; Lemos et al.,

2008; Oral et al., 2011). Among these techniques, solid phase extraction deserves special attention because of several advantages this technique offers. Among these advantages is the reduction or elimination of the use of toxic organic solvents, the achievement of high enrichment factors, a high versatility due to various types of sorbents that are low in cost, decreased operating time and ease of automation. Many substances are used as sorbents for solid phase extraction of trace elements, such as naphthalene, activated carbon, alumina, silica, biosorbents, such as hair, bagasse and peat, and polymeric sorbents, including polyurethane foam (PUF) and polystyrene-divinylbenzene (Beketov et al., 1996; Zhang et al., 2011; Jamshidi et al, 2011; Costa et al., 2011; Matos & Arruda, 2006; Gonzales et al., 2009).

Naphthalene is used in preconcentration procedures because of its ability to extract or form a complex with organic species (Fathi et al., 2011). This sorbent makes it possible to achieve high enrichment factors and is soluble in many solvents, which can facilitate the detection process. However, naphthalene is toxic and possibly carcinogenic, and its use is restricted to batch procedures.

Activated carbon is an excellent material for use in extraction procedures due to its large surface area and its strong interaction with organic species and trace elements (Zhang et al., 2011). The disadvantage of using activated carbon in extraction procedures is that the interaction with some species is so strong it can cause an irreversible sorption. In addition, activated carbon has a heterogeneous surface with active functional groups that often lead to low reproducibility. These sorbents are also very reactive and can act as catalysts of undesirable chemical reactions.

Silica gel is also widely used in solid phase preconcentration procedures due to its mechanical strength, resistance to swelling caused by solvent change and a high adsorption capacity of several species (Tzvetkova et al., 2010). The disadvantages of using this sorbent include a very low selectivity and hydrolysis at basic pH.

Biosorbents are materials of great interest for use in an extraction system because of their availability and versatility (Gonzales et al., 2009). However, some of these sorbents are difficult to recover and the mechanisms of sorption of trace species are not yet fully understood.

Macroporous hydrophobic resins are good supports for developing chelating sorbents. These materials are resins based on polystyrene-divinylbenzene with a high hydrophobic character and no ion-exchange capacity. Several reagents have been incorporated into polystyrene-divinylbenzene to form chelating resins for use in preconcentration procedures (Lemos et al., 2006a).

Among the sorbents used for the extraction of organic and inorganic species for separation and preconcentration, PUF is noteworthy due to the benefits achieved. This material was used for the extraction of aqueous species for the first time in the early 1970s (Braun and Farag, 1978). Since then, PUF has been used in several procedures for separation and preconcentration, and its use has been the subject of several books and reviews (Braun, 1983; Navratil et al., 1985; Lemos et al., 2007). PUF is low in cost, easy to purchase, and provides high enrichment factors. However, this sorbent may swell when treated with solvents (Braun, 1983; Navratil et al., 1985).

The polyurethane-based polymers are widely used for solid phase extraction, with or without treatment, due to their very low cost and simplicity of preparation (Moawed & El-Shahat, 2006; Saeed & Ahmed, 2005). Moreover, the sorbent is resistant to changes in pH and has a reasonable resistance to swelling in the presence of organic solvents. This material can also be used as a support for many reagents in separation and preconcentration. For example, the combination of PUF with chelating reagents has resulted in powerful sorbents for the extraction of metal ions. This association can be effected via impregnation, functionalization or grafting of chelating reagents in PUF. The use of this material to associate with organic reagents in the separation and preconcentration of trace species has many possible applications. However, the potential for grafted or functionalized foam in solid phase extraction has not been fully explored (El-Shahat et al., 2003; Lemos et al., 2006b).

Loading of chelating reagents in PUF is a fairly simple process. In this procedure, the chelating reagent is usually dissolved in an alcoholic solution and passed through a column that is packed with PUF. In batch procedures, the reagent solution is maintained in contact with the foam for a short period, while stirring. After sorption, the sorbent is washed with an alkaline solution to remove excess chelating reagents (El-Shahawi et al., 2011a; El-Shahawi et al., 2011b).

Chelating reagents may be directly introduced into PUF by chemical bonding. This procedure, sometimes called functionalization, produces a very stable sorbent. The binding is based on the participation of amino groups, which are constituents of terminal toluidine groups of PUF, in typical reactions of aromatic amines, such as diazotization or azo coupling reactions (Lemos et al., 2010)]. These reactions result in the incorporation of a ligand by an –N=N– or –N=C– spacer arm (Burhan, 2008; Azeem et al., 2010).

PUFs grafted with chelating reagents have shown excellent characteristics for use in systems of solid phase extraction, such as selectivity, high enrichment factors and stability. Grafted PUF is a material with two or more monomers polymerized through an addition reaction to provide a simple polymer containing different subunits. Grafted PUFs are considered excellent adsorbent materials that have good stability, high extraction capacity of inorganic and organic species, and good flexibility. Grafted PUF has been prepared by mixing PUF with an appropriate reagent prior to the addition of diisocyanate to form the foam material. Several substances, such as Nile blue A (Moawed & El-Shahat, 2006), methylene blue (Moawed et al., 2003; El-Shahat et al., 2007), rosaniline (Moawed, 2004), rhodamine B (El-Shahat et al., 2003) and brilliant green (El-Shahat et al., 2003), have been incorporated into PUF to obtain grafted sorbents. Table 1 summarizes the characteristics of various procedures that use grafted PUF for solid phase extraction of trace species.

In this work, we have synthesized a new sorbent material based on PUF grafted with the chelating reagent 2-[2'-(6-methyl-benzotiazolilazo)]-4-aminophenol (Me-BTAP). The material was characterized by IR spectroscopy and thermogravimetry. The sorbent was used in a solid phase extraction system for preconcentration and determination of cadmium and lead in water samples.

Reagent	Analyte	Limit of detection (μg L^{-1})	Enrichment factor	Sample	Ref
Nile blue A	Zn(II)	-----	-----	Wastewater	(M.F. El-Shahat et al., 2003)
	Cd(II)				
	Hg(II)				
Methylene Blue	Cd(II)	<5,0	40	Wastewater	(Moawed et al., 2003)
	Hg(II)	<5,0	40		
	Ag(I)	<5,0	40		
Rosaniline	Cd(II)	-----	100	Wastewater	(Moawed, 2004)
	Hg(II)				
Methylene blue	U(VI)	-----	-----	Wastewater	(M.F. El-Shahat et al., 2007)
Rhodamine B					
Brilliant green					
Methylene blue	Penicillin G	12	14	Antibiotics	(M.F. El-Shahat et al., 2010)
	Amoxicillin	15	16		
	Ampicillin	19	11		

Table 1. Procedures involving grafted PUF applied in the preconcentration of chemical species by spectrophotometry.

2. Experimental

2.1. Apparatus

A Perkin Elmer (Norwalk, CT, USA) model AAnalyst 200 atomic absorption spectrometer equipped with a deuterium lamp for background correction was used for absorbance measurements. The wavelength value for the hollow cathode lamp for cadmium and lead was 228.8 nm and 283.3 nm, respectively. The flow rate of acetylene and air in the burner were 2.5 and 10.0 L min^{-1}, respectively. The nebulizer flow rate was 8.0 mL min^{-1}.

A Digimed (model DM 20, São Paulo, Brazil) pH meter was used to measure the pH of metal solutions. The preconcentration procedure was performed on a simple on-line system involving two steps: preconcentration and elution. The system consists of a Milan model 204 (Colombo, Brazil) four-channel peristaltic pump operated with silicone tubes and a six-port Rheodyne valve model 5041 (Cotati, USA). Teflon tubes were used in the construction of the system. Polyvinyl chloride (PVC) was used in the construction of the minicolumn (3.50 cm in length and an internal diameter of 4.0 mm). The minicolumn was filled with 400 mg of the synthesized sorbent.

The infrared spectrum was obtained using KBr pellets at 1.0% (w/w) in a Perkin Elmer Spectrum One FTIR spectrometer. Differential thermal analysis (DTA) and thermogravimetric analysis (TGA) curves were obtained in a Shimadzu TGA-50H apparatus in an aluminum cell under air or N_2 (50 mL min^{-1}) and scanned between the temperature values of 0.0 to 800.0 °C at a heating rate of 20 °C min^{-1}.

2.2. Reagents

Ultrapure water from an Elga Purelab Classic was used to prepare all solutions. Working solutions of cadmium and lead at the μg L^{-1} level were prepared daily by diluting a 1000 μg mL^{-1} solution of each element (Merck). Hydrochloric acid solutions were prepared by direct dilution of the concentrated solution (Merck) with ultrapure water. Acetate buffer solutions (pH 4.0-6.0), borate (pH 7.0-8.5) and ammonia (pH 9.0-9.5) were used to adjust the pH. The reagents 2,4- toluene diisocyanate (Aldrich), tin(II) 2-ethylhexanoate (Aldrich), dimethylamino-1-propanol (Aldrich), polyethylene glycol (Aldrich) and silicone oil AP 100 (Aldrich) were used in the synthesis of the sorbent.

2.3. Synthesis of the reagent Me-BTAP

The synthesis of the reagent 2-[2′-(6-methyl-benzothiazolylazo)]-4-aminophenol (Me-BTAP) was performed as described previously (Lemos et al., 2006a). The production of the reagent has been completed in two steps: diazotization of 6-methyl-2-aminobenzothiazole followed by the coupling of the diazotized product with 4-aminophenol. The diazotization reaction was performed by dissolving 6-methyl-2-aminobenzothiazole (3.0 g) in 50 mL of a 6.0 M hydrochloric acid solution. Then, a solution of 2.0 g of sodium nitrite in 20 mL of water at 0-5 °C was added dropwise, and the mixture was stirred at a constant temperature of 0–5 °C for 1 h. The diazotate mixture was added dropwise to a solution of 3.0 g of 4-aminophenol in 20 mL of an 1.0 M sodium carbonate solution at 0–5 °C under vigorous stirring. The system was allowed to stand overnight in a refrigerator at 0–5 °C. The resulting dark-green precipitate was filtered and purified by recrystallization in ethanol. The proposed structure of Me-BTAP is shown in Figure 1.

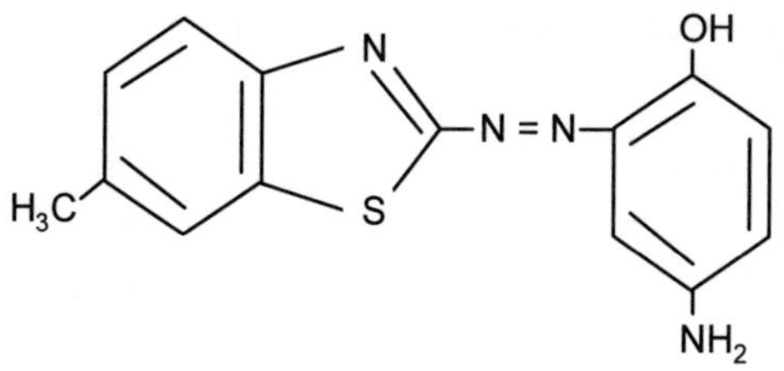

Figure 1. Proposed structure of Me-BTAP.

2.4. Synthesis of the sorbent

Preparation of the grafted sorbent was performed according to the following procedure (El-Shahat et al., 2003; Moawed, 2004): 20.00 g of polyethylene glycol and 0.04 g of dimethylamino-1-propanol were added to 1.0 g of distilled water under vigorous stirring. Next, silicone oil (0.05 g) and tin(II) 2-ethylhexanoate (0.04 g) were added to the mixture. The system was shaken to obtain a homogeneous mixture. The reagent Me-BTAP (0.02 g) was then added, and the mixture was stirred for ten minutes. Afterward, approximately 13.0 g of toluene diisocyanate was added gradually. The resulting polymer was cut into small pieces and washed with 1.0 M HCl, ethanol and water. After washing, the polymer was dried at room temperature.

2.5. Procedure for preconcentration

Solutions containing Cd and Pb were adjusted to pH 7.5 with borate buffer. These solutions were passed through the sorbent minicolumn that contained the PUF-Me-BTAP. At this stage, the elements are sorbed onto the solid phase. After preconcentration (120 s), the position of the six-port valve was changed, and an eluent flow was passed through the minicolumn. The eluent transported Cd (II) or Pb (II) to the nebulizer of the flame atomic absorption spectrometer. The analytical signal was then measured as the peak height (absorbance).

3. Results and discussion

3.1. Characterization of the sorbent

The spectrum in the infrared region of the material PUF-Me-BTAP (Figure 2) shows that the absorptions in the range of 3600-3300 cm^{-1}, centered at 3448 cm^{-1} and 3358 cm^{-1}, can be attributed to the stretches of the -OH and-NH groups, respectively. The bands between 2970 and 2920 cm^{-1} are characteristic of the aliphatic part (-CH_2 and CH_3) of the sorbent structure. The peaks between 1500 and 1448 cm^{-1} are characteristic vibrations of the -CS- group, related to the segment molecular organic reagent for the Me-BTAP. The absorption at 1708 cm^{-1} was assigned to the axial deformation of the urethane carbonyl group conjugated by hydrogen bonding. There were no bands featuring free –NCO of urethane groups (1730-1720 cm^{-1}) (Radhakrishnan Nair, 2008)]. Solubility tests showed that the sorbent is insoluble in the following solvents: chloroform, methanol, ethanol, tetrahydrofuran, benzene, toluene, acetone, diethyl ether, isopropanol, dioxane and acetic acid. In the presence of pyridine and dimethyl sulfoxide, the material swelled. Briefly, the particles formed a gel with a volume greater than 3.2 times the initial volume.

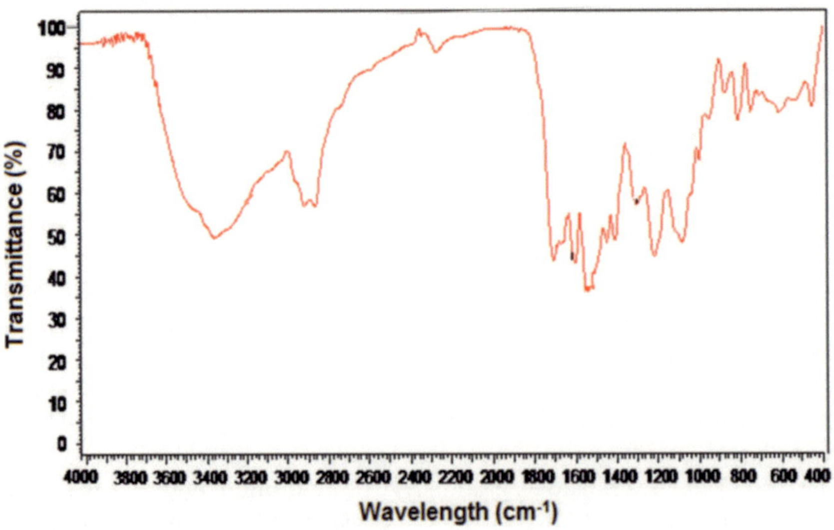

Figure 2. Infrared spectrum of PUF-Me-BTAP.

The graphs of TGA for Me-PUF BTAP in N_2 (Figure 3) and O_2 (Figure 4) show that there are no significant differences related to the atmosphere used during the degradation process for the amount of steps, temperature ranges and loss of mass.

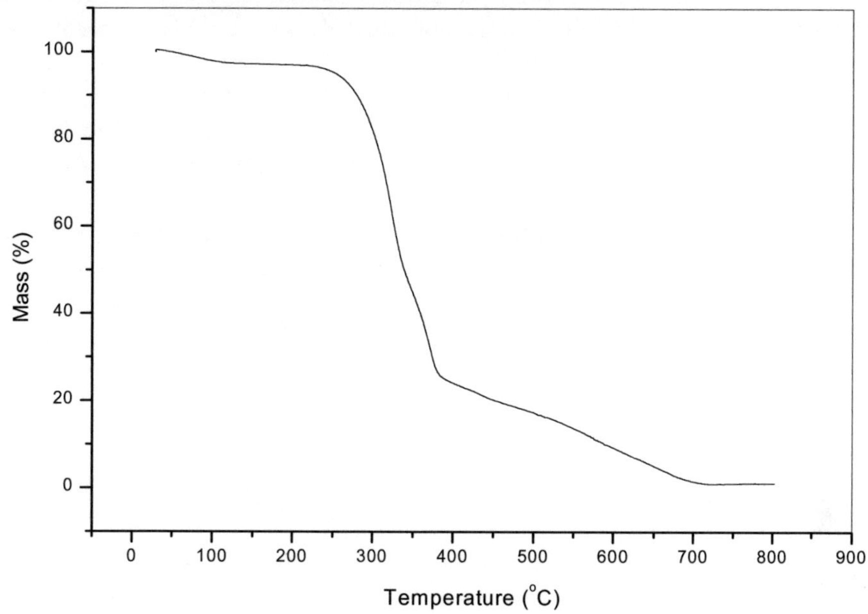

Figure 3. Thermogravimetric curve of the material PUF-Me BTAP under an N_2 environment.

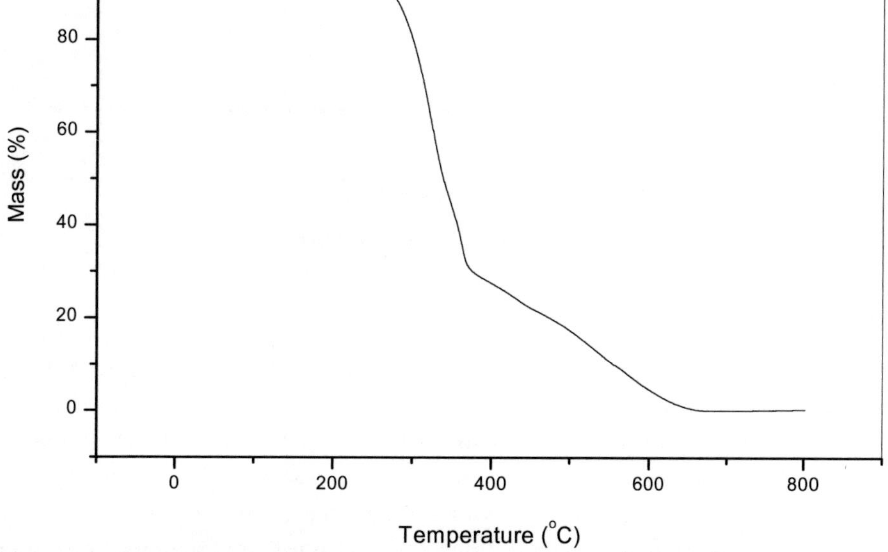

Figure 4. Thermogravimetric curve of the material PUF-Me BTAP under an O_2 environment.

The degradation of the material occurs in two general stages. In Stage I, the degradation is mainly due to the decomposition of rigid segments and involves the dissociation of urethane and the original chain extender, which then form primary amines, alkenes and carbon dioxide. Stage I is influenced by the amount of rigid segments. In the subsequent stage II, depolymerization and degradation of the polyol occur. Therefore, this stage is affected by the content of flexible segments. According to Figure 5, there is a maximum of degradation, indicated by the first derivative curve of DTA at 364 °C in air (O_2). The first stage of degradation occurred concomitantly with a phase transition, possibly because part of the polymer changed from a crystalline to an amorphous phase. An important observation is that the sorbent Me-BTAP-PUF has a high thermal stability at the final temperature for the first stage of degradation in air at 280 °C. The observed mass reduction, which occurred between 0 and 120 °C, was attributed to loss of water.

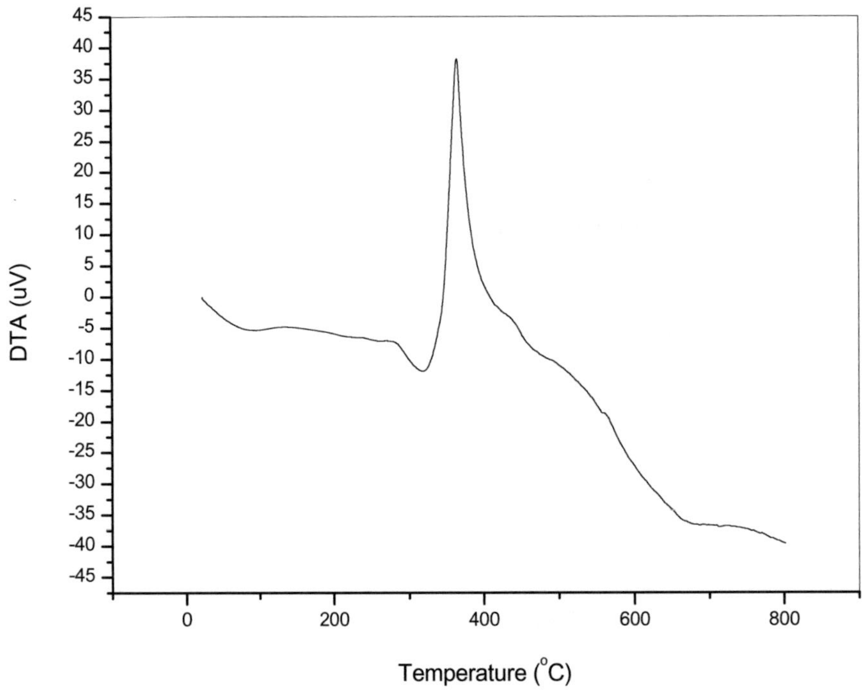

Figure 5. DTA curve of the material PUF-Me BTAP in an O_2 environment.

3.2. Effect of pH

Many complexing agents are Lewis bases (capable of donating electron pairs) and Brönsted bases (capable of receiving protons) and, as such, will be affected by changes in pH. The reaction for the formation of the chelate is influenced by pH because the chelating agent is not presented entirely in the form of a free ion. Thus, the effect of hydrogen concentration was studied to observe the pH range over which the cations cadmium and lead are absorbed by PUF-Me-BTAP. Figure 6 shows the influence of pH on the extraction of cadmium and lead by PUF-Me-BTAP.

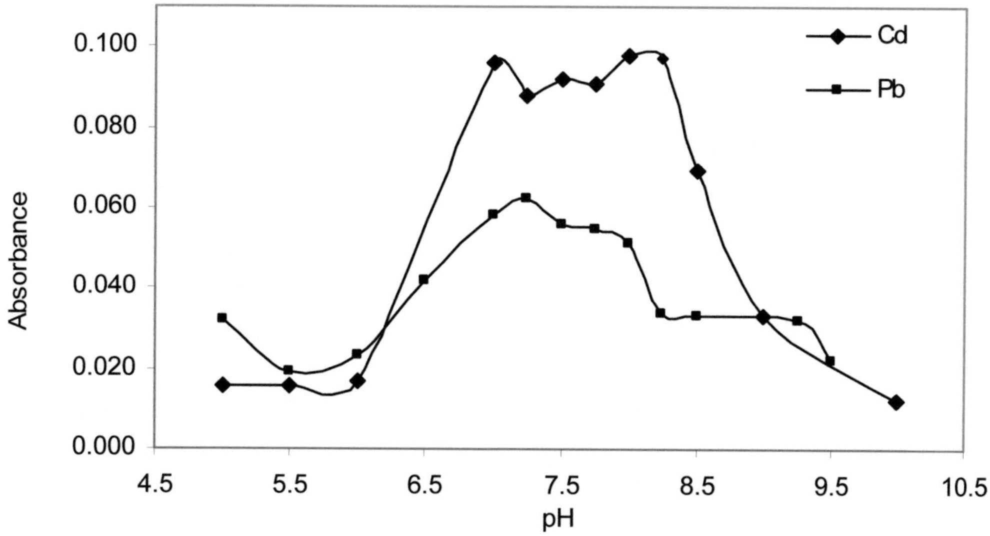

Figure 6. Influence of pH on the determination of Cd and Pb using solid phase extraction.

According to Figure 6, the best pH range for the extraction of cadmium is between 6.8 and 8.2. The extraction of lead is maximal when performed at pH values between 7.0 and 7.8. Thus, the extraction of both metals was performed at pH 7.5 in all subsequent experiments.

3.3. Type and concentration of the eluent

Polyurethane may be dissolved by concentrated sulfuric acid or oxidized by concentrated nitric acid and potassium permanganate solutions. This material was resistant to solvents such as water, hydrochloric acid (up to 6 mol L^{-1}), ethanol and glacial acetic acid (Navratil et al., 1985). Thus, HCl was chosen to prevent a reduction in the lifetime of the PUF that was grafted with Me-BTAP. PUF-Me-BTAP is resistant to ethanol. However, the use of this solvent in the elution of cadmium and lead presented pressure problems in the on-line system in this work. When ethanol was used as the eluent, there was a swelling of the sorbent, which caused backpressure on the minicolumn. This increase in pressure caused leaks throughout the system on-line. Thus, the use of this solvent was discontinued. Hydrochloric acid solutions were then used as the eluent in all further experiments.

Figure 7 illustrates the phenomenon of desorption of cations from the solid phase when the concentration of HCl is varied. It was observed that the hydrochloric acid solutions that provided the highest analytical signals were those at concentrations ranging between 0.01 and 1.00 mol L^{-1} (Cd) and 0.10 and 1.00 mol L^{-1} (Pb). The use of low concentrations of acid is beneficial because it can increase the lifetime of the minicolumn. Moreover, in this work, we chose to use an eluent of identical concentration for both metals, with the aim of simplifying the operation of the on-line system. Therefore, a solution concentration of 0.10 mol L^{-1} for the desorption of both chemical species was used in all subsequent experiments.

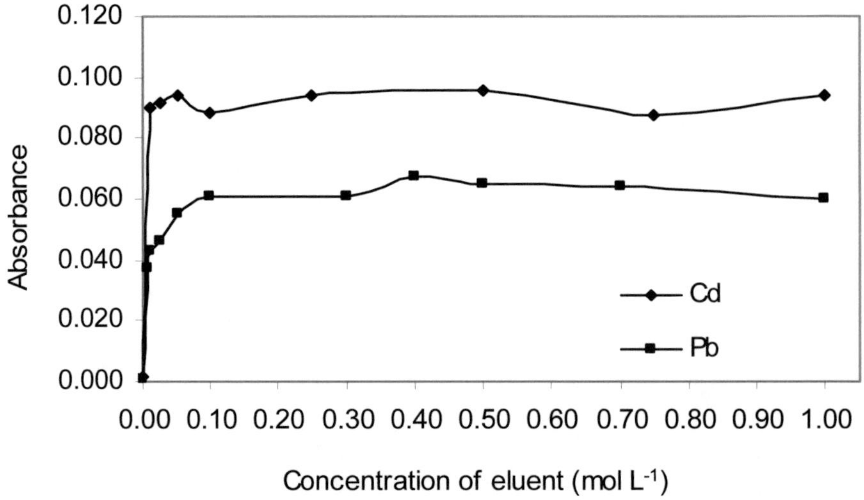

Figure 7. Influence of eluent concentration on the determination of Cd and Pb using solid phase extraction.

3.4. Flow rate of solutions

In on-line preconcentration systems, it is crucial to study the flow of the sample to meet the appropriate speed at which the ions pass through the minicolumn. The results of the influence of flow rate in on-line preconcentration systems of Cd (II), shown in Figure 8, show that the extraction is maximal when the flow rate ranges between 3.3 and 4.6 mL min^{-1}. Values outside this range cause a decrease in the analytical signal.

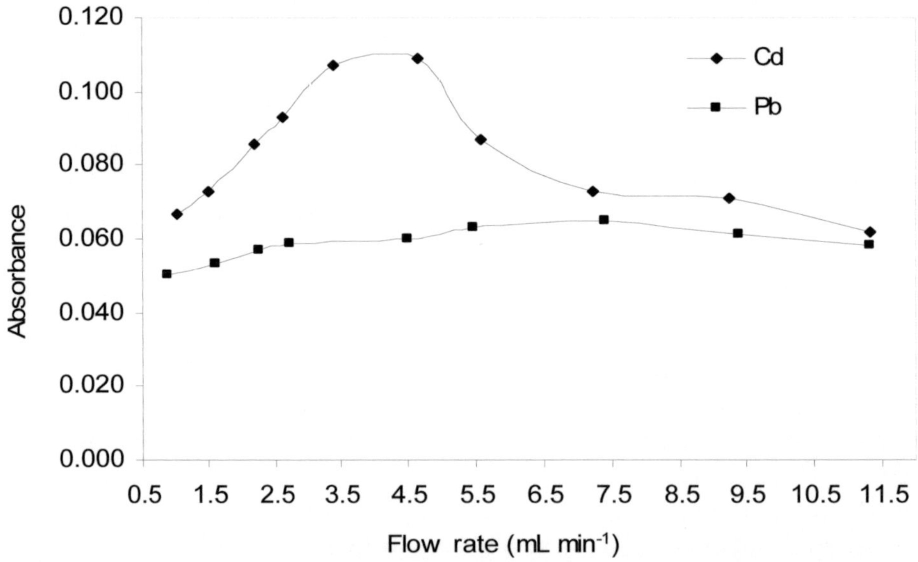

Figure 8. Influence of flow rate of the Cd and Pb solutions for the determination of elements using solid phase extraction.

If the flow rate of the metal solution is too high, there is a possibility that the metal ions can pass through the minicolumn at a speed so quickly that a portion of the analyte passes through without being sorbed. Conversely, an excessively low flow rate of the metal solution can also cause problems with the analytical signal. A solution that passes through the minicolumn with a very low flow rate can result in leaching of the complexed species and significantly increase the analysis time. Considering the curve that corresponds to the lead solution, we observed a similar behavior to that of cadmium. However, the decrease in the amount of extracted metal was smoother. The range of flow rate that produces the maximum extraction of lead was between 5.5 and 7.4 mL min^{-1}. Based on these results, flow rates of 4.5 and 7.0 mL min^{-1} were used in further experiments for solutions of cadmium and lead, respectively.

The inconsistency between the rate of aspiration of the nebulizer of the spectrometer and the flow rate of eluent of the on-line system could result in peak broadening of the analytical signal. This broadening will result in a decrease in the analytical signal (Lemos et al., 2007). Thus, the flow of the eluent for desorption of cadmium and lead ions was adjusted to 8.0 ml min^{-1} to match the flow rates of elution and aspiration of the nebulizer of the spectrometer.

3.5. Preconcentration time

A linear relationship between preconcentration time and analytical signal is dependent on the flow of metal solution and the mass of sorbent. The graph in Figure 9 shows the variation of the analytical signal when the sample is inserted into the on-line preconcentration system at various time intervals. It is observed that the analytical signal is linear for preconcentration periods up to 210 and 120 seconds for cadmium and lead, respectively.

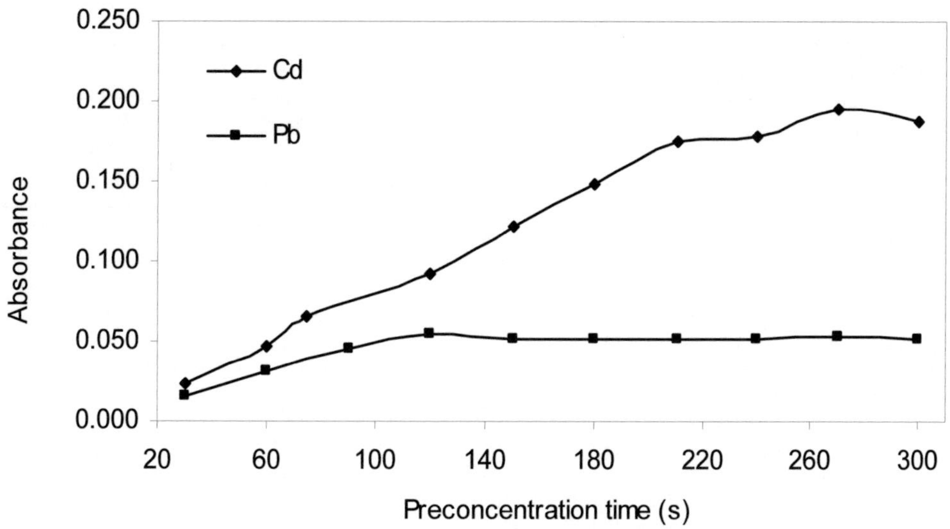

Figure 9. Influence of preconcentration time on the determination of Cd and Pb using solid phase extraction.

3.6. Lifetime of minicolumn

The life of the sorbent was investigated by monitoring the analytical signal corresponding to solutions of Pb (100.0 mg L^{-1}) or Cd (10.0 mg L^{-1}) at the end of a work day and by counting the number of runs. It was observed that the packed minicolumn did not provide a significant change in the extraction, even when used 350 times.

3.7. Analytical characteristics

The analytical characteristics of the method were calculated under the optimized conditions for preconcentration. Table 2 summarizes the analytical characteristics of the method. The analytical curves were constructed using solutions of cadmium and lead ranging from 1.0 to 10.0 and from 10.0 to 100.0 mg L^{-1}, respectively.

Regression curves without preconcentration resulted in the following equations: A = 3.97 x 10^{-4} C + 6.40 x 10^{-3} for Cd and A = 1.98 x 10^{-5} C + 2.38 x 10^{-4} for Pb, where A is the absorbance and C is the metal concentration in solution, in µg L^{-1}. These equations were obtained under optimum conditions of the spectrometer. Enrichment factors (EF) were calculated as the ratio of the slopes of the linear section in the calibration graphs for preconcentration and direct aspiration (Fang et al., 1992). The term concentration efficiency (CE) is defined as the product of EF and the number of samples analyzed per minute (Fang et al., 1992). Therefore, if f is the sampling frequency expressed in samples analyzed per hour, CE = EF x (f/60).

Element	**Cadmium**	**Lead**
Preconcentration time, s	120	120
Enrichment factor	30	35
Concentration efficiency, min^{-1}	13	15
Transfer phase factor	0.91	0.86
Sample volume, mL	9.00	14.0
Consumptive index, mL	0.30	0.40
Sample frequency, h^{-1}	26	26
Limit of detection, µg L^{-1}	0.8	1.0
Limit of quantification, µg L^{-1}	2.7	3.3
Precision, %	4.1	4.8
Calibration function	A= 3.20 x 10^{-3} + 1.19 x 10^{-2} C	A= 7,78 x 10^{-4} + 6.93 x 10^{-4} C

Table 2. Features of the preconcentration system for the determination of cadmium and lead (A, absorbance and C, metal concentration, µg L^{-1}).

The transfer phase factor is defined as the ratio between the analyte mass in the original sample and that in the concentrate. Consumptive index (CI) quantifies the efficiency of an FI on-line column preconcentration system in terms of the sample volume consumed to achieve a defined EF. The term CI is defined as the sample volume (V), in mL, consumed to achieve a unit EF, expressed by the equation CI = V/EF (Fang et al., 1992).

The limits of quantification and detection of the method were also calculated. The detection limit was calculated using the following equation: $3s_b/b$, where s_b is the standard deviation for eleven measurements of the blank, and b is the slope of the analytical curve for each metal. The limit of quantification was calculated as $10s_b/b$.

The accuracy of the proposed procedure was evaluated by determining the amounts of Cd and Pb in a certified reference material. The following material was analyzed: BCR-713, Wastewater (effluent) from the Institute for Reference Materials and Measurements (IRMM, Geel, Belgium). The results were 4.8 ± 0.5 µg L^{-1} for cadmium and 45 ± 4 µg L^{-1} for lead. According to the results, no significant difference was found between the results that were obtained and the certified values of the reference material (5.1 ± 0.6 µg L^{-1} and 47 ± 4 µg L^{-1} for cadmium and lead, respectively).

3.8. Application of the proposed procedure

The preconcentration procedure was applied to determine the metal content in water samples. These real samples were collected at Jequié, Bahia, Brazil. Known concentrations of Cd and Pb were added to the samples to minimize the change in the matrix of the original sample. Recoveries of the spiked samples (2.0 and 30.0 µg L^{-1} for cadmium and lead, respectively) were determined. The results shown in Table 3 demonstrate the applicability of the method. Recoveries (R) were calculated as follows: R (%) = $\{(C_m-C_o)/m\}$ x100, where C_m is the concentration of metal in a spiked sample, C_o is the concentration of metal in a sample and m is the amount of metal spiked. The described procedure can be successfully applied to these matrices for the preconcentration and determination of cadmium and lead.

Amostra	Amount of Cd (µg L^{-1})		R (%)	Amount of Cd (µg L^{-1})		R (%)
	Added	Found		Added	Found	
Tapwater Sample 1	0.0	< LOQ	93.5	0.0	< LOQ	95.7
	2.0	1.87 ± 0.05		30.0	28.70 ± 0.98	
Tapwater Sample 2	0.0	< LOQ	101.5	0.0	15.23 ± 0.98	96.8
	2.0	2.03 ± 0.06		30.0	44.57 ± 2.59	

Table 3. Results for the determination of Cd and Pb using the proposed procedure. LOQ: limit of quantification, R: recovery.

The results obtained from this procedure are comparable to those of other preconcentration methods for Cd and Pb determination. Table 4 summarizes some of these methods and their characteristics.

Element	Process of extraction	Enrichment factor	Limit of detection (μg L^{-1})	Sample volume (mL)	Reference
Cd	Solid phase extraction	30	0.8	9.0	**This work**
Pb		35	1.0	14.0	
Pb	Ionic liquid dispersive liquid–liquid microextraction	40	1.5	20.0	(Soylak & Yilmaz, 2011)
Pb	Solid phase extraction		2.1	-----	Yalcinkaya et al, 2011)
Cd	Dispersive liquid-liquid microextraction	55	0.4	-----	(Rojas et al., 2011)
Cd	Solid phase extraction	19	0.77	-----	(Tang & Hu, 2011)
Cd	Ion-flotation Separation	45	1.2	-----	(Tavallali et al., 2011)
Cd	Cloud point extraction	29	0.1*	-----	(Moghimi & Tajodini, 2010)
Pb	Cloud point extraction	56 and 42	1.14	-----	(Shah *et al.*, 2011)
Pb	Solid phase extraction	50	0.65	-----	(Melek et al., 2006)

Table 4. Analytical characteristics of various procedures for the determination of Cd and Pb by FAAS, * μg g^{-1}.

4. Conclusion

The PUF grafted with Me-BTAP was successfully applied to the preconcentration of cadmium and lead. The sorbent showed characteristics that are desirable for materials used in solid phase extraction systems, such as resistance to swelling and changes in pH, low resistance to flow passage and simplicity in preparation. Moreover, when applied to the preconcentration system, the solid phase provided a simple and sensitive method for the determination of cadmium and lead by FAAS. The synthesized material is a good alternative for the determination of these elements. The extraction of other elements will be tested using this sorbent, and further work in this area is currently being conducted in our laboratory.

Author details

Rafael Vasconcelos Oliveira and Valfredo Azevedo Lemos
Universidade Estadual do Sudoeste da Bahia,
Laboratório de Química Analítica (LQA), Campus de Jequié, Jequié, Bahia, Brazil

5. References

Azeem, S. M. A., Arafa, W. A. A. & El-Shahat, M. F. (2010). Synthesis and application of alizarin complexone functionalized polyurethane foam: Preconcentration/separation of metal ions from tap water and human urine. *Journal of Hazardous Materials*, Vol.182, No.1-3, p.p.286-294.

Beketov, V. I., Parchinskii, V. Z. & Zorov, N. B. (1996). Effects of high-frequency electromagnetic treatment on the solid-phase extraction of aqueous benzene, naphthalene and phenol. *Journal of Chromatography a*, Vol.731, No.1-2, p.p.65-73.

Braun, T. & Farag, A. B. (1978). Polyurethane foams and microspheres in analytical-chemistry - improved liquid-solid, gas-solid and liquid-liquid contact via a new geometry of solid-phase. *Analytica Chimica Acta*, Vol.99, No.1, p.p.1-36.

Braun, T. (1983) Trends in using resilient polyurethane foams as sorbents in analytical chemistry, *Fresenius' Journal of Analytical Chemistry*, Vol. 314, No. 7, p.p. 652-656.

Burham, N. (2008). Uses of 5-Methylresorcin-Bonded Polyurethan Foam as a New Solid Phase Extractor for the Selective Separation of Mercury Ions from Natural Water Samples. *Central European Journal of Chemistry*, Vol.6, No.4, p.p.641-650.

Costa, L. M., Ribeiro, E. S., Segatelli, M. G., Do Nascimento, D. R., De Oliveira, F. M. & Tarley, C. R. T. (2011). Adsorption studies of Cd(II) onto Al2O3/Nb2O5 mixed oxide dispersed on silica matrix and its on-line preconcentration and determination by flame atomic absorption spectrometry. *Spectrochimica Acta Part B-Atomic Spectroscopy*, Vol.66, No.5, p.p.329-337.

El-Shahat, M. E., Moawed, E. A. & Farag, A. B. (2007). Chemical enrichment and separation of uranyl ions in aqueous media using novel polyurethane foam chemically grafted with different basic dyestuff sorbents. *Talanta*, Vol.71, No.1, p.p.236-241.

El-Shahat, M. F., Moawed, E. A. & Zaid, M. A. A. (2003). Preconcentration and separation of iron, zinc, cadmium and mercury, from waste water using Nile blue a grafted polyurethane foam. *Talanta*, Vol.59, No.5, p.p.851-866.

El-Shahawi, M. S., Bashammakh, A. S. & Abdelmageed, M. (2011a). Chemical Speciation of Chromium(III) and (VI) Using Phosphonium Cation Impregnated Polyurethane Foams Prior to Their Spectrometric Determination. *Analytical Sciences*, Vol.27, No.7, p.p.757-763.

El-Shahawi, M. S., Hamza, A., Al-Sibaai, A. A. & Al-Saidi, H. M. (2011b). Fast and selective removal of trace concentrations of bismuth (III) from water onto procaine hydrochloride loaded polyurethane foams sorbent: Kinetics and thermodynamics of bismuth (III) study. *Chemical Engineering Journal*, Vol.173, No.1, p.p.29-35.

Fang Z., Dong L. P. & Xu S. K., (1992). Critical Evaluation of the Efficiency and Synergistic Effects of Flow Injection Techniques for Sensitivity Enhancement in Flame Atomic Absorption SpectrometryJ. *Anal. Atom. Spectrom*. Vol.7 pp. 293-299.

Fathi, M. R., Pourreza, N. & Ardan, Z. (2011). Determination of aluminum in food samples after preconcentration as aluminon complex on microcrystalline naphthalene by spectrophotometry. *Quimica Nova*, Vol.34, No.3, p.p.404-407.

Gonzales, A. P. S., Firmino, M. A., Nomura, C. S., Rocha, F. R. P., Oliveira, P. V. & Gaubeur, I. (2009). Peat as a natural solid-phase for copper preconcentration and determination in a multicommuted flow system coupled to flame atomic absorption spectrometry. *Analytica Chimica Acta,* Vol.636, No.2, p.p.198-204.

Jamshidi, M., Ghaedi, M., Mortazavi, K., Biareh, M. N. & Soylak, M. (2011). Determination of some metal ions by flame-AAS after their preconcentration using sodium dodecyl sulfate coated alumina modified with 2-hydroxy-(3-((1-H-indol 3-yle)phenyl) methyl) 1-H-indol (2-HIYPMI). *Food and Chemical Toxicology,* Vol.49, No.6, p.p.1229-1234.

Lemos, V. A., Baliza, P. X., De Carvalho, A. L., Oliveira, R. V., Teixeira, L. S. G. & Bezerra, M. A. (2008). Development of a new sequential injection in-line cloud point extraction system for flame atomic absorption spectrometric determination of manganese in food samples. *Talanta,* Vol.77, No.1, p.p.388-393.

Lemos, V. A., David, G. T. & Santos, L. N. (2006a). Synthesis and application of XAD-2/Me-BTAP resin for on-line solid phase extraction and determination of trace metals in biological samples by FAAS. *Journal of the Brazilian Chemical Society,* Vol.17, No.4, p.p.697-704.

Lemos, V. A., Santos E. S. & Gama, E. M. (2007). A comparative study of two sorbents for copper in a flow injection preconcentration system. *Separation and Purification Technology,* Vol.56, No.2, pp.212-219.

Lemos, V. A., Santos L. N., Alves, A. P. O. & David, G. T. (2006b). Chromotropic acid-functionalized polyurethane foam: A new sorbent for on-line preconcentration and determination of cobalt and nickel in lettuce samples. *Journal of Separation Science,* Vol.29, No.9, pp.1197-1204

Lemos, V. A., Santos, L. N. & Bezerra, M. A. (2010). Determination of cobalt and manganese in food seasonings by flame atomic absorption spectrometry after preconcentration with 2-hydroxyacetophenone-functionalized polyurethane foam. *Journal of Food Composition and Analysis,* Vol.23, No.3, p.p.277-281.

Lemos, V. A., Santos, M. S., Santos, E. S., Santos, M. J. S., Dos Santos, W. N. L., Souza, A. S., De Jesus, D. S., Das Virgens, C. F., Carvalho, M. S., Oleszczuk, N., Vale, M. G. R., Welz, B. & Ferreira, S. L. C. (2007). Application of polyurethane foam as a sorbent for trace metal pre-concentration - A review. *Spectrochimica Acta Part B-Atomic Spectroscopy,* Vol.62, No.1, p.p.4-12.

Matos, G. D. & Arruda, M. A. Z. (2006). Online preconcentration/determination of cadmium using grape bagasse in a flow system coupled to thermospray flame furnace atomic absorption spectrometry. *Spectroscopy Letters,* Vol.39, No.6, p.p.755-768.

Melek, E., Tuzen, M. & Soylak, M. (2006). Flame atomic absorption spectrometric determination of cadmium(II) and lead(II) after their solid phase extraction as dibenzyldithiocarbamate chelates on Dowex Optipore V-493. *Analytica Chimica Acta,* Vol.578, No.2, p.p.213-219.

Moawed E. A. & El-Shahat M. F. (2006). Preparation, characterization and application of polyurethane foam functionalized with α-naphthol for preconcentration and determination of trace amounts of nickel and copper in cast iron and granite. *Reactive & Functional Polymers,* Vol. 66, No.7, pp. 720-727.

Moawed, E. A. (2004). Separation and preconcentration of trace amounts of cadmium(II) and mercury(II) ions on rosaniline-grafted polyurethane foam. *Acta Chromatographica,* Vol.14, p.p.198-214.

Moawed, E. A., Zaid M. A. A. & El-Shahat, M. E. (2003). Methylene blue-grafted polyurethane foam using as a chelating resin for preconcentration and separation of cadmium(II), mercury(II), and silver(I) from waste water. *Analytical Letters,* Vol.36, No.2, pp.405-422.

Moghimi, A. & Tajodini, N. (2010). Extraction of Cadmium(II) Using Cloud-Point Method and Determination by FAAS. *Asian Journal of Chemistry,* Vol.22, No.7, p.p.5025-5033.

Nair, M. N. R. & Nair, M. R. G. (2008). Synthesis and characterisation of soluble block copolymers from NR and TDI based polyurethanes. *Journal of Materials Science,* Vol.43, No.2, p.p.738-747.

Navratil, J. D., Braun, T. & Farag, A. B. (1985). *Polyurethane foam sorbents in separation science* (first edition), CRC Press, Inc, 0-8493-6597-x, Boca Raton, Florida

Oral, E. V., Dolak, I., Temel, H. & Ziyadanogullari, B. (2011). Preconcentration and determination of copper and cadmium ions with 1,6-bis(2-carboxy aldehyde phenoxy)butane functionalized Amberlite XAD-16 by flame atomic absorption spectrometry. *Journal of Hazardous Materials,* Vol.186, No.1, p.p.724-730.

Rojas, F. S., Ojeda, C. B. & Pavon, J. M. C. (2011). Dispersive liquid-liquid microextraction combined with flame atomic absorption spectrometry for determination of cadmium in environmental, water and food samples. *Analytical Methods,* Vol.3, No.7, p.p.1652-1655.

Saeed M. M. & Ahmed R. (2005). Adsorption modeling and thermodynamic characteristics of uranium(VI) ions onto 1-(2-pyridylazo)-2-naphthol (PAN) supported polyurethane foam. *Radiochimica Acta,* Vol. 93, No. *6,* pp. 333-339.

Serafimovska, J. M., Arpadjan, S. & Stafilov, T. (2011). Speciation of dissolved inorganic antimony in natural waters using liquid phase semi-microextraction combined with electrothermal atomic absorption spectrometry. *Microchemical Journal,* Vol.99, No.1, p.p.46-50.

Shah, F., Kazi, T. G., Afridi, H. I., Naeemullah, Arain, M. B. & Baig, J. A. (2011). Cloud point extraction for determination of lead in blood samples of children, using different ligands prior to analysis by flame atomic absorption spectrometry: A multivariate study. *Journal of Hazardous Materials,* Vol.192, No.3, p.p.1132-1139.

Soylak, M. & Yilmaz, E. (2011). Ionic liquid dispersive liquid-liquid microextraction of lead as pyrrolidinedithiocarbamate chelate prior to its flame atomic absorption spectrometric determination. *Desalination,* Vol.275, No.1-3, p.p.297-301.

Tang, A. N. & Hu, Y. F. (2011). Determination of trace cadmium by flow injection on-line microcolumn preconcentration coupled with flame atomic absorption spectrometry using human hair as a sorbent. *Instrumentation Science & Technology,* Vol.39, No.1, p.p.110-120.

Tavallali, H., Lalehparvar, S., Nekoei, A. R. & Niknam, K. (2011). Ion-flotation Separation of Cd(II), Co(II) and Pb(II) Traces Using a New Ligand before Their Flame Atomic Absorption Spectrometric Determinations in Colored Hair and Dryer Agents of Paint. *Journal of the Chinese Chemical Society,* Vol.58, No.2, p.p.199-206.

Teixeira, L. S. G., Bezerra, M. D., Lemos, V. A., Dos Santos, H. C., De Jesus, D. S. & Costa, A. C. S. (2005). Determination of copper, iron, nickel, and zinc in ethanol fuel by flame atomic absorption spectrometry using on-line preconcentration system. *Separation Science and Technology*, Vol.40, No.12, p.p.2555-2565.

Tuzen, M., Citak, D. & Soylak, M. (2008). 5-Chloro-2-hydroxyaniline-copper(II) coprecipitation system for preconcentration and separation of lead(II) and chromium(III) at trace levels. *Journal of Hazardous Materials*, Vol.158, No.1, p.p.137-141.

Tzvetkova, P., Vassileva, P. & Nickolov, R. (2010). Modified silica gel with 5-amino-1,3,4-thiadiazole-2-thiol for heavy metal ions removal. *Journal of Porous Materials*, Vol.17, No.4, p.p.459-463.

Yalcinkaya, O., Kalfa, O. M. & Turker, A. R. (2011). Preconcentration of Trace Copper, Cobalt and Lead from Various Samples by Hybrid Nano Sorbent and Determination by FAAS. *Current Analytical Chemistry*, Vol.7, No.3, p.p.225-234.

Zambrzycka, E., Roszko, D., Lesniewska, B., Wilczewska, A. Z. & Godlewska-Zylkiewicz, B. (2011). Studies of ion-imprinted polymers for solid-phase extraction of ruthenium from environmental samples before its determination by electrothermal atomic absorption spectrometry. *Spectrochimica Acta Part B-Atomic Spectroscopy*, Vol.66, No.7, p.p.508-516.

Zeeb, M. & Sadeghi, M. (2011). Modified ionic liquid cold-induced aggregation dispersive liquid-liquid microextraction followed by atomic absorption spectrometry for trace determination of zinc in water and food samples. *Microchimica Acta*, Vol.175, No.1-2, p.p.159-165.

Zhang, L., Li, Z. H., Du, X. H. & Chang, X. J. (2011). Activated carbon functionalized with 1-amino-2-naphthol-4-sulfonate as a selective solid-phase sorbent for the extraction of gold(III). *Microchimica Acta*, Vol.174, No.3-4, p.p.391-398.

Chapter 13

Fast, Selective Removal and Determination of Total Bismuth (III) and (V) in Water by Procaine Hydrochloride Immobilized Polyurethane Foam Packed Column Prior to Inductively Coupled Plasma – Optical Emission Spectrometry

M.S. El-Shahawi, A.A. Al-Sibaai, H.M. Al-Saidi and E.A. Assirey

Additional information is available at the end of the chapter

http://dx.doi.org/10.5772/47962

1. Introduction

Bismuth is found in nature in trivalent state as bismuthinite, Bi_2S_3, bismite, Bi_2O_3 and bismuth sulfide- telluric, Bi_2Te_2S. It is also found as a secondary component in some lead, copper and tin minerals [1]. Bismuth (V) compounds do not exist in solution and are important in the view of pharmaceutical analytical chemistry [1]. In the Earth's crust, bismuth presents at trace concentration (8 μg Kg^{-1}) while, bismuth minerals rarely occur alone and are almost associated with other ores [2]. Bismuth appears to be environmentally significant because its physical and chemical properties have led it to be used in different areas of life. Pamphlett et al, 2000 [3] have reported that, bismuth compounds after oral intake enter the nervous system of mice, in particular, in motor neurons [3]. Hence, bismuth species are included in the list of potential toxins [3].

The development of selective, separation, pre-concentration and determination method for bismuth at sub-micro levels is a challenging problem because of its extremely low concentrations in natural samples and of its strong interference from the sample matrices. Several methods e.g. hydride generation atomic absorption spectrometry [4], electro thermal atomic absorption spectrometry [5], atomic fluorescence spectrometry [6], hydride generation atomic absorption spectrometry [7], and cathotic and anodic adsorptive stripping voltammetry [8 - 10] have been reported for bismuth determination. Most of these methods require preconcentration of bismuth for precise determination because most analytical techniques do not possess adequate sensitivity for direct determination.

Solvent extraction in the presence of co-extractant ligands e.g. bis (2, 4, 4-trimethyl pentyl) monothiophosphinic acid [11], pyrolidine dithocarbamate [12] etc has received considerable attention. However, these methods are too expensive, suffer from the use of large volumes of toxic organic solvents, and time-consuming. Thus, recent years have seen considerable attention on preconcentration and/ or monitoring of trace and ultra trace concentrations of bismuth by low cost procedures in a variety of samples e.g. fresh, marine and industrial wastewater [13]. Solid phase extraction (SPE) techniques have provided excellent alternative approach to liquid – liquid extraction for bismuth preconcentration prior to analyte determination step [14 -18].

Polyurethane foams (PUFs) sorbent represent an excellent solid sorbent material due to their high available surface area, cellular and membrane structure and extremely low cost [19]. Thus, several liquid solid separation involving PUFs methods have been employed successfully for separation and sensitive determination of trace and ultra trace levels of metal ions including bismuth (III) [19-29]. The membrane like structure and the available surface area of the PUFs make it a suitable stationary phase and a column filling material [25, 27]. Thus, the main objectives of the present chapter are focused on: i. developing of a low cost method for the removal of bismuth(III) and (V) species after reduction of the latter to tri valence state employing PUFs impregnated $PQ^+.Cl^-$; ii. Studying the kinetics, and thermodynamic characteristics of bismuth (III) sorption by trioctylamine plasticized $PQ^+.Cl^-$ treated PUFs and finally iii. Application of the developed method in packed column for complete removal and / or determination of bismuth (III &V) species in wastewater by $PQ^+.Cl^-$ treated PUFs sorbent.

2. Experimental

2.1. Reagents and materials

All chemicals used were of A.R. grade and were used without further purification. Stock solution (1000 μg mL^{-1}) of bismuth (III) was prepared from bismuth (III) nitrate (Aldrich Chemical Co Ltd, Milwaukee, WC, USA). More diluted solutions of bismuth (III) (0.1 – 100 μg mL^{-1}) were prepared by diluting the stock solution with diluted nitric acid. Stock solutions of procaine [2-(diethylamino)ethyl 4 aminobenzoate] hydrochloride, $PQ^+.Cl^-$ (1.0 %w/v), Fig.1 and KI (10%w/v) were prepared by dissolving the required weight in water (100 mL). A stock solution (1%v/v) of trioctylamine (Aldrich) was prepared in water in the presence of few drops of concentrated HNO_3. Sodium bismuthate

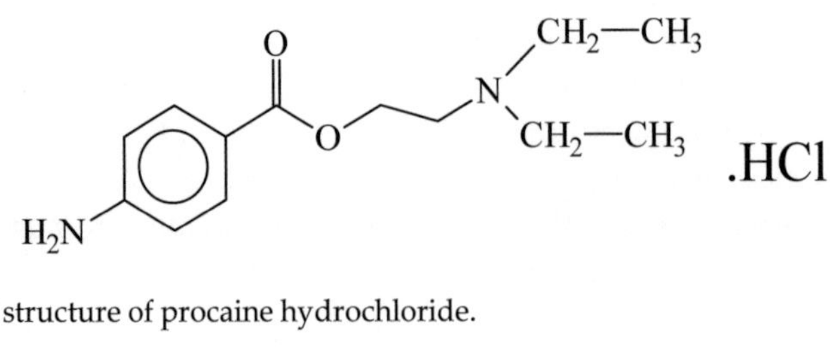

Figure 1. Chemical structure of procaine hydrochloride.

($NaBiO_3$, 85% purity) (BDH, Poole, England) was used for preparation of stock solution (50 μg mL^{-1}) of bismuth (V) in dark bottle [30] as follows: an accurate weight of $NaBiO_3$ was heated in a suitable volume of $HClO_4$ (20 mL, 0.5 mol L^{-1}) filtered and the solution was made up to 250 mL with deionized water and finally analyzed under the recommended conditions of bismuth determination by ICP-OES (Table 1). The measured concentration was taken as a standard stock solution of bismuth (V) in the next work. Bismuth (V) solution was finally stored in low density polyethylene bottles (LDPE) in dark. Stock solutions (0.1-1% w/v) of $PQ^+.Cl^-$ (BDH) and trioctylamine (Merck, Darmstadt, Germany) abbreviated as TOA were prepared in deionized water containing few drops of concentrated HNO_3. Sodium diethyldithiocarbamate (Na-DDTC) and $PQ^+ .Cl^-$(1% w/v) were purchased from Fluka, AG (Buchs, Switzerland). Commercial white sheets of PUFs were cut as cubes (10 -15 mm), washed, treated and dried. The reagent $PQ^+ .Cl^-$ (1.0 % w/v) was dissolved in water, shaken with the PUFs cubes in the presence of TOA (1% v/v) with efficient stirring for 30 min, squeezed and finally dried as reported [21]. The certified reference material (CRM) i.e. trace metal in drinking water standard (CRM-TMDW) was obtained from High-Purity Standard Inc. Sulfuric acid (0.5mol L^{-1}) was used as an extraction medium in the sorption process of bismuth (III) by the PUFs. Commercial white sheets of open cell polyether type polyurethane foam were purchased from the local market of Jeddah City, Saudi Arabia and were cut as cubes (10-15 mm). The PUFs cubes were washed and dried as reported [21, 27]. A series of Britton- Robinson (B-R) buffer (pH 2-11) was prepared as reported [31].

Parameter	
Rf power (kW)	1050 (900.0)
Plasma gas (Ar) flow rate, L min^{-1}	15 (15)
Auxiliary gas (Ar) flow rate, L min^{-1}	0.2 (1.2)
Nebulizer gas (Ar) flow rate, Lmin^{-1}	0.80 (0.93)
Pump rate, mL min^{-1}	1.5
Observation height, mm	15
Integration time, s	10
Wavelength, nm	Bi 223.061

*ICP –MS operational parameters are given in parentheses. Other parameters are: lens voltage =9.0; analog stage voltage 1750 V; pulse stage voltage =750 V; quadrupole rod offset std = =0.0; cell rod offset =-18.0; discriminator threshold =17.0; cell path voltage Std = -13.0 V and atomic mass 208.98 am.

Table 1. ICP-OES operational conditions and wavelength (nm) for bismuth determination*

2.2. Instrumental and apparatus

A Perkin - Elmer (Lambda 25, Shelton, CT,USA) spectrophotometer (190 - 1100 nm) with 10 mm (path width) quartz cell was used for recording the electronic spectra and measuring the absorbance of the ternary complex ion associate $PQ^+. BiI_4^-$ of bismuth (III) at 420 nm before and after extraction with the reagent $PQ^+.Cl^-$ treated PUFs. A

Perkin Elmer inductively coupled plasma – optical emission spectrometer (ICP- OES, Optima 4100 DC (Shelton, CT, USA) was used and operated at the optimum operational parameters for bismuth determination (Table 1). A Perkin Elmer inductively coupled plasma – mass spectrometer (ICP – MS) Sciex model Elan DRC II (California, CT, USA) was also used to measure the ultra trace concentrations of bismuth in the effluent after extraction by the developed PUFs packed column at the operational conditions (Table 1). A Corporation Precision Scientific mechanical shaker (Chicago, CH, USA) with a shaking rate in the range 10 – 250 rpm and glass columns (18 cm x 15 mm i.d) were used in batch and flow experiments, respectively. De-ionized water was obtained from Milli-Q Plus system (Millipore, Bedford, MA, USA). A thermo Orion model 720 pH Meter (Thermo Fisher Scientific, MA, USA) was employed for pH measurements with absolute accuracy limits being defined by NIST buffers.

2.3. General batch procedures

2.3.1. Preparation of the immobilized reagent ($PQ^+.Cl^-$) polyurethane foams.

The reagent $PQ^+.Cl^-$(1% w/v)in water was shaken with the PUFs cubes in the presence of the plasticizer TOA (1% v/v) with efficient stirring for 30 min. The loaded $PQ^+.Cl^-$ PUFs cubes were squeezed and dried between two filter papers [20, 21]. The amount of $PQ^+.Cl^-$ retained onto the PUFs sorbent was calculated using the equation [21]:

$$a = (C_o - C)\frac{v}{w} \tag{1}$$

where, C_o and C are the initial and final concentrations (mol L^{-1}) of the reagent ($PQ^+.Cl^-$) in solution, respectively, v = volume of the reagent solution (liter) and w is the mass (g) of the PUFs sorbent. The reproducibility of $PQ^+.Cl^-$ treated PUFs is fine and the PUFs can be reused many times without decrease in its efficiency.

2.3.2. Batch extraction step

An accurate weight (0.1 ± 0.002 g) of unloaded- or $PQ^+.Cl^-$immobilized PUFs was equilibrated with an aqueous solution (100 mL) containing bismuth (10 μg mL^{-1}) in the presence of KI (10% w/v) , H_2SO_4 (0.5 mol L^{-1})and ascorbic acid (0.1%w/v) to minimize the aerial oxidation of KI. The test solution was shaken for 1 h on a mechanical shaker. The aqueous phase was then separated out by decantation and the amount of bismuth (III) remained in the aqueous phase was then determined spectrohotometrically against reagent blank [32] or by ICP-OES at ultra trace concentrations. The amount of bismuth (III) retained on the foam cubes was then calculated from the difference between the absorbance of $[BiI_4]^-$ in the aqueous phase before (A_b) and after extraction (A_f). The sorption percentage (%E) , the distribution ratio (D), the amount of bismuth (III) retained at equilibrium (q_e) per unit mass of solid sorbent (mol/g) and the distribution coefficient (K_d) of sorbed analyte onto the foam cubes were finally calculated as reported. The %E and K_d are the average of three independent measurements and the precision in most cases was ±2%. Following these procedures, the influence of shaking time and temperature on the retention of bismuth (III) by the PUFs sorbents was fully studied.

2.3.3. Retention and recovery of bismuth (III)

An aqueous solution (100 mL) of bismuth (III) ions at concentration (5 – 100 μg L^{-1}), KI (10%) and H_2SO_4 (1.0 mol L^{-1}) was percolated through the $PQ^+.Cl^-$ loaded PUFs (1.0 ± 0.002 g) column at 2.0 mL min^{-1} flow rate. A blank experiment was also performed in the absence of bismuth (III) ions. Bismuth (III) sorption took place quantitatively as indicated from the analysis of bismuth species in effluent solutions by ICP- OES. After extraction, the ultra trace concentrations of bismuth (III) remained in the test aqueous solutions were estimated by ICP-MS. Bismuth (III) species were recovered quantitatively with HNO_3 (3.0 mol L^{-1}, 10 mL) at 2.0 mL min^{-1} flow rate.

2.3.4. Retention and recovery of bismuth (V)

An aqueous solution (100.0 mL) of bismuth (V) at concentration < 10 μg L^{-1} was allowed to react with an excess of KI (10% w/v) - H_2SO_4 (1.0 mol L^{-1}). The solution was then percolated through PQ^+ .Cl^- loaded PUFs (1.0 ± 0.002 g) packed column at 2.0 mL min^{-1} flow rate of 2.0 mL min^{-1}. The retained bismuth (III) species were recovered with HNO_3 (10.0 mL, 1.0 mol L^{-1}) at 2.0 mL min^{-1} flow rate and analyzed by ICP- OES.

2.3.5. Sequential determination of total bismuth (III) and (V)

An aqueous solution (100 mL) containing bismuth (III) and (V) at a total concentration ≤ 10 μg L^{-1} was analyzed according to the described procedure for bismuth (V) retention and recovery. Another aliquot portion (100 mL) was adjusted to pH 3 - 4 with acetate buffer and then shaken with Na-DDTC (5 0 mL, 1%w/v) for 2-3 min. Bismuth (III) ions were then extracted with methylisobutylketone (5.0 mL) as Bi $(DDTC)_3$after 2 min [24]. Bismuth (V) remained in the aqueous solution was reduced to bismuth (III) by an excess of KI (10% w/v) in the presence of H_2SO_4 (0.5 mol L^{-1}) and then percolated through the $PQ^+.Cl^-$ loaded PUFs column at 2 ml min $^{-1}$ flow rate at the optimum experimental conditions. The retained bismuth species were recovered and finally analyzed following the recommended procedures for bismuth (III). Thus, the net signal intensity of ICP- OES (or ICP-MS) at ultra trace concentrations of the first aliquot (I_1) will be a measure of the sum of the bismuth (III) and (V) ions in the mixture, while the net signal intensity of the of the second aliquot (I_2) is a measure of bismuth (V) ions. The difference (I_1-I_2) of the net signal intensity is a measure of bismuth (III) ions in the binary mixture.

2.4. Analytical applications

2.4.1. Analysis of certified reference material TMDW

The TMDW water sample (2 mL) was digested with nitric acid (10 mL, 3.0. mol L^{-1}) and hydrogen peroxide (10 mL, 10% v/v), boiled for 5 min and diluted by an excess of KI (10% w/v) - H_2SO_4 (1.0 mol L^{-1}) to 100 mL. After cooling, the test solution was percolated through the $PQ^+.Cl^-$ loaded PUFs column at 2 ml min $^{-1}$ flow rate. The retained bismuth species were recovered with HNO_3 (10.0 mL, 1.0 mol L^{-1}) at 2.0 mL min^{-1} flow rate and nalyzed by ICP-OES following the recommended procedures for bismuth (III).

2.4.2. Analysis of total bismuth in wastewater

Wastewater samples (1.0 L) were collected and filtered through a 0.45 μm membrane filter (Milex, Millipore Corporation). The test solution was digested with nitric acid (10 mL, 3.0. mol L^{-1}) and hydrogen peroxide (10 mL, 10% v/v), boiled for 5 min and spiked with different amounts (0.05- 0.5 μg) of bismuth (III) in presence of an excess of KI (10% w/v). After centrifugation for 5 min, the sample solutions were percolated through $PQ^+.Cl^-$ loaded PUFs packed columns at 5 mL min^{-1} flow rate. The concentration of bismuth in the effluent solution was determined by ICP - MS. The retained bismuth (III) species on the PUFs were then recovered and analyzed as described above.

2.4.3. Analysis of total bismuth in seawater

The general procedure for the extraction and recovery of bismuth (III) ions from seawater samples onto $PQ^+.Cl^-$ impregnated PUFs was performed as follow: A 100 mL of water samples were filtered through 0.45 μm membrane filter, adjusted to pH zero with H_2SO_4 (0.5 mol L^{-1}) in the presence of KI (0.1%w/v) and ascorbic acid. The sample solution was then passed through $PQ^+.Cl^-$ impregnated PUFs (1.0 ± 0.001 g) packed column (10 cm x 1.0 cm i.d.) at 5 mL $min.^{-1}$ The retained bismuth(III) species were then recovered and analyzed as described above. The recovered bismuth (III) ions were then determined by ICP-OES.

3. Results and discussion

In recent years [28, 29], PUFs immobilizing some ion pairing reagents have received considerable attention for selective separation, determination and / or chemical speciation of trace and ultra trace metal ions. The non-selective sorption characteristic of the PUFs has been rendered and became more selective by controlling the experimental conditions e.g. pH, ionic strength, etc. Preliminary investigation has shown that, on shaking unloaded PUFs and $PQ^+.Cl^-$ immobilized PUFs with aqueous solutions containing bismuth (III) ions , KI (10%w/v) and H_2SO_4 (0.5 mol L^{-1}), considerable amount of bismuth (III) species were retained onto $PQ^+.Cl^-$ treated PUFs in a very short time compared to the untreated PUFs ones. Thus, in subsequent work, detailed study on the application of $PQ^+.Cl^-$ immobilized PUFS for retention of various bismuth (III & V) species to assign the most probable kinetic model, sorption isotherm models, mechanism and thermodynamic characteristics of retention of bismuth (III) from the test aqueous solutions.

3.1. Retention profile of bismuth (III) from the aqueous solution onto PUFs

Bismuth (III) forms an orange – yellow colored tetraiodobismuthate(III) complex, $[BiI_4]^-$ [32] in aqueous solutions containing sulfuric acid (0.5 mole L^{-1}) and an excess of KI (10%w/v). Thus, the sorption profile of aqueous solutions containing bismuth (III) at different pH by $PQ^+.Cl^-$ loaded foams was critically studied after shaking for 1h at room temperature. After equilibrium, the amount of bismuth (III) in the aqueous phase was determined spectrophotometrically [32]. The results are shown in Fig. 2. The %E and K_d of bismuth (III) sorption onto the PUFs markedly decreased on increasing solution pH and maximum uptake was achieved at pH zero. At pH >1, the sorption of bismuth (III) by $PQ^+.Cl^-$ treated PUFs towards bismuth (III) decreased markedly (Fig.2). This behavior is most likely attributed to the deprotonation of the ether oxygen ($-CH_2 - O- CH_2 -$) and/or urethane nitrogen (- NH– CO–) of PUFs, instability, hydrolysis, or incomplete extraction of the produced ternary complex ion associate of $PQ^+.[BiI_4]^-$ on/ in the PUFs sorbent.

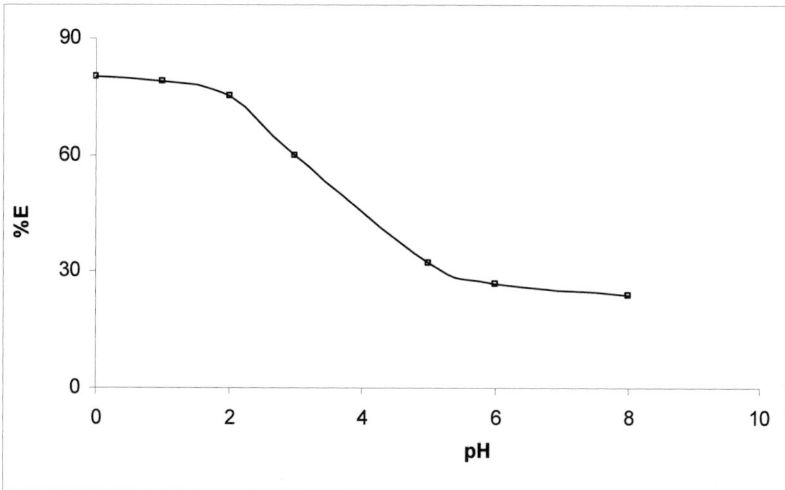

Figure 2. Effect of pH on the sorption percentage of bismuth (III) from aqueous solutions containing KI (10 % m/v) - H_2SO_4 (2.0 mol L^{-1}) onto $PQ^+ .Cl^-$ immobilized PUFs (0.1 ± 0.002 g) at 25 ± 0.1°C.

The retention of bismuth (III) at low pH of aqueous media is most likely attributed to sorbent membranes. The pK_a values of protonation of oxygen atom of ether group ($- CH_2- OH^+- CH_2-$)$_{foam}$ and nitrogen atom of the amide group ($- N^+H_2$ - COO-)$_{foam}$ are - 3 and - 6, respectively [32]. Thus, in extraction media containing H_2SO_4 (0.50 mole L^{-1}) and KI, the complexed species of bismuth $[BiI_4]^-$ are easily retained onto the protonated ether group of the PUFs than amide group of PUFs sorbent. The stability constants of the binding sites of the PUFs with $[BiI_4]^-$ were calculated using the Scatchard equation [33]:

$$\frac{n}{[Bi]} = K(n_i - n) \quad (2)$$

and n is given by the equation:

$$n = \frac{weight \text{ of bismuth bound to foam (g)}}{\text{weight of foam (g)}} \quad (3)$$

where, K = stability constant of bismuth (III) on PUF, n_i = maximum concentration of sorbed bismuth (III) by the available sites onto the PUFs, and [Bi] is the equilibrium concentration of bismuth (III) in solution (mol L^{-1}). The plot of n /[Bi] versus n is shown in Fig. 3. The curvature of the Scatchard plot demonstrated that more than one class of complex species of bismuth (III) has been formed and each complex has its own unique formation constant. The stability constants log K_1 and log K_2 for the sorbed species derived from the respective slopes were 5.56 ± 0.2 and 4.82 ± 0.5, respectively. The values of n_1 and n_2 calculated from the plot were found equal 0.038 ± 0.005 and 0.078 ± 0.01 mol g^{-1}, respectively. The values of the stability constants (log K_1 and log K_2) indicated that, the sorption of bismuth (III) species took place readily on site K_1 that most likely belong to the ether group. The fact that, ether group has a stability greater than the amide group (site K_2) as reported [32]. Moreover, the high values of K_1 and K_2 indicated that, both bonding sites of PUFs are highly active towards $[BiI_4]^-$ species in good agreement with the data reported involving the extraction of the bulky anion $[BiI_4]$ by methyl isobutyl ketone and other solvents that posses ether linkages in their structures e.g. diethyl ether and isopropyl ether [34]. Based on these data and the results reported on the retention of $AuCl_4^-$ and $CdI_4^-$by PUFs [29, 34], a sorption mechanism involving a weak base anion ion exchanger and/ solvent extraction of $[BiI_4]^-_{aq}$ by the protonated ether oxygen or urethane nitrogen linkages of the PUFs as a ternary complex ion associate is most likely proceeded as follows:

Ether group, PUF:

$$(-CH_2-O-CH_2-)\ foam + H^+ \leftrightarrow (-CH_2-HO^+-CH_2-)_{foam} \quad (4)$$

$$(-CH_2-HO^+-CH_2-)_{foam} + [BiI_4]^-_{aq} \leftrightarrow [-CH_2-HO^+-CH_2-].[BiI_4]^-_{foam} \quad (5)$$

Urethane group, PUF:

$$(-NH-COO-)_{foam} + H^+ \leftrightarrow (--NH^+_2-COO-)_{foam} \quad (6)$$

$$(-N^+H_2-COO-)_{foam} + [BiI_4]^-_{aq} \leftrightarrow [-N^+H_2-COO-].[BiI_4]^-_{foam} \quad (7)$$

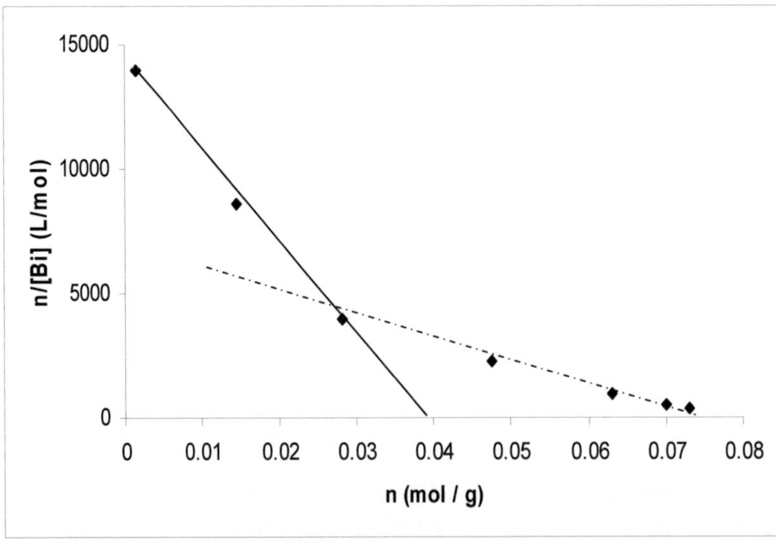

Figure 3. Scatchard plot for the binding of $[BiI_4]^-$ species by $PQ^+.Cl^-$ immobilized PUF (0.1 ± 0.002 g) from aqueous media containing KI (10 % m/v) - H_2SO_4 (0.5 mol L^{-1}) at 25 ± 0.1°C.

The distribution ratio of bismuth (III) onto $PQ^+.Cl^-$ immobilized PUFs showed high retention (D= 6.17 x 10^4 mL g^{-1}) compared to the unloaded PUFs (3.05 x 10^3 mL g^{-1}) due to the formation of the ion associate $([(PQ^+).(BiI_4)]^-_{foam}$ on/in treated PUFs. Thus, the solution pH was adjusted at pH 0.0 – 1.0 and $PQ^+.Cl^{-1}$ treated PUFs was used as a proper sorbent in the subsequent work.

The influence of the plasticizer e.g. tri-n-octylamine (TOA, 0.5 -2.0 %v/v) and tri-n-butyl-phosphate (TBP,0.01%v/v) on the retention of bismuth (III) from the aqueous solutions onto the $PQ^+ .Cl^-$ loaded PUFs was studied. Bismuth (III) sorption onto the PUFs sorbent increased (D = 6.6 X 10^4 mL g^{-1}) in presence of TOA (1% v/v). The formation of the co ternary complex ion associates $TOA^+.BiI_4{}^-$ and $PO^+.\ BiI_4{}^-$ in acidic media may account for the observed increase.

3.2. Kinetic behavior of bismuth (III) sorption onto $PQ^+ .Cl^-$ -TOA loaded PUFs

The influence of shaking time (0 – 60 min) on the uptake of bismuth (III) from the aqueous acidic media at pH zero was investigated. The sorption of bismuth (III) ions onto TOA plasticized $PQ^+ .Cl^-$ immobilized PUFs was fast and reached equilibrium within 60 min of shaking time. This conclusion was supported by calculation of the half-life time ($t_{1/2}$) of bismuth (III) sorption from the aqueous solutions onto the solid sorbents PUFs. The values of $t_{1/2}$ calculated from the plots of -log C/ C_0 versus time for bismuth (III) sorption onto PUFs, where C_0 and C are the original and final concentration of bismuth(III) ions in the test aqueous solution, respectively . The value of $t_{1/2}$ was found 2.32 ± 0.04 min in agreement with $t_{1/2}$ value reported earlier [19]. Thus, gel diffusion is not only the rate-controlling step for $PQ^+ .Cl^-$ immobilized PUFs as in the case of common ion exchange resins [19] and the kinetic of bismuth (III) sorption by $PQ^+ .Cl^-$ immobilized PUFs sorbent depends on film and intraparticle diffusion step where, the more rapid one controls the overall rate of transport.

The sorbed bismuth (III) species onto PUFs sorbent was subjected to Weber–Morris model [35]:

$$q_t = R_d\,(t)^{1/2} \tag{8}$$

where, R_d is the rate constant of intraparticle transport in μ mole g^{-1} $min^{-1/2}$ and q_t is the sorbed Bi (III) concentration (μ mole g^{-1}) at time t. The plot of qt *vs.* time (Fig 4) was linear (R^2= 0.989) at the initial stage of bismuth (III) uptake by TOA plasticized $PQ^+ .Cl^-$ loaded PUFs sorbents was linear up to 10 ± 1.1 min and deviate on increasing shaking time. The rate of diffusion of $[BiI_4]^-_{aq}$ species is high and decreased on increasing shaking time. Thus, the rate of the retention step of $[BiI_4]^-_{aq}$ onto the used solid sorbent is film diffusion at the early stage of extraction [34, 35]. The values of R_d computed from the two distinct slopes of Weber – Morris plots (Fig.4) for bismuth(III) retention by the solid sorbent were found equal 3.076 ± 1.01 and 0.653 m mole g^{-1} with correlation coefficient (R^2) of 0.989 and 0.995, respectively. The observed change in the slope of the linear plot (Fig.4) is most likely attributed to the different pore size [34, 35]. Thus, intra-particle diffusion step is most likely the rate determining step.

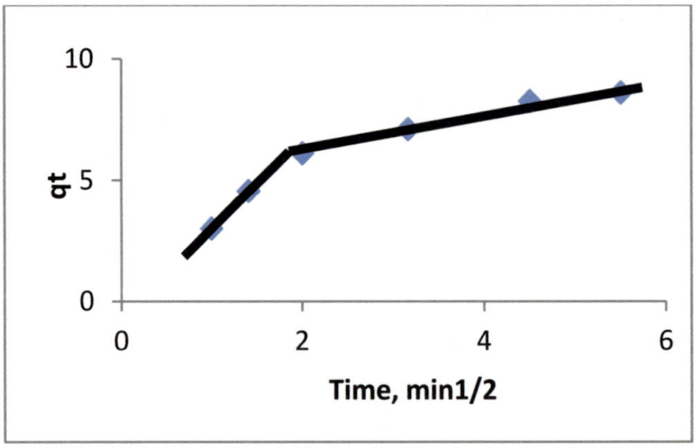

Figure 4. Weber – Morris plot of sorbed bismuth (III) onto $PQ^+.Cl^-$ immobilized PUFs *vs.* square root of time. Conditions: Aqueous solution (100 mL) containing KI (10 % m/v) and H_2SO_4 (0.5 mol L^{-1}), foam doze = (0.1 ± 0.002 g and 25 ± 0.1°C.

The retention step of the $[BiI_4]^-$ species onto the loaded PUFs at 25 ± 1 °C was subjected to Lagergren model [28]:

$$\log (q_e - q_t) = \log q_e - \frac{K_{Lager}}{2.303} t \quad (9)$$

where, q_e is the amount of $[BiI_4]^-$ sorbed at equilibrium per unit mass of PUFs sorbent (μmoles g^{-1}) ; K_{Lager} is the first order overall rate constant for the retention process per min and t is the time in min . The plot of log ($q_e - q_t$) *vs.* time (Fig.5) was linear. The computed value of K_{Lager} was 0.132 ± 0.033 min^{-1} (R^2= 0.979) confirming the first order kinetic model of sorption of $[BiI_4]^-$ species onto the solid sorbent [29]. The influence of adsorbate concentration was investigated and the results indicated that, the value of K_{Lager} increased on increasing adsorbate concentration confirming the first order kinetic nature of the retention process and the formation of monolayer species of $[BiI_4]^-$ onto the surface of the used adsorbent [26, 29].

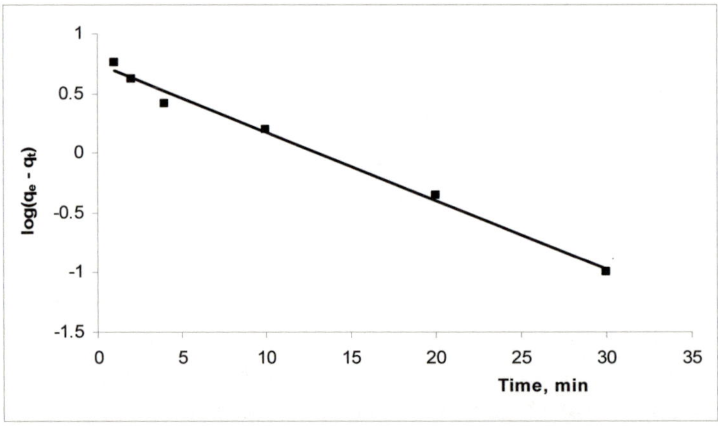

Figure 5. Lagergren plot of bismuth (III) uptake onto $PQ^+.Cl^-$ PUFs from aqueous solutions containing KI (10 % m/v) - H_2SO_4 (2.0 mol L^{-1}) *vs.* time at 25 ± 0.1°C.]

The sorption data was also subjected to Bhattacharya- Venkobachar kinetic model [36].

$$\log\left(1 - U_{(t)}\right) = \frac{-K_{Bhatt}}{2.303}\,t \qquad (10)$$

$$\text{where, } U_{(t)} = \frac{C_0 - C_t}{C_o - C_e},$$

where, K_{Bhatt} = overall rate constant (min^{-1}), t = time (min), C_t= concentration of the bismuth (III) at time t in μg mL^{-1}, C_e= concentration of Bi (III) at equilibrium in μg mL^{-1}. The plot of log (1-$U_{(t)}$) *vs.* time was linear (Fig.6) with R^2= 0.987. The computed value of K_{Bhatt} (0.143 ± 0.002 min^{-1}) from Fig. 6 was found close to the value of K_{Lager} (0.132 ± 0.033 min^{-1}) providing an additional indication of first order kinetic of bismuth (III) retention towards $PQ^+ .Cl^-$ loaded PUFs sorbent.

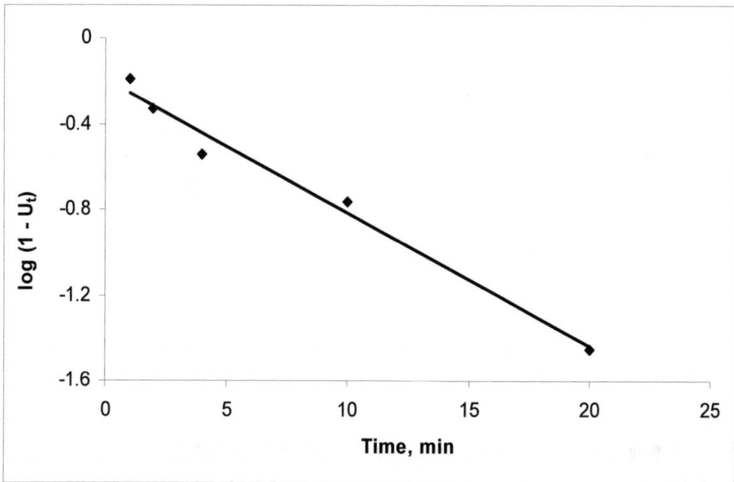

Figure 6. Bhattacharya-Venkobachar plot of bismuth (III) retention from aqueous media containing KI (10 % m/v) - H_2SO_4 (0.5 mol L^{-1}) at 25 ± 0.1°C onto the $PQ^+ .Cl^-$ and TOA loaded PUFs.

The value of BT, which is a mathematical function (F) of the ratio of the fraction sorbed (q_t) at time t and at equilibrium (q_e) in μ mole g^{-1} i.e. F= q_t / q_e calculated for each value of F employing Reichenburg equation [36].

$$BT = -0.4977 - 2.303 \log (1- F) \qquad (11)$$

The plot of Bt versus time at 25 ± 1 °C for TOA plasticized $PQ^+ .Cl^-$ PUFs towards bismuth (III) species was linear(R^2 = 0.990) up to 35 min (Fig. 7) . The straight line does not pass through the origin indicating that, particle diffusion mechanism is not only responsible for the kinetic of $[BiI_4]^-$sorption onto the $PQ^+ .Cl^-$ treated sorbents. Thus, the uptake of $[BiI_4]^-$ onto the employed sorbents is most likely involved three steps: i- bulk transport of $[BiI_4]^-$ in solution, ii- film transfer involving diffusion of $[BiI_4]^-$ within the pore volume of TOA plasticized $PQ^+ .Cl^-$ treated PUFs and/ or along the wall surface to the active sorption sites of the sorbent and finally iii- formation of the complex ion associate of the formula [- CH_2- HO $^+$– CH_2-]. $[BiI_4]^-_{Foam}$ or [-$^+NH_2$ – COO^-]. $[BiI_4]^-_{Foam}$. Therefore, the actual sorption of $[BiI_4]^-$ onto the interior surface of PUFs was rapid and hence particle diffusion mechanism is not the rate determining step in the sorption process. Thus, film and intraparticle transport might be the two main steps controlling the sorption step. Hence, "solvent extraction" and/or "weak base anion ion exchanger" mechanism is not only the most probable participating mechanism and some other processes e.g. surface area and specific sites on the PUFs are most likely involved simultaneously in bismuth (III) retention [37].

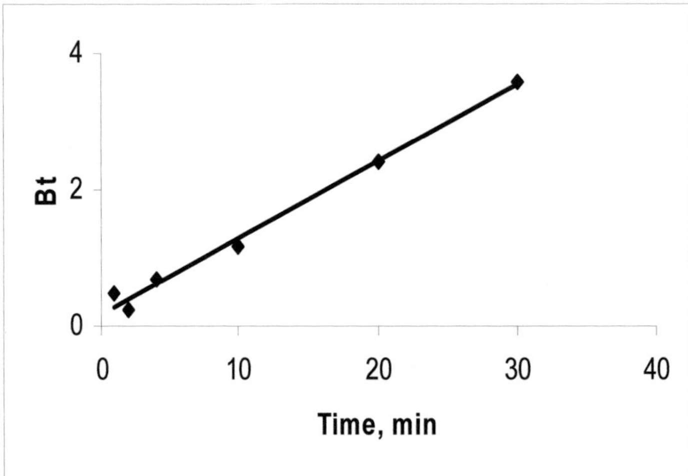

Figure 7. Reichenburg plot of bismuth (III) retention from aqueous media containing KI (10 % m/v) - H_2SO_4 (0.5 mol L^{-1}) at 25 ± 0.1^{0}C onto PQ^+ .Cl^- loaded PUFs.

3.3. Thermodynamic characteristics of bismuth (III) retention onto plasticized PQ^+.Cl^- loaded PUFs

Bismuth (III) retention onto TOA plasticized PQ^+.Cl^- PUFs was studied over a wide range of temperature (293-353 K) to determine the nature of bismuth (III) retention onto solid sorbent at the established experimental conditions. The thermodynamic parameters (ΔH, ΔS, and ΔG) were evaluated using the equations:

$$\ln K_c = \frac{-\Delta H}{RT} + \frac{\Delta S}{R} \tag{12}$$

$$\Delta G = \Delta H - T\Delta S \tag{13}$$

where, ΔH, ΔS, ΔG, and T are the enthalpy, entropy, Gibbs free energy changes and temperature in Kelvin, respectively and R is the gas constant (≈ 8.3 J K^{-1} mol^{-1}). K_c is the equilibrium constant depending on the fractional attainment (Fe) of the sorption process. The values of K_c of bismuth (III) retention from the test aqueous solutions at equilibrium onto the plasticized PQ^+.Cl^- PUFs were calculated using the equation:

$$K_C = \frac{Fe}{1-Fe} \quad (14)$$

Plot of ln Kc *vs.* 1000/T (K^{-1}) for bismuth (III) retention was linear (Fig. 8) over the wide range of temperature range (293- 323 K). The value of Kc decreased on increasing temperature, revealing that, the retention process of $[BiI_4]^-$species onto the sorbents is an exothermic process [21, 22]. The numerical values of ΔH, ΔS, and ΔG calculated from the slope and intercept of the linear plot Fig. 8 were -18.72± 1.01 kJ mol^{-1} , 54.57± 0.5 J mol^{-1} K^{-1} and -2.46 ± 0.1 kJ mol^{-1} (at 298 K), respectively with a correlation factor of 0.998.

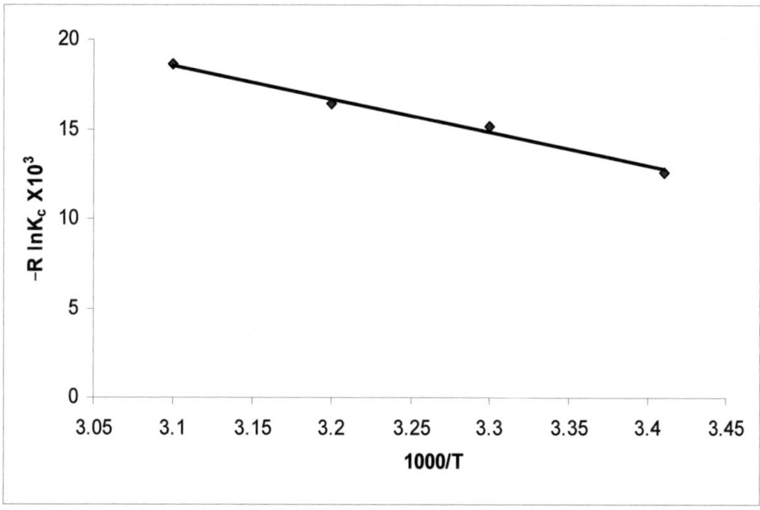

Figure 8. Plot of ln Kc *vs.* 1000/T (K^{-1}) of bismuth (III) sorption from aqueous media containing KI (10 % m/v) - H_2SO_4 (0.5 mol L^{-1}) onto $PQ^+.Cl^-$ treated PUFs.

The retention of bismuth (IIII) by plasticized $PQ^+.Cl^-$ loaded PUFs was also subjected to Vant Hoff model:

$$\log K_d = \frac{-\Delta H}{2.30\ RT} + C \quad (15)$$

where, C is a constant. Vant - Hoff plot of log K_d *vs.* 1000/T (K^{-1}) of bismuth (III) uptake from the test aqueous media of KI (10 % m/v) - H_2SO_4 (0.5 mol L^{-1}) onto plasticized $PQ^+.Cl^-$ loaded PUFs sorbent was linear (Fig. 9). The value of ΔH calculated from the slope of Fig. 9 was -20.1 ± 1.1 kJ mol^{-1} in good agreement with the values evaluated from equations 12 and 13. The ΔS of activation were lower than TΔS at all temperature. Thus, the retention step is entropy controlled at the activation state.

The negative value of ΔH and the data of D and Kc reflected the exothermic behavior of bismuth (III) uptake by the employed solid PUFs and non-electrostatics bonding formation between the adsorbent and the adsorbate. The positive value of ΔS proved that, bismuth (III) uptake are organized onto the used sorbent in a more random fashion and may also indicative of moderated sorption step of the complex ion associate of $[BiI_4]^-$ and ordering of ionic charges without a compensatory disordering of the sorbed ion associate onto the used sorbents. The sorption process involves a decrease in free energy, where ΔH is expected to be negative as confirmed above. Moreover, on raising the temperature, the physical structure of the PUFs membrane may be changing, and affecting the strength of intermolecular interactions between the membrane of PUFs sorbent and the $[BiI_4]^-$ species. Thus, high temperature may make the membrane matrix become more unstructured and affect the ability of the polar segments to engage in stable hydrogen bonding with $[BiI_4]^-$ species, which would result in a lower extraction. The negative of ΔG at 295 K implies the spontaneous and physical sorption nature of bismuth (III) retention onto PUFs. The decrease in ΔG on decreasing temperature confirms the spontaneous nature of sorption step of bismuth (III) is more favorable at low temperature. The energy of urethane nitrogen and/or ether oxygen sites of the PUFs provided by raising the temperature minimizes the interaction between the active sites of PUFs and the complex ion associates of bismuth (III) ions resulting low sorption via "Solvent extraction" [38]. These results encouraged the use of the reagent loaded PUFs in packed column mode for collection, and sequential determination of bismuth (III) and (V) in water samples.

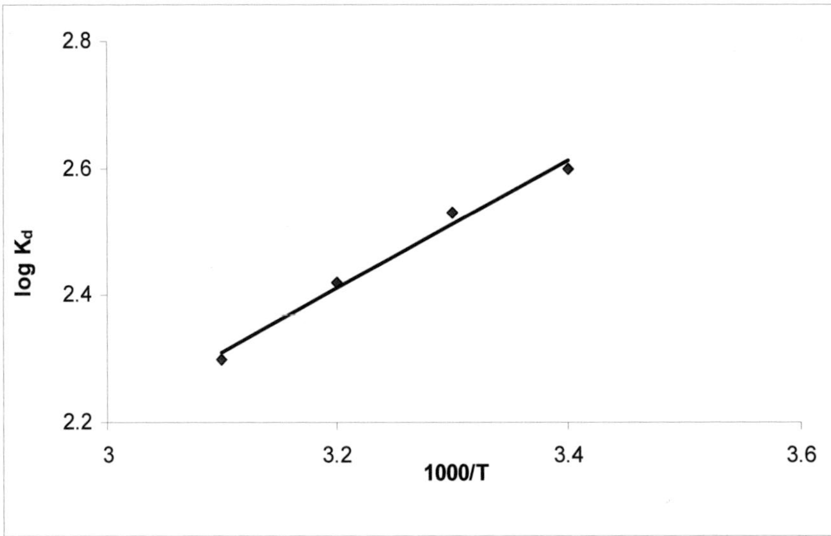

Figure 9. Vant - Hoff plot of log K_d *vs.* 1000/T (K^{-1}) of bismuth (III) retention from aqueous media containing KI (10 % m/v) - H_2SO_4 (0.5 mol L^{-1}) onto $PQ^+ .Cl^-$ loaded PUFs.

3.4. Sorption isotherms of bismuth (III) onto $PQ^+ .Cl^-$ loaded PUFs sorbents

The development of a suitable preconcentration and/ or separation procedures for determination of trace concentrations of bismuth (III) in water is becoming increasingly important. PUFs physically immobilized with a series of quaternary ammonium ion pairireagents e.g. tetraphenyl phosponium chloride, amiloride hydrochloride, tetraheptyl ammonium bromide or procaine hydrochloride was tested for the separation of bismuth (III) from aqueous iodide aqueous media. The results revealed considerable retention of bismuth (III) onto PQ^+ .Cl^- loaded PUFs compared to other onium cations. Thus, the retention profile of bismuth (III) over a wide range of equilibrium concentrations of bismuth (III) ions onto PQ^+ .Cl^- loaded PUFs sorbent from aqueous KI (10%w/v) -H_2SO_4 (1.0 mol L^{-1}) solutions was investigated. The amount of $[BiI_4]^-$ retained onto the PUFs at low or moderate bismuth (III) concentration varied linearly with the amount of bismuth (III) remained in the test aqueous solution (Fig. 10).The equilibrium was approached only from the direction of $[BiI_4]^-$ species-rich aqueous phase confirming a first- order sorption behavior [39]. The sorption capacity of bismuth (III) species towards $PQ^+.Cl^-$ immobilized PUFs as calculated from the sorption isotherm (Fig.10) was 40.0 ± 1.10 mg g^{-1}. The plot of distribution coefficient (K_d) of bismuth (III) sorption between the aqueous solution H_2SO_4 (0.5 mol L-1) and KI (10% w/v) and $PQ^+.Cl^-$ loaded PUFs sorbent is given in Fig. 11. The most favorable values of K_d of bismuth (III) sorption onto PUFs sorbent were also obtained from more diluted aqueous solutions (Fig. 11). The K_d values decreased on increasing the concentration of bismuth (III) ions in the aqueous phase and the PUFs membranes became more saturated with the retained $[BiI_4]^-$ species.

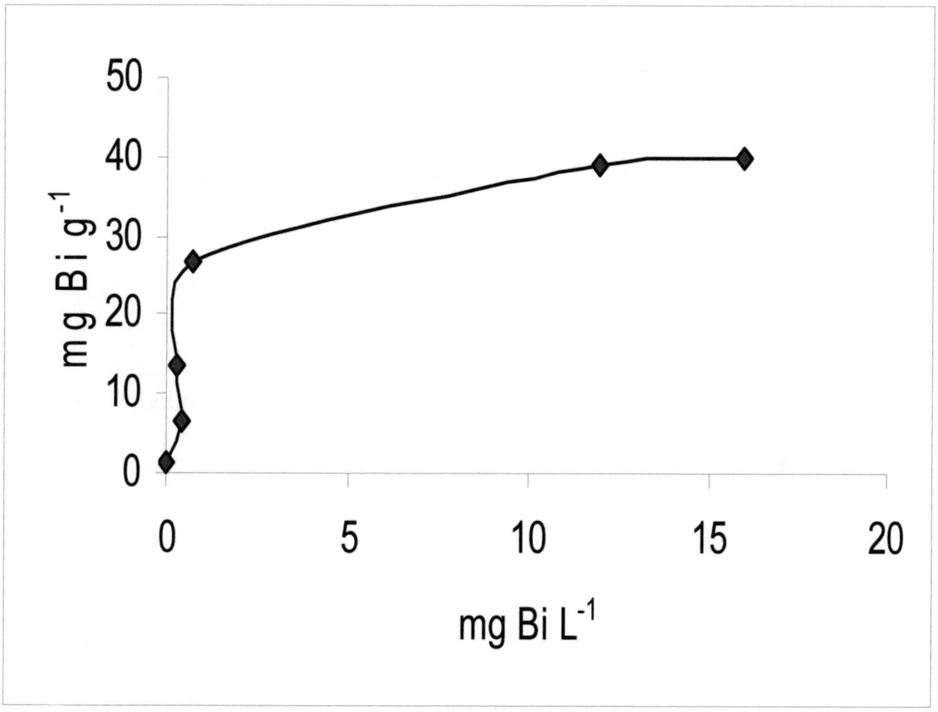

Figure 10. Sorption isotherm of bismuth (III) from aqueous solution of H_2SO_4 (0.5 mol L^{-1}) and KI (10% w/v onto the PQ^+. Cl^- immobilized PUFs.

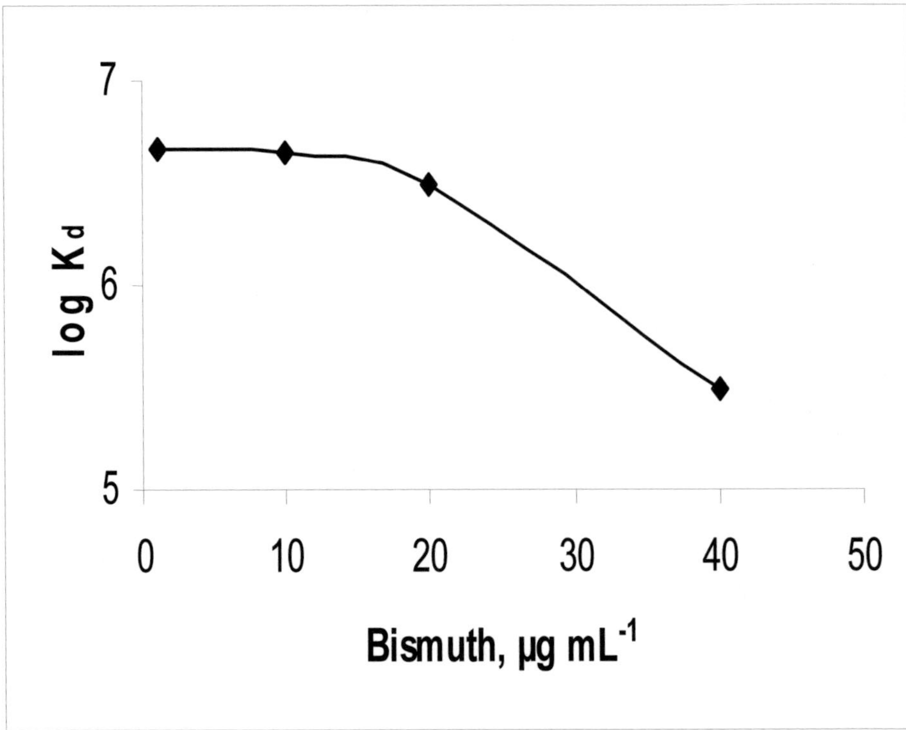

Figure 11. Plot of the distribution coefficient (K_d) of bismuth (III) sorption between the aqueous solution H_2SO_4 (0.5 mol L-1) and KI (10% w/v) and $PQ^+.Cl^-$ loaded PUFs

Sorption of bismuth (III) onto PUFs sorbent was subjected to Langmuir isotherm model expressed in the following linear form [40]:

$$\frac{C_e}{C_{ads}} = \frac{1}{Qb} + \frac{C_e}{Q} \tag{16}$$

where, C_e is the equilibrium concentration (µg mL^{-1}) of bismuth (III) in the test solution, C_{ads} is the amount of bismuth (III) retained onto PUFs per unit mass. The Langmuir parameter Q and b related to the maximum adsorption capacity of solute per unite mass of adsorbent required for monolayer coverage of the surface and the equilibrium constant related to the binding energy of solute sorption that is independent of temperature, respectively. The plot of C_e/C_{ads} *vs.* C_e over the entire range of bismuth (III) concentration was linear (Fig.12) with correlation coefficient of, R^2= 0.998 indicating adsorption of the analyte by PQ^+ $.Cl^-$ treated PUFs sorbents followed Langmuir model. The calculated values of Q and b from the slope and intercept of the linear plot (Fig.12) were 0.21 ± 0.01 m mol g^{-1} and 5.6 ± 0.20 x 10^5 L mol^{-1}, respectively.

Dubinin - Radushkevich (D - R) isotherm model [41] is postulated within the adsorption space close to the adsorbent surface. The D-R model is expressed by the following equation:

$$\ln C_{ads} = \ln K_{DR} - \beta \varepsilon^2 \tag{17}$$

where, K_{DR} is the maximum amount of bismuth (III) retained, β is a constant related to the energy transfer of the solute from the bulk solution to the sorbent and Є is Polanyi potential which is given by the following equation:

$$\varepsilon^2 = RT \ln (1+1/Ce) \tag{18}$$

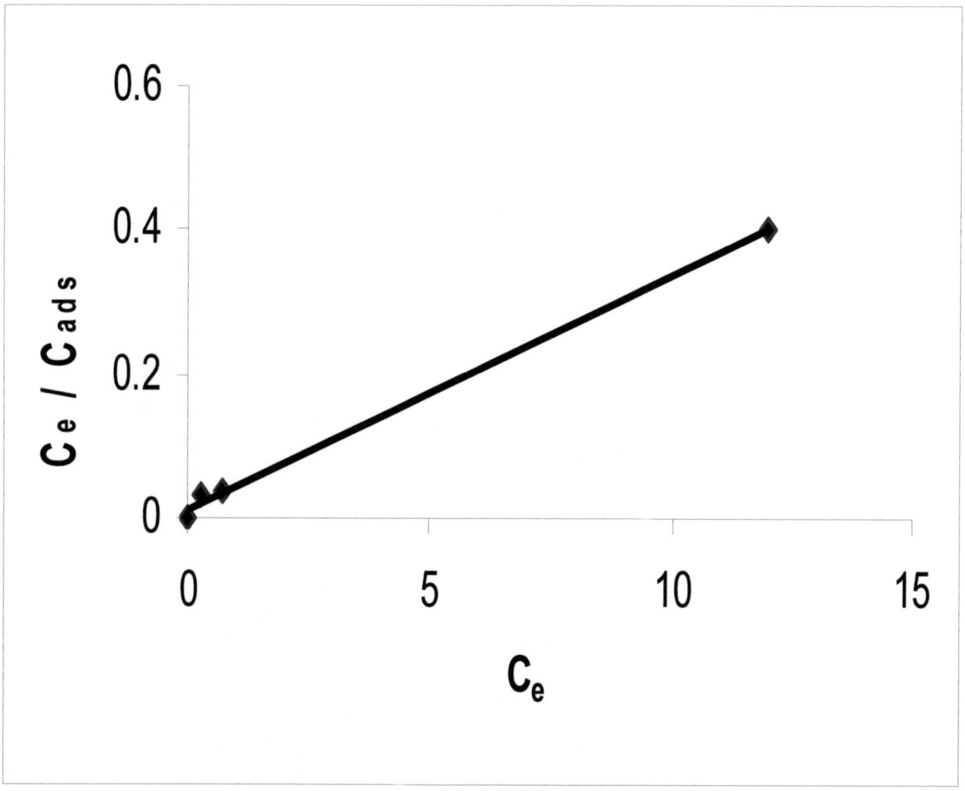

Figure 12. Langmuir sorption isotherm of bismuth (III) uptake from aqueous solution onto $PQ^+.Cl^-$ loaded PUFs at optimum conditions.

The plot of ln C_{ads} versus $Є^2$ was linear with $R^2 = 0.986$ (Fig. 13) for the $PQ^+ .Cl^-$ immobilized PUFs indicating that, the D-R model is obeyed for bismuth (III) sorption over the entire concentration range . The values of β and K_{DR} computed from the slope and intercept were found 0.33 ± 0.01 $mol^2\ KJ^{-2}$ and 171 ± 2.01 μ mol g^{-1}, respectively. Assuming that, the surface of PUFs is heterogonous and an approximation to Langmuir isotherm model is chosen as a local isotherm for all sites that are energetically equivalent, the quantity β can be related to the mean of free energy (E) of the transfer of one mole of solute from infinity to the surface of PUFs. The E value is expressed by the following equation:

$$E = \frac{1}{\sqrt{-2\beta}} \tag{19}$$

The value of E was found 1.23 ± 0.07 $KJmol^{-1}$ for the $PQ^+.Cl^-$ loaded foam. Based on these results, the values of *Q* and *b* and the data reported [42, 43], a dual sorption sorption mechanism involving absorption related to "weak – base anion ion exchange" and an added component for "surface adsorption" is the most probable mechanism for the uptake of bismuth (III) by the used PUFs. This model can be expressed by the equation:

$$C_r = C_{abs} + C_{ads} = DC_{aq} + \frac{SK_L C_{aq}}{1 + K_L C_{aq}} \quad (20)$$

where, C_r and C_{aq} are the concentrations of bismuth (III) retained onto the PUFs and the aqueous solution at equilibrium, respectively. C_{abs} and C_{ads} are the concentrations of the absorbed and adsorbed bismuth (III) species onto the PUFs at equilibrium, respectively and S and K_L are the saturation parameters for the Langmuir adsorption model.

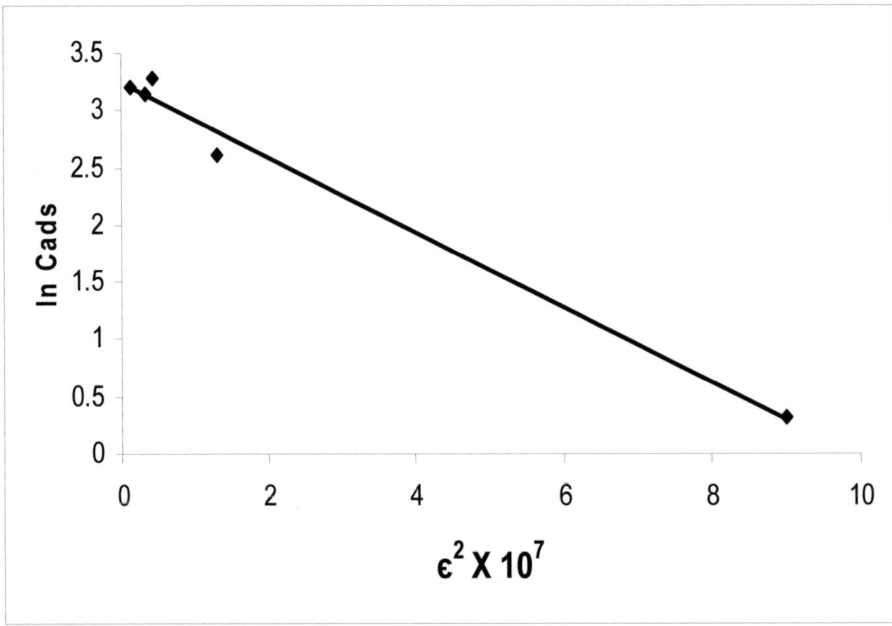

Figure 13. Dubinin-Radushkevich (D-R) sorption of bismuth (III) extraction from aqueous solution onto $PQ^+.Cl^-$ loaded PUFs at the optimum conditions

3.5. Chromatographic behavior of bismuth (III) sorption

The membrane like structures, the excellent hydrodynamic and aerodynamic properties of PUFs sorbent [42, 43], kinetics, capacity and the sorption characteristics of bismuth (III) retention towards plasticized $PQ^+.Cl^-$ PUFs sorbent [39] encouraged the use of the sorbent in packed column for quantitative retention of bismuth (III) from the test aqueous iodide solution. Thus, the test solutions (1.0 L) of the deionized water containing KI (10% w/v) - H_2SO_4 (1.0 mol L^{-1}) was spiked with various trace concentrations (5 -100 μg L^{-1}) of bismuth (III) and percolated through the PUFs packed columns at 5 mL min^{-1} flow rate. ICP-OES measurements of bismuth in the effluent indicated complete uptake of bismuth (III). A series of eluting agents e.g. NH_4NO_3, $HClO_4$ and HNO_3 (1-5. mol L^{-1}) was tested for complete elution of the retained bismuth (III). An acceptable recovery (96.0 ± 2.1) of bismuth (III) was achieved using HNO_3 (10 mL, 3 mol L^{-1}) at 2 mL min^{-1} flow rate. Therefore, HNO_3 (3 mol L^{-1}) was selected as a proper eluting agent for bismuth (III) from the packed columns. With HNO_3, reproducibility data even at ultra trace concentrations (0.5 ng mL^{-1}) of bismuth (III) were successfully achieved. The data of pre concentration and recovery of various concentrations of bismuth (III) are summarized in Table 2. A recovery percentage in the range 98.0 ± 1.5 - 104.2 ± 2.3 was achieved confirming the performance of the developed of $PQ^+.Cl^-$ loaded PUFs.

Bismuth (III) taken, µg L^{-1}	Bismuth (III) found, µg L^{-1}	Recovery, % *
100	98.5	98.0 ± 1.5
50	52	104 ± 2.3
10	10.2	101 ± 1.1

* Average (n=5) ± relative standard deviation.

Table 2. Recovery percentage (%) of bismuth (III) ions from deionized water by the developed PUFs packed columns

The proposed PUFs packed columns was also tested for collection and recovery of bismuth (V) species (< 5 µg L^{-1}) from aqueous solutions after reduction to bismuth (III). A series of reducing agents e.g. H_2S, Na_2SO_3, and KI was tested and satisfactory results were achieved using KI. Thus, in the subsequent work, KI was selected as a proper reducing agent for bismuth (V) to bismuth (III) species. Reduction of bismuth (V) to bismuth (III) was found fast, simple and also form a stable $[BiI_4]^-$ species. The solutions were then percolated through PUFs packed column following the described procedures of bismuth (III) retention. The results are summarized in Table 3. An acceptable recovery percentage of Bismuth (V) in the range 94.0 ± 2.1 – 95.0 ± 3.5 was achieved. The proposed PUFs packed column was also tested for chemical speciation and determination of total bismuth (III) and (V) species in their mixtures. An aqueous solution of bismuth (III) and (V) was first analyzed according to the described procedure for bismuth (V). Another aliquot portion was also adjusted to pH 3 – 4 and shaken with Na-DDTC for 2-3 min and extracted with chloroform (5.0 mL) as Bi $(DDTC)_3$ [33]. The remaining aqueous solution of bismuth (V) was reduced to bismuth (III) with KI (10%w/v) - H_2SO_4 (0.5 mol L^{-1}) and percolated through the $PQ^+.Cl^-$ loaded PUFs column. The retained bismuth species were then recovered and finally analyzed following the recommended procedures of bismuth (III) retention. The signal intensity of ICP- OES of the first aliquot (I_1) is a measure of the sum of bismuth (III) and (V) ions in the mixture, while the net signal intensity of the second aliquot (I_2) is a measure of bismuth (V) ions. The difference (I_1-I_2) of the net signal intensity is a measure of bismuth (III) ions in the binary mixture. Alternatively, bismuth (III) as $Bi(DDTC)_3$ in the methylisobutyl ketone phase was stripped to the aqueous phase by HNO_3 (1 mol L^{-1}) and analyzed by ICP-OES The results are given in Table 4. An acceptable recovery percentage in the 92.5 ± 3.01– 104.3 ± 4.5% of bismuth (III) and (V) ions was achieved.

Bismuth (V) added μg L^{-1}	Bismuth (V) found, μg L^{-1}	Recovery,%
100	95 ± 1.5	95.0± 3.5
250	235 ± 50	94.0 ± 2.1

*Average recovery of five measurements ± relative standard deviation.

Table 3. Recovery (%) of bismuth (V) ions from deionized water by PUFs packed columns

Bismuth (III) and (V) taken, μg L^{-1}		Total bismuth found μg L^{-1}	Recovery,% *
Bi (III)	Bi (V)		
20	25	47 ± 3.5	104 ± 4.5
25	100	118± 5	94.4 ± 2.9
10	10	18.5± 1.5	92.5 ± 3.01

* Average recovery of five measurements ± relative standard deviation.

Table 4. Recovery (%) of total bismuth (III) and (V) in their mixture from aqueous media

3.6. Capacity of the $PQ^{+}.Cl^{-}$ immobilized PUFs

The developed method was assessed by comparing the capacity of the used sorbent towards bismuth (III) sorption with most of the reported solid sorbents e.g. 2, 5- di- mercapto-1, 3, 4-thiadiazol loaded on Silica gel [44] and amionophosphonic dithio-carbamate functionalized polyacrylonitrile [45]. The capacity of the used $PQ^{+}.Cl^{-}$ loaded PUFs sorbent (40.0 ± 1.10 mg g^{-1}) towards bismuth (III) retention was found far better than the data reported by other solid sorbents e.g. 2, 5- dimercapto-1, 3, 4-thiadiazol loaded on Silica gel (3.5 mg g^{-1}) [44] and amionophosphonic dithiocarbamate functionalized poly acrylonitrile (15.5 mg g^{-1}) [45] and some other solid sorbents.[5]

3.7. Analytical performance of the immobilized PUFs packed column

The performance of the PUFs packed column was described in terms of the number (N) and the height equivalent to the theoretical plate (HETP). Thus, aqueous solution (1.0 L) containing bismuth (III) at concentration of 100 μg L^{-1} at the optimum experimental conditions was percolated through the PUFs packed columns (1.0 ±0.001 g) at 5 mL min^{-1} flow rate. Complete retention of $[BiI_4]^-$ was achieved as indicated from the analysis of bismuth in the effluent solution using ICP-MS. The retained bismuth (III) species were then eluted with HNO_3 (10 mL, 3 mol L^{-1}) and a series of fractions (2.0 mL) of eluent solution at 2.0 mL min^{-1} were then collected and analyzed by ICP –OES. The calculated values of N and HETP values from the chromatogram method (Fig. 14) using Gluenkauf equation [14] were equal to 90 ± 3.02 and 0.11± 0.02mm, respectively. The values of N and HETP were also computed from the breakthrough capacity curve (Fig. 15) by percolating aqueous solution (2.0 L) containing bismuth (III) at 100 μgL^{-1} under the experimental conditions through $PQ^{+}.Cl^{-}$ loaded PUFs column at 5 mL min^{-1} flow rate of. The critical and breakthrough capacities [42, 45] calculated from Fig.15 were 1.95 ± 0.1 and 31.25 ± 1.02 mg g^{-1}, respectively. These HETP (97 ± 4) and N (0.13 ± 0.02 mm) values are in good agreement with the values obtaioned from the chromatogram method.

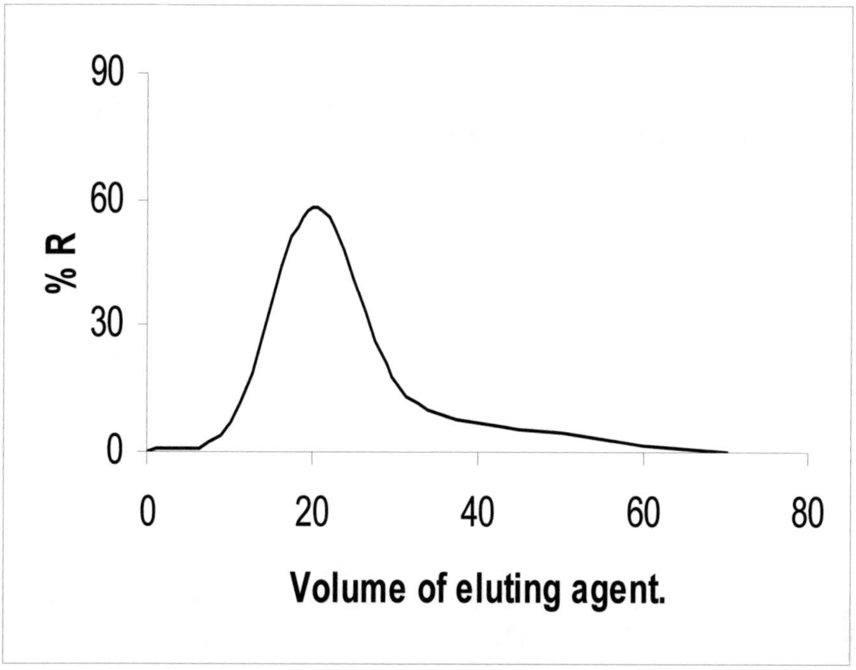

Figure 14. Chromatogram of bismuth (III) recovery from PQ+.Cl- loaded PUFs packed column using nitric acid (5 mol L^{-1}) as eluting agent at flow rate of 2.5 mL min^{-1}.

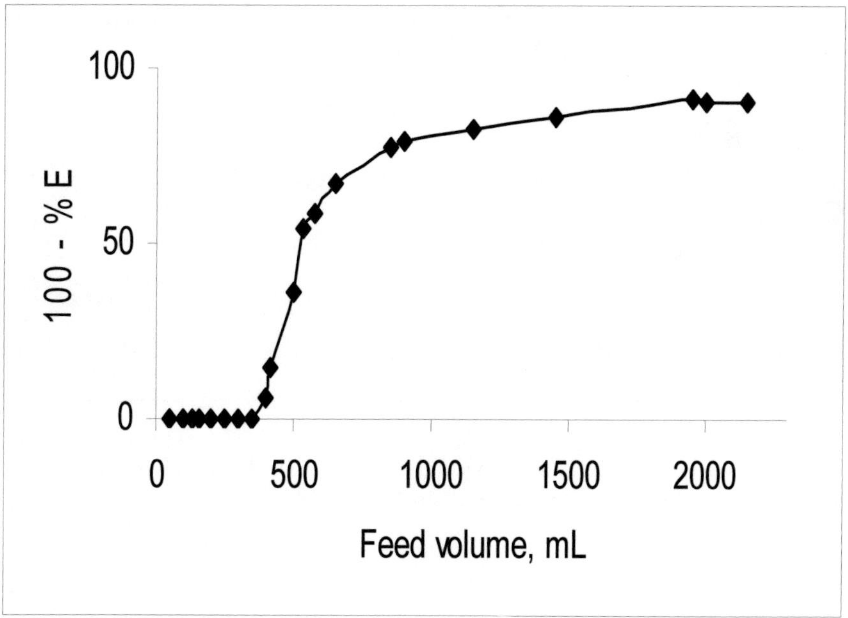

Figure 15. Breakthrough capacity curve for bismuth retention onto PQ+.Cl- loaded packed column at the optimum conditions.

3.8. Figure of merits of the $PQ^{+}.Cl^{-}$ immobilize PUFs packed column

The LOD, LOQ, enrichment and sensitivity factors and relative standard deviation, (RSD) under the optimized conditions were determined. The plot of signal intensity of ICP- OES (I) versus bismuth (III) concentration (C) has the regression equation:

$$I = 4.19 \times 10^{3} C \ (\text{ng L}^{-1}) + 12.96 \ (r = 0.9995) \quad (21)$$

According to IUPAC [46, 47], the LOD = $3S_{y/x}/b$ and LOD = $10S_{y/x}/b$ were 0.9 and 3.01 ngL^{-1}, respectively (V_{sample} = 100 mL) where, $S_{y/x}$ is the standard deviation of *y*- residual and *b* is the slope of the calibration plot [46]. The LOD of the developed method is much better than direct measurement by ICP – OES (5.0 μg mL^{-1}). The enrichment factor ($F_c = V_{s,b}/V_{e,v}$) was defined as the ratio between the volume of analyte sample ($V_{s,b}$ = 1000 mL) before preconcentration and the eluent volume ($V_{e,v}$) after retention and recovery. An average value of F_c of 100 was achieved. The sensitivity factor (the ratio of the slope of the preconcentrated samples to that obtained without preconcentration) was 33.3. The RSD of the method for the determination of standard bismuth (III) solution (50 μg L^{-1}) was ± 2.5% (n= 5) confirming the precision of the method. The figure of merits of the developed method were compared satisfactorily to the reported methods e.g. ICP-OES [45], spectrophotometric [47] and electrochemical [49 -51] (Table 5) in water confirming the sensitivity and applicability of the proposed method. The LOD of the method could be improved to lower values by prior pre concentration of bismuth (III) species from large sample volumes of water (>1.0L). Thus, the method is simple and reliable compared to other methods [50 -52].

SPE	Technique	Linear range,μgL^{-1}	LOD, μgL^{-1}	Reference
Microcrystalline benzophenone	UV – Vis	0 – 2 X10^{4}	—	34
Microcrystalline naphthalene	DPP	180 – 135x10^{2}	55	35
Octylsilane (RP-8) cartridge	ASV	10.5 – 1000	0.73	36
Amberlite XAD-7 resin	HG-ICP-OES	Up to 100	0.02	6
Modified Chitosan	ICP-MS		0.1*	15
$PQ^{+}.Cl^{-}$loaded PUFs	ICP – OES	0.01 – 100	2.7*	Present work

ng L^{-1}

Table 5. Figure of merits of the developed and some of the reported SPE coupled with spectrochemcal and electrochemical techniques for bismuth determination in water

3.9. Interference study

The influence of diverse ions relevant to wastewater e.g. alkali and alkali earth metal ions Ca^{2+}, Mg^{2+}, Cl^{-}, Zn^{2+}, Mn^{2+}, $Cu2^{+}$, Hg^{2+}, Fe^{2+}, Fe^{3+}, Pb^{2+}, Al^{3+}, Ni^{2+}, $Co2^{+}$ and nitrate at various concentrations (0.5 -1.0 mg/ 100 mL sample solution) on the sorption of 10 μg bismuth (III) from a sample volume of 100 mL at the optimum conditions was studied. The tolerance limits (w/w) less than ± 5% change in percentage uptake of bismuth was taken as free from interference. The tested ions except Pb^{2+} did not cause any significant reduction on the percentage (96 -102 ± 2%) of bismuth (III) sorption. Lead ions were found to interfere at higher concentrations (> 0.5 mg/ 100 mL sample solution). Thus, it can be concluded that, the method could applied for the separation and / or determination of bismuth (III) and bismuth (V) after reduction of the latter to trivalence.

3.10. Analytical applications

The validation of the developed method was performed using the certified reference materials (CRM-TMDW). Good agreement between the concentration measured by the proposed method (8.9 ± 0.9 μgL^{-1}) and the certified value (10.0 ± 0.1 μgL^{-1}) of the total bismuth was achieved confirming the accuracy of the method for trace analysis of bismuth in complex matrices.

The method was also applied for the determination of bismuth in wastewater samples (1.0 L) after digestion and percolation through the PUFs packed columns as described. Complete retention of bismuth was achieved as indicated from the ICP-MS analysis of bismuth in the effluent. The retained $[BiI_4]^-$ species were recovered with HNO_3 (10 mL, 3.0 mol L^{-1}) and analyzed by ICP-OES. Various concentrations of bismuth (III) were spiked also onto the tested wastewater samples and analyzed (Table 6). Bismuth (III) determined by the method and that expected (Table 6) in the tested water samples revealed good recovery percentage (98.4± 2.3 – 104 .3 ± 2.8 %) confirming the accuracy and validation of the method.

Bismuth (III) added, (μgL^{-1})	Bismuth (III) found, (μg L^{-1})	Recovery, %*
–	22	–
50	75	104.3 ±2.8
100	120.5	98.4±2.3

* Average recovery of five replicates ± relative standard deviation.

Table 6. Recovery study applied to the analysis of bismuth in wastewater by the developed method

The selectivity of the procedure was further tested for the analysis of bismuth in Red sea water at the coastal area of Jeddah City, Saudi Arabia following the standard addition. as described..The results are summarized in Table 7. An acceptable recovery percentage of 107.01 ±3.5 -108.1 ±2.7 was achieved confirming the selectivity, accuracy and validation of the method.

Bismuth (III) added, (μg L^{-1})	Bismuth (III) found, (μg L^{-1})	Recovery, %*
–	0.07	–
0.30	0.40 ± 0.01	108.1 ±2.7
0.5	0.61 ±0.02	107.01 ±3.5

* Average recovery of five replicates ± relative standard deviation

Table 7. Recovery test for bismuth in sea water by the developed method

4. Conclusion

$PQ^+.Cl^-$ treated PUFs solid sorbent was successfully used for the pre concentration/separation procedures of bismuth (III) and bismuth (V) after reduction of the latter species to bismuth (III). The developed method minimizes the limitations related to sensitivity and selectivity for bismuth determination in various matrices. The intra-particle diffusion and the first order model of bismuth (III) retention onto the tested $PQ^+.Cl^-$ PUFs sorbent are confirmed from the kinetic data. PUFs packed column has shown itself to be a very useful and precise for the analysis of total bismuth (III) & (V) species in water at trace concentrations in water. The PUFs packed column can be reused many times without decrease in its efficiency. Work is continuing for calculating ligation capacity, influence of competitive agents and organic material present in water samples. The LOD of the method is quite close to the concentration of bismuth species reported in marine water. Work is still continuing on developing $PQ^+.Cl^-$ treated PUFs packed column mode for on line determination of bismuth (III) and/ or (V) species at ultra concentrations in aqueous media.

Author details

M.S. El-Shahawi
Department of Chemistry, Faculty of Science, King Abdulaziz University, Jeddah, Saudi Arabia

A.A. Al-Sibaai
Department of Chemistry, Faculty of Science, King Abdulaziz University, Jeddah, Saudi Arabia

H.M. Al-Saidi
Department of Chemistry, University College, Umm Al-Qura University, Makkah, Saudi Arabia

E. A. Assirey
Department of Applied Chemistry, College of Applied Science, Taibah University, Al-Madinah Al-Munawarah, Saudi Arabia

5. References

[1] J. A. Reyes- Aguilera, M. P. Gonzalez, R. Navarro, T.I.Saucedo, M. Avila-Rodriguez, supported liquid membranes (SLM) for recovery of bismuth from aqueous solution, J.Membrane Sci, 2008, 310, 13.
[2] N. Tokman, Anal. Chim. Acta, 2004, 519, 87.
[3] R. Pamphlett, M .Stottenbery. J. Rungby, G. Danscher, Neurotoxicol. Teratol, 2000, 22, 559.
[4] A . S . Ribeiro, M. A. Z. Arruda, S . Cadrore, Spectrochim. Acta Part B, 2002, 57, 2113.
[5] O. Acar, Z. Kilic, A. R. Turker, Anal. Chim Acta, 1999, 382, 329.
[6] L. Rahman, W. T. Corns, D. W. Bryce, P. B. Stockwell, Talanta, 2000, 52, 833.
[7] Y. Zhang, S.B. Adeloju, , Talanta, 2008, 76, 724.
[8] H. Guo, Y. Li, P. Xiao, N. He,Anal. Chim. Acta, 2005, 534, 143.
[9] H.Y. Yang, W.Y. Chen, I.W. Sun, Talanta, 1999, 50, 977.

[10] R. Hajian, E. Shams, Anal Chim Acta, 2003, 491, 63.
[11] S. G Sarkar, P. M Dhadke, Sep. Purif. Technol, 1999, 15, 131.
[12] J. M. Lo, Y. P. Lin, K. S. Lin, , Anal. Sci (Japan), 1991, 7, 455.
[13] M.A. Taher, E. Rezaeipor, D. Afzali, Talanta, 2004, 63,797.
[14] Y. Yamini, M. Chaloosi, H. Ebrahimzadeh, Talanta, 2002, 56,797.
[15] E. M. Thurman, M. S. Mills "Solid Phase Extraction, Principles and Practice" John Wiley and Sons, 1998.
[16] B. Manadal, N. Ghosh, J. Hazard. Materials, 2010,182, 363.
[17] M. Sun, Q. Wu, J. Hazard. Materials, 2010, 182, 543.
[18] M.A. Didi, A.R. Sekkal, D. Villemin, Colllids and Surfaces A: Physicochemical and Engineering Aspects, 2011, 375 (1-3), 169.
[19] S. Palagyi and T. Braun "Separation and Pre-concentration of Trace Elements and Inorganic Species on Solid Polyurethane Foam Sorbents" In Z. B. Alfassi and C. M. Wai "Preconcentration Techniques for Trace Elements" CRC Press, Baca Rotan, FI, 1992.
[20] G.J. Moody, J.D.R. Thomas, "Chromatographic Separation with Foamed Plastics and Rubbers" Dekker, New York, 1982.
[21] M. S. El-Shahawi, M. A. El-Sonbati, Talanta, 2005, 67, 806.
[22] M. S. El-Shahawi, M.A. Othman, M. A. Abdel-Fadeel, Anal. Chim. Acta, 2005, 546, 221.
[23] M. S. El-Shahawi, R. S. Al-Mehrezi, Talanta, 1997, 44, 483.
[24] M. S. El-Shahawi, H. A. Nassif, Anal. Chim. Acta, 2003,487, 249.
[25] M. S. El-Shahawi, H. A. Nassif, Anal. Chim. Acta, 2003, 487, 249.
[26] T. Braun, J. D. Navratil, A. B. Farag "Polyurethane Foam Sorbents in Separation Science" CRC Press Inc, Boca Raton, FL 1985.
[27] D.D. Mello, S.H. Pezzin, S.C. Amico , The effect of Post – consumer PET particules on the performance of flexible polyurethane foams, Polymer Testing, 2009, 28, 702.
[28] A.B. Farag, M.H. Soliman, O.S. Abdel-Rasoul, M.S. El-Shahawi, Anal. Chim. Acta, 2007, 601, 218.
[29] A.B. Bashammakh, S.O. Bahaffi, F.M. Al-shareef, M.S. El-Shahawi, Anal. Sci (Japan), 2009, 25, 413.
[30] F. A. Cotton and G. Wilkinson"Advanced Inorganic Chemistry" Wiley, London. 1972.
[31] A.I. Vogel "Quantitative Inorganic Analysis"3rd edn. Longmans Group Ltd., England, 1966.
[32] Z. Marczenko "Separation and Spectrophotometric Determination of Elements" 2 nd edn. John Wiley and Sons,1986.
[33] M. M. Saeed, M. Ahmed, Anal. Chim Acta, 2004,525, 289.
[34] P. R. Haddad, N. E. Rochester, J. Chromatogr, 1988, 439, 23.
[35] W. J. Weber, and J. C. Morris, , *J. Sanit. Eng. Div. Am. Soc. Civ. Eng.*, 1963, *89, 31.*
[36] A. K. Bhattacharya and C. Venkobachar, *J. Environ. Eng.*, 1984, 110, 1.
[37] D. Reichenburg, *J. Am. Chem. Soc.*, 1972, *75*, 589.
[38] S. Zhi-Xing, P. Qiao-Sheng, L. Xing-yin, C. Xi-Jun, Z. Guang-Yao, R. Feng-Zhi, Talanta, 1995,42, 1127.
[39] M.S. El-Shahawi, A. Hamza, A. A. Al-Sibaai and H.M. Al-Saidi, Chem. Eng. J. 2011, 173 (2), 255.

[40] L. Langmuir, J. Am. Chem. Soc, 1918, 40, 136.
[41] G.A. Somorjai" Introduction to Surface Chemistry and Catalysis" John Wiley& Sons, INC, 1994.
[42] O. D. Sant'Ana, L. G. Oliveira, L. S. Jesuino, M. S. Carvalho, M. L. Domingues, R. J. Cassella and R . E. Santelli, J. Anal. At. Spectrom, 2002, 17, 258.
[43] S. Palagyi and T. Braun "Separation and Pre-concentration of Trace Elements and Inorganic Species on Solid Polyurethane Foam Sorbent" in Z. B, Alfassi and C. Wai "Pre-concentration Techniques for Trace Elements" CRC Press, Boca Roton FL.1992.
[44] K. Terada, K. Matsumoto and Y. Nanao, Anal. Sci. (Japan), 1985,1, 145.
[45] W.X.Ma, F.Liu, K.A.Li, W.Chen and S.Y.Tong, Anal.Chim.Acta, 2000, 416, 191.
[46] M. Filella, J. Environ. Monitoring, 2010, 12, 90.
[47] J.C.Miller, J. N. Miller "Statistics for Analytical Chemistry" Ellis-Horwood, New York, 4th edn., 1994.
[48] C.Lin, H. Wang, Y. Wang, Z. Cheng, Talanta, 2010, 81,30.
[49] D. Thorburn Burns, N. Tungkananuruk and S. Thuwasin, Anal. Chim. Acta, 2000, 419, 41.
[50] A. Bhalotra and B.K. Prui, J.AOAC Intern, 2001,84, 47.
[51] M.H. Pournaghi-Azr, D. Djozan, and H. A. Zadeh, Anal. Chim. Acta, 2001, 437, 217.
[52] I. Kulaa, Y. Arslanb, S. Bakırdere, S. Titretir, E. Kenduzler and O. Y. Atamanb, Talanta, 2009, 80, 127.

Chapter 14

Polyurethane Grouting Technologies

Jan Bodi, Zoltan Bodi, Jiri Scucka and Petr Martinec

Additional information is available at the end of the chapter

http://dx.doi.org/10.5772/35791

1. Introduction

Grouting with polyurethane [PU] resins represents an effective method of improvement of mechanical and sealing properties of soil and rock environment and constructions. The principle of grouting technologies is injection of liquid grouting material into the rock environment or construction under pressure. During the grouting process, fissures and pores are filled with the grouting material, which subsequently hardens and connects the disintegrated parts of the rock mass or grains of loose material. Polyurethane grouting technologies started to be used in the 80s of the 20th century in the mining industry. In the last recent years, PU grouting technologies spread significantly from the mining applications to civil engineering and geotechnics. The application possibilities have a rising tendency and new possibilities occur. Currently, grouting technologies are used mainly in the following fields:

- **Underground constructions, tunneling**
  - filling of caverns and voids
  - protection when crossing fault zones
  - stabilization of loose material in the foreland of excavation
  - securing of excavation during tunnel construction
  - preventive improvement of mechanical properties of the rock mass in the line of the workings
  - sealing and stopping of water inflows into the construction
  - anchoring of soil and rocks
  - strengthening and stabilization of overburden and etc.
- **Mining**
  - strengthening and stabilization of deposit layers before exploitation
  - crossing of fault zones
  - securing of the overburden
  - stabilization of the surrounding of the mine workings
  - lowering of permeability of the rock mass

  - strengthening of coal in areas with rock burst risk
  - limitation of the mine wind blowing
  - anchoring of soil and rocks
  - stabilization and sealing of old mine pits and etc.
- **Geotechnical works**
  - stabilization of slopes, embankments, excavations
  - anchoring of retaining walls
  - construction of underground barriers with low permeability
  - stabilization of unconsolidated soil
  - sealing of dilatation joints
  - micropiling of foundations
  - stabilization of landslides
- **Civil engineering**
  - strengthening of subsoil (also under groundwater level)
  - securing of stability of structures threatened by mining or construction works
  - strengthening of brick or stone masonry
  - restoration of insulation of structures
  - sealing of utility entries into constructions
  - sealing of joints
  - stopping of water inflows into constructions and etc.
- **Foundation of buildings**
  - sealing and anchoring of bottoms and walls of construction pits under groundwater table
  - anchoring of walls of construction pits
  - improvement of subsoil conditions before starting of the construction
  - micropiling in soil with low bearing capacity
  - foundation of buildings in undermined areas
- **Water management works**
  - sealing of joints on dams
  - anchoring and sealing of flood dams, anchoring of bottom of water canals
  - anchoring and strengthening of embankments
  - repair of concrete structures under water
  - limitation of underflowing of dams
- **Bridges and roads**
  - strengthening and sealing of brick and stone masonry on bridges
  - repair of cracks in the constructions
  - improvement of subsoil parameters under pillars (also in rivers)
  - anchoring and micropiling of foundations

This chapter contains brief description of PU grouting technologies and characteristics of basic grouting material types. It further presents practical findings of the authors obtained throughout their long term experimental research, design work and application of PU grouting technologies. The findings are based also on development of PU grouting systems Geopur, Geocream and Supermin from the production of company GME, s.r.o.

2. PU grouting resin types

PU grouting materials can be divided according to their chemistry to three main groups:

1. two-component (PU) organic resins:

component A – polyol in mixture (polyetherpolyol, catalysts, additives),
component B – isocyanates in mixture (methylene diphenyl diisocyanate [MDI], homologes, isomeres).

After curing they form solid PU resins or foam.

2. one-component organic resins:

react with moisture present in the environment or construction and form an organic resin (material is on the basis of prepolymer MDI)

3. two-component organic-mineral resin (OMR):

component A – polysiliceous acid (natrium water glass, catalyst and additives),
component B - isocyanates (MDI, homologes and isomeres).

The main difference between the above materials is, that **material on the basis of polyol – isocyanate react with moisture present in the environment**, while material on the basis of polysiliceous acid – isocyanate are inert to moisture or water.

In case of OMR material, the mixing of the components plays an important role in the grouting process. The component A is inorganic - formed by water glass and additives. It is very different form the component B, which is of organic character on the basis of MDI. During mixing the water glass disintegrates to small drops in the organic phase of MDI and an inhomogeneous system is formed. Two different components A and B are in contact with each other only at the surface of individual drops. Chemical reaction proceeds better, the smaller the drops of component A are (the contact of the components is more intense). The reaction can be influenced also by additives, which lower the surface tension of water glass (e.g. silicones). The best results are achieved when mixing by ultrasound. Formed product of hardening process is a resin with solid closed pores of polysilicious acid gel.

In case of PU material, the intensity of mixing does not have fundamental impact to the reaction proceeding. A homogenous solution is formed by the mixing, which cures quite well.

The hardening process, following the mixing and injection of the PU mixture into the rock mass, takes from several minutes up to few hours, according to the type of used grouting resin. Currently, a wide variety of PU grouting materials of various producers exist on the market. Physical and mechanical properties of individual systems differ and it is often quite difficult to choose the appropriate system for particular application. In table 1 we present for example technical data of universal PU grouting system Geopur® (Bodi, 2003), produced and used since 1994.

Type	Geopur® 082/1000		Geopur® 082/600		Geopur® 082/350		Geopur® 082/290	
Component	**A**	**B**	**A**	**B**	**A**	**B**	**A**	**B**
Volume weight, 20 °C [kg/m³]	1075	1235	1075	1235	1075	1235	1075	1235
Viscosity, 20 °C [mPas]	150-300	170- -230	150-300	170- -230	150-300	170- -230	150-300	170- -230
Mixing ration A/B weight.	100	126	100	126	100	126	100	126
Mixing ratio A/B volume.	100	100	100	100	100	100	100	100
Foaming factor*	1 - 1,2		1,5 - 2		2 - 4		4 - 5	
Volume weight of the foam [kg/m³]	1000 ± 20		600 ± 20		360 ± 20		290 ± 20	
Temperature of the curing reaction max [°C]	do 132		do 132		do 132		do 132	
Beginning of foaming at 20 °C [sec]	120 ± 2		120 ± 2		120 ± 2		120 ± 2	
Type	**Geopur® 082/180**		**Geopur® 082/90**		**Geopur® 230**		**Geopur® 240**	
Component	**A**	**B**	**A**	**B**	**A**	**B**	**A**	**B**
Volume weight, 20 °C [kg/m³]	1075	1235	1075	1235	1090	1235	1075	1235
Viscosity, 20 °C [mPas]	150-300	170- -230	150-300	170- -230	150-300	170- -230	150-300	170- -230
Mixing ratio A/B weight.	100	126	100	126	100	126	100	126
Mixing ratio A/B volume.	100	100	100	100	100	100	100	100
Foaming factor	5 - 6		9 - 11		10 - 15		až 40	
Volume weight of the foam [kg/m³]	180 ± 20		90 ± 20		90 ± 30		35 ± 3	
Temperature of the curing reaction max [°C]	do 132		do 132		do 140		do 132	
Beginning of foaming at 20 °C [sec]	120 ± 2		120 ± 2		120 ± 12		120 ± 3	

Table 1. Technical data of the grouting system Geopur® produced by the company GME

3. Grouting equipment

Injection of grouting material into the rock massive is performed by grouting pumps. Usually piston type pumps with electric or pneumatic drive are used. There are one component and two component pumps available. An example of a grouting pump is presented on Fig. 1 below.

Figure 1. Two component electric grouting pump DV 97.

Grouting elements are used during the injection of the grouting material into the rock mass. These are technically designed to transfer the pressure of the grouting material, preventing back flow of the material out from the borehole. They are usually equipped with a back valve. According to the method of fastening in the borehole, we distinguish mechanically fastened once, hydraulically, drilled, pushed in, vibrated or glued. They are called grouting packers, grouting anchors or bolts, grouting tubes and etc.

4. Grouting technology

Mixing of the PU mixture is made in mixing chamber, which is located behind the pump. This is located as close as possible to the borehole. Grouting pump sucks both components of the grouting resin from separate tanks or the components flow in gravitationally. The pump takes the components in appropriate ratio and delivers them separately to the mixing chamber. In the mixing chamber, components are mixed and subsequently injected through the packer into the rock mass. The resin penetrates under the pressure into surrounding fissures and cavities up to the distance of a few meters from the borehole. As a result sealing and strengthening of the rock mass or construction is achieved. After finishing of the grouting, it is necessary to flush the pump, hoses and accessories and clean the equipment. In case of longer regular use, it is possible to leave the components in the pump and hoses.

The work team is usually formed by a couple of trained workers. Parameters of the grouting works are recorded during the work like e.g. location of boreholes, grouted quantities, grouting pressure and temperature.

Injection of material into the rock environment proceeds:

- without reshaping of the grouted rock mass or
- with reshaping of the grouted rock mass.

Grouting without reshaping of the rock mass may be of penetration or filling character. Penetration grouting works are performed in sandy soil or in constructions. Filling grouting is used in fissured rock and coarse grained soil like sand or gravel.

Geopur® type	foaming factor [-]	Volume weight [kg/m³]	Water intake after 28days [vol. %]	Flexural strength [MPa]	Elasticity modulus [MPa]	Compressive strength [MPa]
82/90	9 - 11	82	2,8	1,6	21	0,5
82/180	5 - 6	185	2,2	2,8	54	2,8
82/290	4 - 5	276	1,7	7,4	192	5,9
82/350	2 - 4	354	1,2	8,3	241	9,5
82/600	1,5 - 2	589	0,9	13,4	443	23,2
82/1000	1 - 1,2	1060	0,4	30,3	985	67,7

Table 2. Physical and mechanical parameters of the grouting system Geopur® produced by the company GME

In case of grouting with reshaping of the rock environment a so called claquage occurs, which is in principle hydraulic fracturing of the rock well known from the oil and gas exploitation. Due to the high hydraulic pressure of the grouting media in the soil a spatial net of fissures is formed, which are subsequently filled with the grouting media. The length and width of fissures depends on the pressure of grouted resin, velocity of penetration and quantity of the grouting resin. Compacting grouting belongs among the grouting methods considered as reshaping the rock mass as well.

5. Behavior of PUR resin in the grouted environment

Grouting PUR resin enters into the borehole as a mixture. The grouting material flows through the rock mass first as a liquid. After curing reaction start, gaseous CO_2 is formed, which causes foaming of the mixture. In case of contact with moisture present in the soil or rock, the foaming is more intense, because the water reacts with the present isocyanate groups. Foaming causes increase of volume of the PUR mixture. The mixture is pushed into open structures of the rock mass and the viscosity of the mixture consecutively increases. The flowing stops, when the viscosity of the material is so high that further pumping is impossible, and the resin becomes hard foam. In case, that the pump is further operated, the pressure increases and the material density increases. In practice, this situation is indicated by significant pressure increase. Increase of the pressure may sometime cause opening of new structures for the grouting and continuing of the grouting. In case of formation of new openings the pressure drops. This may occur repeatedly until full grouting of the surrounding of the borehole.

Volume weight of the grouting material increases from the front of the grouted structure towards the packer. In case of the PUR resins, when the pump is stopped, so called autogrouting continues, which is induced by the reaction of the material and formed CO_2, which induces pressure of 0,1 to 0,3 MPa. In case the grouting process stops before full saturation of the environment by the grouting media, the saturation continues due to the pressure formed by CO_2 until finishing of the chemical reaction. In case, that the fissure had been already filled, the pressure is higher than the pressure of CO_2 and bubbles are not formed - CO_2 remains dissolved in the grouting media and has minimal volume. The texture of the material is in this case compact. Formation of the bubble structure depends therefore on the pressure under which the mixture cures. Usually porous structures are formed with closed or partly closed pores during the grouting.

6. Properties of grouted soil, rock and building material

During pressure grouting of PU grouting resins into soil, rock mass or fissured or defected constructions, new specific materials are formed. These materials have the properties of composite material and, taking into account their components character, are referred to as **geocomposites** (Snuparek & Soucek, 2000).

In rocks or constructions the grouted environment contains discontinuities. The geo-component of the formed geocomposite is formed by blocks of the rock (or masonry), which are defined by combination of bedding surfaces, metamorphic foliation, fissures and etc.

In soil, two basic types of geocomposites are formed by PU grouting: in case of non cohesive soil (sand-gravel), the geo-component of the geocomposite is built by solid grains or their aggregations of various size and shape. These contain grains of minerals and rocks, organic particles (shells of organisms, wood, carboniferous parts of plants and others) or parts of constructions (building material, metals, ash and others). In case of cohesive soil (clay, claystones, or siltstones), the geo-component of the geocomposite is formed by blocks of soil penetrated by a net of so called claquage fissures (fissures caused by hydraulic fracturing during the grouting), which are filled with the binding material.

The binding material is represented in these geocomposites by hardened organic or organic-mineral PU resin with various degree of foaming.

Penetration of the grouting media through the inhomogeneous environment, and thus also the resulting properties of the formed geocomposite, is influenced by many factors. In case of geocomposites of PU resin – rock (soil) and PU resin – building material, the following factors have primary effect (Scucka & Soucek, 2007):

- **properties of the unpolymerized grouting media** –viscosity of the media as a function of temperature and rheology of hardening, velocity of injection (volume per time unit), grouting pressure, the right stoichiometric ratio of input components and sufficient time and intensity of their mixing;
- **properties of the grouted environment** – composition of the rock (building material), shape and size of soil particles and rock blocks, humidity, effective porosity or voids,

type and orientation of discontinuities, temperature of the environment, permeability (plastic+water+gas), adhesion of grouting media to the rock surface, composition of water, pore pressure.

Formed structure and texture of the geocomposite (usually very variable in case of PU geocomposites) is a result of the effect of the above mentioned factors. This variability depends on the bedding conditions of the grouted rock and on the parameters of the grouting process, mainly on the grouting pressure. Grouting pressure together with moisture cause for example significant zonal heterogeneity of the geocomposite in case of grouting of wet or saturated sand (mainly of lower permeability) (Aldorf & Vymazal, 1996).

Structural and textural variability of geocomposites significantly complicate the estimation of physical and mechanical properties of geotechnical constructions formed within the grouting process. Mainly the determination of strength and deformation properties of the geocomposite is problematic, because it is often hard to prepare standard laboratory testing specimens from the samples available and collected in situ by core drilling or excavation. In cases when it is impossible to prepare testing specimens from real in situ samples, model geocomposites are prepared by grouting into pressure tanks in the laboratory (Snuparek & Soucek, 2000). Physical and mechanical properties are subsequently determined on such prepared model samples. Qualitative and quantitative structural-textural parameters of the geocomposite are also analyzed by the methods of image analysis and are subsequently compared with parameters of real samples.

In the following text, basic types of structures and textures of geocomposites (with PU binding material) will be described and examples of determination of mechanical properties on real and laboratory prepared samples will be presented.

6.1. Structure and texture of geocomposites

Table 3 below presents a simple classification system for description of structure and texture of PU geocomposites according to various criteria. Some of the criteria are taken over from the modified system commonly used for analyses of structure and texture of sedimentary rocks in petrography (Pettijohn, 1975). We describe in more detail the categories created by the authors based on their long-term research. These include division of geocomposite textures according to the character of binding material penetration into the grouted soil (rock), division of structures according to quantity of binding material and description of the structure of the binding material in the geocomposite from the point of view of distribution, size and morphology of bubble pores.

6.1.1. Character of penetration of the binder into the grouted rock

According to penetration of the binder into the grouted soil or rock, the following textures or their combination may be distinguished:

- **honeycomb texture I.** - rock particle is surrounded by the binder and this has good adhesion to the rock surface (Fig. 2),

<table>
<tr><td colspan="5">GEOCOMPOSITE TEXTURES</td></tr>
<tr><td colspan="2">According to ordering of building units</td><td colspan="3">According to character of penetration of the binder into the rock</td></tr>
<tr><td colspan="2">- parallel
• linear parallel
• aerial linear parallel
• aerial parallel (bedded)
  • bed type
  • desk type
  • laminar
- massive</td><td colspan="3">- honeycomb type I. (Fig. 2)
- honeycomb type II. (Fig. 3)
- honeycomb type III. (Fig. 4)
- doughy (Fig. 5)
- stringer type (Fig. 6)
- claquague (Fig. 7)
- diffusive (Fig. 8)
- barrier type (Fig. 9)</td></tr>
<tr><td colspan="2">According to distribution of rock particles in the binder</td><td colspan="2">According to the level of filling of the space</td><td>According to spatial distribution of particles and pores</td></tr>
<tr><td colspan="2">- with particles evenly distributed in the binder material
- with particles unevenly distributed in the binder material</td><td colspan="2">- compact
- porous</td><td>- isotropic
- anisotropic</td></tr>
<tr><td colspan="5">GEOCOMPOSITE STRUCTURES</td></tr>
<tr><td>According to rock grain size</td><td colspan="2">According to relative grain size</td><td colspan="2">According to angularity of clastic particles</td></tr>
<tr><td>- pelitic
- aleuritic
- psamitic
(fine, medium, coarse)
- psefitic
(fine, medium, coarse)
- stone type
- boulder type</td><td colspan="2">- evenly grained
- unevenly grained</td><td colspan="2">- breccious
- conglomerate
- angular psamitic
- sub angular psamitic
- sub oval psamitic
- oval psamitic
- perfectly oval psamitic</td></tr>
<tr><td>According to quantity of binder</td><td colspan="2">According to distribution and morphology of bubble pores in the binder</td><td colspan="2">According to size of pores in rock or binder</td></tr>
<tr><td>- basal
- porous
- contact type
- coating type</td><td colspan="2">- type PUR 1 (Fig. 10a)
- type PUR 2 (Fig. 10b)
- type PUR 3 (Fig. 10c)
- type PUR 4 (Fig. 10d)
- type OMR 1 (Fig. 11a)
- type OMR 2 (Fig. 11b)</td><td colspan="2">- with micro pores
- with macro pores
- with cavities</td></tr>
</table>

Table 3. Classification system for description of geocomposite structure and texture.

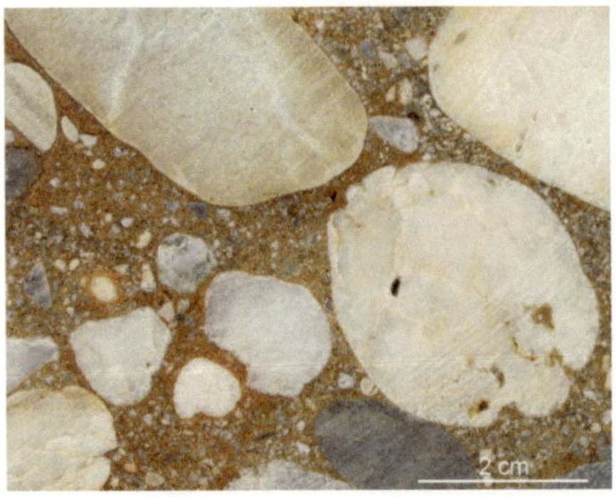

Figure 2. Honeycomb texture I. (PUR is surrounding the rock particle and sticks well to the rock surface).

- **honeycomb texture II.** - rock particle is surrounded by the binder, but the binder sticks only partly to the rock surface (Fig. 3),
- **honeycomb texture III.** - rock particle is surrounded by the binder, but the binder does not stick to the rock surface and is separated from the rock by a gap; free particle may be taken out from the „tissue" of the plastic binder (Fig. 4),
- **doughy texture** - the binder looks like pastry pushed into the gaps between the grains of the aggregate, it does not fill fully the gaps between the grains and does not stick completely to the grains (Fig. 5),

Figure 3. Honeycomb texture II. (PUR is surrounding the rock particles, but sticks only partly to their surface).

Figure 4. Honeycomb texture III. (free rock particle can be taken out of the PUR-binder „tissue").

Figure 5. Doughy texture - OMR-binder has a character of dough pushed into gaps between the conglomerate grains, it does not fill fully the voids and does not stick fully to the grains.

- **stringer texture** - a net of fissures (usually in all directions), not formed due to the grouting, spreads through the rock (masonry) and is filled with the binder (Fig. 6),
- **claquage texture** - a net of fissures, which was formed due to the grouting, spreads through the rock (masonry) and is filled by the binder (Fig. 7),
- **diffusive texture** - the rock is penetrated by the binder "in diffusive way" in pores (Fig. 8),
- **barrier texture** - binder fills only the interconnected cavities and gaps between the grains, it does not penetrate through the barriers formed by the present fine-grained soil (Fig. 9).

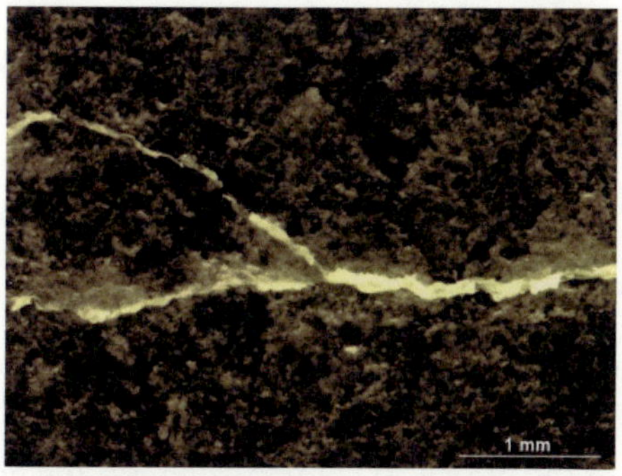

Figure 6. Stringer texture - a net of fissures (usually in all directions), not formed due to the grouting, spreads through the rock (masonry) and is filled with the binder.

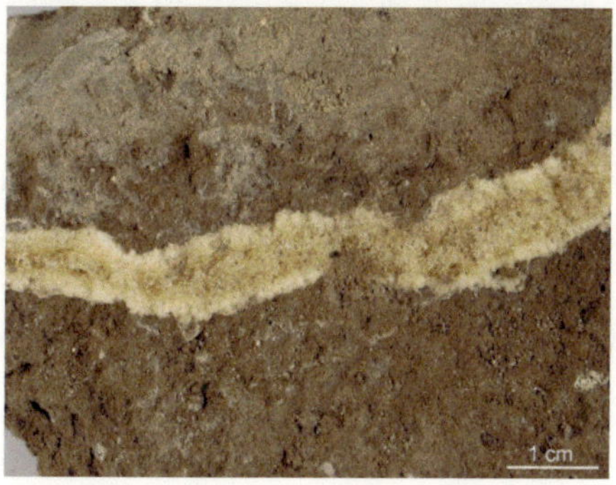

Figure 7. Claquage texture – fine-grained soil fractured hydraulically with claquage fissure, which is filled with PUR-binder.

Figure 8. Diffusive porous zonal texture of geocomposite (crushed brick + PUR). A border formed by penetration of the binder into the pores of the brick fragments is visible on the bigger grain edges. Smaller brick fragments are fully penetrated by the binder.

Figure 9. Barrier texture – the binder fills the interconnected cavities between the grains, it does not penetrate through the barriers formed by the basic mass.

6.1.2. Quantity of the binder in the geocomposite

According to the quantity of the binder in comparison with the quantity of rock component in the geocomposite, the following structures can be distinguished:

- **basal structure** – rock particles are distributed in the abundant binder, particles are separated,
- **porous structure** – binding material fills the pores and voids in between the grains, grains are in contact with each other,
- **contact structure** – binding material is present only in places of grain contact,
- **coating structure** – small amount of binder creates coating around the clastic grains.

6.1.3. Distribution, size and morphology of bubble pores in the binder

A specific feature of most grouting media on the basis of PU is increase of their volume by foaming. In order to describe the relative distribution and morphology of bubble pores in the foamed hardened PU binder, we use the following classification for both micro as well as macro evaluation (Scucka & Soucek, 2007).

- **type PUR 1** – binder is compact, vitreous, bubble pores occur only sporadically or are not present at all (Fig. 10a),
- **type PUR 2** – isolated spherical or ellipsoidal bubble pores of similar size are suspended within the vitreous binder, bubbles have smooth walls, no collapsed walls occur (Fig. 10b),
- **type PUR 3** – partly collapsed bubble pores are suspended within isles of vitreous compact binder, bubbles are in contact, walls are of peel or shell character (Fig. 10c),
- **type PUR 4** – collapsed bubble pores with thin walls are in contact with each other and deform themselves, walls are of peel to honeycomb character. Vitreous compact binder is missing or is sporadic (Fig. 10d).

In case of organic-mineral resins, out of which mainly non foaming types are used in the geotechnics, the structure of the hardened resin has different character. The character strongly **depends on the intensity and time of mixing** of the input components. In case of good mixing, isolated or touching, regular spherical, white drops of polysilicious acid gel are densely distributed within the plastic mass. Irregularly distributed spherical or less regular pores of various sizes are also present in the structure (**type OMR 1**, Fig. 11a). In case of insufficient mixing time and intensity, an inhomogeneous mass is formed containing mineral part, which is irregularly distributed within the plastic mass (**type OMR 2**, Fig. 11b).

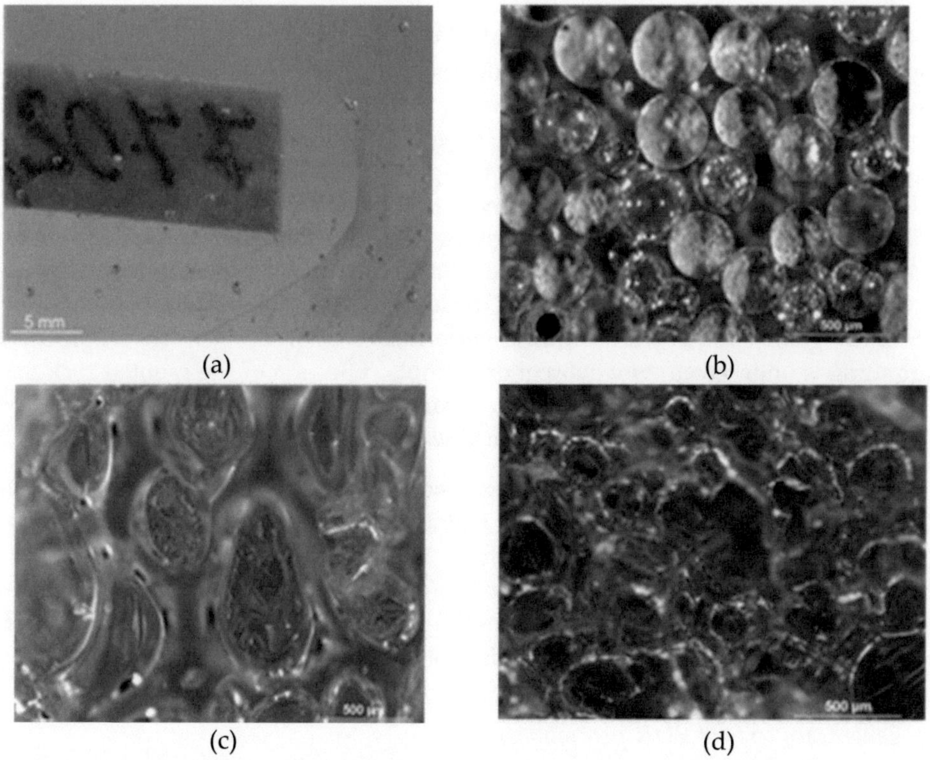

Figure 10. Basic types of plastic binder structure with pores in PUR-geocomposites: (a) type PUR 1, (b) type PUR 2, c) type PUR 3, d) type PUR 4.

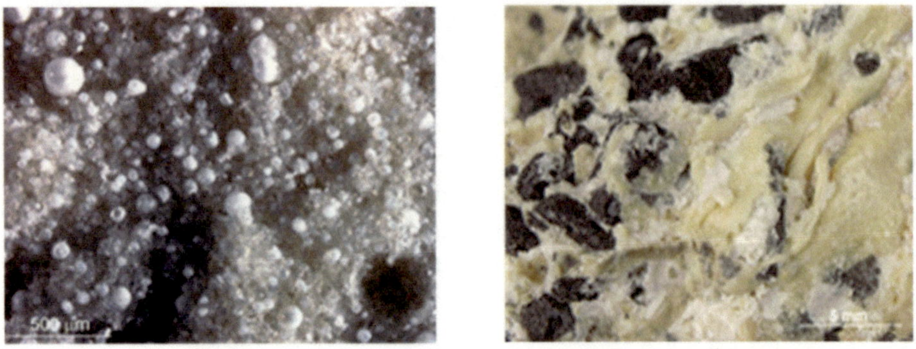

(a) (b)

Figure 11. Basic types of plastic binder structure in OMR-geocomposites: (a) type OMR 1, (b) type OMR 2.

6.2. Determination of mechanical properties of PUR-geocomposites

6.2.1. Preparation of samples and testing specimens

Samples and testing specimens of PUR-geocomposites for laboratory testing of physical-mechanical properties are obtained by the following methods:

1. *by pouring and free foaming* – the simplest method, PUR-mixture is hand mixed with grouted material (sand, gravel, rock debris and others) and is poured into forms of required shape, in which it freely foams. Final shape of the testing specimen is adjusted by cutting off of overfoamed part of the sample (over the volume of the form) (Fig. 12a).
2. *by grouting into pressure tank* – testing samples of required dimensions and shape are drilled or cut from the formed geocomposite (Fig. 12b,d).
3. *by in situ test grouting*– PU mixture is grouted into the rock environment in situ, testing samples of required dimensions and shape are drilled or cut from the formed geocomposite, which is excavated after the test grouting (Fig. 12e).
4. *from real geotechnical projects* – during performance of grouting works in practice, test grouting is undertaken with subsequent sample collection of the grouted rock mass or construction, in some cases also control samples are collected in order to judge the quality and effectiveness of the performed works (Fig. 12c).

The choice of shape and size of the testing specimens is determined by the properties of particular geocomposite type. It depends mainly on the dimensions, shape and textural homogeneity of available geocomposite and also on the possibilities of cutting and machining with cutting or drilling tools. A high-speed abrasive water jet can be well used for cutting of large geocomposite samples (Hlavacek et al., 2009). For shaping of test samples, laboratory drilling machine with diamond bit and diamond saw are used.

6.2.2. Laboratory tests of PUR-geocomposites

There are no standard approaches in the field of laboratory testing of mechanical properties of PUR-geocomposites up to date. Corresponding methods and norms, used in rock mechanics and building material mechanics, are applied for the testing (e.g. ISRM Commision , 1978), and these are adjusted to specific properties of the geocomposites.

Figure 12. Testing specimens of geocomposites prepared by various methods: a) hand mixed mixture of sand + PUR poured into cylinder form with subsequent adjustment of frontal surfaces, b) sand grouted with PUR in pressure tank – cut out specimen of a prism shape of 50mm×50mm×100mm dimensions after uniaxial compressive strength test, c) cylinder shape specimen made from control core drilling, originating from grouted concrete foundation of high voltage pole, d) cube-shaped specimen – coal parts grouted with PU in pressure tank, e) beam type specimen of 40mm×40mm×160mm dimensions during flexural strength test (specimen made of real sample of sand grouted with PUR)

An example of PUR-geocomposite testing is an analysis of sample prepared by grouting in situ with Geopur® 082/90 PU grouting system into saturated sand and shale sandy breccia. Grouting works were performed during construction and excavation of an underground utility tunnel. The underground construction crossed non-coherent strongly saturated sand, where increased water inflows into the construction occurred with subsequent bursting of sand from the working face. Safety of the excavation works at this critical section was secured by creation of a protective "umbrella" above the excavation. This protective "umbrella" was made by the method of PU pressure grouting via perforated steel tubes. During the excavation one of the monolithic geocomposite bodies was dug out for laboratory testing purposes (Fig. 13a). Cross cutting of the geocomposite body showed macroscopically visible zonal heterogeneity of the material (Fig. 13b). Using the methods of image analysis, it was found out, that the degree of foaming of PUR binder increases with the increasing distance from the grouting tube, and that the volume ratio of PUR binder in the geocomposite ranges from 40 to 45% in the various parts of the geocomposite body. Various consistencies of the binder and variable portion of coarse grained breccia grains were identified in the body of the geocomposite. Due to this heterogeneity, the compressive strength values tested on cube-shaped specimens cut from the geocomposite material ranged in relatively wide interval from 5 to 30 MPa (average 12 MPa) and the deformation modulus ranged in interval from 100 to 2000 MPa (average 700 MPa).

(a) (b)

Figure 13. Monolithic geocomposite body formed by GEOPUR grouting into saturated sand and shale sandy breccia (a) and a cross through the geocomposite – zonal heterogeneity of the material is visible (b).

An example of testing of model PUR-geocomposites, prepared in laboratory conditions by grouting into pressure tanks, is an analysis of the effect of grouted environment moisture to the resulting properties of the geocomposite (Scucka & Soucek, 2007). A geocomposite sample, laboratory prepared by grouting into pressure tank filled with loose rock material, is presented in Fig. 14. Grouting was performed into crushed basalt of defined grain size. The material was grouted by the Geopur® 082/1000 resin, which reacts during the curing process with water. Grouting was performed both into dry material and saturated material. Fig. 15 shows macroscopically visible differences in the texture of formed geocomposites. While during the grouting of dry material honeycomb type I texture is formed (good adhesion of binder to the rock particles) with slightly foamed binder *PUR 2* (see sec. 6.1.), in case of saturated grouted material, honeycomb type II texture is formed (only partial sticking of the binder to the rock particles) with strongly foamed binder *PUR 3*. The difference in moisture of the grouted material causes, that the compressive strength of saturated samples is in average lower by approx. 80% and the deformation modulus is lower by approx. 90% compared to the values of samples prepared by grouting into dry material.

Figure 14. Sample of model geocomposite (PUR+basalt aggregate) prepared in laboratory by grouting into pressure tank.

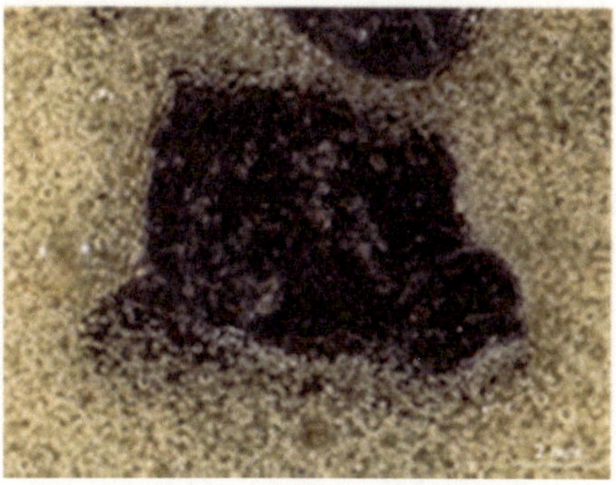

(a)

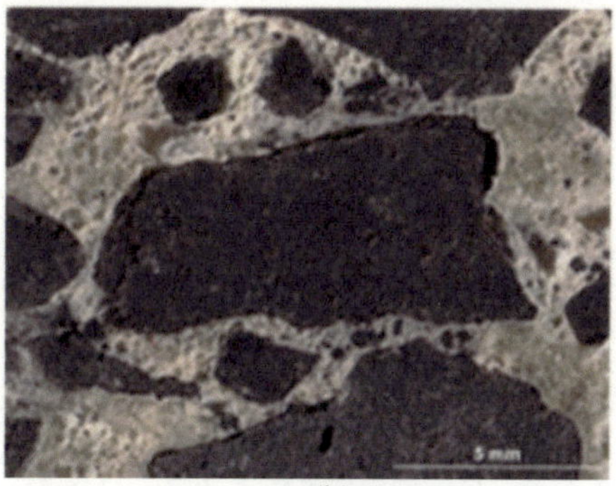

(b)

Figure 15. Different types of textures of laboratory prepared model geocomposite formed due to different moisture level of the grouted material (PUR+basalt aggregate): a) dry aggregate, b) saturated aggregate.

6.2.3. Current knowledge about the mechanical properties of PUR-geocomposites

Data about stress and strain properties of geocomposites with PU binder have not been yet evaluated in summary or statistically. Technical literature or company brochures offer information connected with particular applications under particular geotechnical conditions or from testing and comparison of individual grouting materials. A little bit more complex data and unified interpretation of observed parameters are presented by (Aldorf & Vymazal, 1996), where the properties of laboratory prepared and in situ prepared geocomposites are compared (sand grouted with PU and acrylate resin). Further, we present some conclusions deduced from the results of the above mentioned experiments and from the knowledge of the authors in the field:

- PUR-geocomposites behave in comparison with common rock types extraordinarily, mainly in terms of considerable elasticity and plastic deformations. This feature is observed mainly behind the ultimate strength, when along with the relatively high values of longitudinal deformation (approx. 10 - 20% in case of grouted sand) residual strength of the material remains significantly high.

- The ratio of rock grains to the PU binder, distribution of grains, grain size and the possibility of formation of porous foamed material have significant influence to the values of parameters of physical-mechanical properties of the geocomposite. These factors are always very variable at in situ conditions and depend on the bedding conditions of the rock (local porosity, structure, permeability, moisture and etc.). It is therefore necessary to take into consideration during the laboratory testing mainly the parameters of samples of lower volume weight.
- Samples of lower volume weight contain greater portion of foamed plastic binder. This results in decrease of velocity of longitudinal ultrasound waves spreading through the material and decrease of deformation modulus (higher plasticity).
- Greater portion of rock grains (higher volume weight) positively influences the strength of the geocomposite. Compressive and tensile strengths increase with increasing volume weight.
- Geocomposite deforms within the elastic phase mainly in longitudinal direction, transverse deformations are small. This is indicated also by small values of Poisson's ratio. It is caused probably by high porosity of the binder, which is predominantly elastically deformed in the direction of loading force. High values of tensile strength are probably also result of this.

7. Examples of practical applications of PU grouting

7.1. Reconstruction of Retaining Wall (Prague - Horni Pocernice, Czech Republic)

Task:

The retaining wall (Figs 16-19) is located at the D11 highway at Prague, Horni Pocernice. The highway runs here in a deep trench with walls reaching to 10 m height. The works during the reconstruction of the retaining wall included:

1. stabilization of the fill material behind the current wall by grouting technology from static and safety reasons in order to enable performance of the follow up reconstruction works.
2. anchoring of steel concrete pole prefabricates of the newly built retaining wall.

Solution:

Technology of pressure grouting was used in order to stabilize soil and fill potential cavities behind the retaining wall. Double component PU resin Geopur® 082/90 was used as a grouting medium. Grouting works were performed through drilled or hammered in perforated steel grouting tubes of 16/22 mm diameter. Drillings were drilled approximately perpendicularly to the wall, to the depth of 2 to 3 m in a periodic grid. In total, an area of the wall of approx. 1500 m² was stabilized, total of approx. 10 000 m of grouting rods were drilled, and a total of approx. 46900 kg of Geopur® grouting resin was consumed.

Figure 16. Original state of the retaining wall

Figure 17. Anchoring works

Figure 18. Situation before final completion of the works

Figure 19. Detail of finalized anchor

Drilling of the anchors was made in places of openings of the anchored concrete poles (Figs 17, 19). Self drilling R type rock bolts were used. The length of the anchors was 8 to 10 m, according to the particular geological conditions at the place of the installed anchor (always to reach at least 2 m of stiff rock environment). After drilling in of the anchor rods, pressure grouting was performed through the installed anchor (AR32N) using two component PU resin Geopur® 082/600. The length of the grouted anchor root was approx. 3 to 6 m and approx. 50-60 kg of grouting resin was applied into each anchor.

In total 276 anchors were installed of total length of 1896 m. The consumption of grouting resins Geopur® reached approx. 13 200 kg.

7.2. Hatarmenti water-gate (Zagyva River, Hungary)

Task:

During the time of catastrophic floods on the Tisza River, which occurred from February to March 2001, intensive water seepages occurred around the Hatarmenti water-gate on the Zagyva River by the town of Szolnok (Fig. 20). The dam contains two concrete units with openings of 120cm×120cm. Intensive seepages threatened the stability of the dam. Insulation by PU pressure grouting, using pushed-in grouting tubes at the water-side of the dam, was proposed.

Figure 20. View of the Hatarmenti water-gate

Solution:

Grouting works were launched at 7.00 AM. After preparation of the technical equipment, grouting tubes were pushed in. Pushing in of the 1st tube was followed by immediate start of the grouting works. In total, 9 grouting tubes were made. Grouting was done stepwise. Complete stopping of water seepages was reached at 5 PM. In total, 600 kg of Geopur® 082/350 PU resin material was used.

Figure 21. Grouting material in the surrounding of the water-gate (photo from the control excavation after 10 years)

After 10 years (in February 2011) the river basin authority decided to check the lifetime of the grouting material and its sealing function. An excavation was made down to the concrete construction of the dam. Grouting material was observed on the contact with the concrete structure and in the surrounding (Fig. 21). The material looked undamaged and fulfilled the sealing purpose. It was solid and dry. The efficiency of the performed PU pressure grouting was proved.

7.3. Elimination of contaminated water outflows from the uranium deposit (Rozna, Czech Republic)

Task:

After stopping of mining at the uranium deposit Rozna in 1996, the mine was flooded. The groundwater from the old exploited areas, rich in uranium, infiltrated into the old investigation boreholes, even though the boreholes were sealed in the past (Fig. 22). Contaminated water threatened a drinking water reservoir, and therefore it was necessary to stop the dangerous outflows from these boreholes.

Solution:

The problem was successfully eliminated by application of PU grouting using Geopur® grouting system (Fig. 23). The boreholes were sealed with grouting and the outflows of contaminated water stopped. It was proved that using PU resin of Geopur® type with foaming factor of 12 it is possible to efficiently stop seepages through porous geological environment or boreholes.

Figure 22. Borehole with contaminated water seepages

Figure 23. Sealing of the borehole with PU grouting

7.4. Sealing grouting of sewage collector (Pilsen, Czech Republic)

Task:

During exploitation of a sewage collector in Pilsen – Cernice, drainage in the surrounding area occurred due to drainage effect of the collector. Groundwater disappeared from the wells in the surrounding and drops of the surface occurred, causing even damages to some buildings. In section 4255 – 4350 m of the collector the excavation works ran in close proximity of a residential house. There was a risk of damage to the house and surface drop due to the fast drainage of the groundwater. In order to eliminate inflows into the collector, a hydrogeological survey was performed and a technology of sealing of the environment was proposed.

Solution:

During the exploitation works in the critical section (approx. 100 m), the rock mass was stabilized by PU grouting. Grouting PU materials Geopur® was used. The grouting works were performed always in advance before the exploitation to the distance of 3 m ahead of the face and were followed by exploitation of 2,5 m. Used technology enabled stopping of strong inflows of groundwater and secured higher stability of the rock mass during the exploitation works. Drainage of the surrounding area was eliminated and buildings were not threatened further by the excavation works. Total of 9516 kg of PU grouting material

Geopur® was used. Grouting was performed in total of 275 boreholes. The efficiency of the grouting works was proved by monitoring of the groundwater table level.

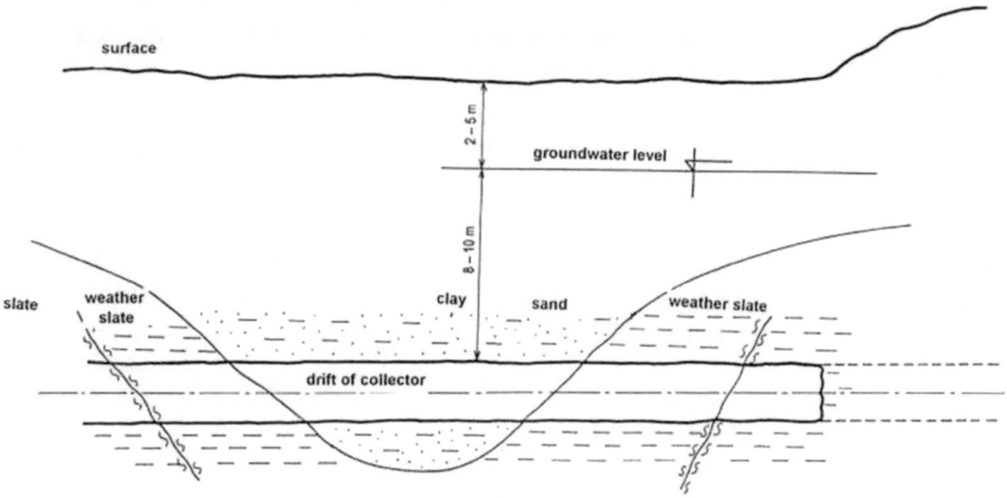

Figure 24. Geological situation of sewage collector construction in Pilsen.

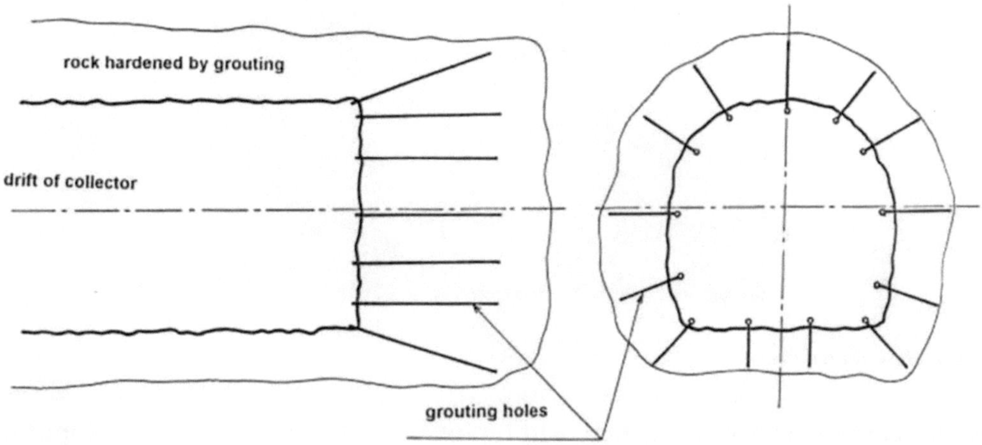

Figure 25. Drift of sewage collector in Pilsen - scheme of grouting works.

7.5. Securing of excavation of underground utility tunnel (Ostrava, Czech Republic)

Task:

In the centre of Ostrava town an underground gallery was exploited (Fig. 26). According to the design of the construction, it was required that the compressive strength of the overburden soil (gravels of the River Ostravice terrace) was minimally 2 MPa.

Solution:

The proposed technology was based on grouting of overlaying rocks ahead of the working-face by PU grouting system Geopur® to the distance of 3 m. In the given geological conditions the technology proved to be a safe and economic solution. The use of PU grouting resin system Geopur® enabled to perform the whole cycle of grouting works in 4 hours. Proposed technology did not require equipment of great size. Borings for grouting were drilled using a light hand drilling equipment; grouting works were performed using transportable pumps. 9-12 grouting tubes were used per one section (3 m), in accordance with local geological conditions. Distance between the grouting tubes was 0,3 m at the face and 0,4 m at the sides of the tunnel. Minimal compressive strength of the formed geo-composite reached in the upper part of the profile 4-6 MPa and 2-4 MPa on the sides. The strength of the rock was reached in 20 min after grouting works.

Figure 26. Grouted gravel in the face of the tunnel

7.6. Repair of cracks in the concrete of highway bridge (Belotin, Czech Republic)

Task:

During the construction of the highway near Belotin, bridge beams were damaged probably due to the frosty weather. Cracks of up to 11 m length formed at the construction. It was necessary to reconnect the cracks using a reinforced grouting technology.

Solution:

Boreholes of 14 mm diameter were drilled into the concrete beam along the cracks, diagonally across the crack. Steel bars of 10 mm diameter were inserted into the drillings and subsequently the drillings were grouted with Geopur® PU resin (Figs 27,28)

Figure 27. Reinforced PU grouting of the crack in the concrete construction

Figure 28. Overall view of the repaired crack

7.7. Lock at the Danube River (Gabcikovo, Slovakia)

Task:

During operation of the lock on the Danube River cavities had been formed on the outer side of the lock in the soil, caused by insufficiently sealed dilatation joints (Fig. 29). Water flowed through the soil embankment, washed out the fine grained particles and cavities were formed. Additional sealing of the dilatation joints was therefore proposed.

Solution:

In 2005 additional sealing of dilatation joints and sealing of cracks in the concrete was performed using PU grouting system Geopur®. In total an area of 42 m^2 of concrete was repaired, 210 m of dilatation joints was resealed and 207 m of cracks in the concrete was sealed. Total of 7 dilatation joints was successfully repaired using total of 700 kg of Geopur® (Fig. 30).

Figure 29. Dilatation joint before the grouting

Figure 30. Dilatation joint after the grouting

7.8. Repair of metro railway (Budapest, Hungary)

Task:

Metro line East-West in Budapest has been operated since 1970. During the operation the railway must withstand a great load. Daily operation represent passing over of 7000 wheels with axis pressure of 7,9 tons. Despite everyday maintenance, signs of damage occurred and repair works were launched.

Solution:

A new method of repair of the concrete sleepers was proposed based on anchoring and use of PU resin. The performance of the works concurrently led also to stopping of water inflows from the subsoil of the railway. Concrete sleepers were stabilized and anchored without the necessity to replace them (Figs 31,32).

Grouting works were performed between 2005 and 2006, mainly during night time and without necessity of putting the metro out of operation. Total of 2842 anchors were made of the following parameters: length of the anchor 500 mm, diameter of the anchor 18 mm, diameter of borehole 25 mm, grouting material Supermin® with reaction start of 2 minutes. Further, total of 2430 grouting boreholes were drilled of the following parameters: borehole diameter 14 mm, length up to 1 m, length of inserted steel bars up to 600 mm, diameter of the bars 8 mm, grouting material Geopur® 082/1000 with reaction time 2 minutes.

Figure 31. Detail of the grouting works

Figure 32. View of the railway after repair works

In addition, water infiltration from below of the railway was stopped in a total length of 58 m of the metro tunnel using material Geopur® 082/350.

Applied methods proved to be very effective and did not disturb the regular operation of the metro line.

8. Conclusion

PU grouting is performed in order to achieve improvement of physical and mechanical properties of the rock, soil, or building material in the construction. It requires experience and complex knowledge from various fields like geology, hydrogeology, structural geology, rock and soil mechanics, geotechnics, mining, underground constructions, construction of foundations, structural stability, defects of constructions and their repair, chemistry of the grouting material, grouting technique (pumps, packers) and etc. Grouting technologies represent an effective technology of solving of various kinds of problems in mining, building industry and geotechnics practice.

Author details

Jan Bodi and Zoltan Bodi
GME, s.r.o., Ostrava, Czech Republic

Jiri Scucka and Petr Martinec
Institute of Geonics AS CR, Institute of Clean Technologies for Mining and Utilization of Raw Materials for Energy Use, Ostrava, Czech Republic

Acknowledgement

This work presents also information and data gathered during a Research project "Výzkum a vývoj nových chemických injektážních materiálů pro zlepšení vlastností hornin, zemin a stavebních konstrukcí v inženýrském stavitelství a stavebnictví 2005-2007 – Research and development of new chemical grouting systems for improvement of rock, soil and constructions at civil engineering and building industry", solved by the companies SG Geotechnika, a.s., Prague and GME Consult, Ostrava. The project was supported by the Ministry of Trade and Industry of the Czech Republic under the programme Impuls no. 47/2007/FI-IM2/072.

In addition this work contains also data gathered within the project of The Institute of clean technologies for mining and utilization of raw materials for energy use, reg. no. CZ.1.05/2.1.00/03.0082 supported by Research and Development for Innovations Operational Programme financed by Structural Founds of Europe Union and from the means of state budget of the Czech Republic.

9. References

Aldorf, J., Vymazal, J. (1996) *Contribution to study of stress and strain properties of sand reinforced using grouts on the basis of PU and acrylate resins.* Proceedings of conference

Application of PUR in mining and underground engineering, Ostrava, February 1996 (in Czech).

Aldorf, J.,Hrubešová,E.,Bódi,J.,Hulla,J. (2003) *Stabilita protipovodňových hrází řeky Dunaje na ostrově Szentendre.* FAST VŠB-TU Ostrava

Bodi, J. (2003). *Instructions for use for double component resin grouting system Geopur®, 2003,* Available from: < www.gmeconsult.cz>

Bodi, J., Bódi, Z. *GME Consult and GME s.r.o. company materials*

Bodi, Z., Ebermann, T. (2009) *Chemical grouting – effective technology for improvement of non-coherent soil and rock environment in the field of foundation of constructions, GME Consult & SG Geotechnika a.s.,* Paper at the Geotechnical Seminar in Stupava, Slovakia, (in Czech).

Bódi, J., Paloncy, L. (1996) *Improvement of the mine workings stability during mining and drivage by the integrated system of bolting and grouting.* World Mine Safety Congress New Delhi, India

Bódi J., (1997) *Insulation injection labours, creation of underground barriers using the PU injection materials during mines closure.* Acapulco World Mining Congress, Mexico

Bódi, J., Pellionis, (1998) *Zajištění stability likvidovaných důlních děl pomocí polyuretanových injektáží a kotvení.,* International conference Děmenovská Dolina Slovakia.

Bódi, J., (1999) *Mining Activities consequences elimination under the Conditions of dense populated areas* 28th World Mine Safety Congress in Mines-Sinai Romania

Bódi, J., (2001) *Repair of defects in hydraulic structures by PU grouting under flood and normal conditions.* International conference on water and nature conservation in the Danube – Tisza river basin Debrecen Hungary

Bódi, J., (2003) *Polyuretanové injektáže, teorie a příklady využití* VŠB TU Ostrava, Fakulta stavební, Habilitation

Bódi, J.,Poštulka,A. (2003) *Injektovatelnost stavebních konstrukcí pomocí polyuretanových pryskyřic.* FAST VŠB-TU Ostrava

Bodi, Z. (1998): *Thesis: Physical properties of geo-composites (soil, rock debris – polyurethane).,* VŠB-TU Ostrava

Hlaváček, P., Valicek, J., Hloch, S., Gregr, M., Foldyna, J., Kozak, D.,Sitek, L., Kusnerova, M., Zelenak, M. (2009). *Measurement of Fine Grain CopperSurface Texture Created by Abrasive Water Jet Cutting. Strojarstvo,* Vol. 51, No. 4, pp. 273-279, ISSN 0562-1887.

ISRM Commision on standardization of laboratory and field tests. (1978) *Suggested methods for determining tensile strength of rock materials.* International Journal of Rock Mechanics and Mining Sciences & Geomechanical Abstracts, Vol. 15, pp. 99-103.

ISRM Commision on standardization of laboratory and field tests. (1978) *Suggested methods for determining the uniaxial compressive strength and deformability of rock materials.* International Journal of Rock Mechanics and Mining Sciences & Geomechanical Abstracts, Vol. 16, pp. 135-140.

Pettijohn, F. J. (1975). *Sedimentary Rocks (Third edition),* Harper & Row, New York

Snuparek, R., Soucek, K. (2000). *Laboratory testing of chemical grouts.* Tunnelling and Underground Space Technology, Vol. 15, No. 2, (April-June 2000), pp. 175-186, ISSN 0886-7798

Scucka, J., Soucek, K. (2007). *Architecture and Properties of Geocomposite Materials with Polyurethane Binders*, Institute of Geonics of ASCR, ISBN 978-80-86407-15-9, Ostrava (in Czech)

Trade mark GEOPUR, no. 148 179, 1999 Czech Republic, Arrangement et Protocole de Madrid no. 864 266 EU, GEOPUR, no. M0002030 , 2000 Hungary

Chapter 15

On the Use of Polyurethane Foam Paddings to Improve Passive Safety in Crashworthiness Applications

Mariana Paulino and Filipe Teixeira-Dias

Additional information is available at the end of the chapter

http://dx.doi.org/10.5772/47996

1. Introduction

The use of cellular materials in general in the automotive industry, and polymeric foams in particular, has been increasing significantly for the last few decades. These materials are used within a particular vehicle for many different purposes, among which are, for example, sound and thermal insulation, vibration damping, fire protection and, of course, crashworthiness. Thus, crashworthiness, safety and protection parameters are strongly influenced by the materials used and, as a consequence, polymeric foams play a major role in the vehicle's crashworthiness levels. In absolute terms, the energy absorption capability of this class of materials can lead to significant improvements on the vehicle's passive safety, better protecting the passengers from aggressive impacts, by absorbing impact energy in a gradual and controlled manner. In addition, design limitations due to environmental constraints are growing steeply as are safety concerns. Whilst the former often leads to a reduction in the weight of the vehicle, the latter will most probably lead to the opposite. Therefore, the combination of properties such as low density, low cost and design flexibility with a great energy absorption capability, is what makes cellular materials so attractive for the automotive industry.

Presently, vehicle structures with high levels of crashworthiness protection are almost always light-weight and must deform in such a way as to dissipate the largest amount of impact energy possible. Several distinct mechanisms may contribute to this, such as, for example, plastic deformation, wrinkling, heat generation, etc. [20, 35]. One way to achieve these effects is to fill tubular or hollow metallic or composite structures with cellular materials, such as foams. During the last few decades, many researchers have been working on these issues [22, 28, 30–32].

Polyurethane foam is nowadays being widely used in many energy absorption engineering applications such as cushioning and packaging [4, 15, 50]. Its use in automotive industry as an energy absorbing material in passive safety mechanisms goes beyond the protection functionality since it also provides more comfort, insulation and sound absorption. Thus, the role of this class of materials in vehicles is of special interest from both the consumer and the manufacturer points of view.

On a microscopical level, most cellular materials, including polyurethane foams, have the ability to absorb energy while deforming due to the mechanics of cell crushing. In the process of absorbing impact energy, cell walls deform plastically and get damaged (e.g. fractured) [7, 18].

Vehicle-to-vehicle side-impacts and vehicle rollover are presently among the most common types of car accidents and collisions. Additionally, these are also frequently the most serious accidents in terms of occupant injuries [6, 8, 39, 48]. Among these, frontal and side impact are the most severe. As a consequence, quite a large effort has been widely focused in improving passive and active safety mechanisms for frontal impact situations for the last decades. However, more recently, the number of serious injuries resulting from side-impacts has brought the attention of many researchers to the importance of developing similar or adapted mechanisms for such collisions [12, 33, 53]. In this type of collisions, the risk and/or the severity of the resulting injuries is frequently a direct consequence of the contact between the occupants and the lateral structure of the vehicle, given the reduced space between the occupant and the door [21]. Pelvic and chest areas have been reported by many authors as the the two areas most affected in this type of car-to-car collision [29, 43, 44].

In the late 1990s Morris *et al.* [37] observed, through a series of numerical simulations of side-impact collisions, that the space available between the structure of the vehicle and the passengers is one of the most important parameters with direct influence on the levels of occupant's injuries. This statement was also supported by many other researches, as can be seen, for example, from the works of Tencer *et al.* [49] and Schiff *et al.* [44]. Morris *et al.* state that the space available not only has influence on the impact velocity but also on the point of the velocity profile at which the door initiates contact with the occupant. These authors also evaluated the benefits of the use of paddings of different sizes in the door interior and of lateral airbags. Lim *et al.* [26] also studied, numerically, the inclusion of padding material for protection of the occupant pelvic area and concluded that it significantly reduced the severity of the resulting injuries. Additionally, Majumder *et al.* [29] studied the dynamic response of the pelvis and established fracture limits in side-impact collisions. These authors supported their conclusions with the results from numerical simulations using finite element modelling software. One of the most important conclusion these researchers derived from their work was that with a more appropriate design of the lateral door and the inclusion of padding material on the level of the pelvic area, the risk and/or severity of occupants' injury could be significantly reduced.

Based on the previous considerations, the authors propose the use of cellular materials, among which polyurethane foams, within an energy absorbing system specifically designed in such a way as to significantly improve passive safety on the event of side-impacts. This system consists of a foam like impact padding confined inside the lateral doors of the vehicle. In order to be efficient, this foam padding must be mounted aligned with the occupant's pelvis, protecting one of the most critical areas in this type of collision. From the experiments and analyses made it is expected that this padding may absorb a significant part of the impact energy, and thus minimise both the forces transmitted to the body of the occupants and, most importantly, the magnitudes of the decelerations experienced, consequently reducing possible injury levels of the occupants of the vehicle.

The design of structures and the choice of materials for crashworthiness and protection systems within a vehicle are also of major importance and relevance for the overall safety of the driver and passengers. These are two of the research fields where there is still quite a large margin for improvement [16, 17, 24, 41, 46]. In the present work, the authors will present, discuss and compare the applicability of several distinct types of cellular materials for impact and energy absorbing paddings.

2. Polyurethane foam as an energy absorbing material

The behaviour of three polymer based structural foams under compressive impact loading — polypropylene, polyamide and rigid polyurethane foam — has been investigated by Avalle *et al.* [4]. As a conclusion, these materials are indicated as players of a very important role in passive safety systems. These authors obtained stress-strain curves in both static and impact loading (dynamic) conditions for the materials examined for different densities at room temperature. They analysed in detail, using energy-absorption and efficiency diagrams, the energy absorption characteristics of each material. Among the materials tested, these authors observed that polyurethane foam is the less sensitive to strain-rate and the one that presents the longest intermediate plateau stage. These facts distinguish this material from the remaining foams studied. The authors also concluded that the rigid polyurethane foam exhibited one of the highest efficiency levels, however, it lost its integrity during compression.

The energy absorption behaviour properties of polyurethane foam were also investigated by Anindya and Shivakumar [2]. The authors evaluated the energy absorption attributes of polyurethane foam in various forms — flexible high resilience, flexible viscoelastic and semi-rigid — as a function of the overall foam density, based on the load-displacement behaviour of the material under compressive loads.

Taher *et al.* [47] investigated the use of polyurethane foam with a density of 47 kg/m^3 as a core filler of a composite keel beam as a way of preventing global buckling and improving crashworthiness performance in aeroplanes and helicopters. The results obtained by these authors revealed that the energy absorbing mechanism can meet the requirements for the purpose desired together with substantial savings.

Likewise, the behaviour of polyurethane foam filled thin-wall structures was investigated by Ghamarian *et al.* [13] in terms of crashworthiness improvement for the aerospace industry. The quasi-static crushing behaviour and efficiency of empty and foam-filled structures was investigated experimental and numerically and the efficiency and the authors were able to demonstrate that the filled tubes presented higher energy absorption capabilities than that of the combined effect of the empty structured and the foam.

Furthermore, applications of polyurethane foam in explosive blast and ballistic energy absorption applications have also been subject of investigation [51, 52], indicating that this material may be a valuable part of protection systems against both generic types of threats.

Later, Shim *et al.* [45] investigated the two-dimensional behaviour of rigid polyurethane foam under low velocity impact loadings in terms of both the deceleration of the impactor and the overall amount of energy dissipated. These authors also proposed suitable stress-strain relations as well as failure patterns, failure criteria and equations of motion for this cellular material.

The foam used by Shim *et al.* [45] was obtained by blending — Daltofoam and Suprasec — in the presence of a blowing agent, producing a final product with a density of 25.6 kg/m^3. Alike typical cellular materials the uniaxial compressive behaviour of this material can be described, in terms of stress-strain, by three distinct stages [14]. The first stage — the elastic deformation stage — is followed by a plastic constant stress stage (also known as "plateau" region) where most of the energy absorption occurs. Finally, the material exhibits densification. The elastic part of the behaviour of these materials is mostly due to the axial compressive resistance of the cell walls. The plateau region is mainly related to the bending, crushing and eventually fracture, of the cell walls. The material starts to densify when all the cells are crushed and the behaviour approaches the behaviour of a monolithic material [19, 42, 54].

The type of polyurethane foam investigated by Shim *et al.* [45] is adopted for the scope of this study and its stress-strain curve is represented on Figure 1, where the three stages are clearly evident. As the overall behaviour of this material can be divided in three distinct stages, its stress-strain constitutive modelling can be defined by the following set of equations:

$$\sigma = \begin{cases} E\varepsilon & \text{if } \varepsilon \in [0, \varepsilon_y] \text{ (elastic behaviour)} \\ E\varepsilon_y & \text{if } \varepsilon \in [\varepsilon_y, \varepsilon_d] \text{ (plastic behaviour, plateau)} \\ E\varepsilon_y \exp \dfrac{a(\varepsilon - \varepsilon_d)}{(\varepsilon_l - \varepsilon)^b} & \text{if } \varepsilon \in [\varepsilon_d, \varepsilon_l] \text{ (densification)} \end{cases}, \tag{1}$$

where E is the material elastic modulus (considered to be $E = 2.78$ MPa), ε_y is the compressive yield strain ($\varepsilon_y = 0.05$), ε_d is the densification strain ($\varepsilon_d = 0.8$), ε_l is the maximum compressive strain ($\varepsilon_l = 0.95$) and a and b are constants which define the shape of the stress-strain curve in the densification regime.

3. Crashworthiness efficiency of polyurethane foam

In previous investigations the authors used a Finite Element Analysis (FEA) approach to study the behaviour of four distinct cellular materials under impact loading in order to evaluate their relative efficiency in terms of crashworthiness applications [40]. The materials tested within these studies were two polymeric foams: polyurethane foam and IMPAXX™ ; a metallic foam: aluminium foam; and a natural cellular material: micro-agglomerated cork. The most relevant mechanical properties of these materials are listed on Table 1, where ρ^* and E^* are the density and elastic modulus of the cellular material, respectively, and E, σ_y and ν are the

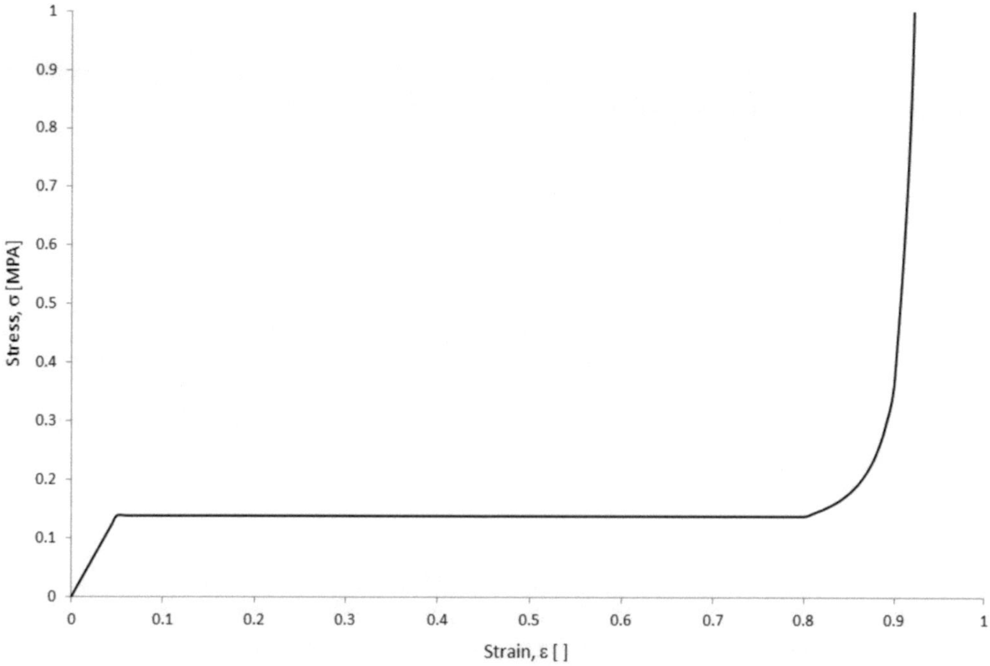

Figure 1. Stress-strain compressive behaviour of rigid polyurethane foam as obtained by [45].

elastic modulus, yield strength and Poisson coefficient of the base material, accordingly. The four materials share great energy absorption capabilities and are used in impact dissipation applications. For this purpose the materials were tested numerically, using a Finite Element Method simulation software, LS-Dyna™ [1], under impact loading in the same conditions and were analysed in terms of acceleration peak and energy absorption.

Mechanical Properties	ρ^* [kg/m^3]	E^* [MPa]	E [MPa]	σ_y [MPa]	ν
Polyurethane Foam [a]	25.6	2.78	1600	127	0.44
Micro-agglomerated Cork [b]	293	15	9000	1	0.30
IMPAXX™	33.661	16.322	3400	80	0.40
Aluminium Foam [c]	470	117	69000	241	0.285

[a] [11]; [b] [23, 45]; [c] [3];
[d] [10, 34].

Table 1. Mechanical properties of the polyurethane foam and of other cellular materials used for comparison purposes.

Paulino and Teixeira-Dias[40] proposed a quantitative procedure that allows a padding or protection system designer to determine the crashworthiness efficiency and performance of specific cellular materials. This procedure assumes that the best method to assess the material value for automotive safety applications is by evaluating the rate at which it dissipates energy. In vehicle impacts, the ideal would be for energy to be dissipated in a gradual and controlled manner. Bearing this in mind, the authors proposed and tested a performance index, ϕ, as an attempt to quantitatively evaluate the energy absorption rate of a certain cellular material. The analytical expression that better describes the dependency of the absorbed energy with time during an impact can be given by:

$$E = E(\bar{t}) = \bar{E}_c^{\min} + \left(\bar{E}_c^{\max} - \bar{E}_c^{\min}\right)\left[\frac{\exp\left(\lambda \frac{\bar{t}-\bar{t}^{\max}}{\bar{t}^{\min}-\bar{t}^{\max}}\right) - 1}{\exp(\lambda) - 1}\right] \quad (2)$$

where $E = E(\bar{t})$ is the analytical function of the energy absorption in time, $E_c^{\min}$ is the minimum kinetic energy of an impacting wall and $E_c^{\max}$ is the maximum kinetic energy of the moving wall. $t^{\min}$ and $t^{\max}$ are the minimum and maximum time values considered for the analysis, respectively. The use of the overbar indicates that the respective variable is normalised. λ is a dimensionless parameter that defines the shape of the energy absorption curve. The methodology and interpretation of this performance is explained in more detail by Paulino and Teixeira-Dias [40].

On a first analysis, the research carried out by the authors showed that for all the impact loading cases studied polyurethane foam was actually not the best performing cellular material among the ones tested, as can be observed from the results on Figure 2. The four materials investigated share a tendency of decreasing performance index with the increase of initial impact kinetic energy. For all the levels of energy studied PUF has a better behaviour than IMPAXX™. However, for low values of initial impact energy PUF exhibits relatively high values of performance index, being overtaken only by micro-agglomerated cork or aluminium foam. Nonetheless, in the same study it was also verified that the specific energy absorption results of the rigid polyurethane foam tested was higher than aluminium foam and micro-agglomerated cork. In fact, the former was the material with the lowest density (see Table 1). It is then reasonable to assume that for denser foams the results in terms of total absorbed energy, maximum acceleration, average force or performance index could be similar — or even better — to those obtained when using other cellular materials.

The passive safety system proposed to protect the occupant pelvic area from side-impact collision consists of a padding confined inside the vehicle's lateral doors, positioned in the direction of the occupants pelvic area, as can be seen in Figure 3. This protection padding should result in lower forces transmitted to the occupant and lower maximum accelerations due to the material energy absorption capabilities, as explained previously.

Standard side-impact crash tests should be performed in order to make a correct evaluation of the efficiency of polyurethane foam as a material dedicated to energy absorption specifically to improve passive safety in side-impacts. Given the complexity associated to the numerical simulation of crash tests and consequently due to the highly expensive procedures, a simplified model was used to replicate a vehicle-to-vehicle side-impact as defined by the European New Car Assessment Programme (Euro NCAP) [38]. A schematic representation of this simplified model is shown in Figure 4. Within this scope, a set of finite element analyses

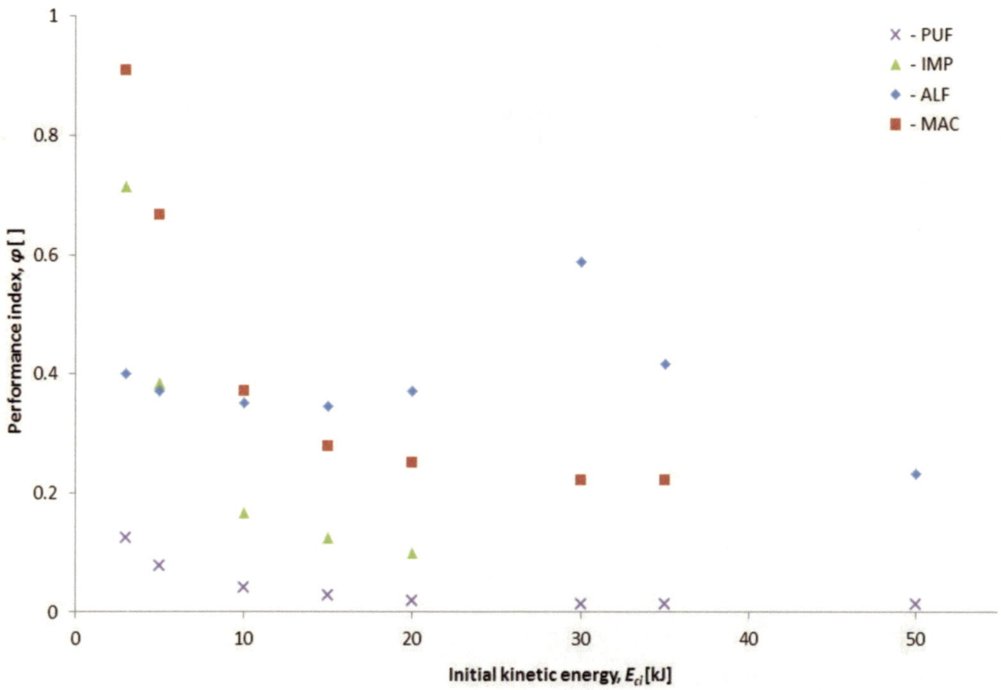

Figure 2. Variation of the performance index ϕ as a function of the initial impact kinetic energy E_{ci} for different cellular materials [40]. PUF: polyurethane foam; IMP: IMPAXX™ ; ALF: aluminium foam; MAC: micro-agglomerated cork.

was performed using LS-Dyna™to assess the benefits of including a padding confined in the vehicle's lateral door and compare the efficiency of different paddings made from different cellular materials, i.e. rigid polyurethane foam (PUF), IMPAXX™ (IMP), micro-agglomerated cork (MAC) and aluminium foam (ALF).

Three key parameters in terms of crashworthiness are defined and analysed, namely the acceleration profiles, the intrusion levels and the loads acting on the vehicle structure.

3.1. Numerical modelling

The use of a simplified model is a common strategy for cost effective preliminary evaluation of crashworthiness situations [9, 12]. Following this line of thought and in the scope of this investigation, an approximate model to crash tests performed by Euro NCAP [38] is developed, implemented and used for the evaluation of the safety performance in vehicle-to-vehicle side-impacts.

The generic vehicle tested in this simplified model consists of a subset of elastic spring elements and the lateral door. The subsets of springs are defined in such a way as to approximately describe the behaviour of the remaining structure of the vehicle. For this purpose, the weight of the vehicle and passengers, the vertical position of the centre of gravity of the car and the friction coefficient between the vehiclet's tires and the asphalt are

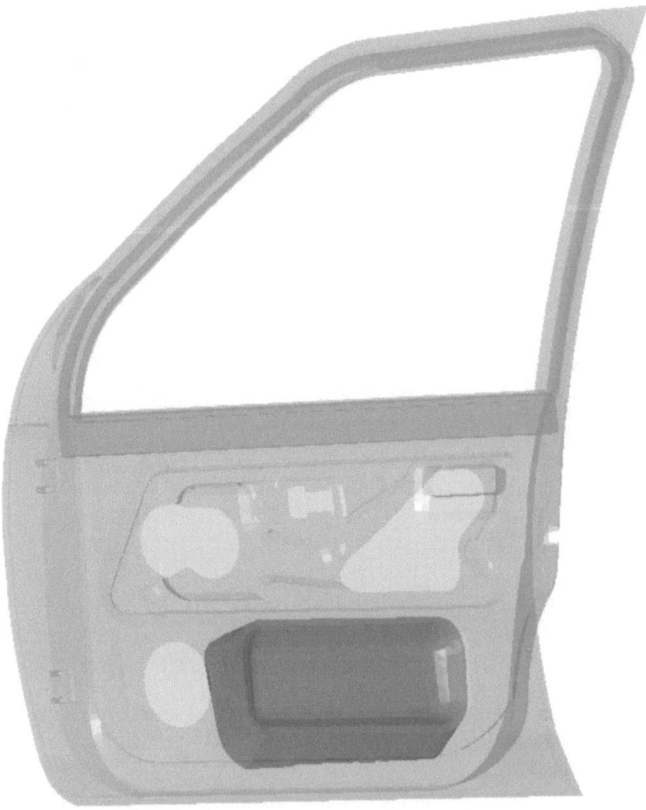

Figure 3. Illustration of the inclusion and position of the protection padding, confined in the lateral door of the vehicle.

considered. This schematic description is illustrated on Figure 5. This approach is considered satisfactory given that the deceleration and intrusion behaviour of a vehicle during a collision are mostly influenced by the two following structural properties: (i) its mass and (ii) its global stiffness [36]. With this approach it should be possible to assess the efficiency of a given structural component without modelling the full vehicle and still assure reasonable results and precision.

The springs representing the global structure of the vehicle were modelled within LS-Dyna™ using two-node discrete elements and `*MAT_SPRING_ELASTIC` stiffness response model. Different material stiffness magnitudes were assigned to the springs in accordance to their relative position to the centre of gravity of the vehicle. The vehicle's lateral door was modelled considering four-node fully integrated shell elements implemented with the Belytschko-Tsay formulation [5].

The material considered for the model of the door was DC06 steel, constitutively described with the `*MAT_PIECEWISE_LINEAR_PLASTICITY` material model, with an elastic modulus $E = 210$ GPa, density $\rho = 7850\ \mathrm{kg/m^3}$ and Poisson ratio $\nu = 0.3$.

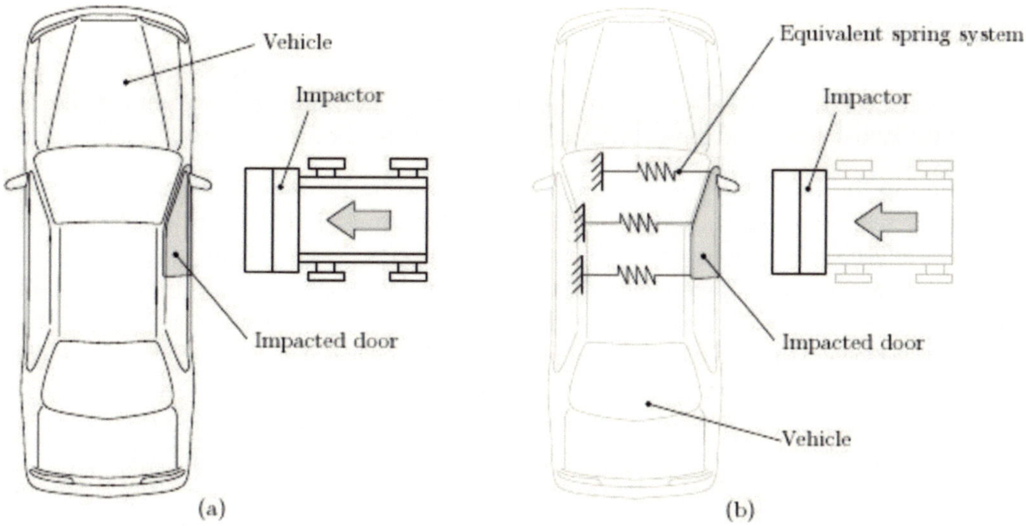

Figure 4. Schematic representation of the simplified model used for finite element implementation of a side-impact crash test according to the EuroNCAP standard [38].

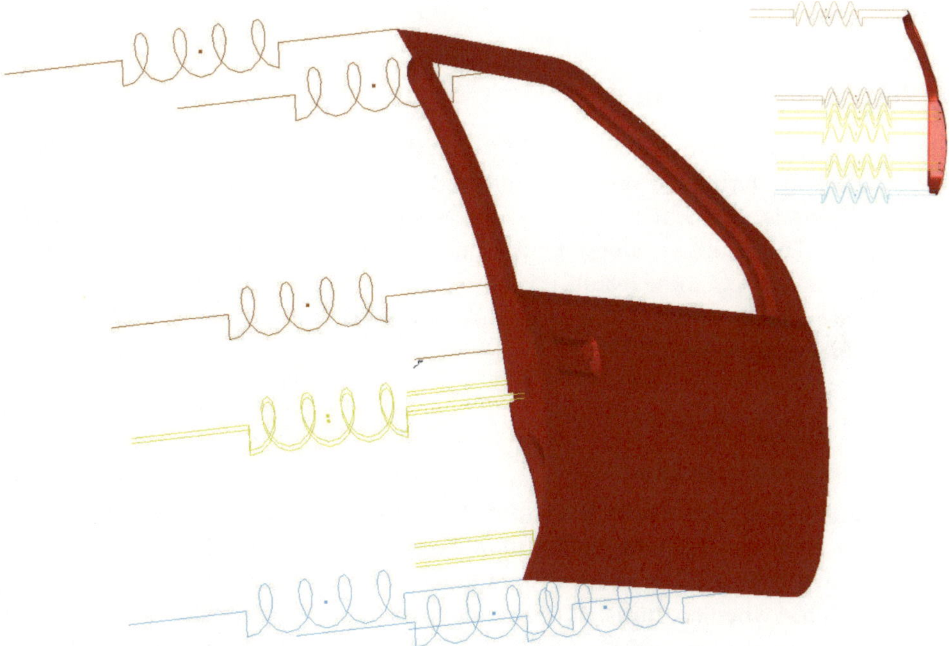

Figure 5. Finite element representation of the simplified model of the vehicle's lateral structure developed and implemented to simulate the side-impact.

The designed and proposed padding system was modelled with four-node tetrahedral elements. The material behaviour was described using the *MAT_HONEYCOMB constitutive approach in order to describe all three different cellular materials. This material is generally adequate for honeycomb and foam materials with anisotropic behaviour [11, 25, 27]. This modelling approach assumes zero value for the Poisson ratio and considers a variable elastic modulus, increasing linearly from the initial value as a function of the relative volume (i.e. the ratio of the actual volume to the initial volume) up to the fully compacted material modulus.

According to the EuroNCAP standards [38] the impacting vehicle (see Figure 4) must be modelled considering deformable 3030 and 5052 aluminium honeycomb blocks. These blocks should be attached to a mobile structure that is to be considered rigid. Within the scope of this work the deformable blocks were modelled using eight-node hexahedral finite elements. The material behaviour was once again described by the `*MAT_HONEYCOMB` constitutive model, for the same reasons stated for the padding materials. The necessary material properties were determined considering both honeycombs [14]. The propeller structure is, however, fairly complex according to EuroNCAP regulations and it is not fully described on the side-impact protocols available. Nonetheless, the total weight of the impactor system is known and, ultimately, the geometry of the impactor moving structure is not significantly relevant for the conditions of the performed tests. For this reason this moving structure was modelled as a moving rigid wall. The most determinant feature is the part of the impactor colliding with the vehicle's lateral door, i.e., the deformable aluminium honeycomb blocks. Hence, mass was added to the impactor's anterior part and the initial velocity of the crash tests ($v_0 = 13.89\ \mathrm{m/s}$) was assigned to the structure.

4. Results and discussion

Once the full model of the vehicle side-impact is defined and implemented in LS-Dyna™ tests were made considering the vehicle door with the paddings made from PUF, MAC and IMP cellular materials. An additional test was made with the door with no padding, for the sake of comparison. The resultant values of kinetic energy, loads, accelerations and absorbed energy were registered. From these results it is possible to evaluate the relative performance of each material used in the side-impact padding. A detailed discussion of these results is presented in the following paragraphs.

4.1. Load distribution

The evolution of the reaction force on the moving rigid wall (the propellant of the deformable barrier in the impactor vehicle) is computed and its dependence with time is registered and plotted on Figure 6. When the rigid wall first contacts the deformable barrier, a force peak is registered at instants $t \approx 0.5$ ms for all the simulations performed. This occurs in an early stage of the crash test when no intrusion has yet happened, not even contact with the vehicle door. Thus, this event shall not be considered relevant for the analysis and should be considered a side-effect of the modelling approach used. After $t \approx 8$ ms, it can be verified that the the reaction force evolution is, as expected, considerably more unstable for the simulation with no padding in the lateral door than for the remaining tests (using paddings with cellular materials). This fact suggests that there is a significant improvement on the behaviour of the vehicle in terms of protection in side-impact collisions when a lateral padding is applied.

Furthermore, from $t \approx 90$ ms on the load almost goes back to zero on all numerical simulations. This is a consequence of the separation of the rigid wall from the impactor. During this stage, however, the system exhibits rigid body motion and, thus, the results obtained for times $t > 90$ ms are not considered relevant for the scope of this research.

It can also be clearly observed that the average load during the considered time interval is considerably lower ($\approx$ 50kN), one order of magnitude, for all the crash tests including a lateral padding when compared the one observed for the test with no padding ($\approx$ 500kN). This leads to the conclusion that the implementation of a padding, either PUF, MAC, IMP or ALF, leads to a much lower and smoother distribution of the load for the whole duration of the impact. The average force obtained for the simulations with the protection padding was around 85% lower than the ones obtained without padding and the maximum force was up to 79% lower.

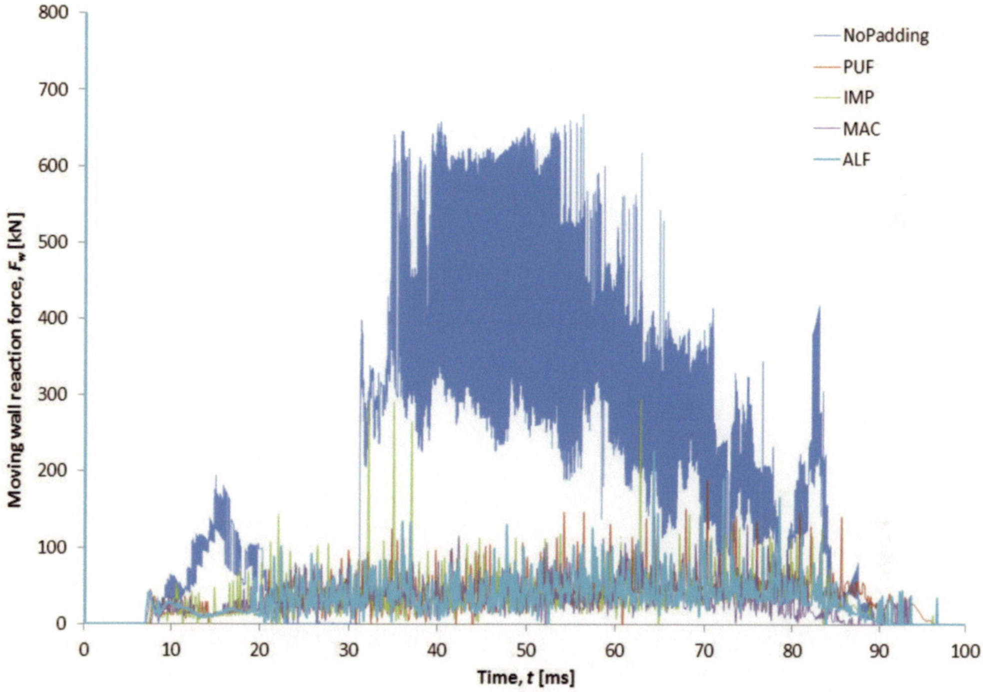

Figure 6. Evolution of the reaction force on the moving rigid wall with time for the simulations of side-impact crash tests.

4.2. Kinetic energy

The evolution of the kinetic energy of the whole system with time is represented on Figure 7. A sudden increase of the kinetic energy can be observed for $t \approx 3$ ms for the crash test with no padding. This corresponds to the instant at which the impactor structure initiates contact with the door of the vehicle. Additionally, for all the simulations performed the curves exhibit an inflection near the final stage of the impact ($t \approx 80$ ms), increasing from this instant until the end of the numerical simulation. From the visualisation of the kinematic results it is possible to relate this inflection to the beginning of rigid body motion of the door (and vehicle) after impact, where incremental deformation ceases to exist [40]. Hence, for the purpose of this investigation the parameters studied will only consider the instants between these marks (that is $3 < t < t_r$ ms), where t_r is the time when rigid body motion initiates. This time instant will be considered different according to each simulation: (i) $t < t_r \approx 80$ ms for the simulation without padding, (ii) $t < t_r \approx 90$ ms when using the polyurethane foam padding, (iii) $t < t_r \approx 85$ ms when using the IMPAXX™ padding, (iv) $t < t_r \approx 75$ ms for the micro-agglomerated cork padding and (v) $t < t_r \approx 90.5$ ms when using the aluminium foam padding.

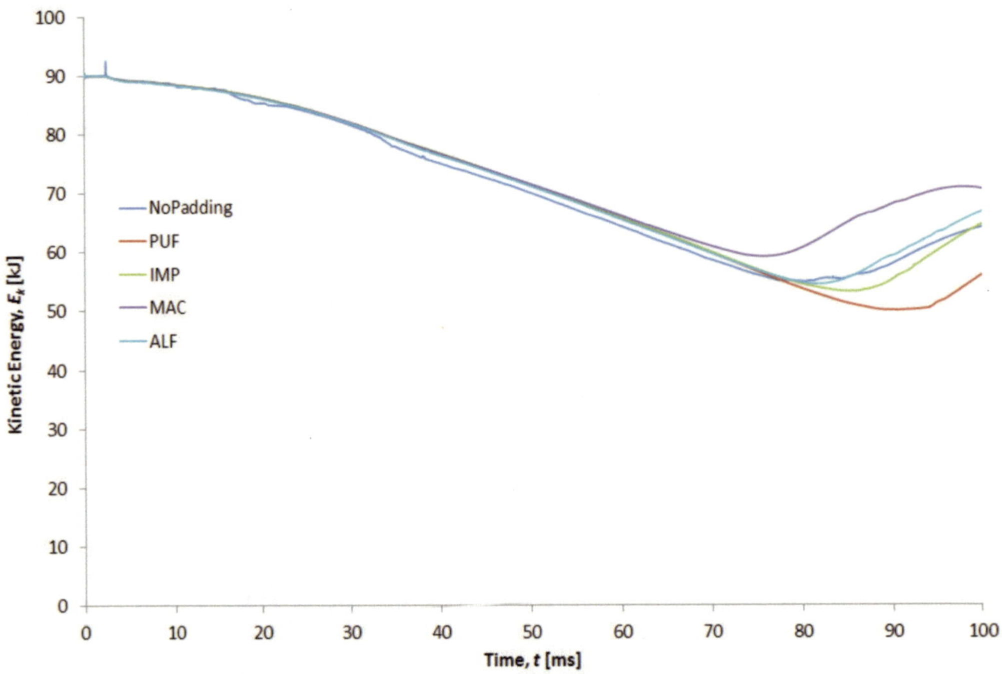

Figure 7. Evolution of the system's kinetic energy with time for the simulations of side-impact crash tests.

4.3. Maximum acceleration

An anthropomorphic dummy was not considered or modelled during the finite element analyses carried out during this research because of the added complexity and Central Processing Unit Time (CPU) time it would bring. Hence, for the purpose of evaluating the maximum acceleration (or maximum deceleration) resulting from the impact felt on the passenger's pelvis area, as well the intrusion level in the passenger compartment, numerical results on nodes on the pelvis direction, on the inner side of the vehicle door, were analysed and are discussed herein.

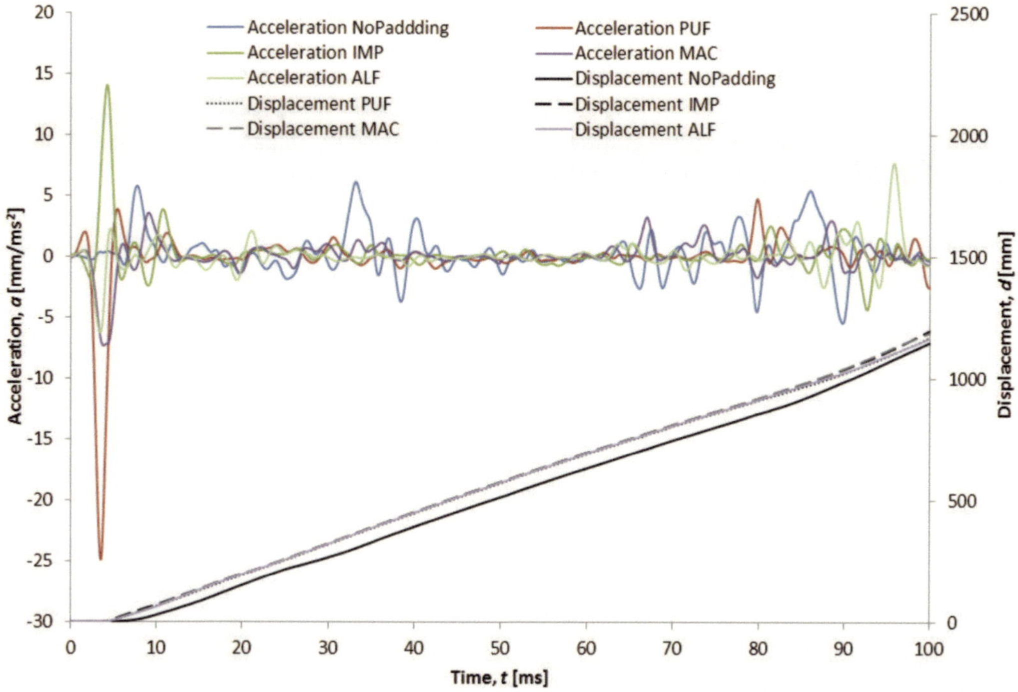

Figure 8. Dependence with time of the acceleration and displacement of the the door's interior structure in the direction of the occupant's pelvic area, obtained from the numerical simulations of the side-impact crash tests.

The resulting evolution in time of the acceleration and displacement measured on the internal structure of the lateral vehicle's door, in the direction of the pelvic area of the occupant, is shown on Figure 8. A SAE-180 filter was used to refine the acceleration results as advised by the protocols used by EuroNCAP [38] for acceleration measurements on the occupants' pelvic area. Very high maximum values of acceleration can be observed at the initial instants of the simulation for all the tests that included a padding, as opposed to the ones observed for the simulations with no padding. This fact is most probably due to the added stiffness of the padding on the first instants of the crash. However, this happens for the early stages of the impact when the displacement of the lateral door's interior is still inexistent and, consequently, there is no contact between the door and the occupant. Thus, the maximum acceleration values were determined only starting from the moment when the displacement of the door initiates.

The results obtained prove that the use of an interior padding can lead to a significant decrease in the maximum acceleration felt by the occupants. This decrease can be as high as 59%. While the maximum acceleration value for the simulations with no padding is $a_0 = 6.1$ mm/ms^2, acceleration peaks of $a_{PUF} = 4.8$ mm/ms^2, $a_{IMP} = 3.9$ mm/ms^2, $a_{MAC} = 3.5$ mm/ms^2 and $a_{ALF} = 2.5$ mm/ms^2 were observed for the crash tests with PUF, IMP, MAC and ALF padding, respectively. Thus, the inclusion of a polyurethane foam padding results in a reduction of roughly 20% even though it is not the best performing material among the ones investigated.

4.4. Energy absorption

The dependence of time of the energy absorbed by the structure of the vehicle is plotted on Figure 9. Analysing these results, it becomes clear that the inclusion of a polyurethane foam padding inside the door structure is the best way to increase the capability to absorb impact energy during a side-impact, when compared to both the crash test numerical simulations with no padding and those using paddings made from other cellular materials. The energy absorbed by the structure with the PUF padding exhibits a higher dissipation capability, leading to an increase in energy absorption of approximately 13% when compared to the structure to no padding.

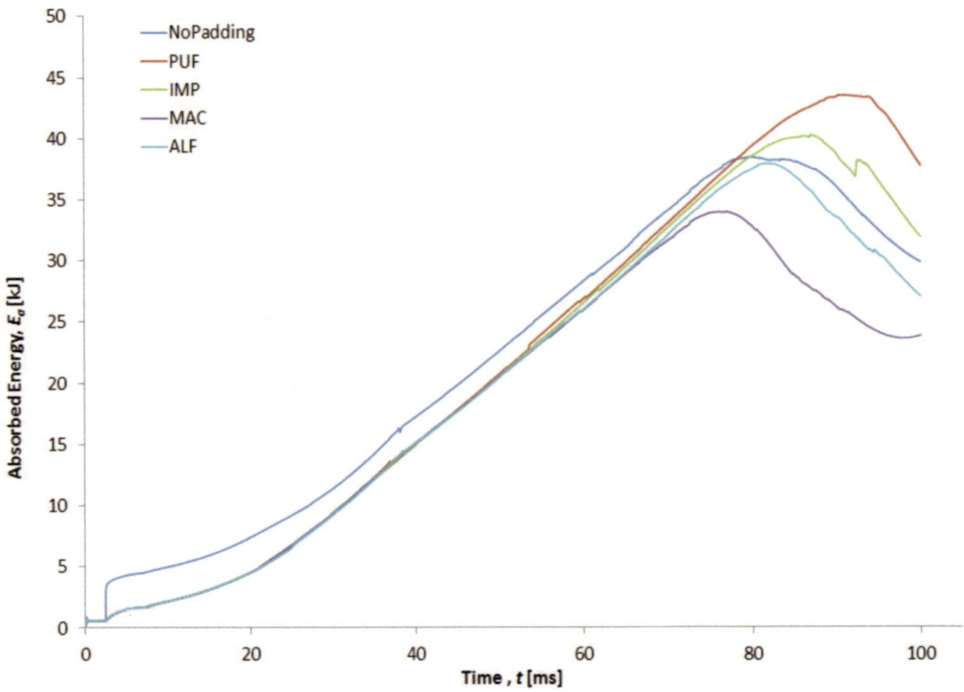

Figure 9. Variation with time of the energy absorbed by the structure of the vehicle during the side-impact crash tests.

5. Conclusions

The introduction of a structural padding made from cellular materials inside the lateral doors of common vehicles is suggested as a passive safety mechanism in side-impact vehicle-to-vehicle collisions. In order to evaluate the viability and efficiency of this safety mechanism crash tests were performed using finite element analysis software LS-Dyna™. The EuroNCAP [38] standards and definitions were considered when the defining and implementing the crash-test models. Rigid polyurethane foam, IMPAXX™ , micro-agglomerated cork and aluminium foam paddings were tested and their performance as energy absorbers was confronted with the results with no padding. The results obtained show that the implementation of a foam like material — a cellular material — as a padding for energy dissipation in lateral doors can, in fact, lead to considerable improvements, mainly in terms of maximum values of deceleration (the direct consequence leading to injury levels) and loads transmitted to the occupants of the vehicle.

Reductions as high as 59% in terms of maximum acceleration values can be observed when comparing the results obtained with and without padding. This reduction was achieved by implementing an aluminium foam padding. This was followed by a cork micro-agglomerate padding, with an improvement of 43%, and IMPAXX™ with 36%. A padding of rigid polyurethane foam, even though it is the one leading to a smaller reduction, can result in maximum accelerations 21% lower when compared to tests without padding.

The average loads in the crash tests with padding are more than 85% lower than the ones from the tests with no padding and its distribution is more balanced. Additionally, the maximum load could also be reduced by up to 79% when including a protective padding, being the best results obtained with cork micro-agglomerate. Polyurethane foam padding enclosed inside the vehicle's lateral door resulted in reductions of 83 and 73% in terms of average and maximum load, respectively.

Furthermore, in terms of energy absorbed by the vehicle's global structure, polyurethane foam was the material exhibiting the best behaviour. The inclusion of this padding, as well as micro-agglomerated cork padding, resulted in improvements of approximately 13%.

Author details

Mariana Paulino
Faculty of Engineering & Industrial Sciences, Swinburne University of Technology, Australia

Filipe Teixeira-Dias
Dept. Mechanical Engineering, University of Aveiro, Portugal

6. References

[1] [n.d.]. LS-Dyna™ (971) [Software]. (2008). Livermore, CA, Livermore Software Technology Corporation.

[2] Anindya, D. & Shivakumar, N. D. [2009]. An experimental study on energy absorption behavior of polyurethane foams, *Journal of Reinforced Plastics and Composites* 28: 3021–3026.

[3] Automotive, D. [2006]. *Tech Data Sheet IMPAXX™ 300 Energy Absorbing Foam*, The Dow Chemical Company.

[4] Avalle, M., Belingardi, G. & Montanini, R. [2001]. Characterization of polymeric structural foams under compressive impact loading by means of energy-absorption diagram, *International Journal of Impact Engineering* 25: 455–472.

[5] Belytschko, T., Lin, J. & Chen-Shyh, T. [1984]. Explicit algorithms for the nonlinear dynamics of shells, *Computer Methods in Applied Mechanics and Engineering* 42(2): 225–251.

[6] Buzeman, D., Viano, D. & Lövsund, P. [1998]. Car occupant safety in frontal crashes: a parameter study of vehicle mass, impact speed, and inherent vehicle protection, *Accident Analysis & Prevention* 30(6): 713–722.

[7] Chen, W. & Wierzbicki, T. [2001]. Relative merits of single-cell, multi-cell and foam-filled thin-walled structures in energyabsorption, *Thin-Walled Structures* 39(4): 287–306.

[8] Coimbra, R., Conroy, C., Hoyt, D., Pacyna, S., May, M., Erwin, S., Tominaga, G., Kennedy, F., Sise, M. & Velky, T. [2008]. The influence of damage distribution on serious brain injury in occupants in frontal motor vehicle crashes, *Accident Analysis & Prevention* 40(4): 1569–1575.

[9] Forsberg, J. & Nilsson, L. [2006]. Evaluation of response surface methodologies used in crashworthiness optimization, *International Journal of Impact Engineering* 32(5): 759–777.

[10] Gama, B., Bogetti, T., Fink, B., Yu, C., Dennis Claar, T., Eifert, H. & Gillespie Jr., J. W. [2001]. Aluminum foam integral armor: a new dimension in armor design, *Composite Structures* 52(3-4): 381–395.

[11] Gameiro, C., Cirne, J. & Gary, G. [2007]. Experimental study of the quasi-static and dynamic behaviour of cork under compressive loading, *Journal of Materials Science* 42: 4316–4324.

[12] Gandhi, U. & Hu, S. [1996]. Data based models for automobile side impact analysis and design evaluation, *International Journal of Impact Engineering* 18(5): 517–537.

[13] Ghamarian, A., Zarei, H. R. & Abadi, M. T. [2011]. Experimental and numerical crashworthiness investigation of empty and foam-filled end-capped conical tubes, *Thin-Walled Structures* 49: 1312–1319.

[14] Gibson, L. & Ashby, M. [1997]. *Cellular Solids: Structure and Properties*, second edition edn, Cambridge University Press.

[15] Henry, F. P. & Williamson, C. L. [1995]. Rigid polyurethane foam for impact and thermal protection, *International Conference on the Packaging and Transportation of Radioactive Materials*.

[16] Hosseini-Tehrani, P. & Nikahd, M. [2006]. Two materials s-frame representation for improving crashworthiness and lightening, *Thin-Walled Structures* 44(4): 407–414.

[17] Hou, S., Li, Q., Long, S., Yang, X. & Li, W. [2009]. Crashworthiness design for foam filled thin-wall structures, *Materials and Design* 30(6): 2024–2032.

[18] Jin, H., Lu, W.-Y., Scheffel, S., Hinnerichs, T. & Neilsen, M. [2007]. Full-field characterization of mechanical behavior of polyurethane foams, *International Journal of Solids and Structures* 44(21): 6930–6944.

[19] Kasparek, E., Zencker, U., Scheidemann, R., Völzke, H. & Müller, K. [2011]. Numerical and experimental studies of polyurethane foam under impact loading, *Computational Materials Science* 50(4): 1353–1358.

[20] Kim, D.-K., Lee, S. & Rhee, M. [1998]. Dynamic crashing and impact energy absorption of extruded aluminum square tubes, *Materials and Design* 19(4): 179–185.

[21] Kim, G. [1995]. Study of safety regulation for occupant protection in side impacts, *Journal of KSME* 35: 525–541.

[22] Klempner, D. & Frisch, K. [1991]. *Handbook of polymericfoams and foam technology*, Hanser Publishers.

[23] Lakes, R. [1986]. Experimental microelasticity of two porous solids, *International Journal of Solids and Structures* 22(I): 55–63.

[24] Lam, K., Behdinan, K. & Cleghorn, W. [2003]. A material and gauge thickness sensitivity analysis on the nvh and crashworthiness of automotive instrument panel support, *Thin-Walled Structures* 41(11): 1005–1018.

[25] Liao, X., Li, Q., Yang, X., Li, W. & Zhang, W. [2008]. A two-stage multi-objective optimisation of vehicle crashworthiness under frontal impact, *International Journal of Crashworthiness* 13(3): 279–288.

[26] Lim, J., Choi, J. & Park, G. [1997]. Automobile side impact modelling using atb software, *International Journal of Crashworthiness* 2(3): 287–298.

[27] Lopatnikov, S. L., Gama, B. A., Haque, M. J., Krauthauser, C. & Gillespie, J. W. [2004]. High-velocity plate impact of metal foams, *International Journal of Impact Engineering* 30(4): 421 – 445.

[28] Lu, G. & Yu, T. [2003]. *Energy absorption of structures and materials*, Woodhead Publishing.

[29] Majumder, S., Roychowdhury, A. & Pal, S. [2004]. Dynamic response of the pelvis under side impact load - a three-dimensional finite element approach, *International Journal of Crashworthiness* 9(1): 89–103.

[30] Mamalis, A., Manolakos, D., Ioannidis, M., Chronopoulos, D. & Kostazos, P. [2009]. On the crashworthiness of composite rectangular thin-walled tubes internally reinforced with aluminium or polymeric foams: Experimental and numerical simulation, *Composite Structures* 89: 416–423.

[31] Mamalis, A., Manolakos, D., Ioannidis, M. & Kostazos, P. [2003]. Crushing of hybrid square sandwich composite vehicle hollow bodyshells with reinforced core subjected to axial loading: numerical simulation, *Composite Structures* 61(3): 175–186.

[32] Mamalis, A., Manolakos, D., Ioannidis, M., Spentzas, K. & Koutroubakisa, S. [2008]. Static axial collapse of foam-filled steel thin-walled rectangular tubes: experimental and numerical simulation, *International Journal of Crashworthiness* 13(2): 117–126.

[33] McIntosh, A., Kallieris, D. & Frechede, B. [2007]. Neck injury tolerance under inertial loads in side impacts, *Accident Analysis and Prevention* 39(2): 326–333.

[34] Mines, R. [2004]. A one-dimensional stress wave analysis of a lightweight composite armour, *Composite Structures* 64(1): 55–62.

[35] Miranda, V., Teixeira-Dias, F., Pinho-da Cruz, J. & Novo, F. [2010]. The role of plastic deformation on the impact behaviour of high aspect ratio aluminium foam-filled sections, *International Journal of Non-Linear Mechanics* 45(5): 550–561.

[36] Mooi, H. G. & Huibers, J. H. A. M. [1998]. Simple and effective lumped mass models for determining kinetics and dynamics of car-to-car crashes, *International Journal of Crashworthiness* 5(1): 7–24.

[37] Morris, R., Crandall, J. & Pilkey, W. [1999]. Multibody modelling of a side impact test apparatus, *International Journal of Crashworthiness* 4(1): 17–30.

[38] NCAP, E. [2011]. European new car assessment programme - side impact testing protocol, *Technical Report version 5.1*, Euro NCAP.

[39] O'Connor, P. & Brown, D. [2006]. Relative risk of spinal cord injury in road crashes involving seriously injured occupants of light passenger vehicles, *Accident Analysis & Prevention* 38(6): 1081–1086.

[40] Paulino, M. & Teixeira-Dias, F. [2011]. An energy absorption performance index for cellular materials – development of a side-impact cork padding, *International Journal of Crashworthiness* 16(2): 135–153.

[41] Pickett, A., Pyttel, T., Payen, F., Lauro, F., Petrinic, N., Werner, H. & Christlein, J. [2004]. Failure prediction for advanced crashworthiness of transportation vehicles, *International Journal of Impact Engineering* 30(7): 853–872.

[42] Ren, X. & Silberschmidt, V. [2008]. Numerical modelling of low-density cellularmaterials, *Computational Materials Science* 43(1): 65–74.

[43] Samaha, R. & Elliot, D. [2003]. Nhtsa side impact research: Motivation for upgraded test procedures, Proceedings of the 18th Conference on the enhanced safety of vehicles, Nagoya, Japan.

[44] Schiff, M., Tencer, A. & Mack, C. [2008]. Risk factors for pelvic fractures in lateral impact motor vehicle crashes, *Accident Analysis & Prevention* 40(1): 387–391.

[45] Shim, V., Tu, Z. & Lim, C. [2000]. Two-dimensional response of crushable polyurethane foam to low velocity impact, *International Journal of Impact Engineering* 24(6-7): 703–731.

[46] Sun, G., Li, G., Hou, S., Zhou, S., Li, W. & Li, Q. [2010]. Crashworthiness design for functionally graded foam-filled thin-walled structures, *Materials Science and Engineering: A* 527(7-8): 1911–1919.

[47] Taher, S. T., Mahdi, E., Moktar, A. S., Magid, D. L., Ahmadun, F. R. & Arora, P. R. [2006]. A new composite energy absorbing system for aircraft and helicopter, *Composite Structures* 75: 14–23.

[48] Tavris, D., Kuhn, E. & Layde, P. [2001]. Age and gender patterns in motor vehicle crashinjuries: importance of type of crash and occupant role, *Accident Analysis & Prevention* 33(2): 167–172.

[49] Tencer, A., Kaufman, R., Mack, C. & Mock, C. [2005]. Factors affecting pelvic and thoracic forces in near-side impact crashes: a study of us-ncap, nass, and ciren data, *Accident Analysis & Prevention* 37(2): 287–293.

[50] Tey, J., Soutar, A., Mhaisalkar, S., Yu, H. & Hew, K. [2006]. Mechanical properties of uv-curable polyurethane acrylate used in packaging of mems devices, *Thin Solid Films* 504(1-2): 384–390.

[51] Uddin, M. F., Mahfuz, H., Zalnuddin, S. & Jeelani, S. [2009]. Improving ballistic performance of polyurethane foam by nanoparticle reinforcement, *Journal of Nanotechnology* 2009: ID 794740, 8 pages.

[52] Woodfin, R. L. [2000]. Using rigid polyurethane foams (rpf) for explosive blast energy absorption in applications such as anti-terrorist defenses, *Technical report*, Sandia National Laboratories.

[53] Yoganandan, N., Pintar, F., Zhang, J. & Gennarelli, T. [2007]. Lateral impact injuries with side airbag deployments—a descriptive study, *Accident Analysis and Prevention* 39(1): 22–27.

[54] Zheng, Z., Liu, Y., Yu, J. & Reid, S. [in press]. Dynamic crushing of cellularmaterials: Continuum-based wave models for the transitional and shock modes, *International Journal of Impact Engineering* .

Chapter 16

Polyurethane Trickling Filter in Combination with Anaerobic Hybrid Reactor for Treatment of Tomato Industry Wastewater

Ahmed Tawfik

Additional information is available at the end of the chapter

http://dx.doi.org/10.5772/47982

1. Introduction

The tomato industry is among the most polluting food industries in its huge amount of water consumption. This wastewater is predominantly loaded with organic wastes and is rich in organic content (Bozinis et al., 1996). Biological treatment processes are widely used for the treatment of agro-industry wastewaters, such as tomato industry wastewater which contain high concentrations of biodegradable organic matter (BOM) (Satyanarayan et al., 2005; Tawfik & El-Kamah 2011). Anaerobic treatment of high strength wastewater is widely accepted in the industry (Tawfik et al., 2008; Mehrdad et al., 2007; Del Pozo et al., 2003; Bernet &Paul, 2006). It has several advantages over aerobic processes, which include the use of less energy due to omission of aeration, the conversion of organic matter to methane which is an energy source by itself and can be used to supply some of the energy requirement of the process. Lower production of sludge, which reduces sludge disposal costs greatly and low level of maintenance, are other benefits of anaerobic processes (Tawfik et al., 2006; Cakira & Stenstromb, 2005; Shin et al., 2005). Moreover, high substrate removal efficiencies would be achieved in anaerobic reactors with short hydraulic retention time (HRT) and high organic loading rate (OLR) (Tawfik et al., 2010). One of the most efficient and quite flexible designs available is an anaerobic hybrid (AH) reactor which combines advantages of both anaerobic filter (AF) and up-flow anaerobic sludge blanket (UASB) designs (Chang ,1989; Hawkes et al., 1995; Wu et al., 2000; Tawfik et al., 2011; Mahmoud et al., 2009). The presence of polyurethane media in the upper portion of AH reactor, in addition to its physical role for biomass retention, also exerts some biological activity which contributes to chemical oxygen demand (COD) reduction in a zone where generally active biomass is lacking in a calssical UASB reactor (Elmitwalli et al., 1999; Tawfik et al., 2009). However; the effluent quality of anaerobic reactor is still not complying in terms of COD, total suspended solids (TSS) and nitrogen for discharge into drainage canals. Therefore, post-treatment is needed. Trickling filter (TF) is one of the aerobic biological treatment systems that are widely used for post-treatment of anaerobically pretreated effluent. However, the TF still has some drawbacks such as difficulty in maintenance of appropriate biofilm thickness under limitation of oxygen supply and biofilm detachment from the packing materials. In this investigation polyurethane trickling filter (PTF) was proposed. It has several advantages compared to conventional TF. In TF of which micro-organisms attach themselves only to a media surface creating a biological filter or slime layer, but in PTF module, the micro-organisms are retained outside and inside the polyurethane media which create long sludge residence time (SRT>100 days) (Tawfik et al., 2006) and consequently, achieve a complete nitrification and produce a very low amount of sludge (Tawfik et al., 2006). Once again, PTF brings the following advantages compared to conventional TF, the amount of active biomass brought in contact with the wastewater is very high and the biological process is very fast, thus very short retention time is needed which gives small plants with no investment cost (Tawfik et al., 2010). The process is in contrast to conventional TF which needs a very low OLR to achieve a nitrification process (Tawfik et al.,2011; Chernicharo &Nascimento, 2001). Moreover, it is easy to enlarge the capacity of PTF system in case the flow and/or the organic load would increase in the future.

The objectives of this investigation are to 1. assess the efficiency of a combined system consisting of AH reactor and PTF for the treatment of tomato industry wastewater at different HRTs and OLRs with emphasis on the COD fractions (COD_{total}; $COD_{soluble}$ and $COD_{particulate}$); TSS and total nitrogen (TN) removal. In addition, the mechanism for the removal COD, TSS and nitrification efficiency along the height of PTF is investigated.

2. Materials and methods

2.1. Tomato wastewater industry

Tomato-processing wastewaters are typical high strength wastewater generated from the food canning industry. Composite wastewater samples from tomato manufacturing company were collected and analyzed for parameters considered necessary for wastewater characterization and system design for a year. Characteristics of the Tomato wastewater industry showed that 73.4 % of the TSS, was volatile organics; and 64% of the COD was insoluble form. Soluble NH_4 –N constituted 74% of TN. The tomato processing industry wastewater were used as substrates for the combined system consisting of AH reactor as a pretreatment and PTF as a post-treatment unit (Fig. 1).

2.2. Anaerobic hybrid (AH) reactor

5 l AH reactor was designed and manufactured from polyvinyl chloride (PVC) as described earlier by Tawfik et al., (2011) and illustrated in Fig. 1. The AH reactor consisted of a sludge blanket at the bottom, and floating polyurethane carriers at the top to overcome washout of sludge from the reactor. The surface area of polyurethane carriers in the AH reactor was

0.57m². $NaHCO_3$ was added to adjust the influent pH in the range of 6–7. The seed sludge was taken from an up-flow anaerobic sludge blanket (UASB) reactor treating juice industry wastewater. The sludge is typically flocculent with mixed liquor suspended solids (MLSS) of 16.8 g/l, mixed liquor volatile suspended solids (MLVSS) of 10.8 g/l and VSS/TSS ratio of 0.64. Three liters of the sludge was pumped into the AH reactor as an inoculum.

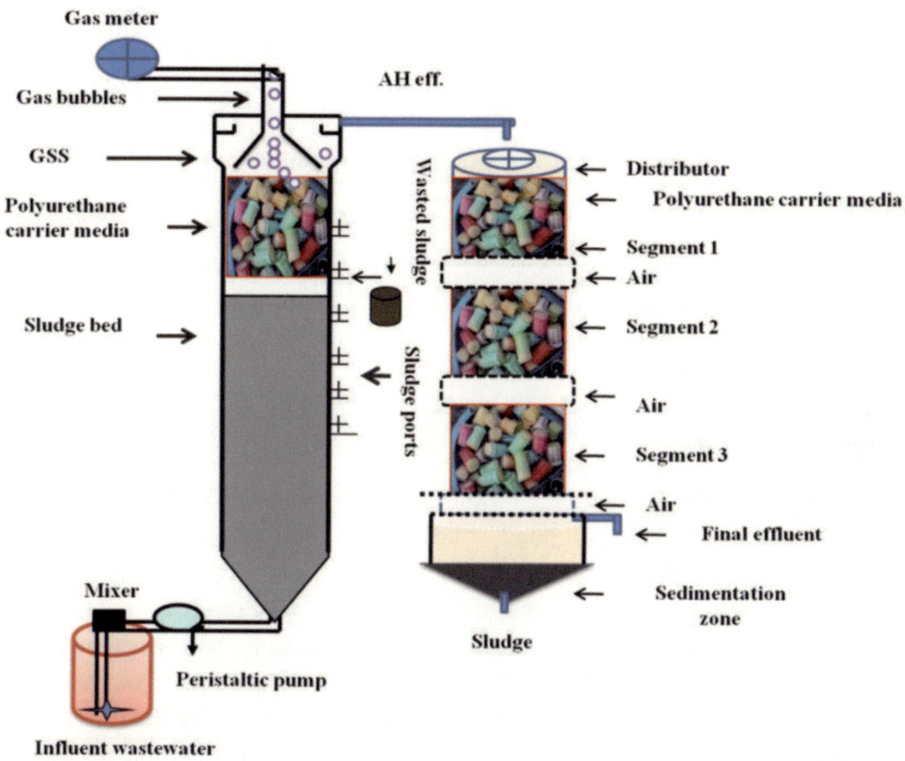

Figure 1. PTF module coupled with AH reactor for treatment of tomato industry wastewater

2.3. Polyurethane trickling filter (PTF) reactor

The PTF is packed with porous polyurethane foam (pore size= 0.63 mm) with relatively high specific surface area (256 m²/m³) to increase both the biofilm mass content and the removal efficiency of the reactor. The PTF module used in this study, consisted of three segments connected vertically in series. The polyurethane represents 15% of the total reactor volume (23l) (Fig. 1). Polyurethane media was warped with perforated polypropylene material (0.5 cm) to avoid clogging of the media and facilitate the air penetration inside the packing material. The PTF was equipped with 276 pieces each 40 mm height and 20 mm in diameter with 90% of porosity. The reactor was operated without inoculums. The distributer is situated on the top of the reactor and operated at 18 rpm for equal distribution of wastewater over the packing material. The air was naturally diffused to the reactor via three windows along the reactor height. There is no need for aeration as well as no backwashing.

2.4. Operating conditions

The operational conditions of the combined AH–PTF are shown in Table 1. Both reactors were operated for 330 days, 30–83; 134–212; and 234–324 days at HRTs of, respectively 14.5, 10 and 7.2 h. The first 30 days of operation were considered as a start-up period, while the periods from day 83 to 134 and from 212 to 234 were considered as acclimatization periods to the new HRT.

Operational conditions/ reactors	Run 1		Run 2		Run 3	
	HRT (h)	OLR (kgCOD/m³.d)	HRT (h)	OLR (kgCOD/m³.d)	HRT (h)	OLR (kgCOD/m³.d)
AHreactor	8.6	2.8	6	3.5	4.3	4.5
PTF reactor	5.9	1.0	4	1.43	2.9	3

Table 1. Operational conditions of AH reactor in combination with PTF for the treatment of tomato industry wastewater

2.5. Analytical methods

Composite samples of the influent wastewater and the treated effluents were biweekly analyzed. COD, TSS, volatile suspended solids (VSS), total Kjeldahl nitrogen (TKj-N), ammonia (NH_4-N), nitrite (NO_2-N), nitrate (NO_3-N) and protein were analyzed according to standard methods (APHA, 2005). Raw wastewater samples were used for COD_{total}, 0.45 μm membrane-filtered samples for $COD_{soluble}$. The $COD_{particulate}$ was calculated by the differences between COD_{total} and $COD_{soluble}$, respectively. Biogas composition was measured using a gas chromatograph fitted with a thermal conductivity detector (TCD) and Poropak Q stainless steel column. The oven, injector, and detector temperatures were set as 40, 60 and 60 °C, respectively and hydrogen was used as the carrier gas. The instrument was calibrated using a mixture of 50% methane and 50% carbon dioxide. Volatile fatty acid (VFA) concentration was measured after centrifuging the samples to remove the suspended solids. A gas-liquid chromatograph equipped with a Flame Ionization Detector (FID) and Chromasorb 101 column was used for the analysis of VFA. The detector, injector and oven temperature were 200, 195 and 180 °C, respectively. The carrier gas used was nitrogen, and a mixture of hydrogen and air was used to sustain the flame in the detector.

2.6. Scanning electron microscope (SEM)

The surface of sponge carriers and the attached microorganism species in the PTF reactor were analyzed by a JSM-5600 LV scanning electron microscope (JEOL, Japan). A sample of the microorganisms attached to the carriers was withdrawn from the PTF and placed in bottles. After drying for 10 h under vacuum at 40 °C, these samples were fixed in 0.1 mol/l phosphate buffer solution (pH 7.3) containing 2.5% glutaraldehyde for 12 h at 4 °C. After fixation, samples were rinsed three times in 0.1 mol/l of phosphate buffer solution (pH 7.3) and dehydrated gradually by successive immersions in ethanol solutions of increasing concentration (30, 50, 70, 80, 90, and 95%). The samples were then washed three times in 100% ethanol. The drying process was then completed by incubating the samples for 2 h at 40 °C. The sponge were then coated with gold powder and attached to the microscope support with silver glue. SEM photographs were taken at 25 and 20 kV.

3. Results and discussion

3.1. Efficiency of AH reactor as a pretreatment of tomato wastewater industry

Figs 2a, b and c show the effect of HRT on the percentage reduction of COD fractions (COD_{total}, $COD_{particulate}$ and $COD_{soluble}$). By increasing the HRT from 4.3 to 8.6 h, the COD_{total} of the effluent significantly reduced from 377±88 to 267±48 mg/l, and the removal efficiency of COD_{total} substantially increased from 51±12 to 71±7%. However, the residual values of COD_{total} in the treated effluent of the AH reactor remained unaffected by increasing the HRT from 6.0 to 8.6 h. Likely, the results in Fig. 2b show that the effluent quality of $COD_{soluble}$ and removal efficiency was maintained at the same level of 117 ±11mg/l and 64±9% respectively by decreasing the HRT from 8.6 to 6 h. This indicates that the AH reactor was operated under substrate limiting conditions at an HRT of 8.6 h. Accordingly it is recommended to apply such a system at OLR 3.5 kgCOD/m^3. d and HRT not exceeding 6.0 h. An increase in the HRT would result in a decrease in the wastewater linear velocity through the support material, improving the mass transfer from the liquid phase to the biomass and, therefore, favoring the process performance (Elmitwalli et al., 2000).

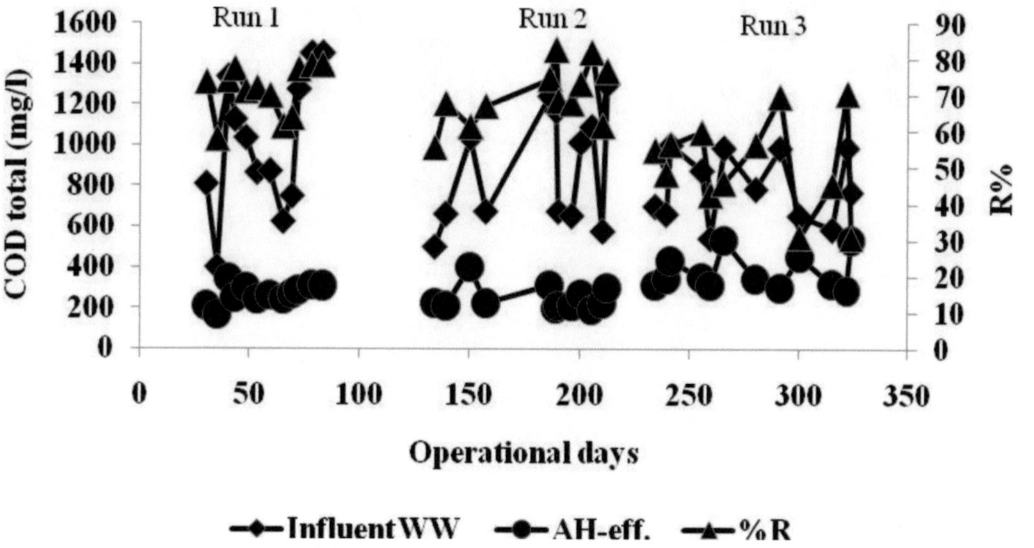

(a)

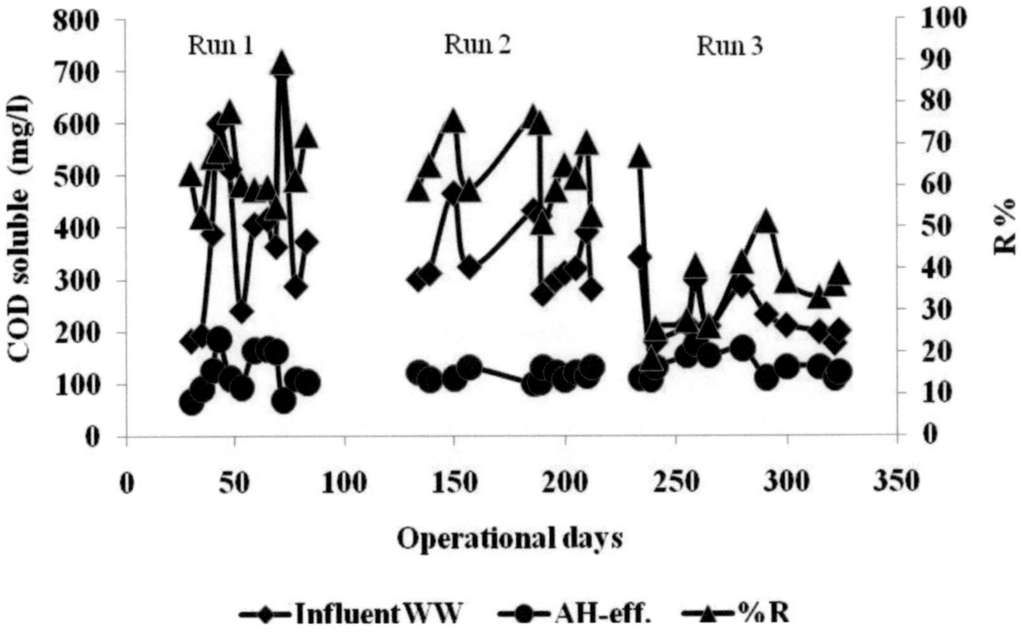

(b)

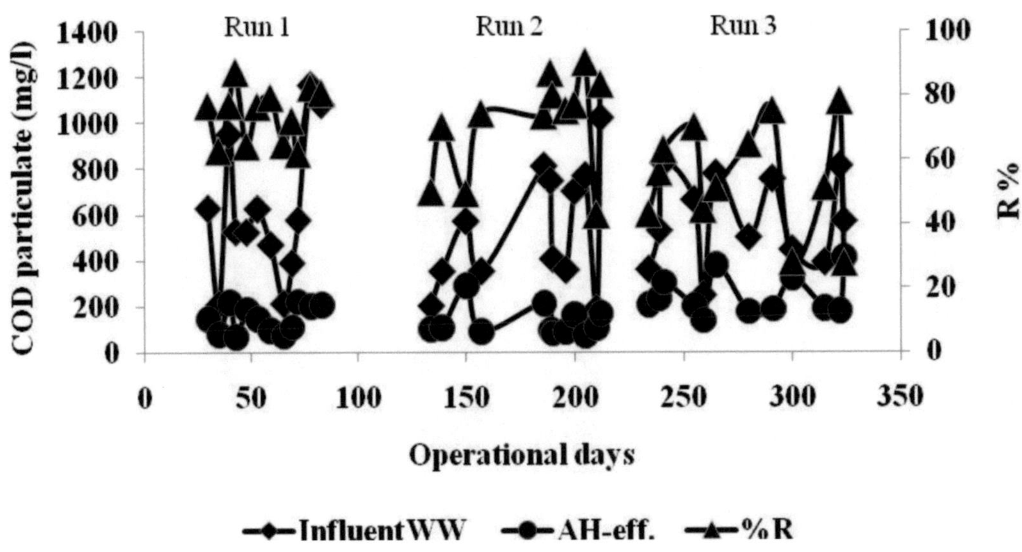

(c)

Figure 2. (a) COD $_{total}$ removal efficiency in an AH reactor treating Tomato industry wastewater at different HRTs ; (b) COD $_{soluble}$ removal efficiency in an AH reactor treating Tomato industry wastewater at different HRTs; (c) COD $_{particulate}$ removal efficiency in an AH reactor treating Tomato industry wastewater at different HRTs

The removal efficiency of COD_{total} in an AH reactor at an HRT of 4.3 h was higher than those obtained by Demirer and Chen, (2005) who used AH reactor for treatment of dairy wastewater at longer HRT of 15 days.Also, Gu¨ngo¨r and Demirer, (2004) achieved a lower COD removal efficiency of 37.9–50% in anaerobic batch reactor treating food industry wastewater.. The improved removal efficiency of COD_{total} in this study was mainly due to a higher removal efficiency of $COD_{particulate}$ as shown in Fig. 2c. In previous studies on opaque beer wastewater with UASB, 57% COD_{total} reduction was achieved at HRT of 24 h (Parawira et al., 2005). Similarly, studies of Cronin and Lo (1998) and Driessen and Vereijken (2003) on UASB with brewery wastewater showed that the COD_{total} reduction of 75–80% with the HRT in the range of 12–36 h. In the present study AH reactor could be optimally operated at an OLR of 3.5 kg COD/m^3.d and HRT not exceeding 6 h with COD_{total} reduction of 71% and methane yield of 0.48 m^3 CH_4/kg COD_{total} reduced. This high efficiency of AH reactor as compared to UASB reactor can be due to the presence of polyurethane carrier material in the sedimentation part which overcome sludge washout and improve the biodegradation process. Moreover,polyurethane carriers provide a much larger surface area for the attachment of biofilm which then leads to an increase of anaerobic biodegradation process.

Variations of VFAs in the influent and effluent of AH reactor are shown in Fig. 3a. Although, there was a significant fluctuation in the VFAs of the feed between 198 and 689 mg/l, the AH reactor showed that VFAs in the feed was effectively utilized by methanogenesis bacteria. VFAs in the effluent was quite low (below 121±23 mg/l) at HRTs of 8.3 and 6 h.However, the residual values of VFAs in the treated effluent was increased at decreasing the HRT(4.3 h) as shown in Fig.3a). Apparently, this can be attributed to limited activity of methanogens in the reactor under these operating conditions. Likely, Amit et al., (2007), found that the VFAs concentration increased in the treated effluent of AH reactor treating industrial cluster wastewater, when the HRT reduced from 12 to 4 h.

The variations of biogas production at different HRTs are shown in Fig. 3b. The biogas production was low (2.6 l/d) at an HRT of 4.3 h. HRT was prolonged up to 6 and 8.3 h and the gas production reached as high as 4.0 l/d, equivalent to 0.48 m^3/ kg COD removed. d. Similarly, Oscar et al., (2008) found that the value of methane yield in an AH treating food industry wastewater increased from 0.07 to 0.18 l CH_4/g COD added when the HRT increased from 1.0 to 5.5 days. The average methane yield in the gas composition was 67% as shown in Fig.3b.

AH reactor was found to be very effective for removal of TSS and VSS as shown in Figs 4a and b. TSS and VSS removal efficiencies increased from 57 ±10 to 70±8 % and from 70±8 to 78±4 % when the HRT rose from 4.3 to 6 h and from 6 to 8.3 h., respectively. The results obtained demonstrate that clogging of the support polyurethane media in the AH reactor was not evident in spite of the high concentration of TSS contained in the influent (Fig.4a). A previous study (Vartak et al., 1997) reported VSS removal efficiencies of up to 91% in up-flow anaerobic attached film reactors with a combination of limestone and polyester as support media treating diluted dairy wastewaters but operating at a longer HRT of 33 d. Lower VSS removal efficiencies (66%) have been achieved in an anaerobic baffled reactor (ABR) fed with dairy wastewater at a HRT of 5 d. (Chen and Shyu, 1996).

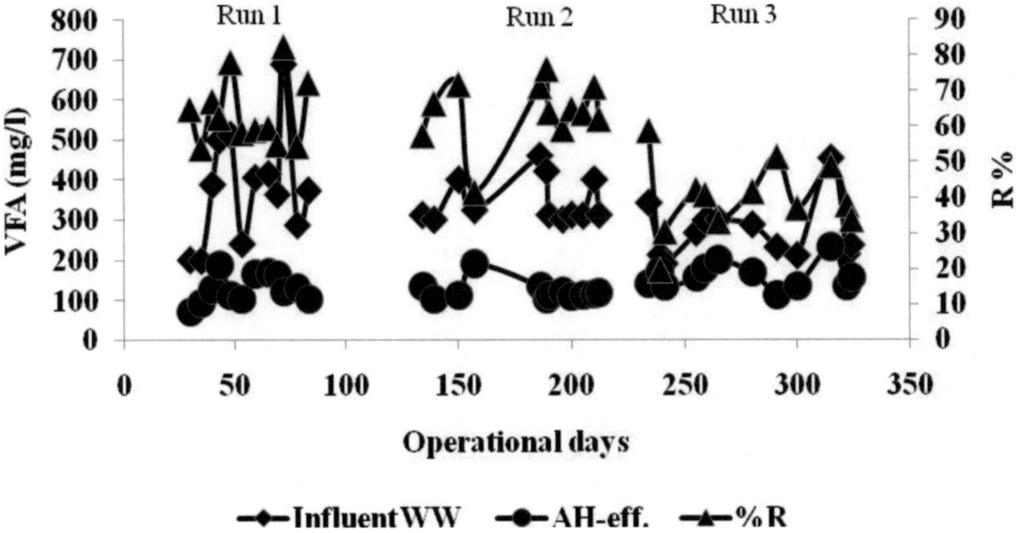

(a)

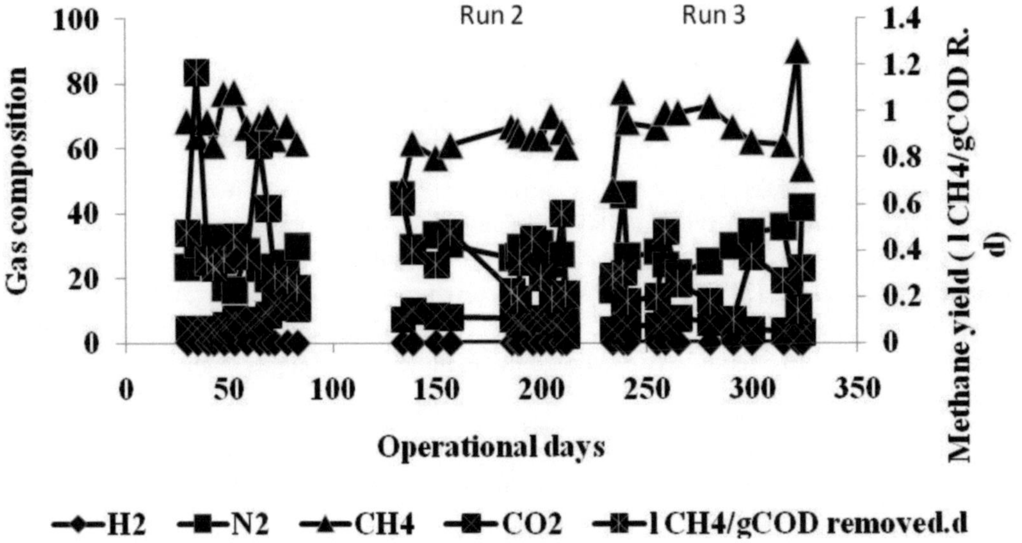

(b)

Figure 3. (a) VFAs removal efficiency in an AH reactor treating Tomato industry wastewater at different HRTs; (b) Biogas production and gas composition in an AH reactor treating Tomato industry wastewater at different HRTs

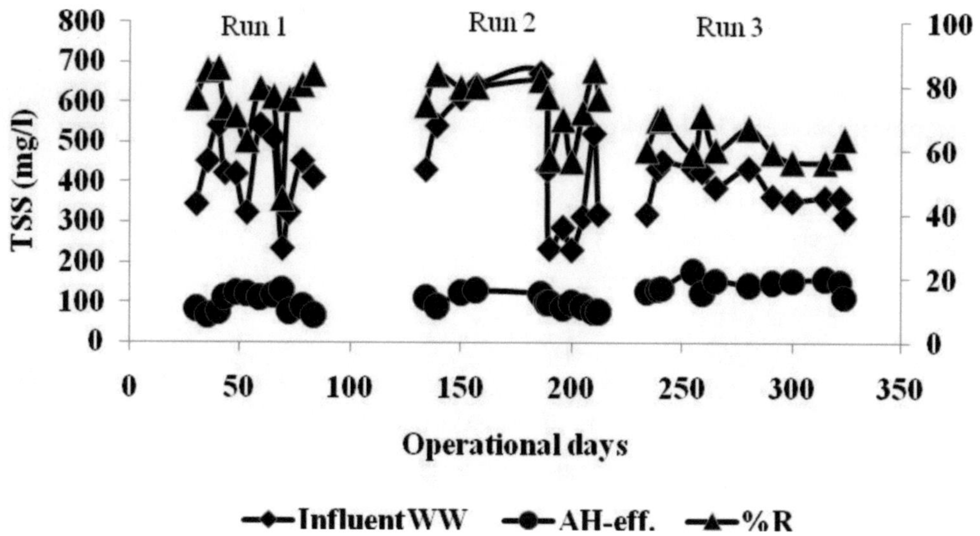

(a)

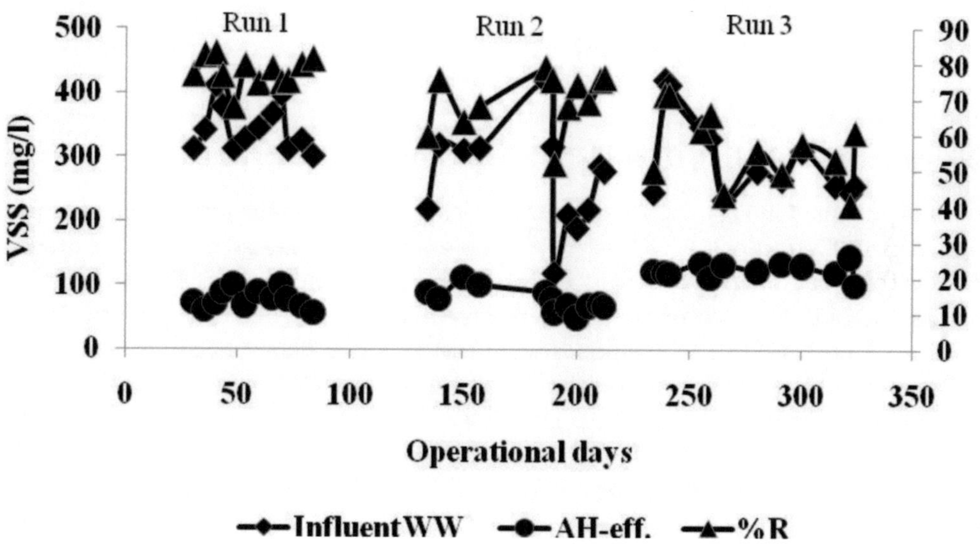

(b)

Figure 4. (a) TSS removal efficiency in an AH reactor treating Tomato industry wastewater at different HRTs ; (b) VSS removal efficiency in an AHreactor treating Tomato industry wastewater at different HRTs

No significant difference was found in the removal of protein in the AH reactor between different HRTs as shown in Fig. 5. The maximum conversion of protein was achieved and accounted for 19.8±8.5% at an HRT of 4.3 h of the protein content. The conversion of protein dropped at an HRT of 8.6 and 6 h (14±5%). The drop in protein hydrolysis might be due to chemical precipitation of NH_4-N (Miron et al., 2000).

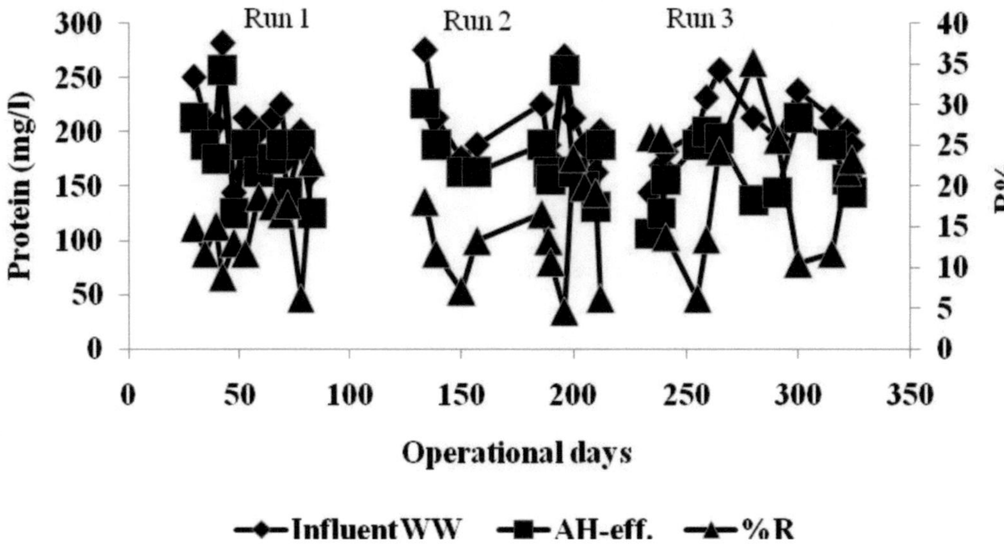

Figure 5. Protein removal efficiency in an AH reactor treating Tomato industry wastewater at different HRTs

3.2. Polyurethane trickling filter (PTF) as a post-treatment system

The results presented in Figs. 6a, b and c show the effect of OLR on the removal efficiency of the different COD fractions (COD_{total}, $COD_{soluble}$ and $COD_{particulate}$) in the PTF system treating AH reactor effluent. The results reveal a significantly improved COD_{total} removal at decreasing the OLR. The system provided a mean effluent quality of 35±9 mg/l for COD_{total} at an OLR of 1.0 kgCOD/m^3.d., which is significantly lower than that at an OLR of 3.0 kgCOD/m^3.d (86±16 mg/l). The improved removal efficiency of COD_{total} was mainly due to a higher removal efficiency of $COD_{soluble}$ and $COD_{particulate}$ (Figs. 6b and c). This excellent performance towards the removal of $COD_{soluble}$ and $COD_{particulate}$ matter can be attributed to entrapment or/and adsorption followed by hydrolysis and degradation in the polyurethane packing material. Low removal efficiency of COD_{total} at an OLR of 3 kgCOD/m^3.d can be explained by excess biofilm accumulation, filling in pores of the polyurethane packing material and reducing the mass transfer capabilities (Chen et al., 2006; Tawfik & Klapwijk, 2010) and DO concentration dropped from 5.2 to 3.2 mg/l in the PTF as the OLR increased from 1.0 to 3 kgCOD/m^3. d. However, the results presented in Fig. 6a show that the residual value of COD_{total} in the treated effluent of the PTF system remained unaffected by decreasing the OLR from 1.43 to 1.0 kgCOD/m^3.d, as a result of increasing the HRT from 4.0 to 5.9 h. Accordingly it is recommended to apply such a system at loading rate of 1.43

kgCOD/m³. d., and HRT not exceeding 4.0 h. The results obtained in this investigation were higher than those obtained by El-kamah et al., (2010 & 2011) who used down flow hanging sponge (DHS) system for post treatment of anaerobically pretreated onion industry wastewater. The system was operated at an OLR of 5.1 kgCOD/m³.d. and a similar HRT of 4.2 h. The system provided an effluent quality of 80 mg/l for COD and 30 mg/l for TSS.

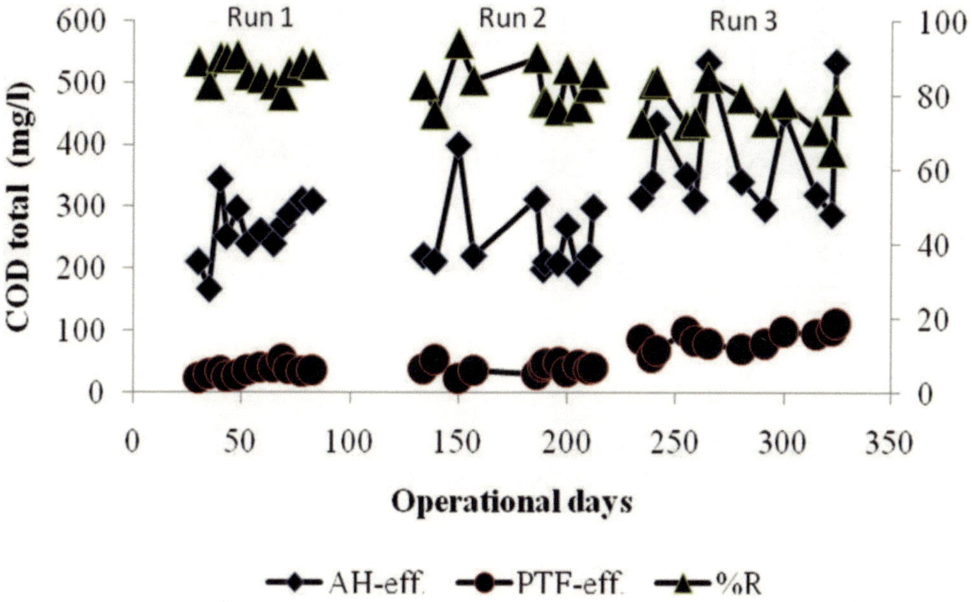

(a)

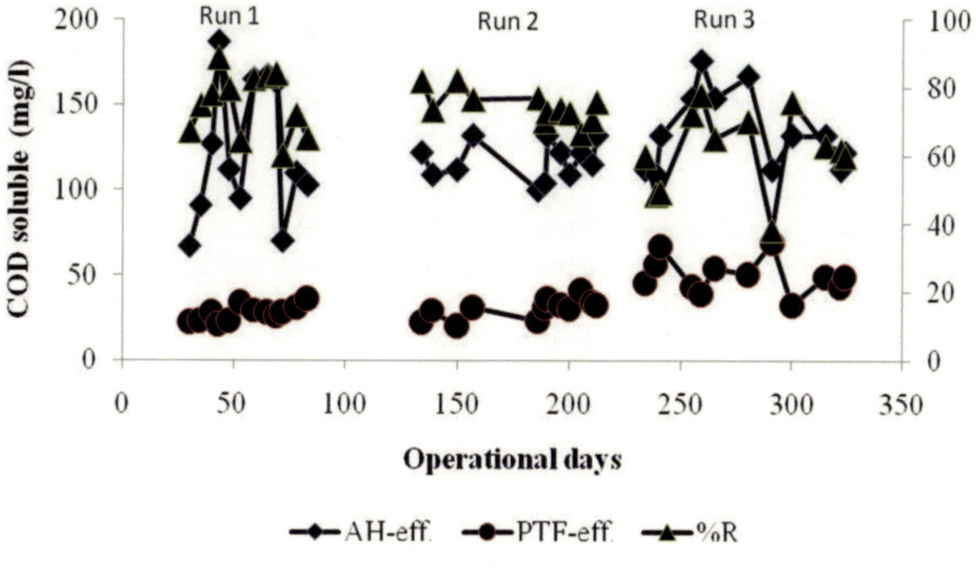

(b)

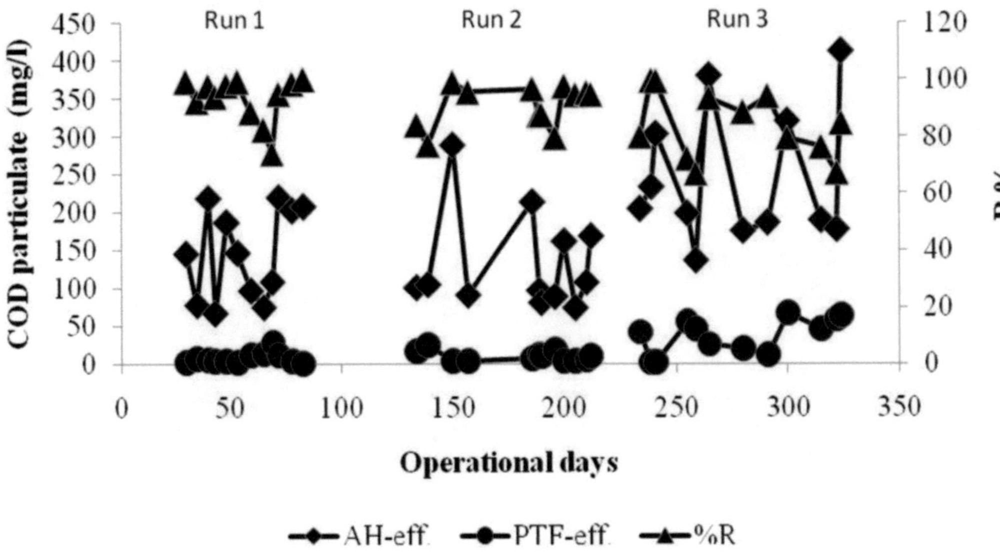

(c)

Figure 6. (a) The efficiency of PTF for removal of COD $_{total}$ at different OLRs; (b) The efficiency of PTF for removal of COD $_{soluble}$ at different OLRs; (c) The efficiency of PTF for removal of COD $_{particulate}$ at different OLRs

The results in Figs. 7a and b revealed that the removal efficiencies of TSS and VSS in the PTF reactor significantly decreased at increasing the OLR from 1.43 to 3.0 kgCOD/m^3.d., while decreasing the OLR from 1.43 to 1.0 kgCOD/m^3.d did not affect seriously on the removal efficiencies. The reactor achieved removal efficiencies of 87.1; 87 and 78.6% for TSS and 89.3; 88.5 and 79.5 % for VSS at OLRs of 1,1.43 and 3.0 kgCOD/m^3.d. respectively. This high removal efficiency for coarse suspended solids in PTF reactor were mainly due to the high entrapment capacity, high specific surface area and porosity of the polyurethane packing material. Tawfik &klapwijk, (2010) found that polyurethane is better than polystyrene packing media for removal of TSS and VSS.

The nitrification efficiency in the PTF treating AH reactor effluent at different OLRs is shown in Fig. 8a. The results show that increasing the OLR from 1.0 to 1.43 and from 1.43 to 3.0 kg COD/m^3.d, results in an increase of the ammonia concentration in the final effluent from 2.7±1.3 to 2.8±1.3 mg/l and from 2.8±1.3 to 17.8±3.7 mg/l, respectively. At OLR of 1.1, 1.43, and 3.0 kg COD/ m^3.d, ammonia was removed by values of 89.4±5.9%, 89.7±4.6 % and 25 ±10 %, while at the same time 17.7 ±3.5, 17±4.4 and 1.7±0.9 mg/l of nitrate were, respectively produced as shown in Fig. 8a. Based on these results, it can be concluded that the OLR imposed to the PTF reactor should remain below 3 kg COD/m^3.d to achieve a high nitrification efficiency as also found by El-kamah et al., (2011) for down flow hanging sponge (DHS) system treating anaerobically pretreated onion industry wastewater.

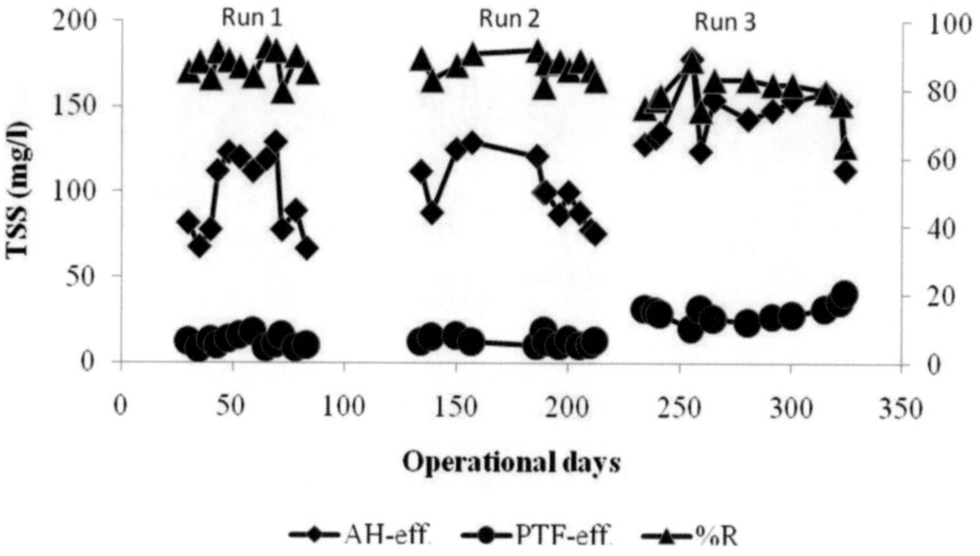

(a)

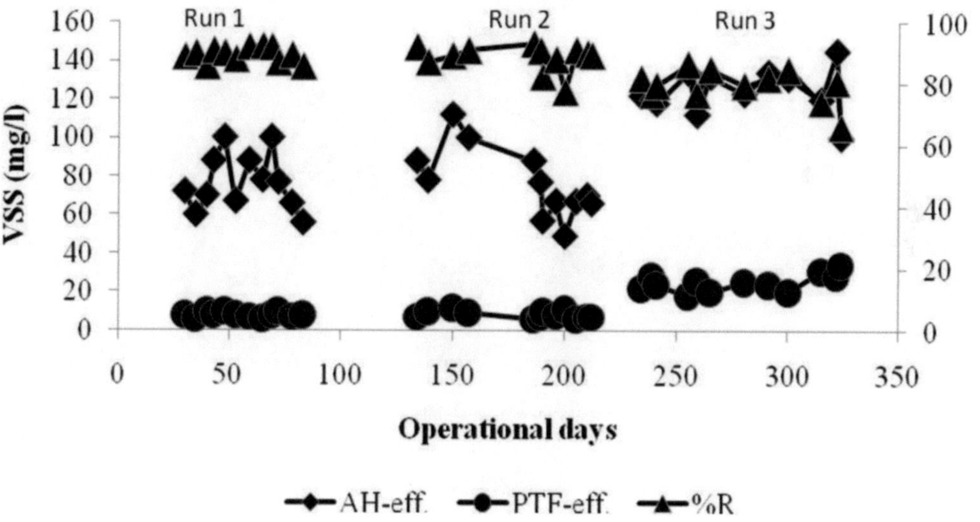

(b)

Figure 7. (a) The efficiency of PTF for removal of TSS different OLRs; (b) The efficiency of PTF for removal of VSS at different OLRs

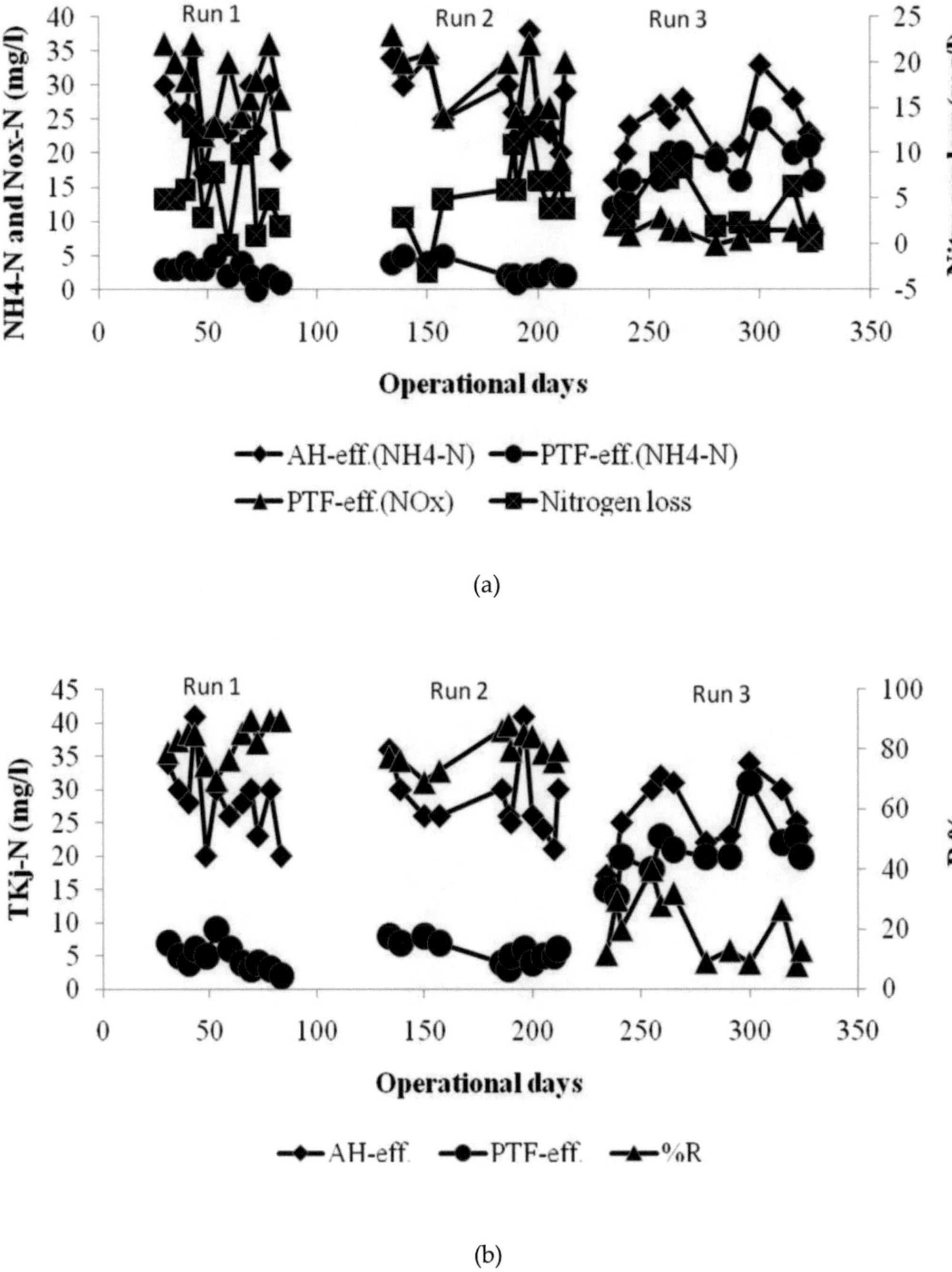

Figure 8. (a) Nitrification efficiency and total nitrogen removal in PTF at different OLRs; (b) The efficiency of PTF for removal of TKj-N at different HRTs and OLRs

The results revealed that the nitrification rate in PTF was strongly dependant on VSS/ TN ratio. A low nitrification rate was achieved in the PTF at the high influent VSS/TN ratio of 5±1, the nitrification rate was 0.013 kg NO_x-N/m^3.d as compared to VSS/N ratio of 2.8, the nitrification rate amounted to 0.1 kgNO_x-N/m^3.d. This can be attributed to the attachment and degradation of volatile suspended solids on the surface of the nitrifying biofilm where they take away oxygen which otherwise would have been available for nitrifiers (Tawfik et al., 2010). The TKj-N removal in the PTF treating AH reactor effluent was 82.8 ±6.4% at an OLR of 1.0 and 1.43 kg COD/m^3.d as compared to 20 ±10% at higher OLR of 3.0 kg COD/m^3.d (Fig. 8b). The nitrogen loss amounted to 20% (Fig. 8a) which can be due to (1) assimilation of biomass (2) denitrification occurring in the anoxic zone of the biofilm (Holman & Wareham, 2005).

3.3. Profile of polyurethane trickling filter (PTF) reactor

Profile of dissolved oxygen (DO) concentration along the height of PTF shows a gradual increase in the concentration of DO as the wastewater flows down. DO in the final effluent was in the range of 4-4.6 mg/l as shown in Fig.9a. The profile results of PTF showed that in the upper part of the PTF system, mainly COD was oxidized while nitrification was taken place in the lower part of the system, where nitrifiers are available. The results in Fig. 9 b,c and d show that most of the COD fractions (COD_{total}, $COD_{soluble}$ and $COD_{particulate}$) were removed in the 1st and 2nd segment of PTF reactor.

The 3rd segment provided a little additional removal of COD fractions as shown in Figs. 9 b, c and d. This can be explained by the fact that the most of the coarse and soluble organic matter were adsorbed and degraded in the segments 1 and 2. Likely TSS and VSS concentrations were gradually decreased from segment 1 to 3 as shown in Figs. 10 a and b. The results in Figs. 11a and b show that nitrification was very limited in the 1st segment of PTF system at OLR of 4.2 kg COD/m^3 .d. This was due to the presence of an insufficient ammonia oxidizer population at high loading rate as they cannot compete with heterotrophs for space and oxygen. In the 2nd and 3rd segment of PTF system, a high nitrification rate was achieved at lower OLRs of 2.1, and 1.4 kg COD/m^3 .d.

These results demonstrate that at OLR exceeding 4.2 kg COD/m^3 d heterotrophic bacteria still prevail in the 1st segment of PTF system, but the nitrifying bacteria promoted in the 2nd , and 3rd segment of PTF system when the OLR drops to 2.1 and 1.4 kg COD/m^3.d, respectively. The ammonia oxidation and TKj-N removal (Fig. 11a and c) was virtually approximately complete, only 1.7 mgNH_4-N/l and 4.0 mg TKj-N/l provided in the final effluent of PTF system.

3.4. Efficiency of the combined system (AH+PTF) treating tomato industry wastewater at different OLRs and HRTs

The results presented in Table 2 revealed that decreasing the total HRT from 14 to 10 h was not significantly affected on the removal efficiency of COD fractions (COD_{total}, $COD_{soluble}$

and COD particulate). However, decreasing the total HRT from 10 to 7.2 exerted a negative impact on the removal efficiency of the total process as shown in Table 2.

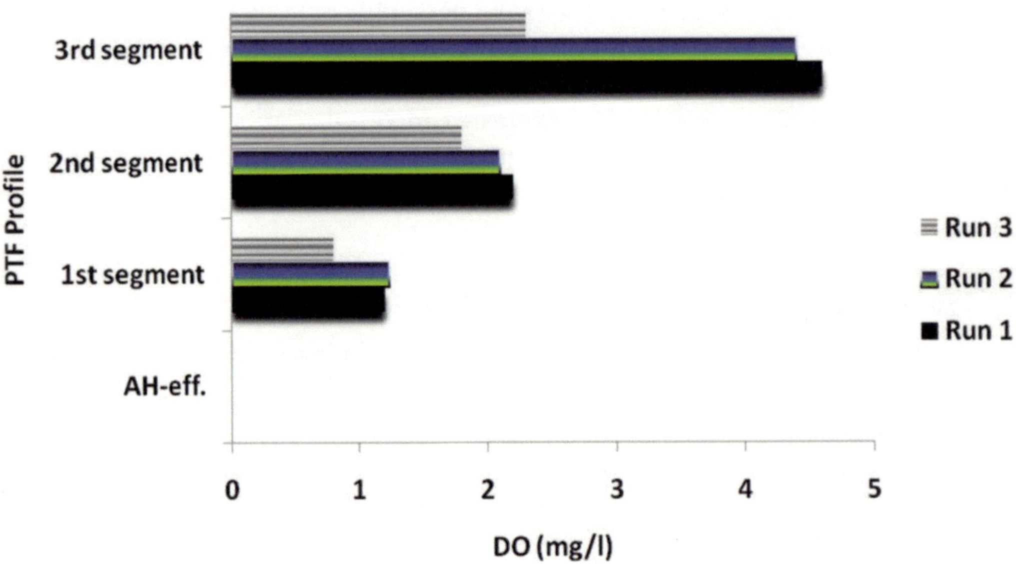

(a)

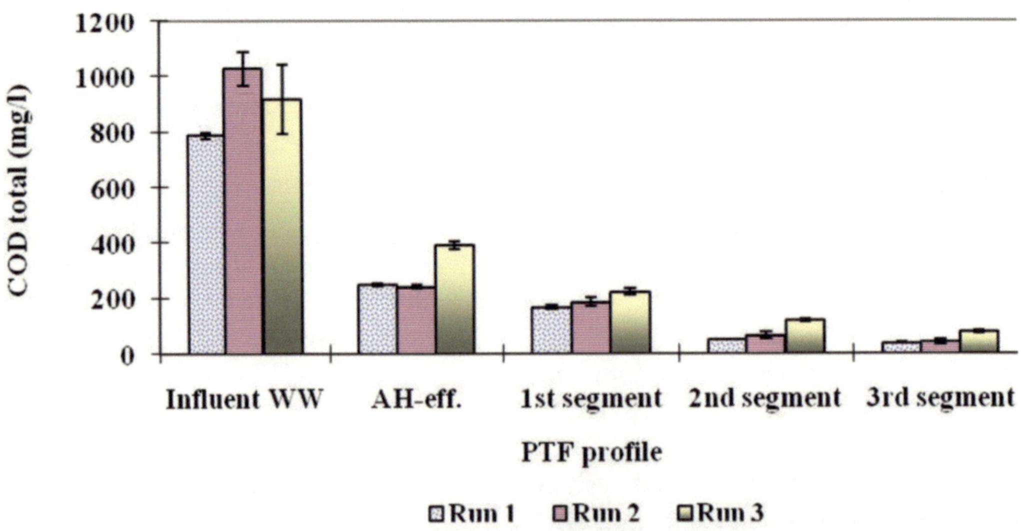

(b)

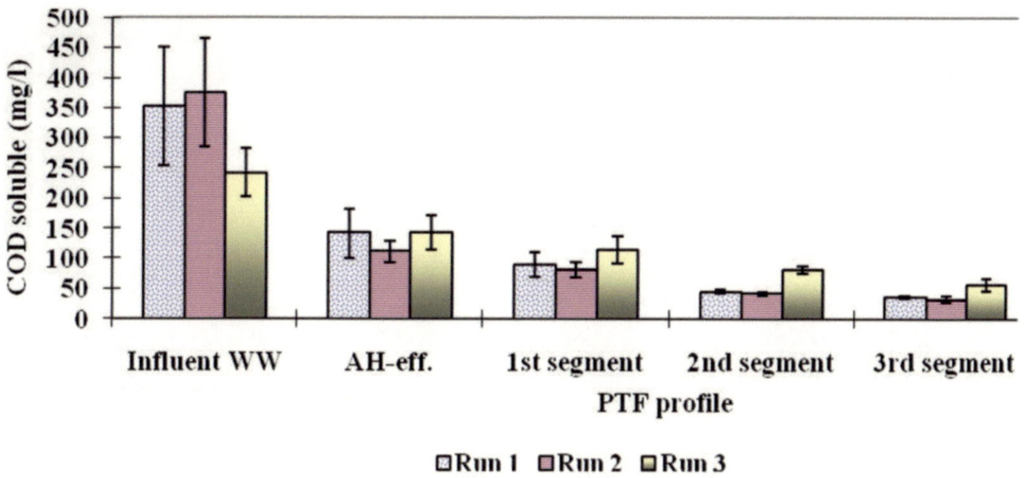

(c)

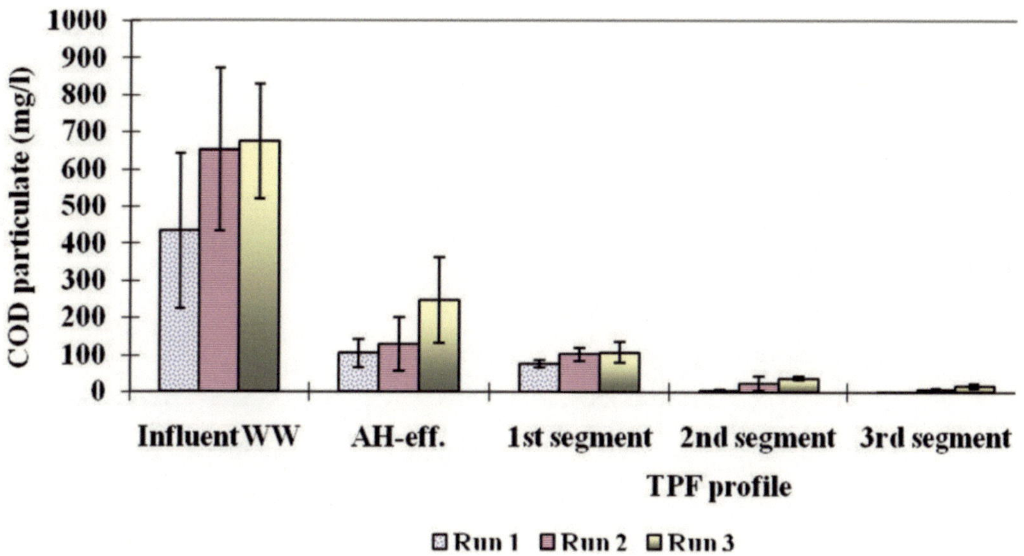

(d)

Figure 9. (a) DO concentration along the height of PTF reactor treating AH reactor effluent; (b) COD $_{total}$ removal efficiency along the height of PTF reactor treating AH reactor effluent; (c) COD $_{soluble}$ removal efficiency along the height of PTF reactor treating AH reactor effluent; (d) COD $_{particulate}$ removal efficiency along the height of PTF reactor treating AH reactor effluent

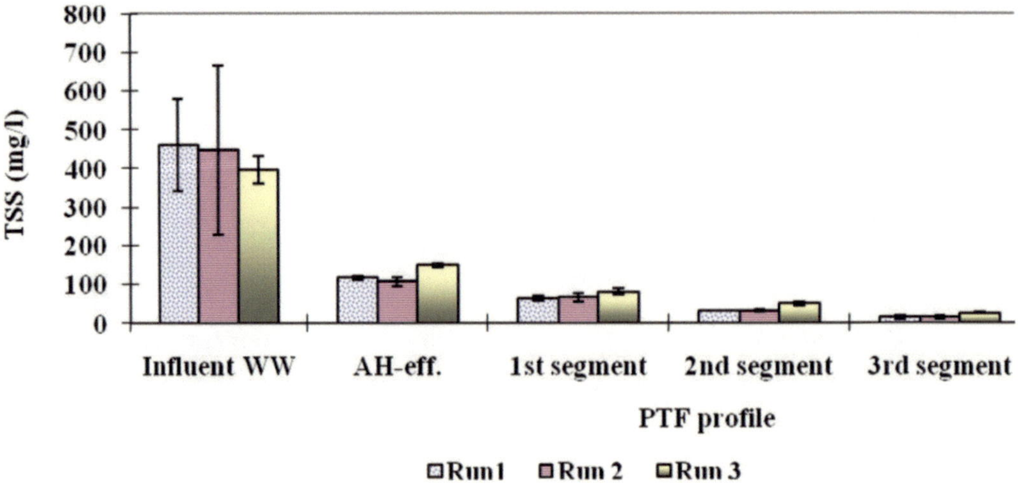

(a)

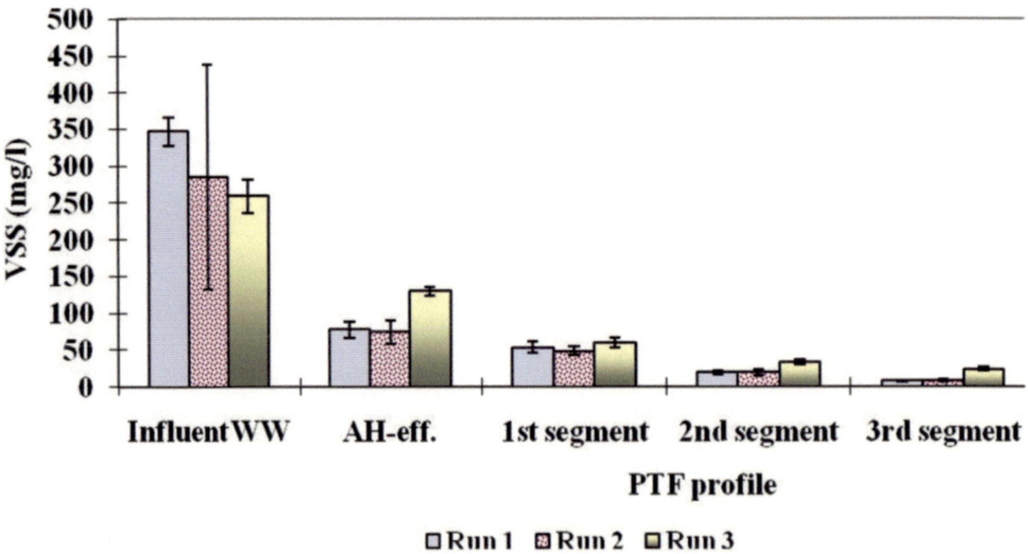

(b)

Figure 10. (a) TSS removal efficiency along the height of PTF reactor treating AH reactor effluent; (b) VSS removal efficiency along the height of PTF reactor treating AH reactor effluent

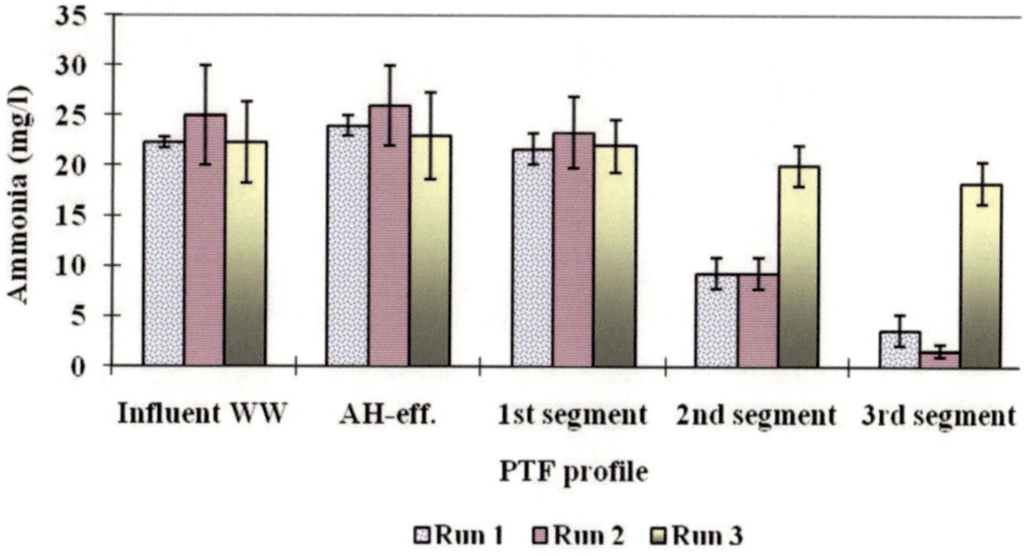
35
30
25
20
15
10
5
0
Ammonia (mg/l)
Influent WW
AH-eff.
1st segment
2nd segment
3rd segment
PTF profile
Run 1
Run 2
Run 3

(a)

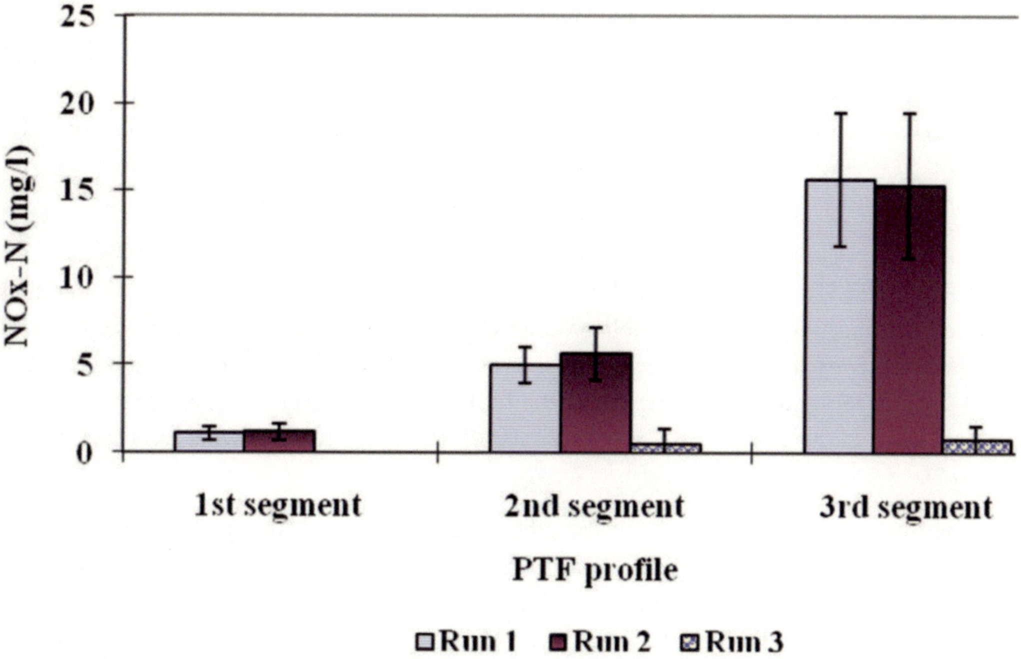
25
20
15
10
5
0
NOx-N (mg/l)
1st segment
2nd segment
3rd segment
PTF profile
Run 1
Run 2
Run 3

(b)

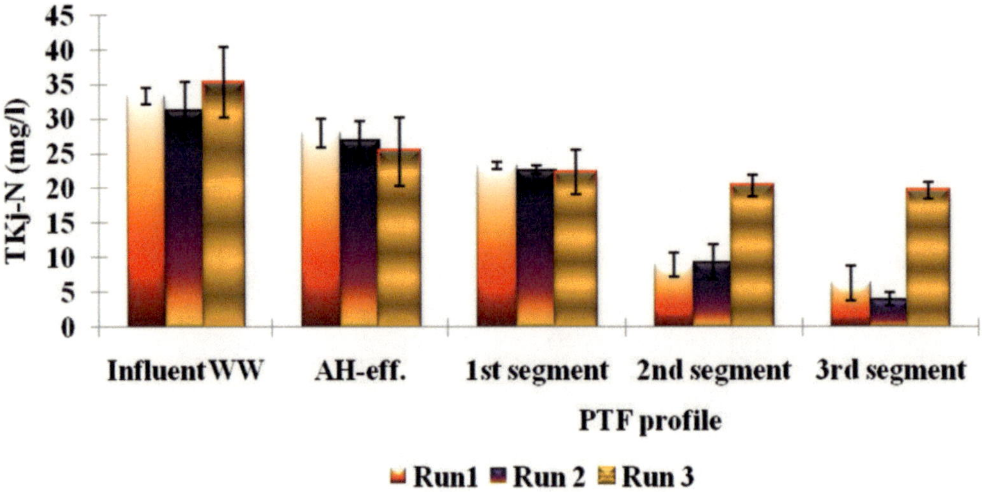

(c)

Figure 11. (a) NH_4-N removal efficiency along the height of PTF reactor treating AH reactor effluent; (b) NOx –N production along the height of PTF reactor treating AH reactor effluent; (c) TKj-N removal efficiency along the height of PTF reactor treating AH reactor effluent

At a total HRT of 14 and 10 h, the combined system (AH+PTF) provided an overall removal efficiencies of 96±2% and 95±2.2% for COD_{total}, 92±3.5% and 91±3% for $COD_{soluble}$ and 98±2.4% and 97.4±2.6 for $COD_{particulate}$ respectively. The overall removal efficiency of COD fractions was dropped at a total HRT of 7.2 h., i.e. 88.7±3.3% for COD_{total}; 76±9.1% for $COD_{soluble}$ and 92±6% for $COD_{particulate}$. The major part of TSS and VSS was removed in the AH reactor, and little additional removal occurred in the PTF system (Table 2).

The total process achieved an overall removal efficiency of 97±1.2%; 96±1.4% and 92.1±2.3% for TSS at total HRTs of 14, 10 and 7.2 h, respectively. The available data indicates that unique contributions of each technology component to the efficiency of the total treatment system i.e. AH reactor was effective for removal of COD fractions (COD_{total}, $COD_{soluble}$ and $COD_{particulate}$), TSS and VSS . By capturing the COD and suspended particles early in the AH process, most of the volatile and oxygen-demanding organic matters were removed in PTF (Table 2).

The removal of COD_{total} and TSS in the AH reactor, improved the nitrification efficiency in PTF as shown in Table 2. This is particularly important in food industry wastewater treatment systems because as shown in Table 2, the effluent after AH system contained significant amounts of TKj-N (28 mg/l), mostly soluble forms of NH_4-N (26 mg/l). The NH_4 –N was efficiently oxidized in the PTF module resulting a removal efficiency of 86±6.5% .

Parameters	COD fractions (mg/l)			Nitrogen species (mg/l)			Solids (mg/l)	
Run 1	Total	Soluble	Particulate	TKj-N	NH_4-N	NO_x-N	TSS	VSS
Wastewater	999.7±337	388±157	612±309	33±6	24.3±5	-	416±95	344±37
AH- effluent	267±48	121.4±40	145±59	28±6	26±5	-	98±23	77±14
PTF- effluent	34.7±8.5	27.4±4.7	7.3±7	4.8±2	2.7±1.4	18±3.6	12.2±3.2	8±1.4
Overall removal efficiency	96±2	92±3.5	98±2.4	85±6	88.6±6.5	-	97±1.2	98±0.5
Run 2								
Wastewater	883±285	344±65	539±267	33±5.7	26.1±5.4	-	437.2±160	266±79
AH- effluent	248±61	118±11	131±65	28.4±6	28±6	-	100.3±18	77±18
PTF- effluent	40±8.5	31±6	10±7	6±1.8	2.8±1.3	17.1±4.5	13±2.8	8.3±2
Overall removal efficiency	95±2.2	91±3	97.4±2.6	83±5.3	88.9±5.2	-	96±1.4	96.5±2
Run 3								
Wastewater	795±168.4	223±59	572±193	32.3±5	23±4.6	-	387.4±48	300±65
AH- effluent	377±89	134.5±23	242.8±89	26±5	24±4.6	-	143.2±17.7	124±12
PTF- effluent	86±16	49.7±10.7	36±23	21±4	17.8±3.7	1.7±0.9	30±5	25±5
Overall removal efficiency	88.7±3.3	76±9.1	92±6	36.3±9	21.2±9.4	-	92.1±2.3	91.3±2.5

Table 2. Overall removal efficiencies of the total process (AH + PTF) treating tomato industry wastewater

3.5. Scanning electronic microscope (SEM) observation

Typical SEM images of porous polyurethane of PTF reactor are shown in Fig. 12. Microorganisms were attached to the porous polyurethane packing material (Fig. 12). The presence of microorganisms in the PTF not only oxidizes ammonia, but also improves the adsorbent and oxidization capability of e organic matter in the wastewater.

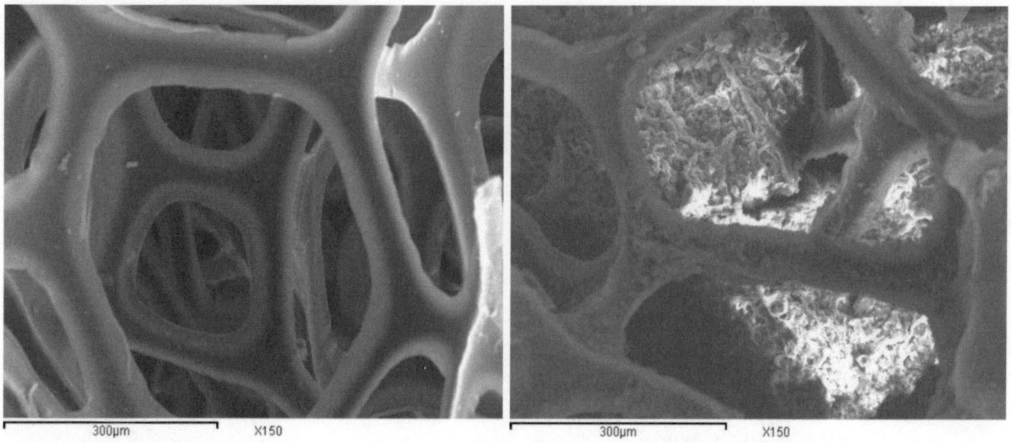

Figure 12. SEM photographs of the microorganisms forming the biofilm in the bioreactor. (a) The clean polyurethane media before attachment of microorganisms; (b) the same surface of the polyurethane after the attachment of microorganisms.

4. Conclusions

- The results obtained revealed that the combined system (AH+ PTF) is very effective for the treatment of tomato industry wastewater at a total HRT not exceeding 10 h. The total process removed 96% of COD_{total}, 92% of $COD_{soluble}$, 98% of $COD_{particulate}$, 85% of TKj-N, 89% of NH_4 –N, 97% of TSS, and 98% of VSS. The effluent quality is complying for reuse and /or discharge according to Egyptian standards for discharge.
- The experimental results obtained here demonstrated that AHreactor and PTF was capable of operating efficiently at short HRT and high values of OLR in the treatment of tomato industry wastewater. Therefore, the volume of the reactor could be reduced five times in comparison with that used in conventional treatment systems without affecting the organic matter removal and nitrification efficiency.

Author details

Ahmed Tawfik
Egypt-Japan University of Science and Technology (E-Just); School of Energy Resources and Environmental Engineering, New Borg El Arab City, Alexandria, Egypt

National Research Center (NRC), Water Pollution Research Dept., Dokki, Cairo, Egypt

Acknowledgement

The author is very grateful for Prof. M. Salem, Prof. R. Abdel Wahab, Dr. A. Al-Asmer, A. Elmitwalli and Prof. Maei for their support and help.

Abbreviations

Trickling filter: TF
Total suspended solids: TSS
Chemical oxygen demand: COD
Up-flow anaerobic sludge blanket: UASB
Anaerobic filter: AF
Anaerobic hybrid: AH
Biodegradable organic matter: BOM
Hydraulic retention time: HRT
Organic loading rate: OLR
Polyurethane trickling filter: PTF
Anaerobic baffled reactor: ABR
Volatile suspended solids: VSS

Dissolved oxygen: DO
Total Kjeldahl nitrogen: (TKj-N),
Thermal conductivity detector: TCD
gCOD removed: gCOD R
Sludge residence time: SRT
Total nitrogen: TN
Mixed liquor suspended solids: MLSS
Mixed liquor volatile suspended solids: MLVSS
Polyvinyl chloride: PVC
Influent wastewater: Influent WW
anaerobic hybrid effluent: AH-eff.
Percentage removal: %R
Volatile fatty acids: VFAs
Influent expression: Inflow wastewater
Effluent expression: Treated wastewater
Anaerobic hybrid effluent: AH-eff
Polyurethane trickling filter effluent: PTF-eff
Scanning electron microscope:SEM
Flame ionization detector: FID
NOx-N: NO3-N + NO2-N

5. References

Amit, K.; Asheesh, K.; Sreekrishnan, T.R.; Santosh, S.; & Kaushik, C.P.; (2007). Treatment of low strength industrial cluster wastewater by anaerobic hybrid reactor Bioresource Technology, 129(1-3):349-57 .

APHA, 2005. Standard methods for the examination of water and wastewater, 20th ed. Washington, DC, USA.

Bernet, N.; & Paul, E.; (2006). In: Cervantes, F.J., Pavlostathis, S.G., van Haandel, A.C. (Eds.), Application of Biological Treatment Systems for Food-Processing Wastewaters. IWA Publishing, London, pp. 237–266 (Chapter 7).

Bozinis,N.A.; Alexiou,I.E.; & Pistikopoulos, E.N.; (1996). A mathematical model for the optimal design and operation of an anaerobic co-digestion plant,Water Sci. Technol. 34, 383–392.

Cakira, F.Y.; & Stenstromb, M.K.; (2005). Greenhouse gas production: a comparison between aerobic and anaerobic wastewater treatment technology. Wat. Res. 39, 4197–4203.

Chang, J.E.; (1989). Treatment of landfill leachate with an up-flow anaerobic reactor combining a sludge bed and a filter. Water Sci. Technol.21, 133–143

Chen, T.H.; & Shyu, W.H.; (1996). Performance of four types of anaerobic reactors in treating very dilute dairy wastewater. Biomass Bioenergy 11 (5), 431–440.

Chen, T.H.;& Shyu, W.H.; (1996). Performance of four types of anaerobic reactors in treating very dilute dairy wastewater. Biomass Bioenergy 11(5), 431–440.

Chernicharo, C.A.I.; & Nascimento, M.C.P., (2001). Feasibility of a pilot –scale UASB/Trickling filter system for domestic sewage treatment. Wat. Sci. Tech., vol. 44, no. 4, pp 221-228.

Cronin, C.; & Lo, K.V.; (1998). Anaerobic treatment of brewery wastewater using UASB reactors seeded with activated sludge. Bioresour. Technol. 64, 33–38.

Del Pozo, R.; Tas, D.O.; Orhon, D.; &Diez, V.; (2003). Biodegradability of slaughterhouse wastewater with high blood content under anaerobic and aerobic conditions. J. Chem. Technol. Biotechnol. 78 (4), 384–391.

Demirer, G.N.; & Chen, S.; (2005). Anaerobic digestion of dairy manure in a hybrid reactor with biogas recirculation. World J. Microbiol. Biotechnol. 21 (8-9), 1509–1514.

El-Gohary, F.; Tawfik, A.; Badawy, M.; El-Khateeb, M.; (2009). Potentials of anaerobic treatment for catalytically oxidized olive mill wastewater (OMW) Bioresource Technology 100 (2009) 2147–2154

El-Kamah, H.; Mahmoud, M.; & Tawfik, A.; (2011). Performance of down-flow hanging sponge (DHS) reactor coupled with up-flow anaerobic sludge blanket (UASB) reactor for treatment of onion dehydration wastewater.Bioresource Technology 102 (2011) 7029–7035

El-Kamah, H.; Tawfik, A.; Mahmoud, M.;& Abdel-Halim, H.; (2010). Treatment of high strength wastewater from fruit juice industry using integrated anaerobic/ aerobic system. Desalination 253, 158–163.

Elmitvalli, T.A.; van Dun, M.; Bruning, H.; Zeeman, G.; & Lettinga, G., (2000). The role of filter media in removing suspended and colloidal particles in an anaerobic reactor treating domestic sewage. Bioresour. Technol. 72, 235–242.

Elmitwalli, T.; Sklyar, V.; Zeeman, G.; & Lettinga, G.; (1999). Low temperature pre-treatment of domestic sewage in an anaerobic hybrid and an anaerobic filter reactor. In: Proceedings of the 4th IAWQ Conference in Biofilm System, New York, October 16–20

Gu¨ngo¨r, G.; & Demirer, G.; (2004). Effect of initial COD concentration, nutrient addition, temperature and microbial acclimation on anaerobic treatability of broiler and cattle manure. Bioresour. Technol. 93 (2004), 109–117.

Hawkes, F. R.; Donnellyh, T.; & Anderson, GK.; (1995). Comparative performance of anaerobic digesters operating on ice-cream wastewater. Water Res. 29,522–533

Holman, J.B.; & Wareham, D.G.; (2005) COD, ammonia and dissolved oxygen time profiles in the simultaneous nitrification/denitrification process. Biochem Eng J 22(2):125–133

M. Mahmoud, A. Tawfik, F. Samhan, F. El-Gohary Sewage treatment using an integrated system consisting of anaerobic hybrid reactor (AHR) and downflow hanging sponge (DHS) Desalination and Water Treatment 4 (2009) 168–176

Mehrdad, F.; Mehdi B.; & Valentina V. U.; (2007). Treatment of beet sugar wastewater by UAFB bioprocess Bioresource Technology 98 (2007) 3080–3083

Miron, Y.; Zeeman, G.; Lier, J.B.; & Lettinga, G.; (2000). The role of sludge retention time in the hydrolysis and acidification of lipids, carbohydrate and protein during digestion of primary sludge in CSTR systems. Wat. Res. 34(5), 1705-1713

Oscar, U.; Svetlana, N.; Enrique, S.; Rafael, B.; & Francisco, R.; (2008). Treatment of screened dairy manure by upflow anaerobic fixed bed reactors packed with waste tyre rubber and a combination of waste tyre rubber and zeolite: Effect of the hydraulic retention time Bioresource Technology, 99 (15), pp.7412-7417

Parawira, W.; Kudita, I.; Nyandoroh, M.G.; & Zvauya, R.; (2005). A study of industrial anaerobic treatment of opaque beer brewery wastewater in a tropical climate using a full-scale UASB reactor seeded with activated sludge. Process Biochem. 40, 593–599.

Satyanarayan,S.; Ramakan, T.;& Vanerkar, A.P., (2005). Conventional approach abattoir wastewater treatment, Environ. Technol. 26, 441–447.

Shin, J.; Lee, S.; Jung, J.; Chung, Y.; & Noh, S.; (2005). Enhanced COD and nitrogen removals for the treatment of swine wastewater by combining submerged membrane bioreactor (MBR) and anaerobic upXow bed Wlter (AUBF) reactor. Process Biochemistry 40, 3769–3776.

Tawfik, A.; El-Gohary, F.; & Temmink, H., (2010). Treatment of domestic wastewater in an up-flow anaerobic sludge blanket reactor followed by moving bed biofilm reactor Bioprocess Biosyst Eng., 33:267–276

Tawfik, A.; & El-Kamah, H.; (2011). Treatment of fruit juice industry wastewater in a two stage anaerobic hybrid (AH) reactors followed by sequencing batch reactor (SBR). Environmental Technology. DOI: 10.1080/09593330.2011.579178

Tawfik, A.; & Klapwijk, A.; (2010). Polyurethane rotating disc system for post-treatment of anaerobically pre-treated sewage Journal of Environmental Management 91, 1183–1192

Tawfik, A.; Badr, N.; *Abou Taleb, E.; El-Senousy, W.; (2011).* Sewage treatment in an up-flow anaerobic sponge reactor followed by moving bed biofilm reactor based on polyurethane carrier material. In press desalination and water treatment journal.

Tawfik, A.; El-Gohary, F.; Ohashi, A.; & Harada, H.; (2008). Optimization of the performance of an integrated anaerobic–aerobic system for domestic wastewater treatment Wat. Sci. & Tech., 58 (1), pp 320-328

Tawfik, A.; Ohashi, A.; & Harada, H., (2006). Sewage treatment in a combined up-flow anaerobic sludge blanket (UASB)-down-flow hanging sponge (DHS) system. Biochemical Engineering Journal 29, 210-219

Tawfik, A.; Ohashi, A.; Harada, H.; (2010). Effect of sponge volume on the performance of down-flow hanging sponge system treating UASB reactor effluent. Bioprocess Biosyst Eng. 33:779–785

Tawfik, A.; Sobhey, M.; Badawy, M.; (2008). Treatment of a combined dairy and domestic wastewater in an up-flow anaerobic sludge blanket (UASB) reactor followed by activated sludge (AS system). Desalination 227; 167–177

Vartak, D.R.; Engler, C.R.; McFarland, M.J.;& Ricke, S.C.; (1997). Attached-film media performance in psychrophilic anaerobic treatment of dairy cattle wastewater. Bioresour. Technol. 62, 79–84.

Wu, M.; Wilson, F.; & Tay, J.H.; (2000). Influence of media-packing ratio on performance of anaerobic hybrid reactors. Bioresource Technol., 71, 151–157

Chapter 17

Polyurethane as an Isolation for Covered Conductors

Žiga Voršič

Additional information is available at the end of the chapter

http://dx.doi.org/10.5772/35083

1. Introduction

The designers of electro energetic system are thinking, to conserve the routes of 220kV transmission lines, and the transition to 400kV lines. The easiest way seems to be the placement of new overhead power lines. The other option is the use of covered conductors. Covered conductors are conductors with insulation made from two dielectrics: the first is the insulation mantel and the other is air. The covered conductors consist of a conductor which is a metal electrode (a cable), and the covering mantel which is made from a dielectric with a greater dielectric constant and higher breakdown voltage. The other dielectric with a lower dielectric constant is the surrounding air. The conductor should not be touched despite the insulation.

According to the usual labelling of conductors, we named the suggested conductor PUAC 2150/490/65 mm^2. Here 2150 mm^2 stands for the cross section of the polyurethane mantel, 490 mm^2 for the aluminium and 65 mm^2 for the core made from carbon fibbers. The saved weight (instead of steel we use carbon fibbers) can be used for the insulation. The insulation has to be thick enough so that the electric field intensity on the edge of the insulator does not exceed the critical electric field intensity of air. Such a conductor, hanged on a typical 220 kV transmission tower meets the electric load exerted. Simplified analytical calculations of the electric field intensity on the edge of the insulation match exact calculation, using the finite element method. This fact encourages new research, both theoretical and practical.

2. The current usage of covered conductors in the middle to high voltage networks

Covered conductors for overhead power lines are meant to replace the existing bare cable power lines, especially in wooden areas where the risk of falling trees is high. Another concern is the weighing down of cables from sticking snow and ice. Reasons for using covered conductors are better safety, ecology (fewer disturbances in the nature, especially less clearing of trees), better operational reliability and lower operating costs.

2.1. Covered conductors

The most widespread structure of covered conductors consists of the core made from hard, compact aluminium alloy and a watertight/waterproof mantel made from a cross-linked polyethylene (XLPE). The use has shown the reliability of this type of conductor in very difficult conditions. It will withstand the weight of a fallen tree for days, mechanically and electrically.

Because of the outer mantel, the covered conductors are not that vulnerable when touching each other, or in contact with tree branches. This enables the spacing between phases/cables to shrink to one third of the space between normal over ground power lines. The platform for an over ground power line in a wooden area can therefore be smaller/narrower.

2.2. The use of covered conductors

Slovenia began to introduce covered conductors in 1992. In the same year Elektro Gorenjska performed reconstruction of the transmission line Savica – Komna, where they replaced the bare conductors with covered ones. All other parts of the power lines stayed the same. In the year 1939, Elektro Ljubljana build the first power line based on finish technology. They decided for the finish 20kV system because of positive experience from a more than 10000 km of build power lines with covered conductors. These experiences show us that the use of such system reliability has increased by 5 times. The number of failures per year on a 100 km long section is 4.5 for bare conductors and 0.9 for covered conductors in the SN network (Tičar I., Zorič T., 2003). Because the two mentioned power lines successfully passed the adverse weather conditions (sleet, an additional burden of winter), the importer and representative (C&G d.o.o. Ljubljana) for the company (Ensto), obtain an expert opinion on the imported equipment. It was found that the covered conductors, the corresponding hanging and insulation materials and the overvoltage protection used in the 20kV distribution network conform to the current JUS standard. The material and equipment that is not included in the JUS standard was covered and conformed in the international IEC standards.

2.3. High voltage

In the Scandinavian countries, the development of overhead lines that use covered conductors began in the mid-eighties. The first country to introduce covered conductors was Finland. The finish distribution company FingridOjv with the cooperation of Eltel Networks began with the use of covered conductor in high voltage networks. They build a 110 kV DV power line Mätäkivi-Sula with a length of 6 km, half of which was equipped with covered conductors. In the construction of the transmission lines, composite insulators were used.

In General Public Utilities/ Pennsylvania Electric Company they researched the electric and mechanic effects of covered conductors insulated with polyethylene. Those cables had high specific weight and were installed on several insulators and clamps. They conducted two experiments with the voltage of 70 kV. The first was to determine the feasibility of the existent spacer (a device for spacing bundled conductors). The second was done on a distribution cable with higher voltage, to determine the criteria for building new compact high-voltage transmission lines in small corridors, especially in densely populated areas.

Since 1980, the use of covered conductors increased worldwide. The reason for this is that the covered conductors are more compact and environmentally friendlier than traditional non-insulated conductors. Also the number of failures is much lower. This development has also an impact on the characteristics of voltage drops and is an important aspect of working with clients, sensitive to such decrease in voltage.

3. Using covered conductors at the highest voltage levels

The space around an electrically charged body is in a special state. This special state acts only on particles that have an electric charge. If we introduce a small electric charge that does not significantly alter the state of that space, we find that there is force acting on that small charge. This force is proportional to the electric charge q and the vector quantity that defines the state of the space. It is denoted as E. The vector E of the electric field intensity has the same direction as the force.

3.1. The electric field: Cylinder of charge

For the understanding of the distribution of voltage and electric field intensity in a covered conductor, we first look at the electric field in a cylinder of charge (Voršič J., Pihler J., 2005). The electric field intensity E(r) of an isolated cylindrical Gaussian surface is shown in figure 1.

In an infinitely long charged cylinder two things are true:

a. the electric charge is evenly distributed over the surface of the cylinder,
b. the electrostatic field is a flat radial field.

From the equation,

$$Q = q \cdot l = D \cdot 2 \cdot \pi \cdot r \cdot l, \tag{1}$$

where:
q –is the electric charge, gathered on the length of the cylinder,
r – is the radius of the equipotential surface,
l – is the length of the cylinder,
D–is the electric displacement field and
Q– is the electric charge;
we obtain the absolute value for the electric displacement field,

$$D = \frac{Q}{A} = \frac{q \cdot l}{2 \cdot \pi \cdot r \cdot l} = \frac{q}{2 \cdot \pi \cdot r} \tag{2}$$

And it's vector quantity

$$\vec{D} = D \cdot \vec{1}_r = \frac{q}{2 \cdot \pi \cdot r} \cdot \vec{1}_r. \tag{3}$$

The corresponding electric field intensity is

$$\vec{E} = \frac{\vec{D}}{\varepsilon} = \frac{q}{2 \cdot \pi \cdot \varepsilon \cdot r} \cdot \vec{1}_r \tag{4}$$

The Electric potential $V(r)$ on an equipotential surface (concentric cylinder) with a radius r is

$$V(r) = \int_r^{r_0} E \cdot \mathrm{d}r = \frac{q}{2 \cdot \pi \cdot \varepsilon} \cdot \int_r^{r_0} \frac{\mathrm{d}r}{r} = \frac{q}{2 \cdot \pi \cdot \varepsilon} \cdot \ln \frac{r_0}{r}, \tag{5}$$

where r_0 is the radius of the equipotential surface, on which we chose the potential s starting point. Because we assume that the cylinder is infinitely long, we cannot select the starting point of the potential to be in infinity.

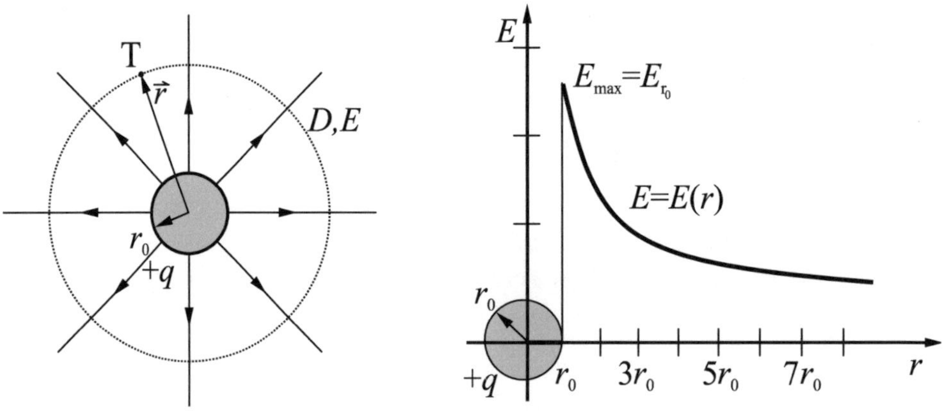

Figure 1. Electric field intensity E(r) of an isolated cylindrical Gaussian surface

Between two equipotential surfaces with the radius r_1 and r_2 there is a potential difference

$$U_{12} = U = V_1 - V_2 = \frac{q}{2 \cdot \pi \cdot \varepsilon} \cdot \ln \frac{r_2}{r_1} \tag{6}$$

Two concentric cylinders with the length l, radius r_1 and r_2 and a dielectric between them they form a cylindrical capacitor with the capacitance

$$C = \frac{Q}{U} = \frac{q \cdot l}{U} = \frac{2 \cdot \pi \cdot \varepsilon \cdot l}{\ln \frac{r_2}{r_1}} \tag{7}$$

Because we don't know the exact electric charge, but only the charge between the electrodes, we denote the expression for the electric field intensity at any random point between the two cylindrical electrodes (figure 2) in the form of

$$E(r) = \frac{U}{r \cdot \ln \frac{r_2}{r_1}} \tag{8}$$

We get the highest value of the electric field intensity on the surface of the inner cylinder.

$$E(r_1) = \frac{U}{r_1 \cdot \ln \frac{r_2}{r_1}} = E_{max} \tag{9}$$

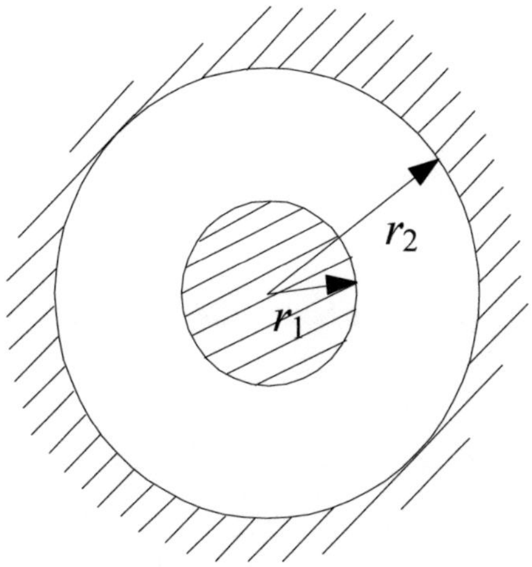

Figure 2. The electric field of two concentric cylinders

3.2. Double-layer single-wire cable

The double-layer single-wire cable (figure 3) is a typical example of the use of double-layer dielectrics. The voltage between the core of the cable and the mantel is distributed over the layers of the dielectric.

$$U_1 = \frac{q}{2 \cdot \pi \cdot \varepsilon_0 \cdot \varepsilon_{r1}} \cdot \ln \frac{r_2}{r_1} \tag{10}$$

$$U_2 = \frac{q}{2 \cdot \pi \cdot \varepsilon_0 \cdot \varepsilon_{r2}} \cdot \ln \frac{r_3}{r_2} \tag{11}$$

$$U = U_1 + U_2 \tag{12}$$

$$U = \frac{q}{2 \cdot \pi \cdot \varepsilon_0} \left(\frac{1}{\varepsilon_{r1}} \cdot \ln \frac{r_2}{r_1} + \frac{1}{\varepsilon_{r2}} \cdot \ln \frac{r_3}{r_2} \right) \tag{13}$$

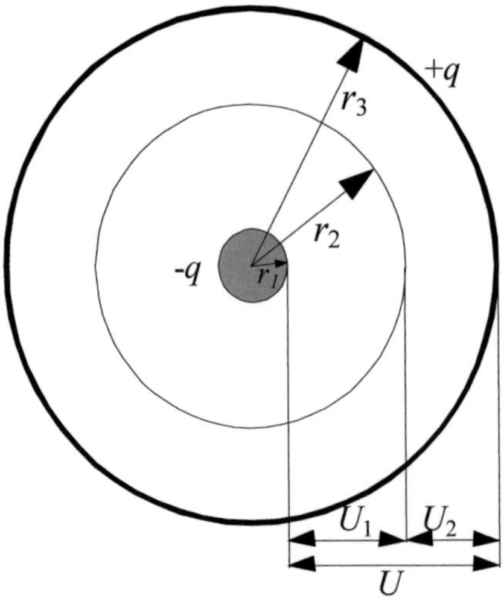

Figure 3. A double-layer single-wire cable

We calculate the electric charge and then both voltage levels.

$$U_1 = \frac{U \cdot \ln \frac{r_2}{r_1}}{\varepsilon_{r1} \cdot \left(\frac{1}{\varepsilon_{r1}} \cdot \ln \frac{r_2}{r_1} + \frac{1}{\varepsilon_{r2}} \cdot \ln \frac{r_3}{r_2} \right)} \quad (14)$$

$$U_2 = \frac{U \cdot \ln \frac{r_3}{r_2}}{\varepsilon_{r2} \cdot \left(\frac{1}{\varepsilon_{r1}} \cdot \ln \frac{r_2}{r_1} + \frac{1}{\varepsilon_{r2}} \cdot \ln \frac{r_3}{r_2} \right)} \quad (15)$$

We get the highest electric field intensity in material 1 at the radius r_1,

$$E_{1\max} = \frac{q}{2 \cdot \pi \cdot \varepsilon_0 \cdot \varepsilon_{r1}} \cdot \frac{1}{r_1} = \frac{U \cdot \frac{1}{r_1}}{\varepsilon_{r1} \cdot \left(\frac{1}{\varepsilon_{r1}} \cdot \ln \frac{r_2}{r_1} + \frac{1}{\varepsilon_{r2}} \cdot \ln \frac{r_3}{r_2} \right)}. \quad (16)$$

and the highest electric field intensity in material 2 at the radius r_2

$$E_{2\max} = \frac{q}{2 \cdot \pi \cdot \varepsilon_0 \cdot \varepsilon_{r2}} \cdot \frac{1}{r_2} = \frac{U \cdot \frac{1}{r_2}}{\varepsilon_{r2} \cdot \left(\frac{1}{\varepsilon_{r1}} \cdot \ln \frac{r_2}{r_1} + \frac{1}{\varepsilon_{r2}} \cdot \ln \frac{r_3}{r_2} \right)}. \quad (17)$$

3.3. Polarized conductors and the electric field intensity

A covered conductor can be regarded as a two layered insulated conductor. In this case the inner electrode is insulated and the space to the outer electrode is air. The insulation has a much higher dielectric strength than air. Because of that, it is irrelevant that the electric field intensity is small on the inner electrode (figure 4). The important thing is that the electric field intensity in the air is as small as possible: $E_{air} = E_2(r_2)$.

Symbols in figure represent:

r_1–the radius of the core,
r_2–the radius of the mantel,
r_3–the radius of insulation,
E_2–the electric field intensity of air.

Taking into account that $\varepsilon_{r1} = \varepsilon_r$ and $\varepsilon_{r2} = 1$ in the equation (20), we get:

$$E_2 = \frac{U}{r_2 \cdot \left(\frac{\ln \frac{r_2}{r_1}}{\varepsilon_r} + \frac{\ln \frac{r_3}{r_2}}{1} \right)} \tag{18}$$

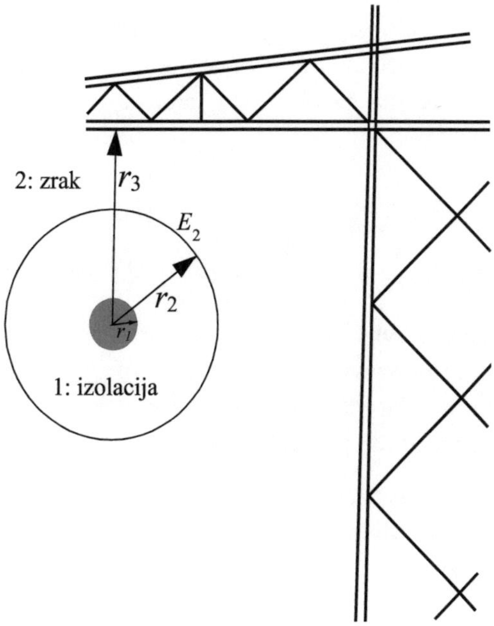

Figure 4. Coaxial cylindrical arrangement of partial isolation

We get the minimum field intensity E_2 as a function of r_2, when the denominator in equation (18) is at its highest value (Beyer M., 1986). When we denote the denominator with y and the variable r_2 as x we derive the function $y(x)$ from x.

Denominator:

$$y(x)=\left(\frac{\ln\frac{x}{r_1}}{\varepsilon_r}+\frac{\ln\frac{r_3}{x}}{1}\right)=x\cdot\left(\frac{1}{\varepsilon_r}\cdot\ln\frac{x}{r_1}-\ln\frac{x}{r_3}\right)=\frac{x}{\varepsilon_r}\cdot\left(\ln\frac{x}{r_1}-\varepsilon_r\cdot\ln\frac{x}{r_3}\right) \tag{19}$$

We find the maximum value of the denominator and get the optimal value of the radius,

$$r_{2\,opt}=r_1\cdot\frac{1}{e}\cdot\left(\frac{r_3}{r_1}\right)^{\frac{\varepsilon_r}{\varepsilon_r-1}}, \tag{20}$$

where the electric field intensity in the air is at its smallest value.

$$E_2=E_{2\,min}=\frac{U\cdot e}{r_1\cdot\left(\frac{r_3}{r_1}\right)^{\frac{\varepsilon_r}{\varepsilon_r-1}}\cdot\left(1-\frac{1}{\varepsilon_r}\right)} \tag{21}$$

With the transmission tower, the chain of insulators (the vertical string of discs, l=2.25 m) and the cable (490/65 Al/Fe, r = 15.3 mm) of the existing 220kV over ground power line (Figure 5) the geometry is already set. The only remaining variable is the relative permittivity. If we attempt to calculate the relative permittivity of polyurethane ($\varepsilon_r = 3.4$), we get the optimal radius where the electric field intensity in the surrounding air is at its smallest value:

$$r_{2\,opt}=r_1\cdot\frac{1}{e}\cdot\left(\frac{r_3}{r_1}\right)^{\frac{\varepsilon_r}{\varepsilon_r-1}}=0{,}0153\cdot\frac{1}{2{,}71828}\cdot\left(\frac{2{,}25}{0{,}0153}\right)^{\frac{3{,}4}{3{,}4-1}}=6{,}62\text{ m}$$

But this is unrealistic. This is why we try to use other materials with higher values of relative permittivity. Figure 6 shows the dependence of the optimal radius (equation 20) from the relative permittivity. Despite the clear advantages of materials with larger values of relative permittivity we try to use polyurethane.

If we use the same radius as the radius in the current 220kV overhead conductor (r = 15.3mm), the relative permittivity $\varepsilon_r = 3.4$ and an insulation that is 15 mm thick, together with the voltage of 400kV, we get:

$$E_{air}=\frac{U}{r_2\cdot\varepsilon_{r2}\cdot\left(\frac{1}{\varepsilon_{r1}}\cdot\ln\frac{r_2}{r_1}+\frac{1}{\varepsilon_{r2}}\cdot\ln\frac{r_3}{r_2}\right)}=2{,}39\text{ MV/m} \tag{22}$$

The value is smaller from the Dielectric Strength of air in normal conditions (3 MV/m).

Figure 5. The sketch of a typical 220 kV transmission tower for an overhead power line

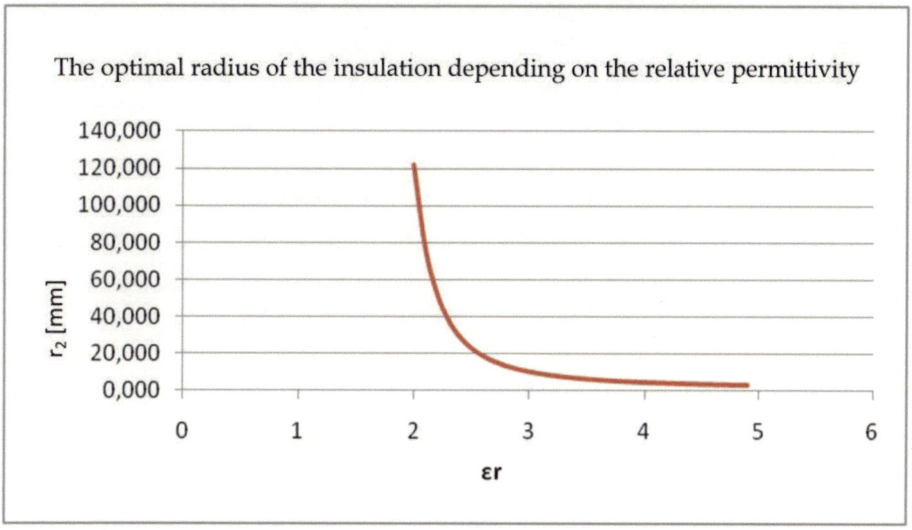

Figure 6. The optimal radius of the insulation depending on the relative permittivity

For other insulation thicknesses, the electric field intensity (according to equation 21) is shown in figure 7. Figure 8 shows the highest electric field intensities on the edge of the insulation for different values of relative permittivity. The difference becomes apparent only at greater thicknesses of the mantel. You can also see that better dielectric displace more electric field in the worse dielectric - the air.

Figure 9 shows the electric field intensity at the edge of the conductor as the function of the insulation thickness. As expected it is substantially lower than the dielectric strength, the reason being that the majority of the electric field is displaced into the worse dielectric – the surrounding which is also more abundant (the length of the string of insulating discs).

According to survey results, we find that the use of covered conductors with insulation made from polyurethane enables the preservation of the existing 220 kV power line platforms and the transition to 400kV power lines. Furthermore we decided to reduce the weight by using the thinnest possible insulation that still meets all requirements – thickness of 15 mm.

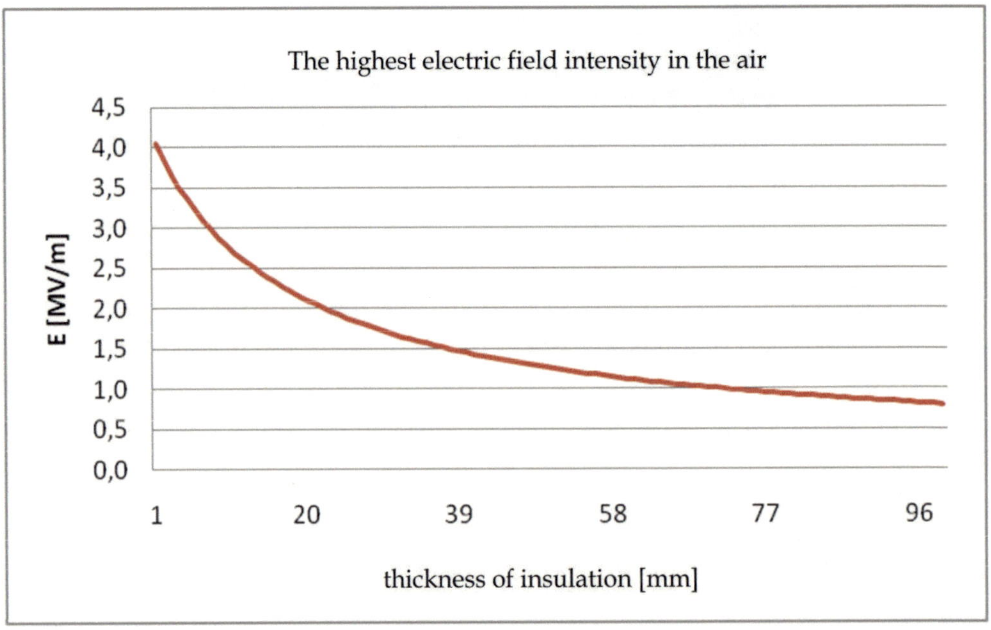

Figure 7. The highest electric field intensity in air

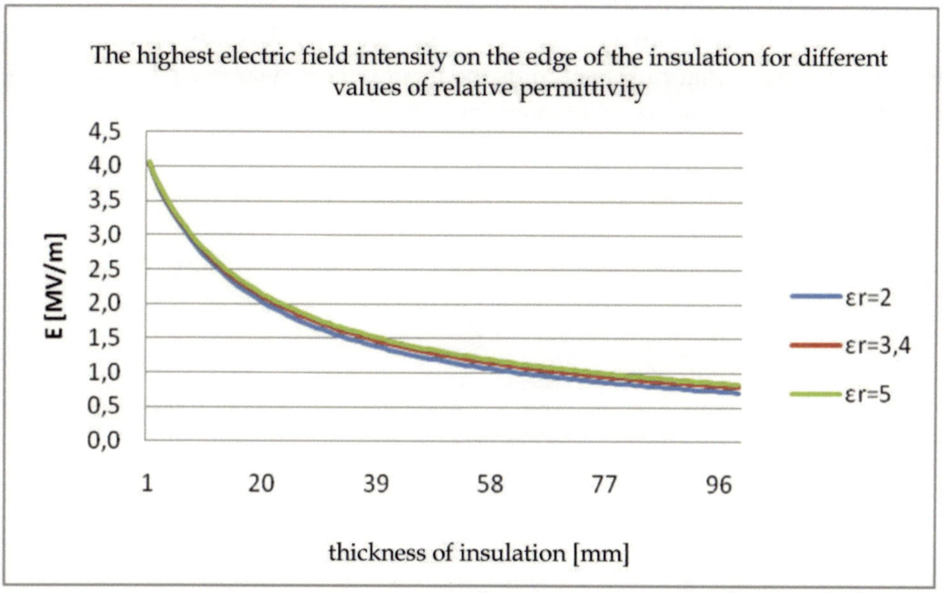

Figure 8. The highest field intensity on the edge of the insulation for different values of relative permittivity

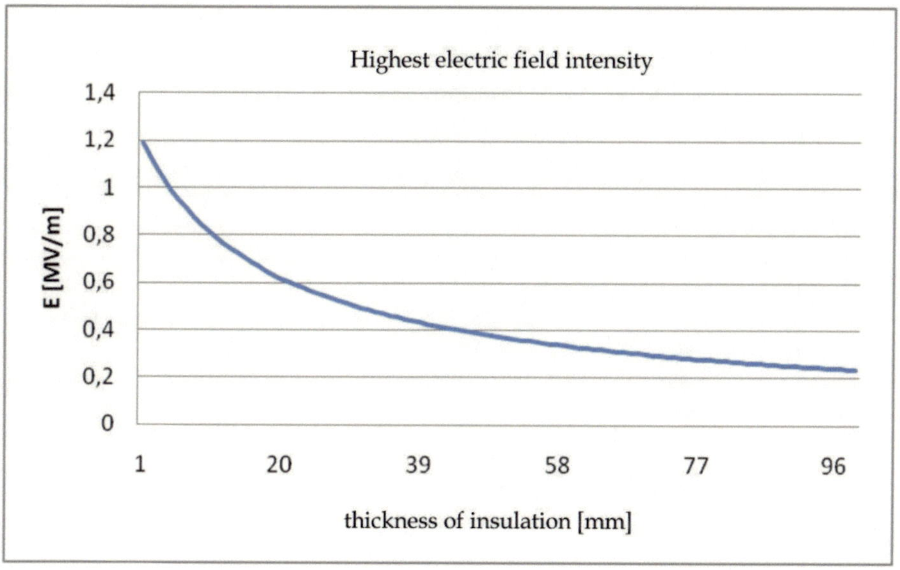

Figure 9. The highest electric field intensity on the edge of the conductor as the function of insulation thickness

4. The mechanics of the PUAC 2150/490/65 power line

Mechanical calculations showed that the proposed cable (figure 10) meets the requirements. With the calculation of the formed catenary we get the height at which the conductor needs to be over the ground (figure 11) and thus the input data to determine the environmental impact of power lines by means of non-ionizing radiation.

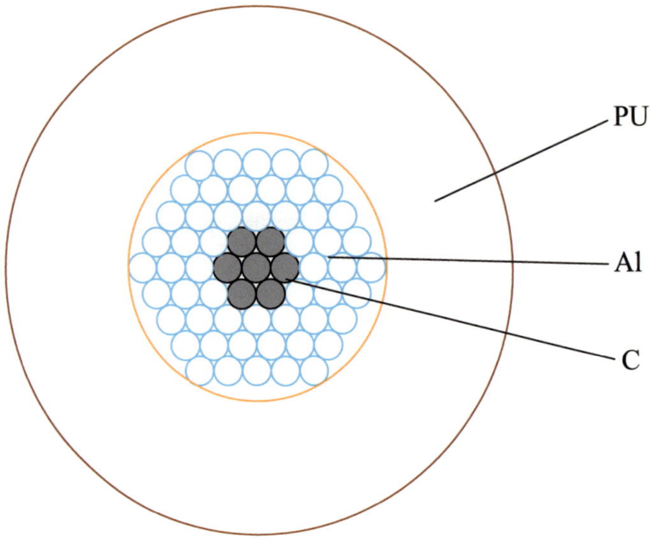

Figure 10. Cross section of the proposed covered conductor

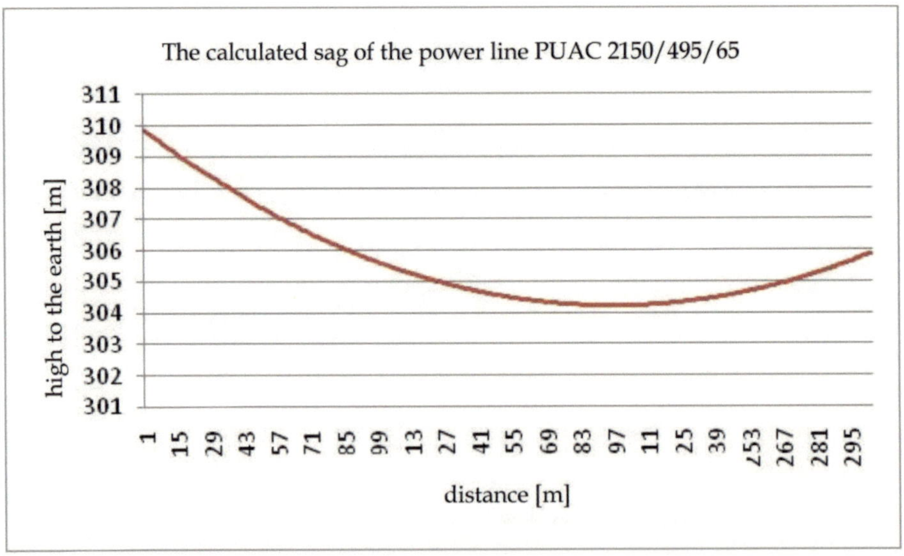

Figure 11. The calculated sag of the power line PUAC 2150/490/65

5. The electric field in the surrounding of a 400 kV PUAC 2150/490/65 power line

We deal with electric field in the surroundings of electrically charged bodies, for example in the vicinity of electric energy transmission lines, transmission antennas of telecommunications equipment. Electric fields are everywhere, where electric charge is present. Every electric conductor under voltage creates an electric field around itself. The field exists even when no current is flowing through the conductor, so even when the power line is not laden with users. The higher the voltage is the greater the electric field. Electric fields have the highest intensity close to the source and decrease very rapidly with distance. Metal shields them very well, but other material weaken it as well. The intensity of electric fields of power lines is greatly reduced by walls, buildings and trees. The electric fields of underground cables are also reduced by the soil.

We have determined that with the use of insulation made from polyurethane we can reduce the electric field intensity on the surface of covered conductors and that it is possible to operate at 400 kV. Due to the increased voltage it is necessary to examine the impact of such above ground power lines on the environment in accordance with our regulations.

5.1. Electromagnetic radiation

Exposure to electromagnetic radiations is not something new. It accompanies us from the very beginning of human existence. Here we think of natural sources of radiation. Another story is artificially generated sources, which are much stronger in intensity and more recently also increased in number. Man is now, unlike in the past, at home and at work, exposed to a complex mix of electric and magnetic fields.

The main sources of electric and magnetic fields of low frequencies 50 Hz are artificial sources of electromagnetic radiations, namely those caused by man, which are devices for transmitting and distributing electricity, electrical substations, and all devices that use electricity for their operation. The intensity of electromagnetic radiations emitted by artificial sources, in comparison with natural resources (the Earth's static magnetic field, electric field caused by the discharge in the atmosphere - lightning) is much higher. When an electrical device is plugged in, an electric field is generated in its surroundings. The higher the voltage, the stronger the electric field at a certain distance from the device. The electric field is present even when the device is not working because there is no need for the electric current to flow to create voltage. The magnetic field on the other hand requires a flow of electrons, so it occurs only when the device is plugged in and the current is flowing. Under these conditions both fields exist in the room. The greater the power consumption and thus the electric current, the stronger the magnetic field is.

5.2. Evaluation of electromagnetic fields in the Slovenian legislation

The Slovenian government adopted a regulation on electromagnetic radiations in the natural and living environment, which specifies the maximum allowed threshold of radiations. The regulation protects the most sensitive areas (EMR protection zone I, which includes living environment, schools, kindergartens, hospitals) with an additional preventive factor.

These areas demand increased protection against radiation therefore they are subject to ten times more severe limitations than in the European Union. For EMR protection zone II (areas with no residential building), the restrictions for magnetic fields are the same as in the European Union but for the intensity of the electric field two times higher values are allowed.

The maximum levels of radiation for networks with a frequency of 50 Hz are 500 V/m and 10 μT for EMR protection zones I and 10000 V/m and 100 μT for EMR protection zones II. Radiation with frequencies of 50 Hz includes electromagnetic fields from distribution substations, over ground and underground power lines, high voltage transformers and others. This is described under the Slovenian Regulation (2nd paragraph of article 2). At this frequency we distinguish two fields:

- The electric field, which is described with the effective value of the electric field intensity E [V/m] and depends on the voltage of the radiation source or the element
- The magnetic field, which is described with the Magnetic flux density B [T] which depends on the electric current passing through the source of radiation or the element.

When calculating the effects of electromagnetic radiation we have to consider the most unfavorable impact on nature that can occur in normal operations.

5.3. Electric field intensity in the vicinity of an overhead power line

For the straight infinitely long conductor we assume that the electric charge is evenly distributed over the whole surface (uniform linear charge density). The charge on such a conductor can be described with an infinite line charge, which in any given point of T (x, y) (Fig. 12) leads to the following vector of electric field intensity:

$$\vec{E} = \vec{1}_r \cdot \frac{q_+}{2 \cdot \pi \cdot \varepsilon_0} \cdot \frac{1}{\left|\vec{r}_+\right|} \tag{23}$$

Where:

$\vec{1}_r$ is the unit vector of distance,

q_+ is the positive value of the line charge,

$\left|\vec{r}_+\right|$ is the absolute value of the distance vector between the electric charge and the point of observation and

ε_0 is the vacuum permittivity.

To calculate the electric field intensity of a conductor above a conductive surface we use the method of equivalent charges. Its main idea is the exchange of the surface charge near a conductive surface (in our case soil) with a charge opposite in sign but equal in quantity that is projected over a conductive surface.

$$[q_+] = [C] \cdot [V],$$

where:

$[q_+]$	is the columnar vector of positive charge,
$\left[V\right]$	is the columnar vector of conductor potentials,
$\left[C\right]$	is the square matrix of the capacitance.

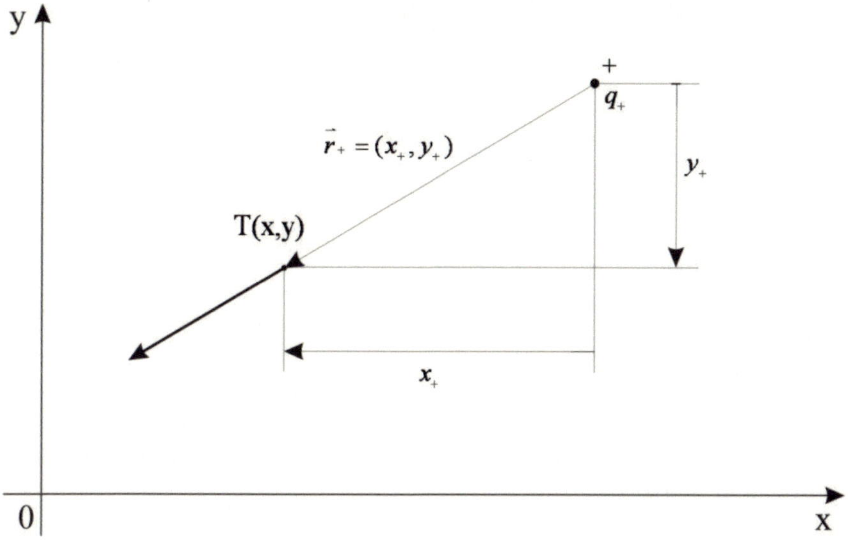

Figure 12. Electric field of a line charge

5.3.1. Voltage on conductors

To determine the electric field intensity we have to determine the charge on the phase conductors. These are obtained from the current values of the voltage taking into account the capacitance. In Figure 13 current value of tension on the 400 kV overhead line are shown.

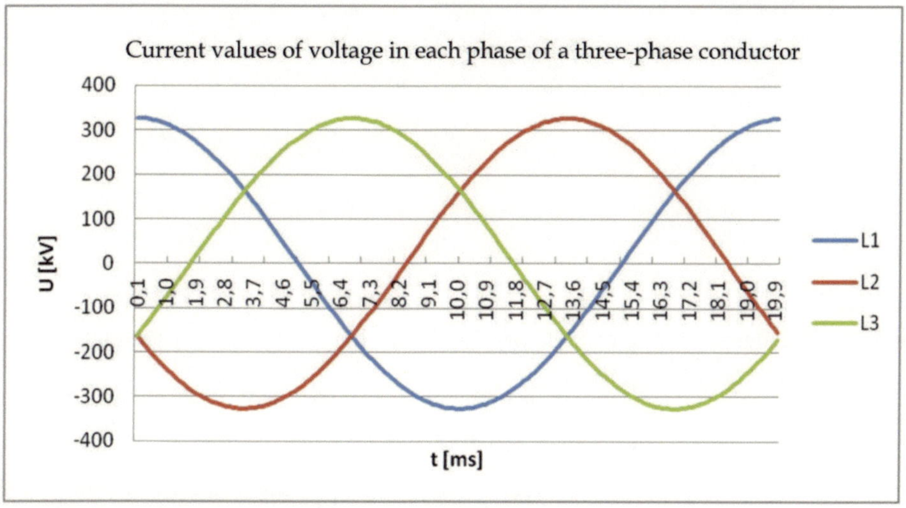

Figure 13. Current values of voltage in each phase of a three-phase conductor

5.3.2. Charge on conductors

We get the matrix of capacitance $[C]$ by first determining the elements of the potential coefficients of the conductor $[p]$ (Tičar I., Zorič T., 2003):

$$p_{ii} = 18 \cdot 10^9 \cdot \ln \frac{H_{ii}}{r} \quad \left[\frac{m}{F}\right] \tag{24}$$

$$p_{ij} = 18 \cdot 10^9 \cdot \ln \frac{H_{ij}}{d_{ij}} \quad \left[\frac{m}{F}\right], \tag{25}$$

where:

p_{ii} is the individual potential coefficient,

p_{ij} is the mutual potential coefficient,

ε_0 is the vacuum permittivity,

d_{ij} is the distance between multi-phase conductors,

H_{ii} is the distance between the multi-phase conductors ant there mirror projections,

H_{ij} is the distance between the multi-phase conductors and the mirror projections of other multi-phase conductors.

Distances established like that apply to bare conductors (line charge) in the air with a constant relative permittivity ε_0. In our case, where we are dealing with insulation around the conductors, the electric charge gathers on the edge of the insulation and we have to consider that the distance between conductors is reduced by the thickness of the insulation. So when we consider these reductions in distance and the designations on figure 14 we get the new:

- »Individual« potential coefficient:

$$p_{ii} = \frac{\frac{1}{\varepsilon_{r2}} \cdot \ln \frac{r_2}{r_1} + \ln \frac{H_{ii}}{r_2}}{2 \cdot \pi \cdot \varepsilon_0} \tag{26}$$

- »Mutual« potential coefficient

$$p_{ij} = \frac{\frac{1}{\varepsilon_{r2}} \cdot \ln \frac{r_2}{r_1} + \ln \frac{H_{ij}}{d}}{2 \cdot \pi \cdot \varepsilon_0} \tag{27}$$

We deal with three potential of conductors, four line charges and nine potential coefficients, which we can combine into a matrix.

$$\begin{bmatrix} V_{L1} \\ V_{L2} \\ V_{L3} \end{bmatrix} = \begin{bmatrix} p_{L1L1} & p_{L1L2} & p_{L1L3} \\ p_{L2L1} & p_{L2L2} & p_{L2L3} \\ p_{L3L1} & p_{L3L2} & p_{L3L3} \end{bmatrix} \cdot \begin{bmatrix} q_{L1} \\ q_{L2} \\ q_{L3} \end{bmatrix} \tag{28}$$

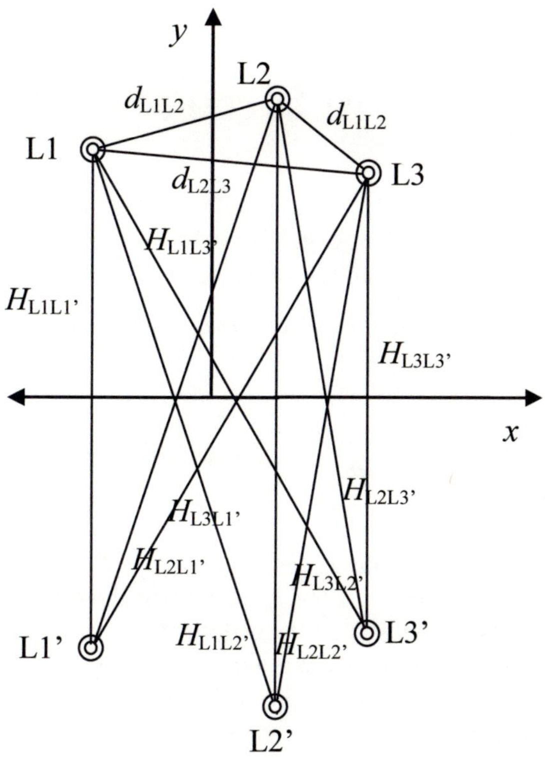

Figure 14. Mirror projections over the conductive surface plane

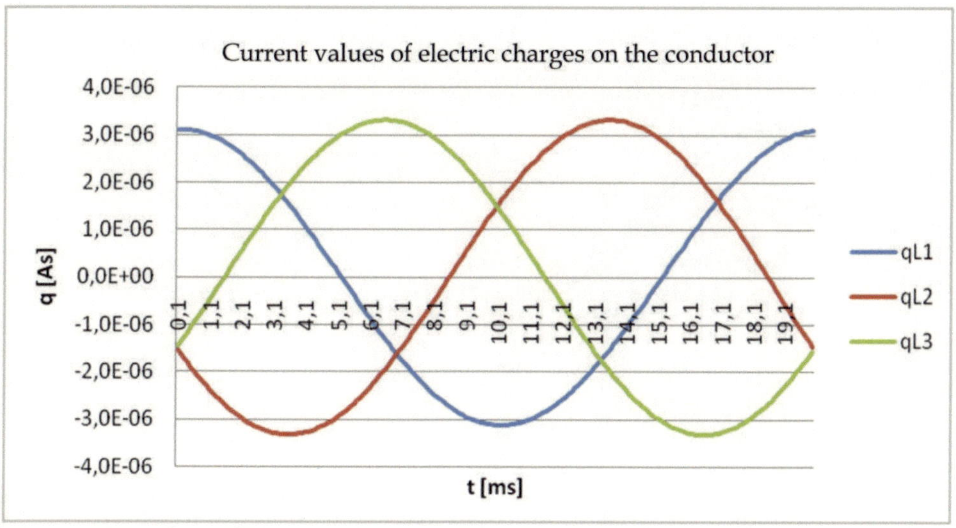

Figure 15. Current values of electric charge on the conductor

From the equation $\left[C\right]=\left[p\right]^{-1}$ and by considering the geometry on figure 15 we get the current value of the electric charge on the conductors:

$$\left[q\right]=\left[p\right]^{-1}\cdot\left[U\right] \tag{29}$$

5.3.3. Electric field intensity

With the current values of electric charge on the conductor we calculated the components of the electric field intensity that exists because of the charge on all three phases.

We get the greatest electrical field intensity in the substance 2 at a radius r_2 (figure 4):

$$E_{2\max}=\frac{q}{2\cdot\pi\cdot\varepsilon_0\cdot\varepsilon_{r2}}\cdot\frac{1}{r_2}=\frac{U\cdot\frac{1}{r_2}}{\varepsilon_{r2}\cdot\left(\frac{1}{\varepsilon_{r1}}\cdot\ln\frac{r_2}{r_1}+\frac{1}{\varepsilon_{r2}}\cdot\ln\frac{r_3}{r_2}\right)}, \tag{30}$$

We calculated the electric field intensity in several points of the space around the conductor. The points were selected at the edge of the insulation (at the radius r_2) of the multi-phase conductors. The radius was chosen so that the distance to the neighbouring conductors was as small as possible. As seen in figure 5, the point for the phase L1 is at (4.6785; 27.9785), for the phase L2 at (3.9215; 30.9785) and for the phase L3 at(5.4785; 25.0215). We got the electric field intensity at the edge of the insulation for each individual phase conductor (figure 16, figure 17 and figure 18) with vector addition of the contribution of all the electric charges (equation 23). The geometric sum of the current values (the current value of the electric field intensity) is a periodic quantity, but not a sinus one.

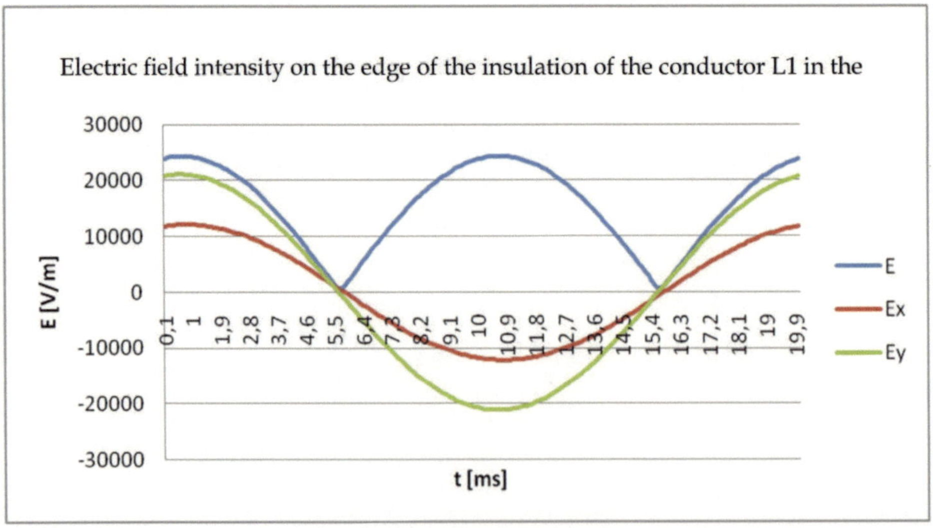

Figure 16. Electric field intensity on the edge of the insulation of the conductor L1 in the point (-4.6785; 27.9785)

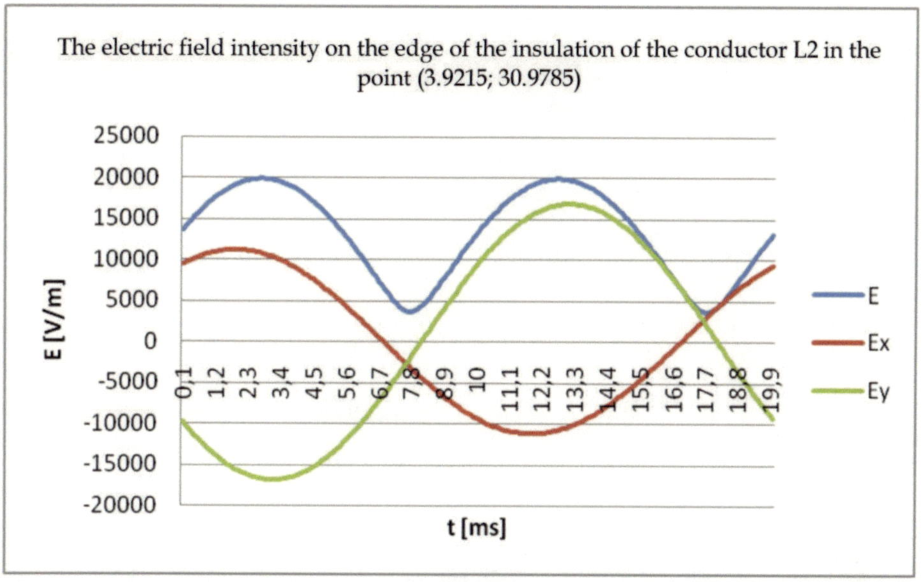

Figure 17. The electric field intensity on the edge of the insulation of the conductor L2

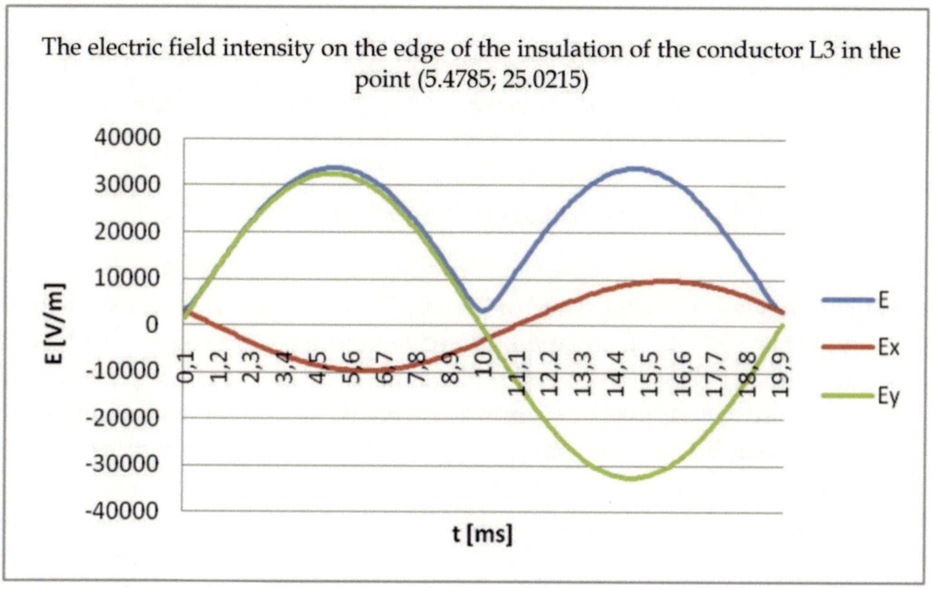

Figure 18. The electric field intensity on the edge of the insulation of the conductor L3

We get the effective value of the electric field intensity by summing over the whole period. By definition, the effective value of the periodic quantity is the one which makes the same effect as the corresponding one-way quantity. In our case the effective value of the electric field intensity is as follows.

$$E_{ef} = \frac{1}{T} \cdot \int_{0}^{T} E^2(t) \cdot \mathrm{d}t \qquad (31)$$

here:

E_{ef} is the effective value of the electric field intensity,

T is the period,

E is the vector sum of all the contributing charges.

5.3.4. The electric field intensity on the edge of the insulation

The highest current value of the electric field intensity in the point (-4.6785; 27.9785) – on the edge of the insulation of the conductor L1 is 24.366e+003 V/m with the effective value of 17.235e+003 V/m.

The highest current value of the electric field intensity in the point (3.9215; 30.9785) – on the edge of the insulation of the conductor L2 is 19.865e+003 V/m with the effective value of 14.287e+003 V/m.

The highest current value of the electric field intensity in the point (5.4785; 25.0215) – on the edge of the insulation of the conductor L3 je 33.813e+003 V/m with the effective value of 24.023e+003 V/m.

From the calculations and figures, we see that the electrical field intensity at the edge of the insulation of any of these three conductors does not exceed the critical dielectric strength of air.

5.3.5. The electric field strength perpendicular to the bisector of the span at the point of the maximum sag

In the end we examine the consistency of the over ground 400 kV power lines that use covered conductors with the Regulation. We have examined the electric field strength perpendicular to the bisector of the span at the point of the maximum sag, 1 meter above the ground. We use the vector addition in each point, to add the contributions of all phase conductors and their mirror images together. We did this in the same manner as we determined the electrical field intensity at the edge of the insulation of each individual phase conductor. We recorded the highest value of the electric field intensity at each point in the time of one period. We then determined the effective value in accordance with the equation (31). We calculated the points within a distance of 100 meters left and right from the bisector of the overhead power line in steps of 1 m. The results are shown in figure 19.

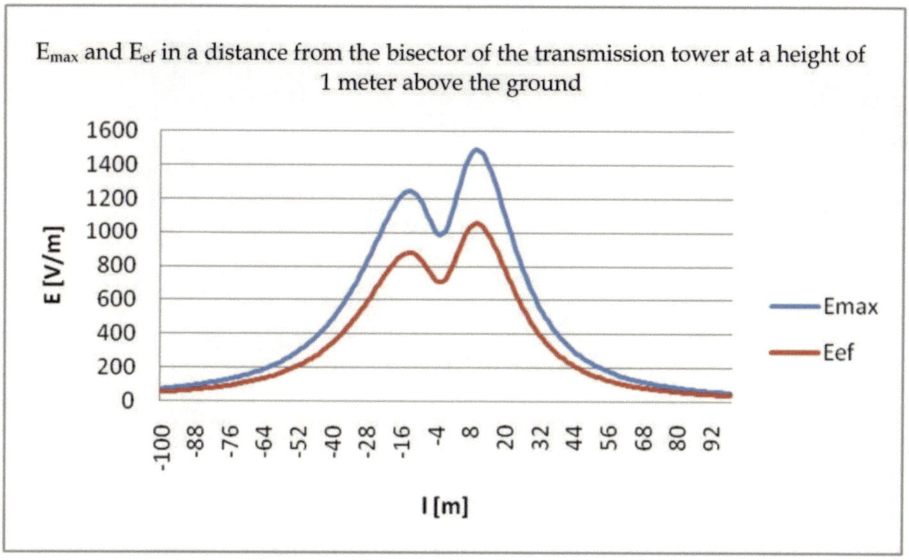

Figure 19. The electrical field intensity perpendicular to the bisector of the span in the point of the greatest sag.

The value of the electric field intensity falls under the permitted limit of the regulation at a distance of 30 m away from the bisector of the power line.

6. The calculation of the electric field intensity with the finite element method

The planned covered conductor does not float in the air by itself, but is mounted in a three-phase electric system and hanged on a steel construction. Because of that, an analytical approach is not sufficient and the approximate results have to be checked with the finite element method.

6.1. Calculation of the electric field intensity at the edge of the insulation of the conductor

Based on the measurements on figure 5, we calculated the maximum electric field intensity of the phase L1, using the computer program ELEFANT®. For the basic harmonic current this holds true when the voltage of the phase L1 is 326 kV and 163 kV for the phase L2 and L3. The phase to phase voltage from phase L1 to the other two phases is then at an amplitude value of 400 kV – 565.69 kV. From figure 20 you can see that the electric field intensity does not exceed the value of 0.1 MV/m, which is less than what we got from our analytical calculation (equation 22).

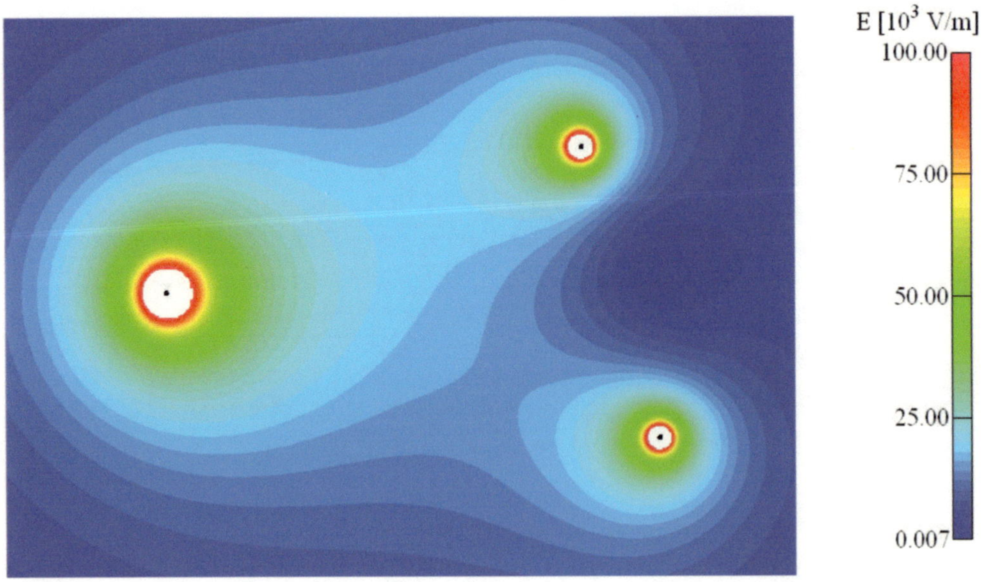

Figure 20. The electric field intensity calculated with the computer program ELEFANT®

6.2. Calculation of the electric field intensity perpendicular to the conductor

As a result, a figure is shown that depicts the calculation of the effective value of the electric field intensity perpendicular to the conductor in the point of the greatest sag (figure 21). The calculation results of the programs Matlab and Elefant are compared (figure 22).

In comparison, we see that both calculations give approximately the same result and that using both methods gives a value that is under the maximum allowed effective value permitted by the law.

With the use of the computer program we calculated the phase voltage of line conductors as a function of time (in a time period of 20 ms).

We calculated the potential coefficients of the capacitance (the inverse value of the matrix of potential coefficients), based on the geometry (arrangement of conductors and the insulation on them). We also calculated the current electric charge on the phase conductors based on the current values of voltage. With the known charges we calculated all three vectors of the electric field intensity, which are caused by the current values of charge on the individual line conductors in the point of interest. In the point of interest the current values of the vectors were added together, to get the total vector of electric field intensity. This total vector is dependent upon the three current values of charge on the line conductors. The resulting electric field intensity is not sinusoid quantity (figure 18) but it is a periodic quantity. We calculate the effective value in accordance with equation (31).

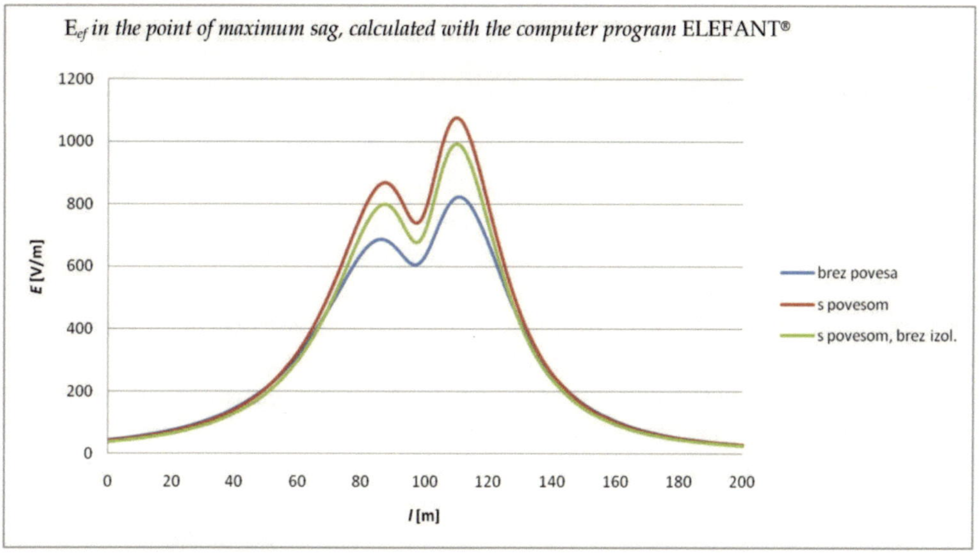

Figure 21. E_{ef} in the point of maximum sag, calculated with the computer program ELEFANT®

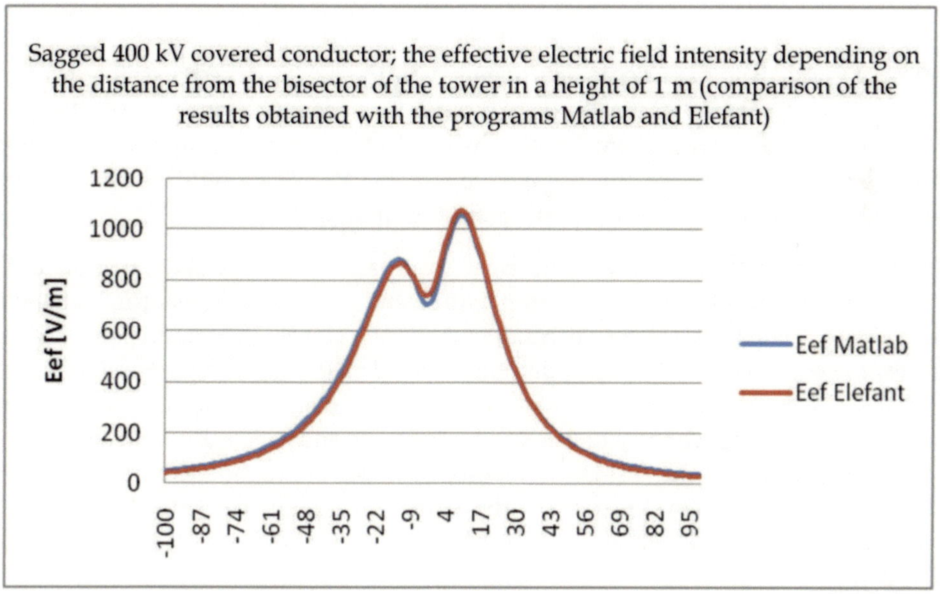

Figure 22. Comparison of E_{ef} perpendicular to the conductors in the point of maximum sag calculated with the programs Matlab and Elefant.

Figure 22 shows the calculation for points 1 m above the ground transverse to the power line in the area of the maximum sag of the cable using the analytical method and the finite element method.

7. Conclusion

The Slovenian power system designers tend to reduce the number of voltage levels. In the future, only four levels will probably exist: 0.4; 20; 110 and 400 kV. The 10 kV levels are in the middle of the range and are present only in large towns (Ljubljana, Maribor). One of the major problems is abandonment of the 220 kV voltage level in the transmission network. The designers are thinking about preservation of the 220 kV power line platforms and the transition to 400 kV conductors. The simplest solution seems to be the erection of new overhead power lines, yet this would involve substantial funds and new permissions. The proposition is the use of covered conductors. The purpose of this chapter was to determine whether it is possible to use the existing platforms and transmission towers of the 220 kV power lines with the new 400 kV conductors. We proposed a covered conductor with a carbon fibber core and a conductive layer made from aluminium, surrounded by in insulation made from polyurethane. The insulation thickness was calculated as the double insulation of the conductor was made from two layers, the one being polyurethane and the other air. We determined a radius at which the electric field intensity at the edge of the insulation is not high enough to cause breakdown of the surrounding air (the electric field intensity has to be lower than the dielectric strength of air). For the reduction of weight of the conductor we assume that we can replace the steel core with a core made from carbon fibbers.

We attempt to calculate the electric field intensity in the air (at the edge of the insulation) with the following values. For the radius of the conductor we take the radius of the current 220 kV conductor, which is 15.3mm. For the thickness of the insulation we use 15mm and its dielectric strength $\varepsilon_r = 3.4$. For the voltage we use 400 kV. The result we get with these values is 2.39 MV/m, which is less than the dielectric strength of air.

The proposed conductor will have a core made from carbon, a conductive layer made from aluminium and the insulation made from polyurethane. According to the usual labelling of conductors we named the suggested conductor PUAC 2150/490/65 mm^2. Here 2150 mm^2 stands for the cross section of the polyurethane mantel, 490 mm^2 for the aluminium and 65 mm^2 for the core made form carbon fibber. The electrical resistance of the covered conductor doesn't change in comparison with a normal conductor and is $R = 0.0592$ Ω/km. Likewise the dielectric strength of the insulation mantle does not affect the electrical reactance of the conductor, which is = 0,414 Ω/km. For the proposed conductor PUAC 2150/490/65 mm^2 with a 15 mm thick insulation layer made from polyurethane ($\varepsilon_r = 3.4$) the capacitance is $C = 12.6$ nF/km.

The over ground conductors are used to transfer electricity between two points and they lead through various parts of the area. We calculated the mechanical properties of the proposed cable and the sag in the middle of the imaginary span and over obstacles. With this data we calculated the impact of non-ionizing radiation that over ground lines exert on the environment.

The installation of an over ground power line is disruptive to the environment. The frequency that we use for the transfer of electricity in the distribution network is 50 Hz, and it causes a magnetic field with the same frequency. This electromagnetic field falls in to the category of low frequency fields. As negotiated at an international level it actually belongs to electromagnetic fields with very low frequencies (ELFF), with frequencies ranging from 30-300 Hz. This is the range at which we talk about electric and magnetic fields separately, instead of electromagnetic fields. The electric field is the result of electric charge on the conductor and in the ground. It is also indirectly linked to the voltage between conductor and ground, the higher the voltage the higher the electric field is. If we look at the limit values that are determined in the Slovenian legislation the electric field is more problematic than the magnetic field. We calculated the electric field intensity in the critical points, and found that it is smaller than the value that is allowed under the regulation about non-ionizing radiation. We also calculated the electric field intensity perpendicular to the axis of the over ground conductor in the point of the greatest sag. It fall on the specified value determined by the regulation for new buildings in a distance of 75 m from the axis of the over ground conductor. We checked the results with a calculation using the finite element method.

We found that the proposed covered conductor does not need a wider corridor as it is already set for the 220 kV overhead power line with bare conductors and allows the transfer of energy at 400 kV.

As future work we propose the construction of a prototype of such a conductor, laboratory experiment of these theoretical calculations and an economic analysis: cost of new conductors and the replacement of these on the existing transmission towers - the price of building the new above ground power line with all the necessary permits.

Author details

Žiga Voršič
University of Maribor, Slovenia

8. References

R.J. Bacha, GPU/PENELEC Compact 115 kV Covered Conductor Study, Minutes of the Meeting – Pennsylvania Electric Association, Engineering Section, 1981

Manfred Beyer, Wolfram Boeck, Klaus Möller, Walter Zaengl, Hochspannungstechnik, Theoretische und praktische Grundlagen; Springer-Verlag 1986.

Ray Elford: Covered Conductors – making the right choice, Electrical engineer, februar 1995

Michèle Gaudry, Francis Chore, Claude Hardy, Elias Ghannoum: Increasing the ampacity of overhead lines using homogeneous compact conductors, Pariz 1998

Damjan Miklavčič: Vpliv elektromagnetnih polj na biološk esisteme, Zbornik 2, Konference slovenskega komiteja CIGRE, Maribor, 7. - 9. junij 1995: 445 - 452

Jože Pihler, Igor Tičar: Design of systems of covered overhead conductors by means of electric field calculation, IEEE, April 2005

Hans Prinz, Hochspannungsfelder, R. Oldenbourg Verlag, Munchen 1969

Heine P., Pitkänen J, Lehtonen M, Sag Characteristics of Covered Conductor Feeders, 38th International Universities Power Engineering Conference, UPEC 2003, Thessaloniki, Greece, September 2003

Recueil CEI / IEC, Symboleslittérauxet conventions, Zajednica JEK, Beograd 1986.

F. Sato, H. Ebiko: Development of a low sag aluminum conductor Carbon fiber reinforced for transmission lines, Pariz 2002

Janko Šarman, Električno in magnetnopolje v okolici visokonapetostnih daljnovodov 110,220 in 400 kV, diplomska naloga, UM FERI, 1999.

Sašo Škorjanc: Izračunelektromagnetnegapoljavodnikov,diplomskodelo, UM FERI, Maribor 2005

Simon Tajnšek, Magistrska naloga: Nov sistem nadzemnih vodov s polizoliranimi vodniki, UM Fakulteta za elektrotehniko, računalništvo in informatiko, Maribor, december 2005

Igor Tičar, Oszkar Biro, Kurt Preis, Uporaba 2D in 3D metode končnih elementov v bielektromagnetnih raziskavah, Tretja konferenca slovenskih elektroenergetikov, Nova Gorica, 3.-5.junij 1997.

Igor Tičar, Tine Zorič, Osnove elektrotehnike 1. zvezek Elektrostatična in tokovna polja, Univerza v Mariboru, FERI, Maribor 2003

M. J. Tunstall, S.P. Hoffmann: Maximising the ratings of national grid's existing transmission lines using high temperature, low sag conductor, Pariz 2000

Stane Vižintin, Tadeja Babnik, Franc Jakl: Možnost uporabe sodobnih tehnologij vodnikov pri novi generaciji 400 kV nadzemnih vodov, Elektroinštitut Milan Vidmar, Ljubljana, januar 2007

Jože Voršič, Jože Pihler: Tehnika visokih napetosti in velikih tokov, UM FERI, 2005

Žlahtič, Franc, Cestnik, Breda, Grajfoner, Slavko, Jakl, Franc. Vpliv uredbe o elektromagnetnem sevanju na parametre 110 in 400 kV daljnovodov = Effects of the newly enforced law controlling electromagnetic radiation on parameters of 110 and 400 kV power lines. V: http://www.eles.si/za-poslovne-uporabnike/razvoj-in-uporaba-prenosnega-omrezja/strategija-razvoja-elektroenergetskega-sistema-rs.aspx

Pravilnik o tehničnih normativih za graditev nadzemnih elektroenergetskih vodov z nazivno napetostjo od 1kV do 400 kV, Ljubljana 2009

SIST EN 50341-3-21; Nadzemni električni vodi za izmenične napetosti nad 45 kV, februar 2009

Terminološkakomisija (urednik Anton Ogorelec), Slovenski elektrotehniški slovar, Področje elektroenergetika, Sloko CIGRÉ, Ljubljana 1996.

Tretja konferenca slovenskih elektroenergetikov, Nova Gorica, 3.-5.junija, 1997. Zbornik. Ljubljana: Slovenski komite Mednarodne conference za velike elektroenergetske sisteme, 1997, str. 36/23-28

Uredba o elektromagnetnem sevanju v naravnem in življenjskem okolju, Uradni list RS, št. 70/1996

Section 4

Bio-Based Polyurethanes

Chapter 18

Seed Oil Based Polyurethanes: An Insight

Eram Sharmin, Fahmina Zafar and Sharif Ahmad

Additional information is available at the end of the chapter

http://dx.doi.org/10.5772/50085

1. Introduction

Seed oils [SO] are cost-effective, eco-friendly and biodegradable in nature. They bear functional groups such as carboxyls, esters, double bonds, active methylenes, hydroxyls, oxirane rings and others, amenable to several derivatization reactions. Their abundant availability, non-toxicity and rich chemistry has established SO as focal point of polymer production, e.g., production of polyesters, alkyds, epoxies, polyols, polyethers, polyesteramides, polyurethanes and others. The escalating prices of petro-based chemicals, environmental and health concerns have further beckoned the enhanced utilization of SO as polymer precursors. SO have attracted enormous attention as potential source of platform chemicals, at both laboratory and industrial scale. Today, oil-seed bearing crop plants are being raised and modified for uses in areas covering biodiesel, lubricants, folk medicines, cosmetics, plastics, coatings and paints.

SO based polyurethanes [PU] occupy an important position due to their simple preparation methods, outstanding properties and versatile applications in foams, coatings, adhesives, sealants, elastomers and others [1-4]. In general, PU are prepared by chemical reaction of a diol, polyol or any oligomer or polymer containing hydroxyl groups, with an aliphatic, cycloaliphatic or aromatic isocyanate. SO serve as green precursors to diols, polyols and other –OH containing polymers offering greener raw materials in PU synthesis, replacing their petro-based counterparts. The choice of SO polyol or isocyanate is governed by the end-use application of SO PU ranging from soft and flexible to rigid PU foams, thermoplastic to thermosetting PU, flexible films to tough, scratch-resistant, impact resistant coatings and paints. It is well known that the structure of a triglyceride is very complex. Every SO has a characteristic fatty acid profile. Amongst the same species, the composition of triglycerides in a particular SO varies by weather conditions of crops and soil. Triglycerides vary by their fatty acid chain lengths, presence as well as the position of double bonds and degree of unsaturation of the constituent fatty acids. The structures of natural SO and their derivatives, i.e., epoxies, diols, polyols, polyesters and alkyds are complex. Thus, the properties of PU obtained from SO derivatives depend on a number of factors such as (i) the type, composition and distribution of fatty acid residues in the constituent triglycerides, (ii) the number, distribution, site of hydroxyls (in the middle or closer to the end of the triglyceride chain) and level of unsaturation in the fatty triester chains of the parent diol or polyol, (iii) the type, position and structure of isocyanates used and (iv) the urethane content of the final PU [5-8].

The ingredients for the preparation of SO based PU generally comprise of a diol or polyol (containing active hydrogens) derived from SO and an isocyanate, aliphatic and aromatic such as 1,6-hexamethylene diisocyanate [HMDI], isophorone diisocyanate [IPDI], cyclohexyl diisocyanate [CHDI], L-Lysine Diisocyanate [LDI], toluylene 2,4-diisocyanate or toluylene 2,6-diisocyanate [TDI], 4,4′- methylenediphenyl diisocyanate [MDI], naphthylene 1,5-diisocyanate [ND]. PU are prepared by polyaddition reaction between the two components, often in presence of a chain extender, cross-linker or a catalyst. Today, several environment friendly routes for the production of PU have cropped up. Research has been focussed on the preparation of PU from non-isocyanate routes, and also on the use of fatty acid based isocyanates for PU production, considering the persisting challenges of polymer industry and immediate attention sought towards environmentally benign chemicals and chemical routes for the same [9-18].

SO based PU are generally flexible in nature. Generally, these PU show low Tg due to the presence of long aliphatic chains and also poor mechanical properties (owing to the presence of dangling chains). The thermal stability of SO based PU is also lower since the decomposition of urethane bonds starts at 150-200°C. Javni et al have studied the decomposition of PU from Olive, Peanut, Canola, Corn, Soybean, Sunflower, Safflower and Castor oils [7]. The decomposition involves the dissociation of urethane bonds to isocyanate, alcohol, amine, olefin and carbon dioxide. The properties of PU depend upon their crosslinking density as well as chemical composition. In an execellent review, Petrovic has highlighted the different properties of PU prepared from polyols obtained by different methods. As the properties of polyols depend upon the methods of preparation, so also the properties of PU derived therefrom. He has presented a brief outline of the effect of polyols prepared by epoxidation, hydroformylation, ozonolysis, effect of crosslinking density, and type of isocyanate on the properties of PU. He described the effect of the structure of polyols prepared by epoxidation followed by ring opening with methanol, HCl, HBr, and by hydrogenation of epoxidized Soybean oil, and showed that PU obtained from these polyols showed relatively higher glass transition temperatures and improved mechanical properties. Halogenated polyols obtained via ring opening by HCl and HBr gave PU that were less stable than ones without halogens, and had higher Tg (70°C –80°C) than the latter. Polyols obtained via hydroformylation crystallize below room temperature while those derived through hydrogenation reveal crystallization at temperature higher than room temperature. PU from non-halogenated polyols had higher thermal stability than brominated (100°C) or chlorinated polyols (160°C). Polyols with primary hydroxyls give more stable PU than their counterparts with secondary hydroxyls. PU with high crosslink density show higher thermal stability. Hydrolytic stability of PU also depends on the degree of crosslinking, temperature, and physical state of PU. In SO based PU, although SO have ester groups susceptible to hydrolysis, long hydrophobic fatty acid chains cause shielding effect. Under highly humid conditions, urethane bonds undergo hydrolysis producing amine and carbon dioxide [1, 19].

SO are rich in various functional groups such as double bonds, active methylenes, esters, hydroxyls and others. These undergo several chemical transformations yielding numerous derivatives with versatile applications as inks, plasticizers, lubricants, adhesives, coatings and paints. Amongst various SO derivatives, those containing hydroxyl groups are used in the production of PU. These include SO based diols, triols, polyols, polyesters, alkyd, polyesteramide, polyetheramide and others (Figure 1).

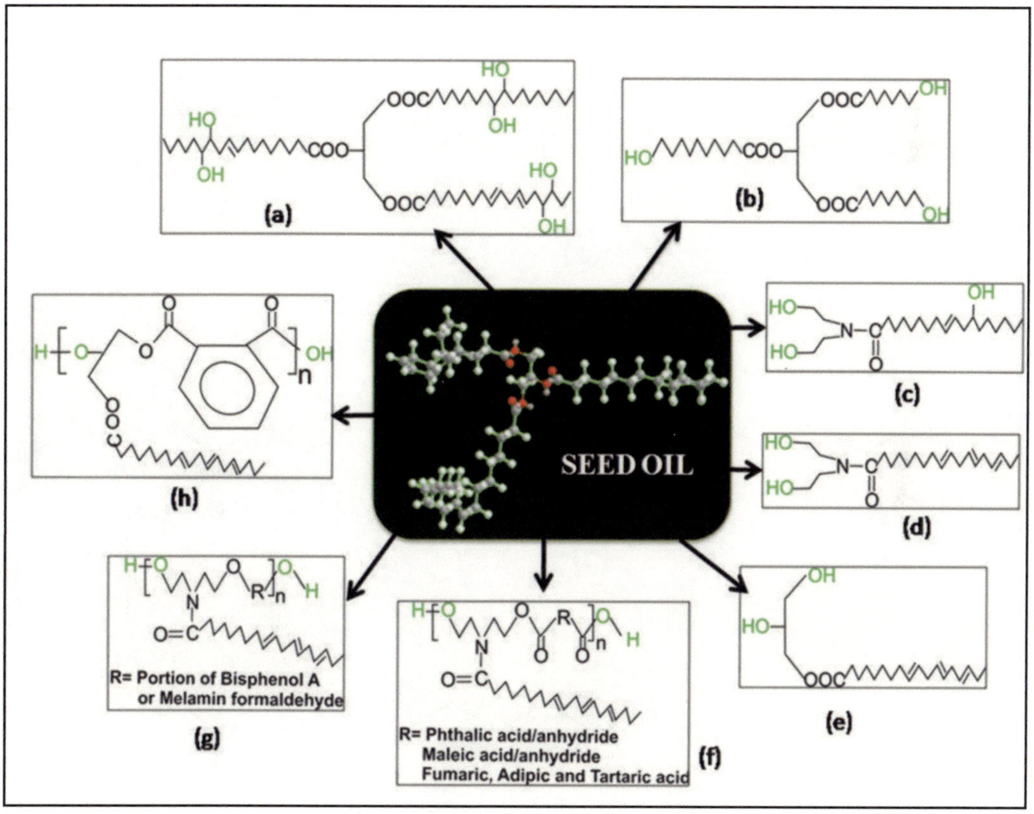

Figure 1. SO derivatives used in the production of PU (a) polyol, (b) triol, (c) fatty amide triol, (d) fatty amide diol, (e) monoglyceride, (f) polyesteramide, (g) polyetheramide, and (h) alkyd.

The chemistry of SO based PU is very vast, governed by several factors such as the type of diol or polyol, type of isocyanate, preparation method of diols or polyols, type of chain extender, cross-linker, reaction temperature and other reaction conditions. In this chapter we have focussed on the preparation, structure and properties of PU obtained from diols, triols and polyols derived by amidation of SO termed as "SO alkanolamides". In the proceeding sections, we have also discussed the modifications of the said SO alkanolamides based PU at the forefront of PU chemistry such as SO based metal containing PU, PU hybrids, composites for applications mainly in coatings, paints and foams.

2. SO based diols

The most excessively used SO based diol in PU production is fatty amide diol or fatty alkan-diol-amide [FAD] (Figure 2). FAD is obtained by the base catalysed amidation of SO with diethanolamine. The structure of FAD is determined by the fatty acid composition of the starting SO.

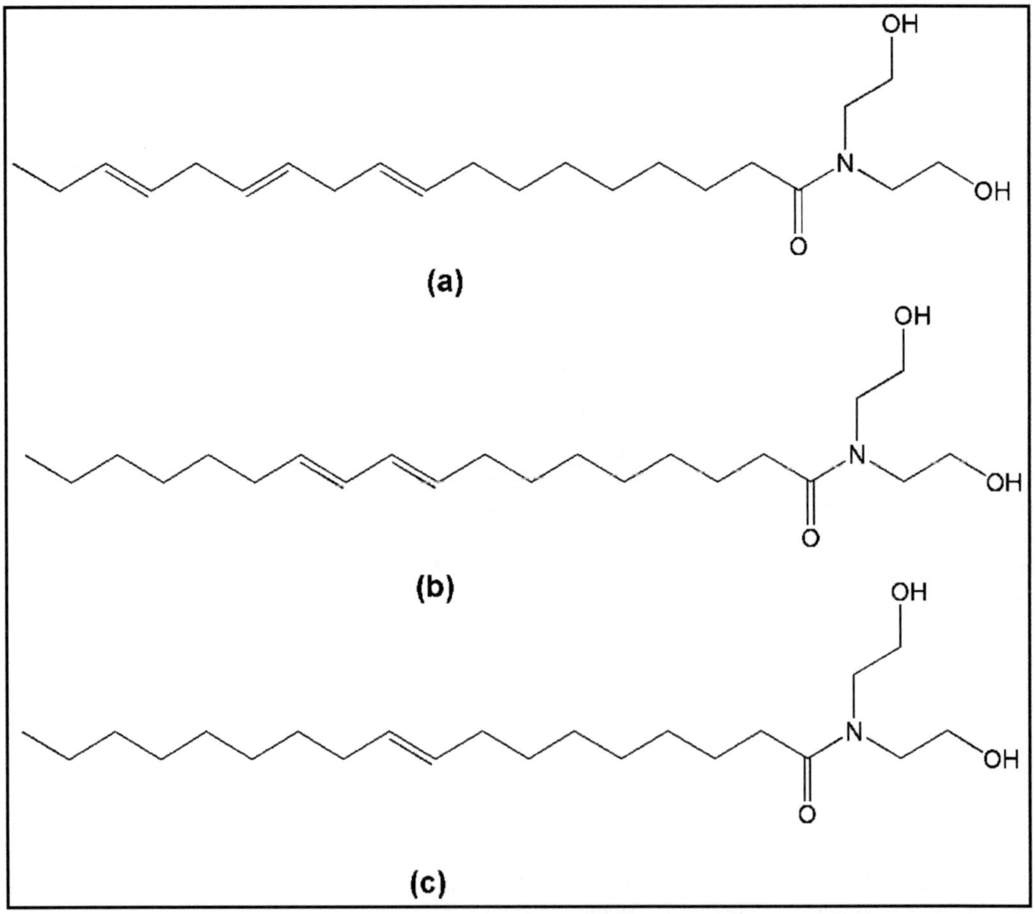

Figure 2. Figure 2. FAD from (a) Linseed (35.0-60.0 % Linolenic acid), (b) Soybean (43.0-56.0 % Linoleic acid) and (c) Karanj (44.5–71.3 % oleic acid), Nahor (55-66% oleic acid) , Jatropha (37-63 % oleic), Olive (65-80 % oleic acid)

The reaction proceeds by nucleophilic acyl substitution bimolecular (SN2) reaction mechanism. As the name suggests, FAD contains an amide group, with two hydroxyl ethyl groups directly attached to amide nitrogen along with the pendant aliphatic chain of SO. FAD is derived from different SO such as Linseed (*Linum ussitassimum*), Soybean (*Glycine max*), Karanj (*Pongamia glabra*), Nahor (*Mesua ferrea*), Jatropha (*Jatropha Curcus*), Olive (*Olea europea*), Coconut (*Cocos nucifera*) and others [20-30] (Figure 2). FAD is used as raw material for various polymers such as PU, poly (esteramide) and poly (ether amide).

3. PU from SO FAD

FAD can be treated with an isocyanate such as TDI, IPDI, HMDI, MDI, ND, CHDI and LDI forming poly (urethane fatty amide) (Figure 3) [FADU] [31].

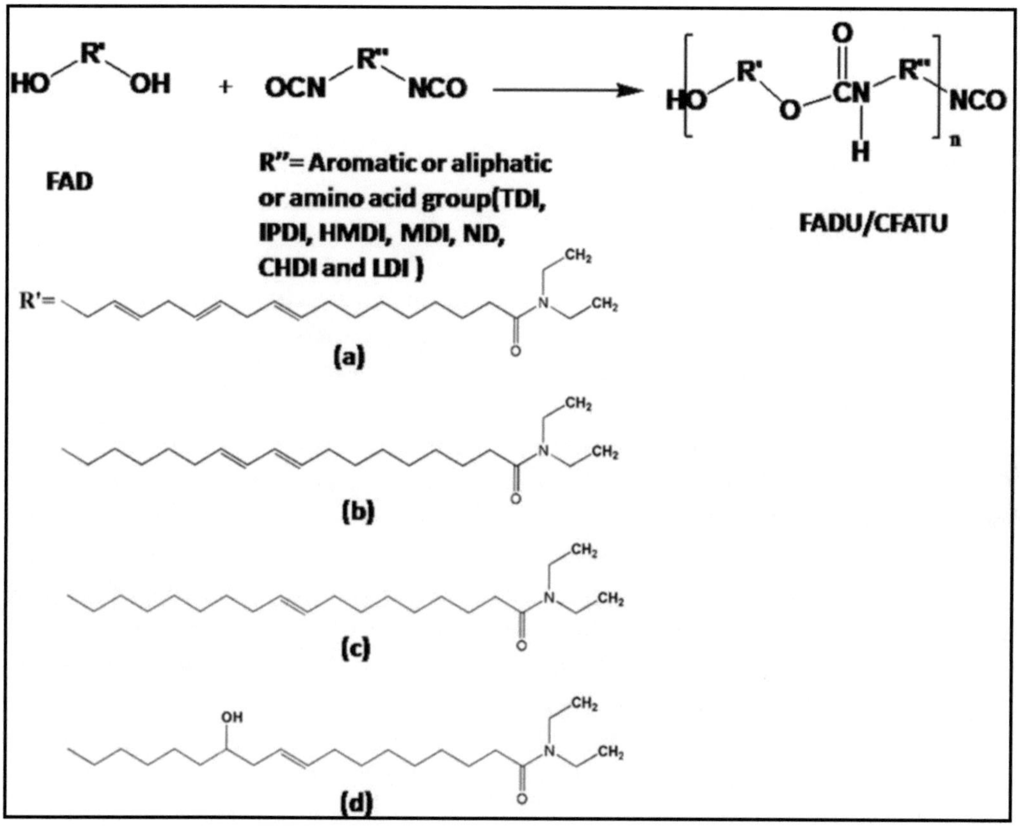

Figure 3. FADU from (a) Linseed, (b) Soybean (c) Karanj, Nahor, Jatropha, Olive and (d) Castor

For the first time, Linseed oil [LO] derived FAD [LFAD] was treated with TDI by one-shot technique to prepare PU [LFADU] as introduced by Ahmad et al [32] (Figure 4).

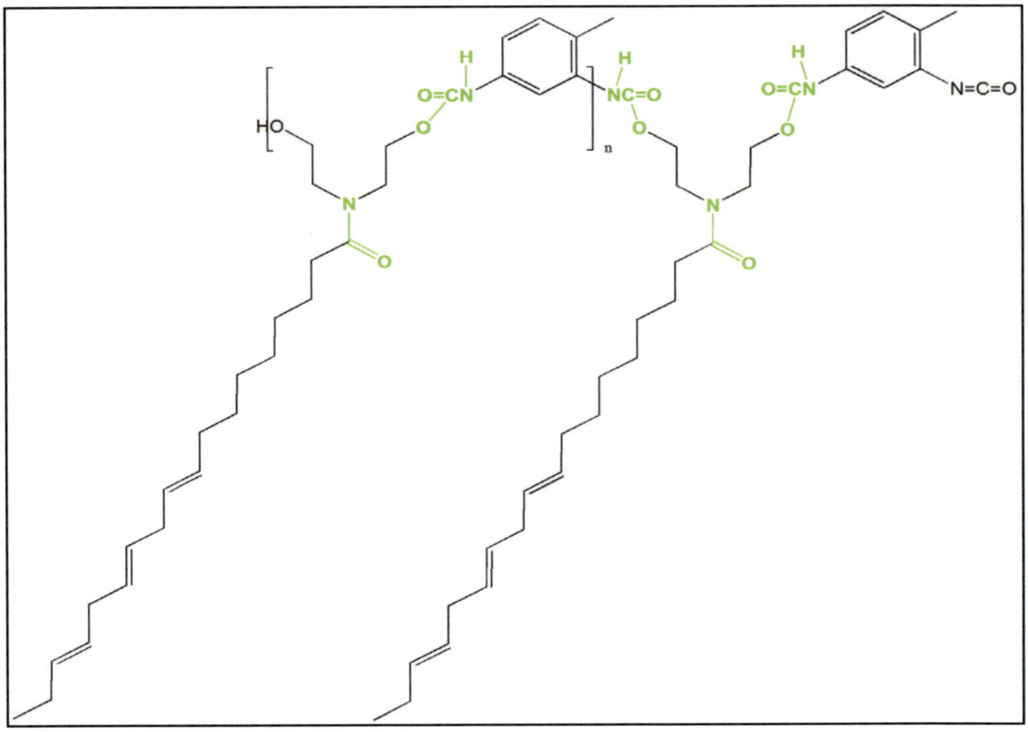

Figure 4. Chemical structure of LFADU.

The structure of LFADU was established by spectral analyses. FTIR, ^{1}H NMR and ^{13}C NMR spectra showed the presence of the main characteristic absorption bands of parent SO [32]. Along with these bands, additional absorption bands are observed supporting the presence of urethane groups in the backbone of LFADU such as those at 3375cm^{-1} for hydroxyl groups, 1716.11 cm^{-1} for urethane carbonyl (str), 1227.56cm^{-1} for C–N of urethane groups, 1735cm^{-1} typical for carbonyl (str) of TDI. The characteristic peaks for hydrogen bonded and non-hydrogen bonded protons of –**H**NCOO– appear at 7.99–7.82ppm and 7.1–6.9ppm, respectively. The aromatic ring protons of TDI occur at 7.5–7.22 ppm. The peaks of –**H**NCOOCH$_2$– are observed at 4.1–3.9ppm and for C**H**$_3$ of TDI appear at 2.25ppm. ^{13}C NMR spectrum also shows the presence of characteristic peaks of LFADU at 17ppm (**C**H$_3$ of TDI), 143.97ppm {–NH–(**C** O)–O–} and 137.46, 136.2, 134.4, 125.94, 125.4, 116.0 ppm (aromatic ring carbons of TDI). TGA thermogram of LFADU has shown four step degradation pattern, at 260 °C (27% weight loss), 360 °C (21% weight loss), 505 °C (40% weight loss), 640 °C (9% weight loss) corresponding to the degradation of urethane, ester, amide and hydrocarbon chains, respectively.

PU from Karanj or *Pongamia glabra* [PGO] oil [PFADU] has also been prepared by similar method. PU obtained from both FAD showed similar structure except for the difference in the structure of pendant fatty amide chains attributed to the variation in the structure of the parent SO chain [33] (Figure 5).

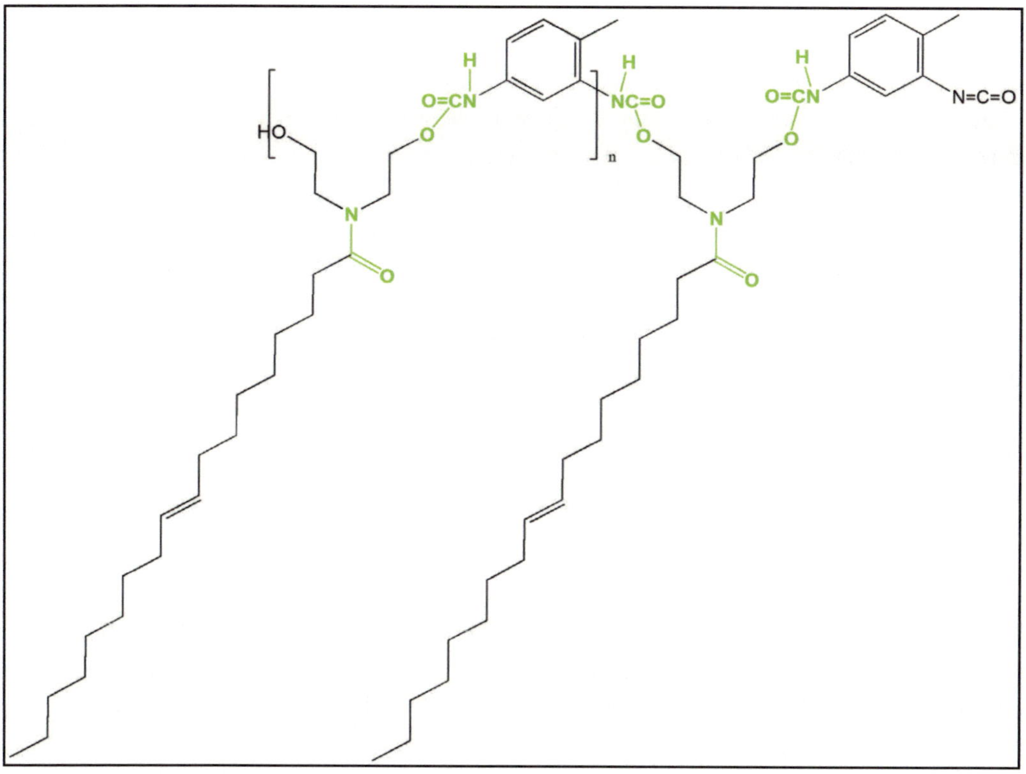

Figure 5. Chemical structure of PFADU or NFADU

The thermal degradation of PFADU was observed at 177°C and 357°C, with 5% weight loss occurring at 200°C attributed to the entrapped solvent and moisture, 10wt% loss at 225 °C, attributed to the decomposition of urethane moieties, 50wt% and 80wt% losses at 390 °C and 455 °C, respectively, attributed to the degradation of the aromatic ring and aliphatic pendant fatty alkyl chains, respectively.

It was observed that in both LFADU and PFADU, beyond 1.5moles loading of TDI, formation of some lumpy aggregates occurred. Upto 1.5 moles of TDI addition, it is speculated that the isocyanate groups of TDI react with hydroxyl groups of FAD forming PU linkages. Beyond this amount, any additional isocyanate added reacts with the urethane groups of LFADU or PFADU forming allophanate groups (secondary reaction). The final PU attains very high viscosity and crosslinking, so much so that the formation of lumpy aggregates occurs and PU is deprived off its free flowing tendency.

Karak and Dutta have reported the production of PU by amidation and urethanation of methyl ester of *M. Ferrea* or Nahor oil [NO], rich mainly in triglycerides of linoleic, oleic, palmitic and stearic acids. They investigated the structure and physico-chemical aspects of FADU from NO [NFADU] [26].

3.1. PU as coatings

LFADU has free –OH, –NCO, aliphatic hydrocarbon chains (from parent LO), amide and urethane groups, which make it an excellent candidate for application in protective coatings (Figure 4). LFADU coatings undergo curing at ambient temperature (28-30°C) by three stage curing phenomenon, (i) solvent evaporation (physical process), (ii) reaction of free –NCO groups of LFADU with atmospheric moisture, and (iii) auto-oxidation. These coatings show good scratch hardness (2.5kg), impact resistance (200lb/inch), bending ability (1/8inch) and chemical resistance to acid and alkali. PU from PGO [PFADU] has shown moderate antibacterial behavior against *Salmonella* sp. with good scratch hardness (1.9kg), impact resistance (150lb/inch), bending ability (1/8inch), and gloss (46 at 45°) [33]. LFADU coatings have shown superior coating properties than PFADU owing to the fatty acid composition of parent oils (PGO, a non-drying oil has higher content of oleic acid while LO, a drying oil, is rich in linolenic acid).

Karak and Dutta have reported the use of NFADU coatings with very good alkali resistance (Figure 5)[32].

3.2. PU as hybrids

Organic-inorganic hybrid materials have been developed with FADU as organic constituent and metal/metalloid as inorganic component to improve the performance and broaden the applications of PU (Figure 6).

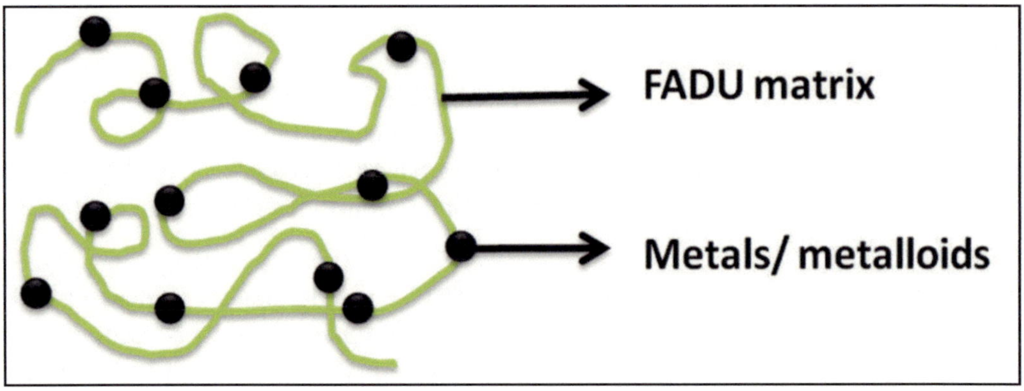

Figure 6. PU as hybrids

In one report, Zafar et al. have prepared organic-inorganic hybrids by using boric acid as inorganic content and PFADU as organic matrix [B-PFADU] [34]. B-PFADU was characterized by standard spectral techniques and physico-chemical methods. B-PFADU performed well as protective coatings in terms of physico-mechanical and chemical resistance tests. B-PFADU showed no change in water and xylene upto 15 days. However, slight deterioration in performance was observed in alkali and acid media, correlated to the presence of -O-B-O- which is susceptible to hydrolysis on exposure to these media. B-

PFADU showed high activity against *E. coli* (Zone of inhibition: 21-30 mm) and moderate activity against *S. aureus* (Zone of inhibition: 16-20 mm). The reason can be the presence of urethane, amide, and hydroxyl groups in the polymer backbone, which can presumably interact with the surface of microbes, causing antibacterial action. B-PFADU can be used as an antibacterial agent as well as coating material.

In another work, Ahmad and co-workers developed LFADU hybrid material with tetraethoxy orthosilane [TEOS] as inorganic constituent in the hybrid material [Si LFADU] by in situ silylation of LFAD with TEOS (at 80 ◦C) followed by urethanation with TDI (at room temperature) [35]. Along with the typical absorption bands for LFADU, additional absorption bands were observed at 484 cm^{-1} (Si–O–Si bending), 795 cm^{-1} (Si–O–Si sym str) and 1088 cm^{-1} (Si–O–Si assym str) in FTIR due to the presence of -Si–O–Si- bond in the hybrid backbone. Hydroxyl value decreases while refractive index and specific gravity increase with the loading of TEOS in Si LFADU, supporting the formation of the hybrid materials by insitu siylation and urethanation reaction. Optical micrograph of Si LFADU showed the presence of SiO_2 particles surrounded by LFADU (Figure 7).

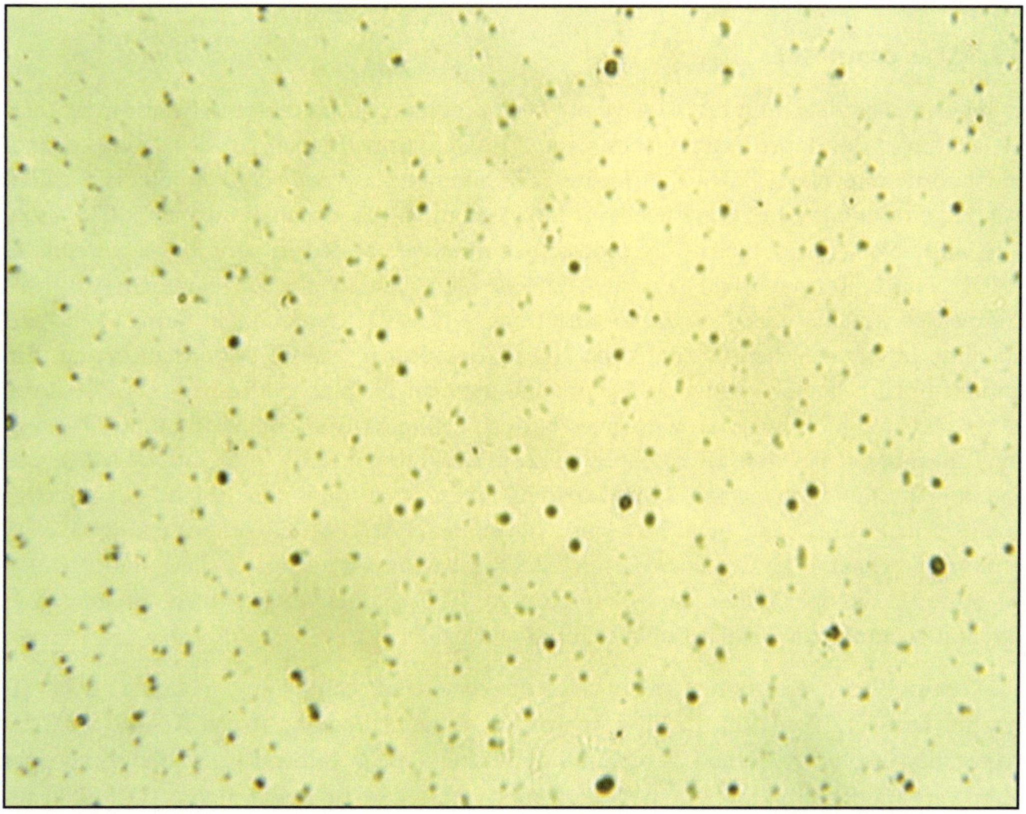

Figure 7. Optical micrograph of Si LFADU

Si LFADU formed hybrid coatings by simple curing route at ambient temperature, over mild steel panels of standard sizes with improved gloss and scratch hardness. SiO2 domains also improved adhesion with the penal surface exhibiting good scratch hardness, bending ability (1/8 inch) and impact tests (150 lb/inch) correlated to the synergism showed by both the components, LFADU backbone imparting flexibility and gloss, while the inorganic domains conferring excellent adhesion and hardness [36].

The corrosion rate (CR) of Si LFADU is much lower (3.08×10^{-4} mm per year) relative to LFADU (3.124 mm/year) In 3.5wt% HCl, with inhibition efficiency (IE%) 99.77. In 3.5% NaOH, CR and IE% were found as 1.26×10^{-3} mm per year and 99.34, respectively. Si LFADU formed uniform and well adhered coating over the metal substrate which prohibits the permeation of corrosive media. The protection mechanism is purely through barrier action attributed to the hydrophobic inorganic content [37, 38]. Coating remained intact when subjected to corrosive media for 192 h as supported by the constant value of polarization resistance (Rp = 1.22 × 104 Ohm in NaOH and 7.7 × 105 Ohm for HCl). Thermal studies showed four step degradation, thermal stability increasing with higher inorganic content, with two glass transition temperatures (Tg) as observed at 115 ◦C and 155 °C in DSC thermogram with safe usage upto 200 °C.

3.3. PU as composites

Composite materials from FADU have not been prepared yet. In their recent research, Zafar et al have developed composites from FADU using metal oxides and organo-montmorillonite clay [OMMT] (Cloisite 30B; modified by an alkyl ammonium cation bearing two primary hydroxyl functions, alkyl group is tallow containing ≈65% C18, ≈30% C16, and ≈5% C14, Southern Clay product) as modifiers added in very lower amounts to FADU matrix (unpublished work). The sole aim behind the development of these composites was the improvement in performance of FADU materials in terms of thermal stability, physico-mechanical and corrosion or chemical resistance performance, and also antimicrobial behavior relative to the pristine material for high performance applications. MMT occurred as nano-sized aggregates with size ranging from 37 to 100 nm as observed by Transmission Electron Micrography [TEM]. The thermal stability of FADU/ MMT was found to increase with increased MMT loading. These composites may be used as protective coatings in future. Zafar et al have also developed FADU/MnO composites, with good antifungal behavior. MnO occurred as needles self-assembled in "lemon slices" morphology as observed in optical micrograph (Figure 8). The said composite material may find application as antimicrobial agent in coatings and paints.

PU composites were prepared by the dispersion of conducting polymer poly (1-naphthylamine) [PNA] in LFADU matrix in amount ranging from 0.5–2.5 wt% by conventional solution method as reported by Riaz et al [39]. At lower levels, the composites were very fragile in nature. The polymerization of PNA occurred through N–C(5) linkages of 1-naphthylamine units as confirmed by FTIR spectra. PNA also reacted with free isocyanate groups of TDI forming urea linkages, as also supported by spectral analysis. UV visible spectra also confirmed the formation of urea linkages between LFADU and PNA. TEM micrographs showed the average particle size as 17–27 nm. Nanoparticles appeared as smaller aggregates which later on formed larger aggregates. XRD analysis revealed purely amorphous nature of composites. With the increase in the loading of PNA in the composites, the distortion and torsional strain increased in the composites due to higher urea linkages. It was found that as the percent loading of PNA in the composites increased, their electrical conductivity values also increased; however, these values fell in the semi-conducting range, which was much higher relative to the conductivity values obtained with very high loading of PNA in previously reported composites. The improved electrical conductivity values of LFADU/PNA composites can be correlated to the hydrogen bonding and urea type linkages formed between the two polymers, which provide the path to charge conduction [40,41].

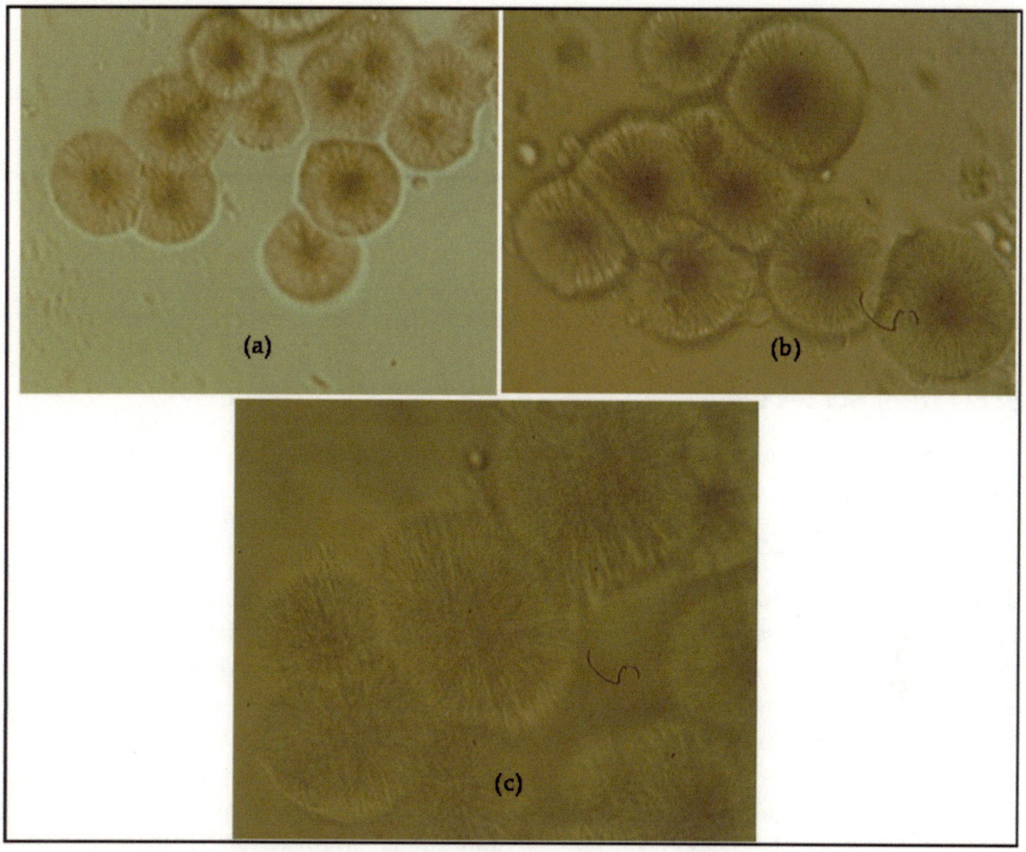

Figure 8. Optical micrographs of FADU/MnO (a) 100 X, (b) 200 X, (c) 500 X

4. SO based triol

Castor oil (CO) is obtained from seeds of *Ricinus communis* or Castor belonging to the family *Euphorbiaceae*. It is non edible oil. The crop is cultivated around the world because of the commercial importance of its oil. India is the world leader in castor production and dominates the international CO trade. Worldwide castor production was about 1.4 million metric tons during the year 2009 with an average yield of about 956 kg ha^{-1}. Ricinoleic acid

(12- hydroxy -9- octadecenoic acid), hydroxyl containing fatty acid, is the major component of CO, and constitutes about 89% of the total fatty acid composition. Castor seed products have widespread applications in many industries like paints, lubricants, pharmaceuticals and textiles. Today, the importance of these products has grown manifolds because of their biodegradable and eco-friendly nature.

Due to the inherent hydroxyl functionality, CO stands as an excellent natural raw material for the development of PU. CO derived PU are flexible due to long aliphatic triglyceride dangling chains [42]. CO has three hydroxyl moieties in one triglyceride molecule. On amidation, CO yields fatty amide triol or alkan-triol-amide [FAT], which bears two hydroxyl ethyl groups directly attached to amide nitrogen, as well as one hydroxyl group in the pendant fatty chain obtained from the parent CO. Thus, CO derived FAT [CFAT] acts as SO derived triol (Figure 9). Rao et al prepared acrylated CFAT as a multifunctional amide for photocuring [31, 43, 44].

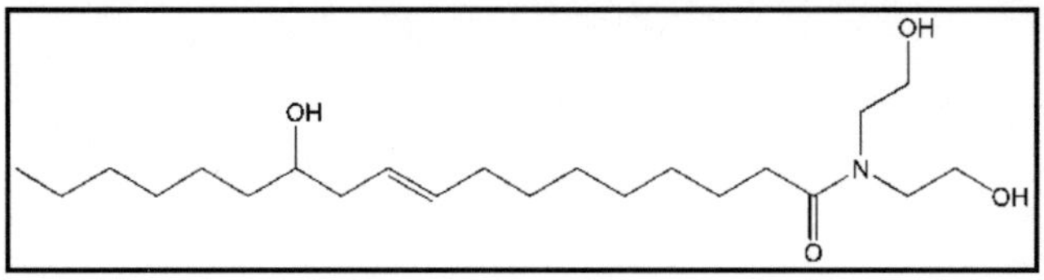

Figure 9. CO (87-90 % Ricinoleic acid) derived FAT [CFAT)

5. PU from SO FAT

CFAT on chemical reaction with TDI by one shot technique yields CFATU (Figure 10), with an additional crosslinking site (hydroxyl group) conferred by parent CO containing 89% ricinoleic acid. Contrary to LFADU and PFADU, where the permissible limit of TDI addition is 1.5 moles, in CFATU, at 1.2.moles of TDI addition, CFATU becomes highly viscous. The difference prevails due to the additional hydroxyl functionality in CFATU, which presumably makes possible higher inter and intra crosslinking sites relative to LFADU and PFADU. As also observed in LFADU and PFADU, the physico-chemical characteristics such as specific gravity, inherent viscosity and refractive index increase, while hydroxyl and iodine values decrease on increasing the content of TDI in PU. The thermal degradation occurred in the temperature range of 150–390 °C. The decomposition observed at earlier temperature range may be attributed to the thermal degradation of urethane linkages followed by the volatilization of the decomposition products while that at higher temperatures is correlated to the degradation of amide bond, aromatic ring and aliphatic alkyl chain of the fatty acid, respectively, followed by the volatilization of the decomposition products [45, 46].

5.1. PU as coatings

CFATU coatings have been prepared and their physico-mechanical, thermal as well as corrosion resistance behavior has also been investigated [31]. CFATU have been further modified for improvement in their performance as discussed in proceeding sections.

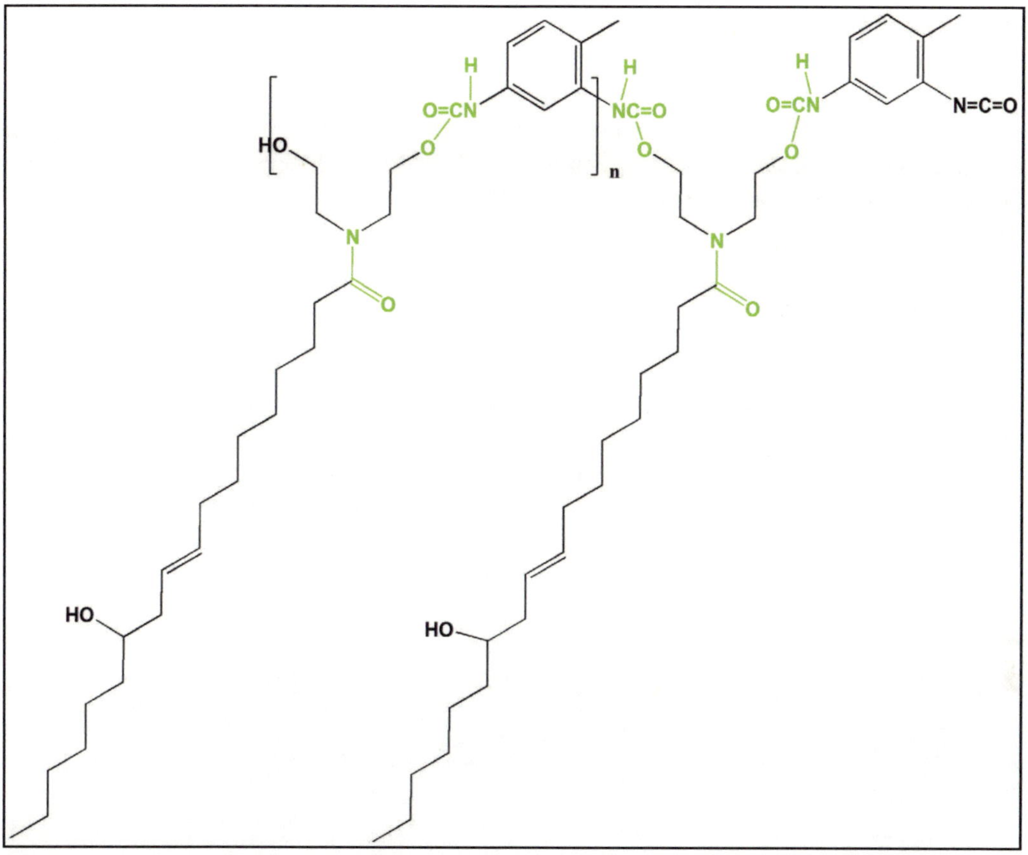

Figure 10. Chemical structure of CFATU

5.2. PU as hybrids

Ahmad et al have reported the preparation and characterization of metal containing CFATU [47]. They treated CFAT with zinc acetate (5, 10 and 15 wt%) and TDI (25–75 wt%) to prepare metal containing CFATU [MCFADU] "in situ" by microwave [MW] assisted preparation method in a domestic MW oven. During the preparation, it was observed that MCFATU with > 5wt% zinc acetate formed gel. While the conventional preparation method of LFADU, CFATU, PFADU and NFADU requires 8-12 hours, CFATU is obtained in 4 minutes by MW assisted preparation method. By molecular interactions with the electromagnetic field and heat generated by molecular collision and friction, the reaction is facilitated to occur in much reduced time periods under the influence of MW irradiations. In this reaction, hydroxyl groups of CFAT react with zinc acetate and TDI successively. Curing of MCFATU is a two step process occurring by solvent evaporation (physical phenomenon) followed by the second stage curing (a chemical process) where free –NCO groups of MCFATU react with the atmospheric moisture to form urethane and amino groups through addition reaction. MCFATU acted as good corrosion protective coatings for mild steel. The best physico-mechanical properties (scratch hardness 3.5 kg, impact resistance

150lb/inch, and bending ability 1/8 inch) were achieved at 5wt% loading of zinc acetate and 55 wt% of TDI, when evaluated by standard methods and techniques. The corrosion protection efficiency of the same was evaluated by potentiodynamic polarisation measurements [PDP] in aqueous solution of 3.5wt% HCl (CR 4.51 × 10-3 mm/year; IE% 96.23), 3.5wt% NaOH (CR 1.36 × 10-3 mm/year; IE% 90.81); 3.5wt%NaCl (CR 2.25 × 10-3 mm/year; IE% 94.50) and tap water (Cl- ion 63 mg/L; Conductivity 0.953 mS/A) (CR 3.56 × 10-3 mm/year; IE% 98.35).

5.3. PU as composites

CFATU composites were developed by the introduction of MMT clay [48] and nano sized ZnO by Zafar et al [unpublished work]. Morphology of the composites as studied by TEM revealed the presence of nanosized globules of size ranging between 15-120 nm in CFATU/OMMT (Figure 11), and ZnO in CFATU/ZnO composites occurring as nano-sized spindles of diameter 50-60nm (Figure 12). Both the composite materials may find application as corrosion protective coatings for mild steel. CFATU/ZnO composites also act as excellent antifungal agents against common fungal strains such as *Candida albicans, Candida krusei, Candida glabrata* and *Candida tropicalis.*

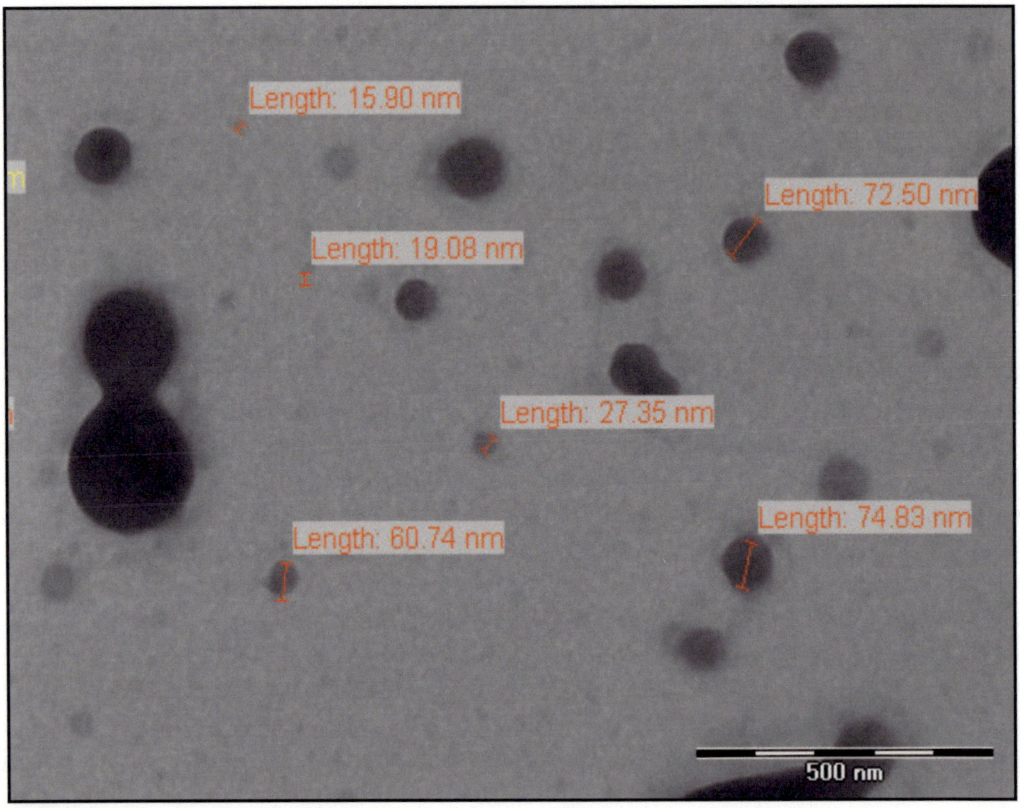

Figure 11. TEM of CFATU/OMMT

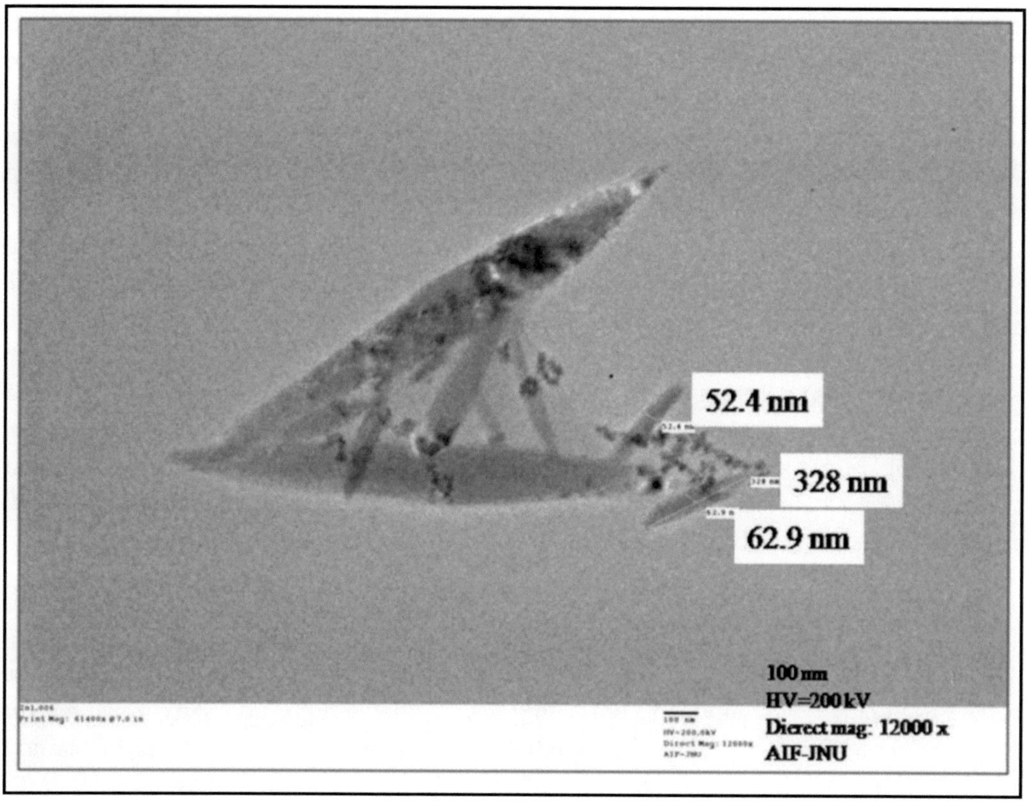

Figure 12. TEM of CFATU/ZnO composite

6. SO based polyols

SO derived polyols serve as the most important oleochemicals for PU production. Some of the SO polyols are derived through various chemical reactions such as epoxidation followed by hydration, ring opening by methanol, acids such as HBr, HCl, hydroformylation, ozonolysis of SO, others are naturally available polyols such as *Ricinus communis* or CO and Lesquerella oil. As discussed previously, the properties of PU also depend upon the type of polyol and the method by which the polyol is derived. In SO polyols, in general, hydroxyl groups are present in the middle of the triglyceride chains. Due to this, in cross linked polyols, the pendant or dangling chains provide steric hindrance to cross-linking, they do not support stress under load and act as plasticizers. In SO polyols, there is difference in the length of elastically active network chains (EANC) and elastically inactive network chains or dangling chains (DC). This variation is also passed onto their respective PU. The properties of PU thus also depend upon the content of EANC and DC, i.e., the number and position of hydroxyl groups. The number of hydroxyls on each chain in SO polyols and their stereochemistry are also variable depending upon the hydroxylating agents, hydroxylating method and other reaction conditions. For example, polyols obtained by ozonolysis and hydrogenation bear hydroxyl groups at terminal position.

SO polyols may also undergo amidation with diethanolamine to yield fatty amide polyols (FAP) [49]. Similar to FAD and FAT, FAP also house two hydroxyl ethyl amides directly attached to amide nitrogen and multiple hydroxyl groups located on the pendant fatty alkyl chains, which were part of the triglyceride molecule in parent SO bearing double bonds at the site of hydroxyl groups in FAP. The properties of SO polyols obtained by either method mentioned previously, also influence the properties of FAP. Hydroxylated, hydroxymethylated, carboxylated SO followed by their amidation yield polyols with higher number of hydroxyl groups with improved distribution [50-56]. These are ideal candidates to produce PU foams. The approach has been accomplished on CO, LO, PO, Rapeseed, Safflower, Soybean oils and refined bleached deodorized Palm Kernel Olein. These polyols have been used as non-ionic surfactants in the household and cosmetic industries and also to produce PU foams on treatment with suitable isocyanates. Such polyols prove to be advantageous over CO as they can be incorporated in higher amounts during PU formulations. Foams obtained show improvement in terms of high close cell contents, good dimensional stability and compression strength.

6.1. PU foams from SO FAP

Alkanolamide polyols serve as excellent starting materials for PU foams [50-56]. The variations in fatty acid components of starting SO, number and position of hydroxyl groups and also the presence of dangling chains in the polyol confer differences in performance and cellular structures in PU foams. The hydroxyl content of PU determines the suitability of PU foams ranging from flexible to rigid foams. A. Palaniswamy et al. produced PU foams from FAP derived from PO and Polymeric Diphenylmethane Diisocyanate (PMDI) by hand foaming. It was found that the decreased FAP content led to increase in compressive strength and density of PU foam [50]. In another research work, they have produced PU foams from PMDI and CO, in the presence of stannous octoate as catalyst and Tegostab by hand mixing process with carbon dioxide as the blowing agent generated from reaction between excess PMDI and water. PU foams with varying FAP content, catalyst and molecular weight of poly propylene glycol were studied with respect to their effect on density and compression strength [50, 51].

6.2. PU foams from SO based FAD epoxies

The epoxidized oil based alkanolamides are also classified as polyols for PU foams [56, 57]. PU show low thermal stability, thus with view to improve the thermal stability and mechanical properties of PU, heterocyclic groups such as isocyanurate, imide, phosphazene and oxazolidone, are incorporated in SO PU. The latter is formed by the chemical reaction between an oxirane ring and isocyanate in presence of a catalyst, the approach improving both thermal stability as well as stress-strain properties of the modified product with respect to the pristine material. PU foams derived from epoxidised alkanolamides show better compression strength, thermal conductivity, close cell contents and dimensional stability relative to plain alkanolamide PU. However, in some examples, a part of epoxy content is lost during amidation reaction occurring at higher temperatures. Thus, it became imperative to determine ideal reaction conditions for amidation to retain maximum number of epoxidized rings, which was attempted by Lee et al and characterized by high performance liquid chromatography and gas chromatography [55, 56].

7. SO based polymers for PU production

As discussed earlier, SO undergo numerous transformations yielding various derivatives. Some of these derivatives bearing (inter or intra located) hydroxyls serve as excellent starting materials for PU production. A large number of PU are prepared from SO polymers such as polyesters, alkyds, polyesteramides, polyetheramides [Figure 1], which find profound applications in paints and coatings.

8. Summary

FAD, FAT and FAP serve as good starting materials for PU production. LFADU, PFADU, CFATU have similar structural characteristics; the difference being due to the pendant fatty amide chains attributed to the fatty acid composition of the parent SO. LFADU, PFADU,and CFATU are formed at a particular NCO/OH ratio. An astonishingly abnormal rise in viscosity was observed in LFADU and PFADU above 1.5moles and in CFATU above 1.2moles of addition of TDI, followed by the formation of lumpy aggregates. According to the general chemistry of PU, a particular NCO/OH ratio is required for a particular application. The best properties in PU are achieved when this ratio is kept as or closer to 1 or 1.1, i.e., when one equivalent weight of isocyanate reacts with one equivalent weight of polyol, to achieve the highest molecular weight. In certain applications this ratio is kept well below the stoichiometry (higher hydroxyl content relative to isocyanate) to obtain low molecular weight PU for applications as adhesives and coatings as described in the chapter. NCO/OH ratio is varied by the formulator based on the type of end use application of PU. Properties of PU mentioned here also depend on the chemical route of raw materials (polyol, isocyanate), functionality and type of the raw materials (diol, triol, polyol and isocyanates-aliphatic, aromatic), the number of urethane groups per unit volume, non-isocyanate PU, as well as other structural differences such as the presence of modifiers (acrylics, metals, nanosized metal oxides, MMT clay).

The preparation through MW technique offers advantages of reduced times and improved yield. Most of these PU are used for coatings and foams. The incorporation of inorganic constituent led to improved thermal and hydrolytic stability as well as coating performance of PU. Another area that is presently being explored is the preparation of green PU from fatty isocyanates or non-isocyanate PU. Due to their numerous applications and advantages SO PU have been extensively studied and extensive research is still going on.

Author details

Eram Sharmin and Fahmina Zafar*
Department of Chemistry, Jamia Millia Islamia (A Central University), New Delhi, India

Sharif Ahmad
Materials Research Lab, Department of Chemistry,
Jamia Millia Islamia (A Central University), New Delhi, India

Acknowledgement

Dr Fahmina Zafar (Pool Officer) and Dr.Eram Sharmin (Pool Officer) acknowledge Council of Scientific and Industrial Research, New Delhi, India for Senior Research Associateships against grant nos. 13(8385-A)/2010-POOL and 13(8464-A)/2011-10 POOL, respectively. They are also thankful to the Head, Department of Chemistry, Jamia Millia Islamia(A Central University), for providing support to carry out the work.

9. References

[1] Petrović Z. S. Polyurethanes from vegetable oils. Polymer Reviews 2008; 48:109-155.
[2] Lligadas G., Ronda J.C., Galia`M., Cadiz V. Plant oils as platform chemicals for polyurethane synthesis:current state-of-the-art. Biomacromolecules 2010; 11: 2825-2835.
[3] Desroches M., Escouvois M., Auvergne R.,Caillol S., Boutevin B. From vegetable oils to polyurethanes: synthetic routes to polyols and main industrial products. Polymer Reviews 2012; 52 (1): 38-79.
[4] Pfister D.P., Xia Y., Larock R.C. Recent advances in vegetable oil-based polyurethanes. Chem Sus Chem 2011; 4(6):703-17.
[5] Zlatanic A., Petrovic Z. S., Dusek K. Structure and properties of triolein-based polyurethane networks. Biomacromolecules 2002; 3 (5): 1048-1056.
[6] Guo A., Cho Y., Petrovic Z. S. Structure and properties of halogenated and nonhalogenated soy-based polyols. J Polym Sci Part A: Polym Chem. 2000; 38 (21): 3900-3910.
[7] Javni I., Petrovic Z. S., Guo A., Fuller R. Thermal stability of polyurethanes based on vegetable oils. Journal of Applied Polymer Science 2000; 77 (8): 1723-1734.
[8] Ligadas G., Ronda J. C., Galia M., Cadiz V. Novel silicon-containing polyurethanes from vegetable oils as renewable resources. Synthesis and properties. Biomacromolecules 2006; 7 (8): 2420-2426.
[9] Bähr M., Mülhaupt R. Linseed and soybean oil-based polyurethanes prepared via the non-isocyanate route and catalytic carbon dioxide conversion. Green Chemistry 2012;14: 483-489.

* Corresponding Author

[10] Guan J., Song Y., Lin Y.,Yin X., Zuo M., Zhao Y., Tao X., Zheng Q. Progress in study of non-isocyanate polyurethane. Industrial Engineering Chemistry Research 2011; 50: 6517-6527.

[11] Gonzalez-Paz R.J., Lluch C., Lligadas G., Ronda R.C., Galia M., Cadiz V. A Green approach toward oleic and undecylenic acid-derived polyurethanes. Journal of Polymer Science Part A. Polymer Chemistry 2011; 49: 2407-2416.

[12] Parzuchowski P.G., Jurczyk-Kowalska M., Ryszkowska J., Rokicki G. epoxy resin modified with soybean oil containing cyclic carbonate groups. Journal of Applied Polymer Science 2006; 102: 2904-2914.

[13] Javni I., Hong D.P., Petrovic Z.S. Soy-based polyurethanes by nonisocyanate route. Journal of Applied Polymer Science 2008; 108: 3867-3875.

[14] Hojabri L., Kong X., Narine S.S. Fatty acid-derived diisocyanate and biobased polyurethane produced from vegetable oil: synthesis, polymerization, and characterization. Biomacromolecules 2009; 10 (4): 884-891.

[15] Hojabri L., Kong X., Narine S.S. Biomacromolecules 2010; 11: 911-918

[16] Palaskar D.V., Boyer A., Cloutet E., Alfos C., Cramail H. Synthesis of biobased polyurethane from oleic and ricinoleic acids as the renewable resources via the AB-type self-condensation approach. Biomacromolecules 2010; 11: 1202-1211.

[17] Tamami B., Sohn S., Wilkes G.L. Incorporation of carbon dioxide into soybean oil and subsequent preparation and studies of nonisocyanate polyurethane networks. Journal of Applied Polymer Science 2004; 92: 883-891.

[18] Matsumura S., Soeda Y., Toshima K. Perspectives for synthesis and production of polyurethanes and related polymers by enzymes directed toward green and sustainable chemistry. Appllied Microbiology Biotechnology 2006; 70: 12–20.

[19] Gast, L.E., Schneider W.J., Mc Manis G.E., Cowan, J.C. Polyesteramides from linseed and soybean oils for protective coatings: Diisocyanate-modified polymers. *Journal of the American Oil Chemists' Society* 1969; 46 (7): 360-364.

[20] Gast L.E., Schneider W.J., Cowan, J.C. Polyesteramides from linseed oil for protective coatings. *Journal of the American Oil Chemists' Society* 1966; 43(6): 418-421.

[21] Gast, L.E., Schneider W.J., Cowan, J.C. Polyesteramides from linseed oil for protective coatings low acid- value polymers. *Journal of the American Oil Chemists' Society* 1968; 45(7): 534-536.

[22] Ahmad S., Ashraf S M., Yadav S., Hasnat A. A polyesteramide from Pongamia glabra oil for biologically safe anticorrosive coating. Progress in Organic Coatings 2003; 47 (2): 95-102.

[23] Zafar F., Sharmin E., Ashraf S. M., Ahmad S. Studies on poly (styrene-*co*-maleic anhydride)-modified polyesteramide-based anticorrosive coatings synthesized from a sustainable resource. Journal of Applied Polymer Science 2004; 92: 2538-2544.

[24] Zafar F., Ashraf S.M., Ahmad S. Studies on zinc-containing linseed oil based polyesteramide. Reactive & Functional Polymers 2007; 67: 928-935.

[25] Raval D.A., Patel V.M., Parikh D.N. Streptomycin release from N,N-bis(2-hydroxyethyl) fattyamide modified polymeric coating. Reactive and Functional Polymers 2006; 66 (3): 315-321.

[26] Dutta S., Karak N., Synthesis, characterization of poly (urethane amide) resins from Nahar seed oil for surface coating applications. Progress in Organic Coatings 2005; 53: 147-152.

[27] Khan N.U., Bharathi N. P., Shreaz S., Hashmi A.A. Development of water-borne green polymer used as a potential nano drug vehicle and its in vitro release studies. Journal of Polymers and the Environment 2011; 19 (3): 607-614.

[28] Bharathi N. P., Khan N. U., Alam M., Shreaz S., Hashmi, A. A. Edible oil-based metal-containing bioactive polymers: synthesis, characterization, physicochemical and biological studies. Jouranl of Inorganic and Organometallic Polymers and Materials 2010; 20:839–846.

[29] Alam M, Alandis N.M., Microwave Assisted Synthesis Of Urethane Modified Polyesteramide Coatings From Jatropha Seed Oil. Journal of Polymers and the Environment 2011; 19 (3): 784-792.

[30] Alam M, Alandis N.M., Microwave assisted synthesis and characterization of olive oil based polyetheramide as anticorrosive polymeric coatings (communicated).

[31] Kashif M. Development and characterization of poly (urethane-amide) protective coating materials from renewable resource. Thesis submitted to Jamia Millia Islamia (A Central University), New Delhi, India.

[32] Yadav S., Zafar F., Hasnat A., Ahmad S., Poly (urethane fatty amide) resin from linseed oil—A renewable resource. Progress in Organic Coatings 2009; 64 (1): 27-32.

[33] Kashif M., Zafar F., Ahmad S., Pongamia glabra seed oil based poly(urethane–fatty amide). Journal of Applied Polymer Science 2010; 117: 1245–1251.

[34] Zafar F., Kashif M., Sharmin E., Ahmad S. Studies on boron containing poly(urethane fattyamide). Macromolecular Symposis 2010; 290: 79-84.

[35] Ahmad S., Zafar F., Sharmin E., Garg N., Kashif M. Synthesis and characterization of corrosion protective polyurethanefattyamide/silica hybrid coating material. Progress in Organic Coatings 2012; 73 (1): 112-117.

[36] Ooij W.J.V., Zhu D., Stacy M., Mugada T., Gandhi J., Puomi P., Corrosion protection properties of organofunctional silanes—an overview. Tsinghua Science & Technology 2005;10 (6): 639-664.

[37] Phanasgaonkar A., Raja V.S., Influence of curing temperature, silica nanoparticles- and cerium on surface morphology and corrosion behaviour of hybrid silane coatings on mild steel. Surface and Coatings Technology 2009; 203(16): 2260-2271.

[38] A.S. Vuc, M. Fir, R. Jese, A. Vilcnik, B. Orel, Structural studies of sol–gel urea/polydimethylsiloxane barrier coatings and improvement of their corrosion inhibition by addition of various alkoxysilanes. Progress in Organic Coatings 2008; 63 (1): 123-132.

[39] Ashraf S.M., Ahmad S., Riaz U., Development of novel conducting composites of linseed-oil-based poly(urethane amide) with nanostructured poly(1-naphthylamine). Polymer International 2007; 56:1173-1181.

[40] Rodriguesa P.C., Akcelrud L. Networks and blends of polyaniline and polyurethane: correlations between composition and thermal, dynamic mechanical and electrical properties. Polymer 2003; 44 (22):6891-6899.

[41] Chiang L.Y., Wang L.Y., Kuo C.S., Lin J.G. and Huang C.Y. Synthesis of novel conducting elastomers as polyaniline-interpenetrated networks of fullerenol-polyurethanes, Synthetic Metals 1997; 84 (1-3):721-724 .

[42] Mutlu H., Meir M.A.R. Castor oil as a renewable resource for the chemical industry. European Journal of Lipid Science and Technology 2012; 112 (1): 10-30.

[43] Rao B.S., Palanisamy A. Synthesis, photo curing and viscoelastic properties of triacrylate compositions based on ricinoleic acid amide derived from castor oil. Progress in Organic Coatings 2008; 63: 416-423.

[44] Rao B.S., Palanisamy A. Photo-DSC and dynamic mechanical studies on UV curable compositions containing diacrylate of ricinoleic acid amide derived from castor oil. Progress in Organic Coatings 2007; 60:161-169.

[45] Somani K, Kansara S, Parmar R, Patel N. High solid polyurethane coatings from castor oil based polyester polyols. International Journal of Polymer Materials 2004; 53:283-293.

[46] Szycher M (1999) Szycher's Handbook of polyurethane, 2nd edn.CRC Press, Sterling Biomedical, Inc, Lynnfield MA, Michael Szycher, Cardio-Tech International, Woburn, Massachusetts.

[47] Zafar F., Mir M.H., Kashif M., Sharmin E., Ahmad S. Microwave assisted synthesis of bio based metallopolyurethaneamide. Journal of Inorganic and Organometallic Polymers and Materials 2011; 21 (1): 61-68.

[48] Zafar F., Sharmin E., Zafar H., Ahmad S. Synthesis and characterization of bio-nanocomposites based on polyurethanefattyamide/ organo-montmorillonite. 2011; communicated

[49] Sharmin E., Akram D., Ahmad S. Polyol from linseed oil for waterborne coatings: synthesis and characterization. International conference. Polymer Science & Technology: Vision & Scenario (APA-2009) at New Delhi, India on Dec. 17-20, 2009

[50] Palanisamy A, Karuna M. S. L., Satyavani T., Rohini Kumar D. B., Development and Characterization of Water-Blown Polyurethane Foams from Diethanolamides of Karanja Oil. Journal of the American Oil Chemists' Society 2011; 88 (4): 541-549.

[51] Palanisamy A, Rao B. S., Mehazabeen S., Diethanolamides of castor oil as polyols for the development of water-blown polyurethane foam. Journal of Polymers and the Environment 2011; 19:698-705.

[52] Khoe T.H., Otey F., Frankel E.N., Cowan J.C. Polyurethane foams from hydroxymethylated fatty diethanolamides. Journal of the American Oil Chemists' Society 1973; 50:331-333.

[53] Khoe T.H., Frankel E.N. Rigid polyurethane foams from diethanolamides of carboxylated oils and fatty acids. Journal of the American Oil Chemists' Society 1976; 53:17-19.

[54] Shapiro SH (1968) Commercial nitrogen derivatives of fatty acids. In: Pattison ES (ed) Fatty acids, their industrial applications. Marcel Dekker, New York, pp 77-154
[55] Lyon C.K., Garret V.H., Frankel E.N. Rigid urethane foams from hydroxymethylated castor oil, safflower oil, oleic safflower oil, and polyol esters of castor acids. Journal of the American Oil Chemists' Society 1974; 51:331-334.
[56] Lee C.S., Ooi T.L., Chuah C.H., Ahmad S. Synthesis of palm oil-based diethanolamides. Journal of the American Oil Chemists' Society 2007; 84:945-952.
[57] Badri K.H., Othman Z., Ahmad S.H. Rigid polyurethane foams from oil palm resources. Journal of Materials Science 2004; 39:5541-5542.

Chapter 19

Polyglucanurethanes: Cross-Linked Polyurethanes Based on Microbial Exopolysaccharide Xanthan

Nataly Kozak and Anastasyia Hubina

Additional information is available at the end of the chapter

http://dx.doi.org/10.5772/48007

1. Introduction

Considering environmental protection and resolution a number of ecological problems (including problem of recourses for chemical synthesis depletion) synthesis of the biodegradable polymer materials becomes one of the most actual tasks of modern polymer chemistry. Among ways of environmental protection from polymer waste (keeping on waste deposits, burials, incineration, pyrolysis, recycling) there can be distinguished the method of minimization of ecological pollution due to creation of polymers able to be destructed under influence of natural factors such – chemical (oxygen, air, water), physical (sun light, heat), biological (bacteria, fungi) etc. These factors are very effective and lead to fragmentation of polymer as a result of macromolecule degradation and turning it into low-molecular compounds which become part of natural circuit of substance. In other words biodestruction is reliable and comparatively fast method of utilization. Usually it can be achieved by implication of natural compounds fragments into polymer structure. Other promising method is biopolymers modification with further creation of new synthetic polymers able for degradation under biological factors. Development of this method in future allows to resolve one of the most actual modern problems and to substitute petroleum refining products as the base of chemical synthesis with renewable source. It is also relevant using as reagents economically effective products which are cheaper than oil refining raw materials.

Purpose of our study was to create new polymerizing systems possessing above metioned attractive features. Therefore new polyglucanurethane (PGU) networks were obtained on the base of microbial polysaccharide xanthan and blocked polyisocyanate (PIC) using environment friendly method. Biopolymer application as reagent provides both preserving advantages of initial materials and developing new advanced properties of obtained biodegradable materials due to chemical modification. Replacement of toxic compounds with latent isocyanate-containing reagent blocked polyisocyanate is also a prominent advantage of developed technique. PGU were obtained via interaction of xanthan hydroxyl groups and isocyanate groups of deblocked above 125°C PIC (Kozak & Nizelskii, 2002).

Microbial polysaccharide xanthan is produced by xanthomonas campestris pv. Campestris bacteria (Gzozdyak et al., 1989). Xantan is well known and most widely used microbial exopolysaccharide. It is used in light industry (textile), heavy industry (drilling and oil production) and food industry as well as in agriculture, forestry, pharmaceutics, medicine and cosmetics. The water solutions of xantane have unique rheological properties due to metal cation complexing ability and formation of primary, secondary and higher levels of structure. The main chain of the polysaccharide is alike to cellulose and its side-chains (pendants) consist of glucose, mannose and glucuronic acid residues.

Blocked polyisocyanate is latent reagent which is able to produce reactive isocyanate groups under elevated temperature. It is multifunctional latent reagent that can consist of 40 to 70 % of dimeric compound , 20 to 60 % of tetramer and 1 to 5 % of trimer and hexamer. Melting temperature interval of blocked PIC is from 80 to-95 °C, NCO-group unblocking temperature ranges from 125 to 130 °C. PIC is soluble in most of organic solvents and can be used both in powder and liquid form.

2. Synthesis

The reaction was provided in solid. Fig. 1 describes the scheme of PGU synthesis. There occur both deblocking of polyisocyanate groups and urethane bonds formation.

There were obtained powdered PGU, hot-pressed samples and reinforced PGU with calculated degree of polysaccharide hydroxyl group substitution of 20, 40, 60, 80 and 100%. As far as obtained polymer is quite new and unexplored polymer material, not full range of the hydroxyl/urethane ratio was studied by methods presented in this Chapter. The obtained materials are acid-, alkali- and thermo resistant.

The reaction path was controlled using sampling procedure and sample analysis with FTIR spectroscopy.

Characteristic band at 2276 cm^{-1} which appears after heating the reactive mixture up to 130°C demonstrates the process of isocyanate groups deblocking. On the initial stages of reaction all of PGU samples show increasing intensity of this band due to active isocyanate group formation.

Interaction of N=C=O groups with the nearest primary and secondary hydroxyl groups of polysaccharide leads to decreasing intensity of characteristic band 2276 cm^{-1} during next 10 min. Diffusion limitations of this reaction are determined by heterogeneity of reactive mixture and results in retarding of urethane bonds formation. Mechanic stirring of reactive mixture allows improve reactive centers contact and leads to total disappearance of isocyanate groups in the system. (Fig 2.a). Consumption of hydroxyl groups is accompanied by disappearance of the band at 3165 cm^{-1} and decreasing of intensity band at 1209 cm^{-1} (valence vibrations and deformation vibrations of O-H bond in glucuronic acid residue respectively) and by intensity reduction of the band at 3215 cm^{-1} (valence vibrations of primary OH-groups in mannopyranose cycles) (Fig 2 c).

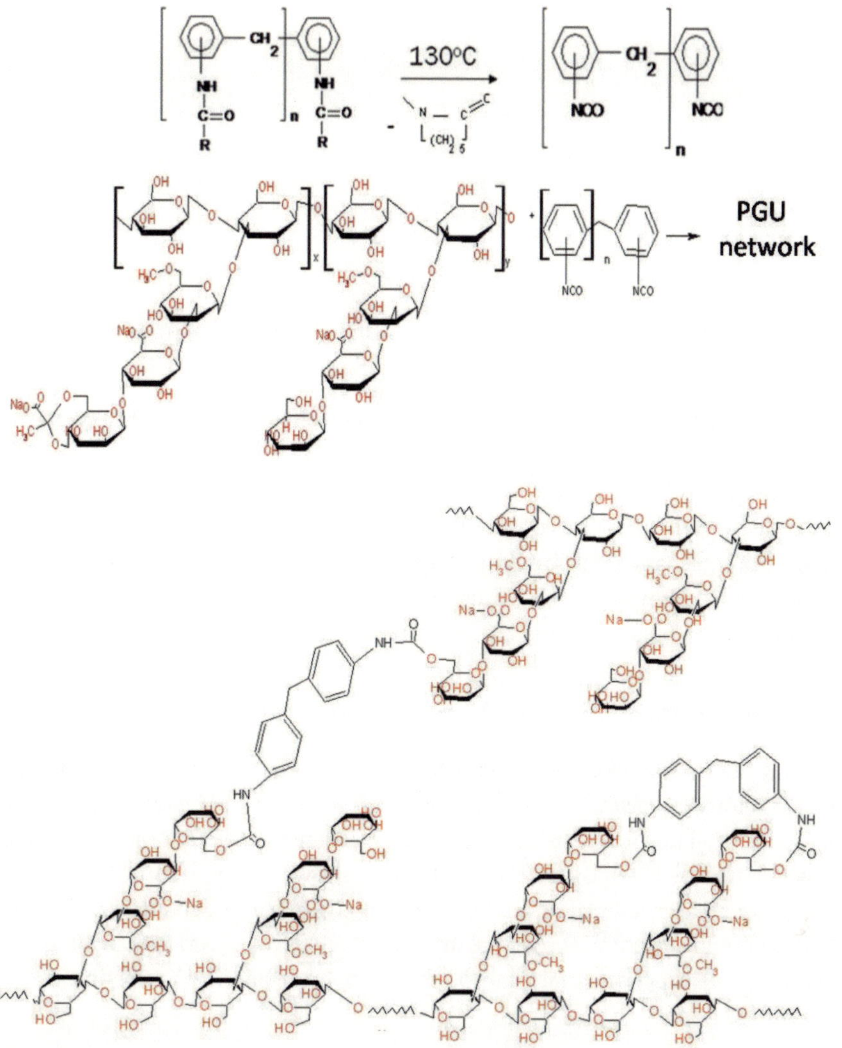

Figure 1. Scheme of PGU synthesis

Increasing intensity of the band at 3364 cm^{-1} (in characteristic doublet of N-H valence vibrations), appearance of the 1635 cm^{-1} band in the region of NH deformation vibrations (amide II) and changes of intensity of 1650 and 1590 cm^{-1} bands respond to formation of urethane bonds and releasing of blocking agent (Fig 2b).

In the wave numbers range from 3000 to 3500 cm^{-1}redistribution is observed of the intensities of absorption bands corresponding to hydrogen linked OH-groups. That points on redistribution of intermolecular bonds in the system during polysaccharide cross-linkage and PGU formation.

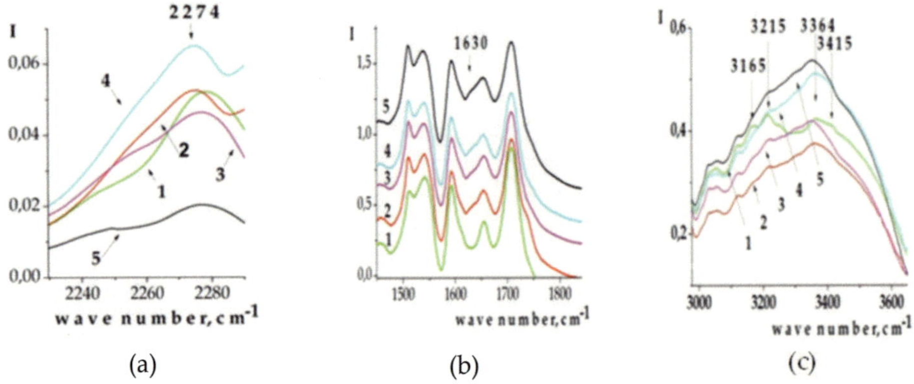

Figure 2. IR spectra of reactive mixture for PGU60 at temperature 130°C(1), 150°C (2), at 150°C after 10 min (3), after 20 min (4) after 30 min(5)

According to FTIR spectra of various PGU (PGU40, PGU80 and PGU100) the number of isocyanate groups released at the initial stage of reaction correlates with the polyisocyanate content in the system. During the first 10 min the process of polyisocyanate deblocking dominates. At the same time the urethane group formation occurs via interactiom of NCO groups and polysaccharide hydroxyl groups in acid residue of glucuronic acid and/or hydroxyl groups of mannose. The time when the urethane group formation begins to dominate depends on balance of the reagents in reaction mixture.

3. Polyglucanurethane chemical and thermal resistance

Obtained PGU networks possess advanced thermal and chemical (both alkali and acid) resistance. Chemical resistance of PGU was analyzed using standard method...[State Standart 12020-72]. Thermal resistance of initial reagents and PGU of various composition were studied with the thermogravimetry. Table 1 shows the results of PGU20 exposure in water, acid- and alkali media. Fig 4. illustrates the TGA curves: mass loss (TG), differential mass loss (DTG) and differential thermal analysis (DTA) that characterize the dependence of thermooxidative destruction character of PGU on the degree of substitution of xanthan hydroxyl groups.

As it can be seen from table 1 data the mass of PGU20 samples remains practically unchanged after the 7 and 13 days exposure in aggressive alkali end acid medium. The initial stages of PGU interaction with water, alkali- and acid media are characterized with ignificant swelling of polysaccharide component.

According to TGA data thermooxidative destruction of the systems analyzed consists of several stages. TGA curves of xanthan are typical for polysaccharides. TGA curves character for PGU networks and number of stages in temperature interval from 20 to 700°C depend on the balance of hydroxyl and urethane groups in PGU. Stage of weight loss in temperature interval 60-140°C is accompanied with endothermal peak on DTA curve and responds to absorbed water removal.

Time, day	Sample weight, g		
	H_2O	*H_2SO_4*	*NaOH*
1	0,081	0,085	0,081
2	0,213	0,205	0,431
3	0,217	0,213	0,489
7	0,250	0,224	0,485
13	0,425	0,241	0,489

Table 1. The weight change of PGU20 exposed in deionized water, concentrated sulphuric acid (V=20 ml, 30% wt.) and concentrate alkali solution (V=20 ml, 40% NaOH).

Presenting mass of absorbed water in modified and non-modified samples of exopolysaccharide as mass loss at the first stage (temperature interval 45 – 150°C) we can see that amount of absorbed water correlates with balance of hydroxyl and urethane groups in the system (Fig.3). It corresponds with the fact that system hydrophilic properties correlate with amount of hydroxyl groups. Weight loss at this stage is 1,5; 2,5; 8,5 % wt. for PGU80, PGU40, PGU20, respectively.

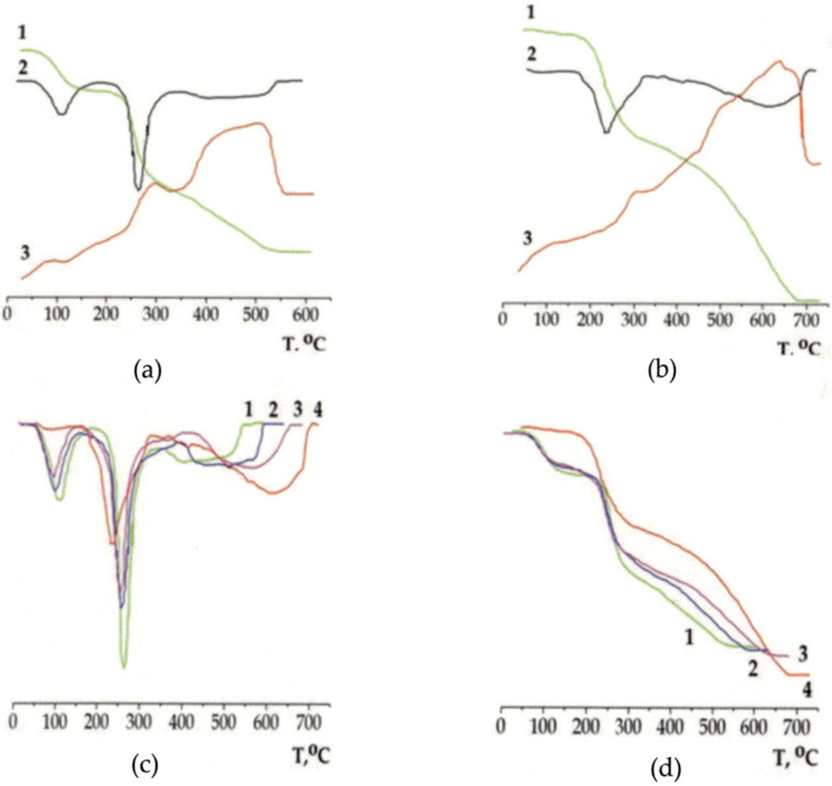

Figure 3. TGA curves - TG (1), DTG (2), DTA (3) of initial reagents: xanthan (a), blocked PIC (b); curves DTG (c) and TG (d) for initial reagents and PGU of various composition: xanthan (1), 2- PGU20 (2), PGU40, PIC (4)

Intensive thermal-oxidative destruction is observed in 200 – 400°C temperature interval. The characteristic temperature responding to maximum speed of weight loss at the stage shifts toward the higher temperatures with increasing of urethane groups amount in the system. Weight loss at this stage is 35, 30, 35 % wt. for PGU80, PGU40, PGU20, respectively.

Wide peak on DTG curves in temperature interval from 500 to 700°C mainly corresponds to destruction of carbon base of polymer. With growth of urethane group content in the system the carbon residual decreases.

As it can be seen thermal resistance of PGU grows with increasing of urethane group amount. For instance, for PGU40 high temperature stage is allocated 40°C lower than for the PGU 80. The degree of hydroxyl substitution also influences the system capacity of water absorbance. With growth of urethane group content in the system the amount of absorbed water declines

4. PGU interaction with water solutions of phenol and transition metal salts

It is known that microbial polysaccharides are considered as prospective raw materials for obtaining effective sorbents for extracting organic compounds and metal ions from water solution (Crini, 2005). Application of water-soluble polysaccharides (eg. xanthan) as sorbents is difficult. Using of PGU allows both keep sorption properties of polysaccharide and eliminate a number of disadvantages (water solubility, low chemical resistance etc.). Study of properties of cross-linked PGU (Hubina, 2009) revealed its ability to quantitative extraction of phenol and bivalent metal ions from water solution, while controlling capacity of the material with cross-linking degree.

To analyze ability of PGU to phenol sorption from its water solution the pollutant concentration was controlled before and after exposure of PGU20 and PGU60 films in 10^{-4}M phenol water solution during 24, 48 and 120 hours. Phenol concentration was monitored using UV electron spectroscopy by the change of band intensity near 256 nm. Experiment conditions were as follows: T=18 °C, $m_{sorbent}$=7 g, $V_{solution}$=50 ml. Calibration curve was plotted for 0,05; 0,075; 0,1; 0,15 and 0,2 M phenol water solutions.

According to UV-spectroscopy data after 24 hours of exposure PGU in phenol solution increasing of phenol concentration is observed both for PGU20 and PGU60 (Fig.4). Further exposure of PGU in solution leads to essential decreasing of phenol concentration. Such phenol concentrating during PGU interaction with phenol solution can be explained by the preferable swelling of polysaccharide component of the polymer in water.

The same character of concentration change is observed while extracting Cu^{2+} ions from copper sulfate solution. (Fig.5). Its concentration change was controlled using electron spectroscopy in Vis-region by the change in band intensity near 810 nm. That band corresponds to d-d transitions in $[Cu(H_2O)6]^{2+}$aqua ion. Exposure of PGU60 sample in copper sulfate water solution during 10 min. is accompanied by sufficient growth of solution absorbance and shows copper ions concentrating. Increasing time of interaction with sorbent to 60 min leads to absorbance reduction.

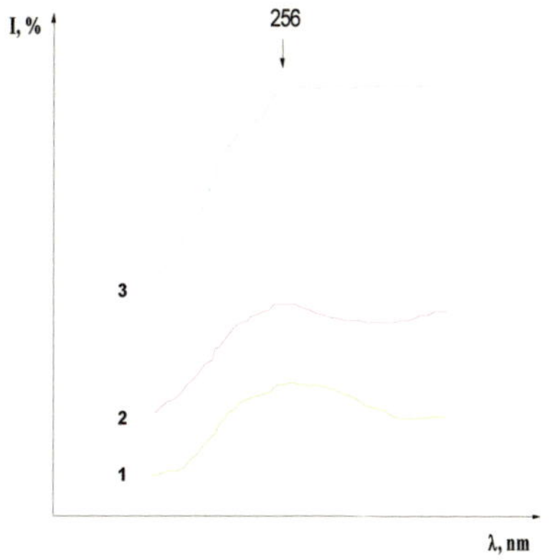

Figure 4. The electron spectra of phenol water solutions: 1 – initial; 2 – PGU60 after 10 min exposure; 3 – PGU20 after 10 min exposure

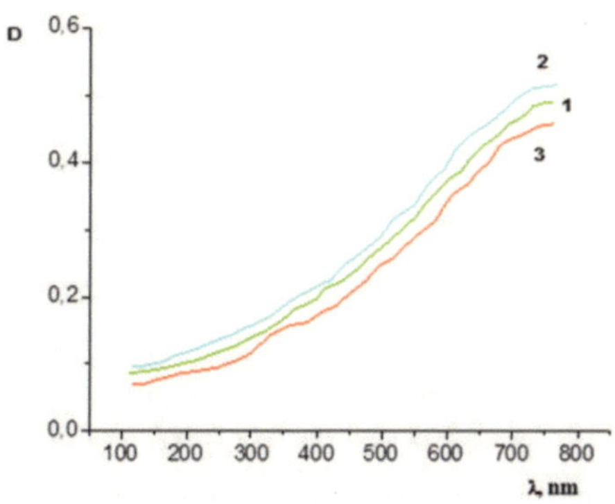

Figure 5. The electron spectra of copper salt water solutions: 1 - initial, 2 - after 10 min of PGU exposure and 3 - after 60 min of PGU exposure

Interaction of PGU40 and PGU80 with water solution of Cu^{2+} and Co^{2+} salts with concentration of 50 and 500 mg/dm^3 was studied in static conditions using conductometry. Solution conductivity was fixed after 2, 12 and 74 hours of PGU exposure. Experiment conditions were as follows m_{PGU}=1,25 g; $V_{solution}$=50 cm^3 ; permanent stirring frequency=2 Hz; room temperature. Intermediate and final concentrations were calculated from calibration curve. Fig 7 demonstrates conductivities of cobalt and copper salt solutions of various concentrations (50 and 500 mg/dm^3) depending on the time of interaction with PGU40 and PGU80. In the table 2 there are the results of concentration changes of copper(2+) sulfate and cobalt(2+) chloride solutions respectively *vs* time of exposure PGU40 and PGU80. Fig 6 shows that conductivity of cobalt chloride and copper sulfate solutions grows on the initial stages of sorbent exposure for both high (500 mg/dm) and low (50 mg/dm) concentrations.

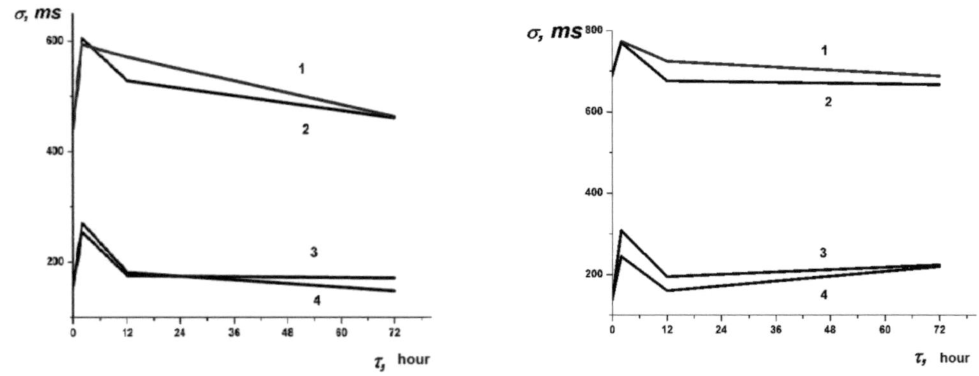

1 – $CuSO_4$ 500 mg/ dm³ in PGU80,
2 - $CuSO_4$ 500 mg/ dm³ in PGU40,
3 - CuSO4 50 mg/ dm³ in PGU80,
4. - $CuSO_4$ 50 mg/ dm³ in PGU40

1 - $CoCl_2$ 500 mg/ dm³ in PGU80,
2 - $CoCl_2$ 500 mg/ dm³ in PGU40
3. - $CoCl_2$ 50 mg/ dm³ in PGU80,
4. - $CoCl_2$ 50 mg/ dm³ in PGU40

Figure 6. Dependence of solution conductivity *vs* time of sorbent exposure

Similar effect was observed for hot-pressed samples of PGU while interacting with water solutions of phenol and transition metal salts due to predominant swelling of polysaccharide component on the initial stages. Next 12 and 74 hours of PGU exposure result in sufficient solution concentration decreasing due to metal ions capture by the functional groups of sorbent. Equilibrium was achieved both for PGU40 and PGU80 in solutions of high cobalt salt concentration (500 mg/ dm³) while for copper salt solutions equilibrium was achieved for low concentration (50 mg/ dm³).

As table 2 and Fig 6 show the efficiency of metal ions extraction depends on both concentration of solution and the balance of hydroxyl and urethane groups in PGU. The highest concentration fall was observed for PGU80. The solutions of low concentration aren't sensible to hydroxyl and urethane balance in PGU. Besides, the best ion extraction of cobalt ions from low concentration solutions is achieved with short-time exhibition of sorbent. For cobalt salts solutions of high concentration and copper salts solutions of low concentration the best effect is achieved during 74 hours of exhibition.

PGU	C, mg/dm³	Cu SO_4			$CoCl_2$		
		ΔC_{2h} , %	ΔC_{12h} , %	ΔC_{74h}, %	ΔC_{2h} , %	ΔC_{12h} , %	ΔC_{74h} , %
40	50	20,0	45,3	55,0	11,2	43,4	33,1
80	50	24,2	47,2	48,3	29,2	53,6	36,4
40	500	73,2	84,7	86,4	55,0	58,0	60,0
80	500	63,2	83,2	86,3	55,2	60,7	61,2

Table 2. Concentration change of cobalt chloride and copper sulfate solutions.

The results of PGU sorptive properties research in static conditions correspond with the research results obtained under dynamic conditions. Ability of synthesized powdered PGU materials to sorb metal ions in dynamic conditions was examined for $CoCl_2$ and $CuSO_4$ water solutions of various concentration. PGU60 sorbent was used. Concentration of initial and final solutions was controlled using electron spectroscopy in Vis-region by the change of band intensity responding to d-d transitions in aqua ion $[Co(H_2O)_6]^{2+}$ and d-d transitions in aquaion $[Cu(H_2O)_6]^{2+}$ (near 510 nm and 810 nm respectively). Calibration graphs were plotted for 0,05; 0,075; 0,1; 0,15, 0,2 M concentrations of $CoCl_2 \times 6H_2O$ and $CuSO_4 \times 5H_2O$. Experiment was carried out at 18 °C. Solutions of copper and cobalt salts with initial concentrations of 0,1 M and 0,05 M were pumped through the column filled with the sorbent powder. m_{PGU}=7 g; $V_{solution}$=50 cm^3. Under dynamic conditions the time of interaction ranged from 60 to 120 seconds. Fig. 7 illustrates the character of Co^{2+} and Cu^{2+} ion sorption.

After dynamic contact of copper salt with sorbent during 60 seconds absorbance of filtrate (D) falls to 1,08 in comparison with the initial solution absorbance value of 1,4. It responds to 40% decreasing of copper ions in solution. For 0,1M solution of cobalt chloride after dynamic contact with PGU40 there is observed absorbance fall from 0,52 to 0,48 that responds to extracting of 20% metal ions. For 0,05M solution of cobalt chloride is achieved 40% decreasing of cobalt ion concentration after 60 sec contact. Calculated concentration of cobalt ions in final solutions was 0,08M for initial 0,1M and 0,03M for initial 0,05M. The dynamic sorption of metal ions with PGU40 sorbent is more effective for solutions with lower concentration.

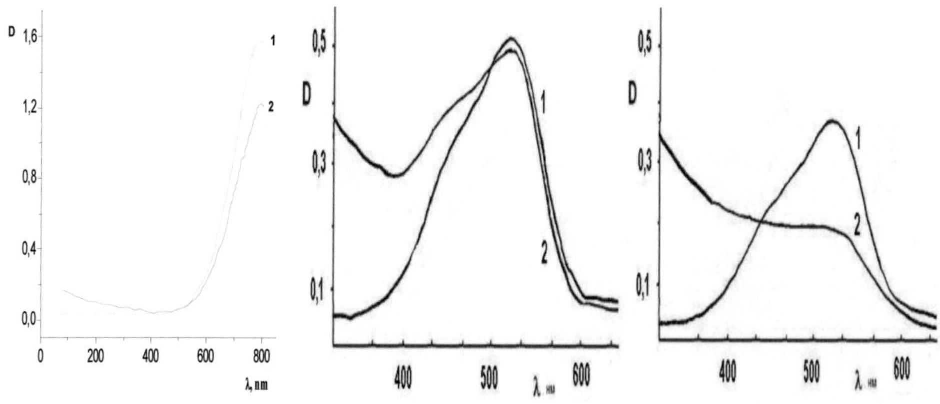

(a) (b) (c)

Figure 7. The electron spectra of water solutions of metal salts before (1) and after (2) passing through the column filled with powdered PGU40

According to (Bergmann et al., 2008) the mechanism of ion capture by water soluble polysaccharides mainly related to formation of complexes of chelate structure. Research of complex formation of PGU with metal ions allows conclude of chelate structure of formed complexes. In particular chelate structure of copper ion (2+) complexes with PGU is confirmed in (Hubina et al., 2010). Using results of (Bergman et al., 2008; Hubina et al., 2010) we can assume that cobalt ions also form chelate structures with PGU.

Analysis influence of cross-linking degree of PGU on complexing metal ions with functional groups of PGU demonstrated that variation of hydroxyl and urethane groups balance in the system allows to achieve effective control over holding metal ions in PGU matrix.

Peculiarities of "PGU-copper ion" complex formation were studied using electron paramagnetic resonance method (EPR). Copper ions were introduced into PGU matrix via pumping of 0,1M water solution of CuSO4 through column filled with PGU60 (the way it described above for dynamic sorption conditions). Then metal-containing PGU sample was dried and the EPR spectra of bivalent copper were recorded. Fig. 8 demonstrates EPR spectrum of dried PGU60 after interaction with copper sulfate water solution.

Recorded spectra are characterized with anisotropy of g-factor and appearance of hyperfine structure in the region of $g_{\|}$ that is usually concerned to tetragonal chelate complexes of bivalent copper. Hyperfine structure components are broadened as a result of superposition of signals from tetragonal copper complexes which can differ both in symmetry distortion and in nearest chemical surrounding. Integral intensity of obtained EPR spectra depends on the balance of hydroxyl and urethane group in PGU while electron spin parameters are almost unchanged. That points on preferable interaction of metal cation with one of the components of PGU. This conclusion corresponds with the regularities of bivalent copper ion complexing with mono saccharides.

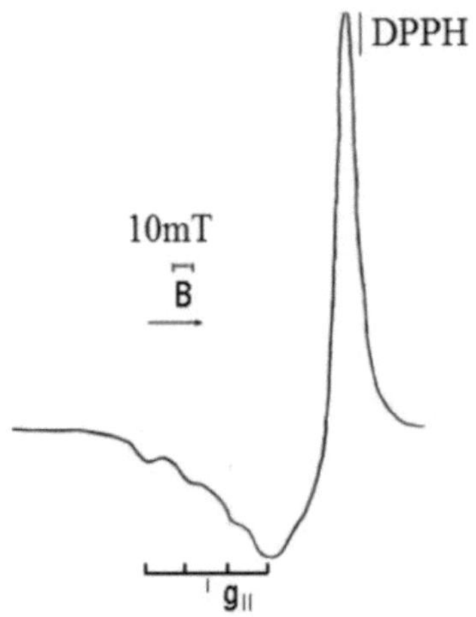

Figure 8. The representative EPR spectrum of copper containing PGU

Experimental analysis of influence of metal ions on water molecules self diffusion process in swelled polysaccharide gel was carried out by the method of quasi-elastic neutron scattering.

For description of experimental dependences there was calculated the values of general coefficient of water self diffusion in swelled gel $D = D_{singl} + D_{col}$ as well as values of its D_{singl} (single particle) component and D_{col} (collective) component. (Fig 9b).

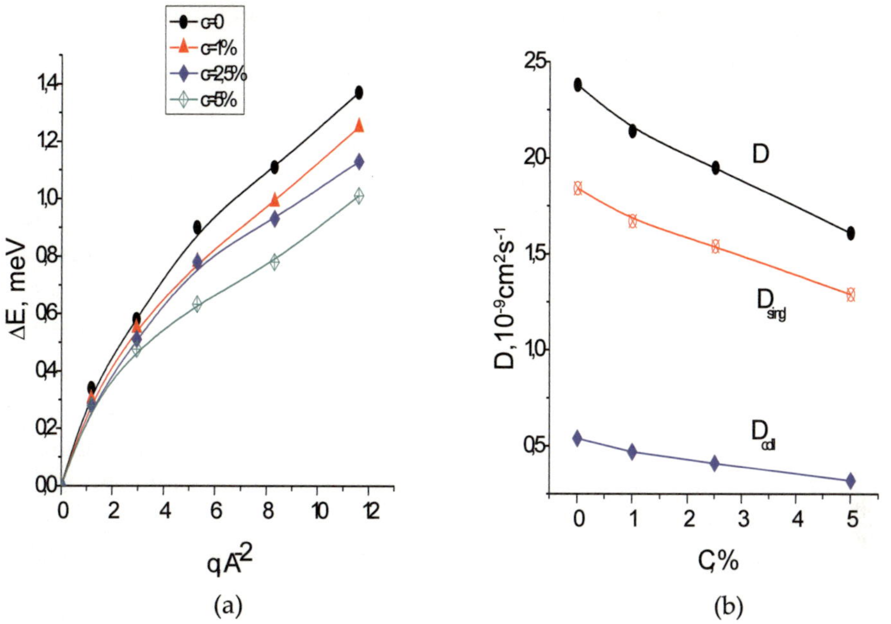

Figure 9. Dependence of energy broadening of quasi-elastic peak ΔE *vs* squared transferred kinetic moment *q* for various polymer–metal salt ratio (a) and concentration dependencies of D, D_{singl} and D_{col} (b).

Obtained data demonstrate that addition of metal salt to water and increasing of solution concentration decreases the coefficient of water molecules self diffusion in swelled xanthan gel both for general coefficient and for its components.

That indicates the swelled polymer density growth in presence of metal ions. Such condensation effect can be explained taking into account ability of transition metal chelates to form charge-transfer complexes with electron-donor centers of polymer resulting in creation of additional 'coordination juncs' both in swelled linear polysaccharide and in chemically cross-linked PGU.

Thus, polyglucanurethanes based on water soluble polysaccharide xanthan and blocked polyisocyanate are able to extract heavy metal ions from their salt water solutions. Variation of hydroxyl and urethane groups balance at the cross-linking stage provides effective regulation of complexing and capturing of metal ions with polyglucaneurethanes.

5. Biodestruction

Used plastics utilization had become an important problem. One of the ways of polymer and in particular of polyurethane waste reclamation is biodegradation of plastic. The ability to biodegradation was analized on the PGU exposed in the medium of common soil microorganisms association as well as into the medium of resistive microbial association isolated from soils polluted by chlorine-organic pesticides. (Hubina et al., 2009). There were explored biodegradable properties of two types of PGU20: based on microbial polysaccharide xanthan (PGU20) and methylcellulose (PGU20-cellulose).

The resistant microbial association with working name "Micros" was isolated from soil polluted by chlorine-organic pesticides. This association has high destructive activity with respect to chlorine-organic, organic-phosphorous, simm-triazine and other pesticide groups. It was supposed, that "Micros" is able to specialize to exotic substratum and could decompose polyglukanurethane systems due to utilization of this polymer as carbon or nitrogen source. To compare destruction of the PGU by common soil microorganisms the soil native-born microbial association was isolated from pollution-free chernozem soil. It has never contacted with pollutants in soil.

Changes in polymer structure after exposure in microbial medium were analyzed by FTIR spectroscopy, optical microscopy and thermogravimetry. Fig presents IR – spectra of PGU20 after contact with aggressive and natural microbial medium. Evaluation of redistribution of bonds in destructed PGU was provided by analysis of the location, width and intensity of valence vibrations band of C-O-C intercycle group of polysaccharide (805 cm^{-1}) and also by analysis of location and intensity of band amide II) in PGU urethane bridges (1600 ÷ 1660 cm^{-1}).

In PGU spectrum after the sample contact with aggressive medium (spectrum 3 fig 10) it is seen resolving of complex band at 1600 – 1660 cm-1 (N-H vibrations) comparing to initial PGU20 spectrum (Fig.10 spectrum 1). This may be caused by destruction of essential part of urethane bonds of PGU in aggressive medium.

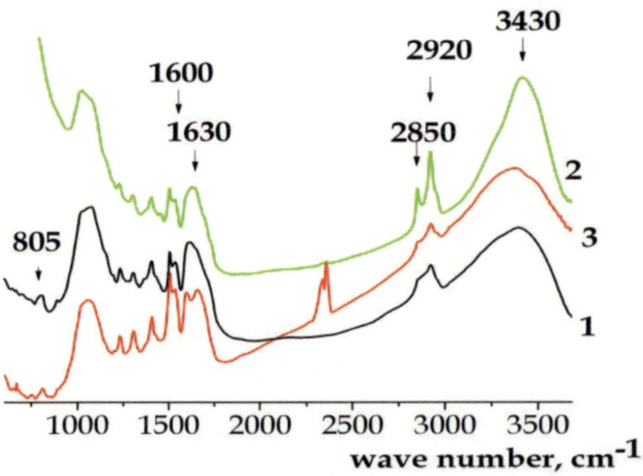

Figure 10. IR spectra of PGU20 based on xanthan 1 – initial PGU, 2 – PGU after contact with natural microbial medium, 3 – PGU after contact with aggressive microbial medium

Other type of PGU destruction is observed in natural medium (fig. 10 Spectrum 2). It is approved by redistribution of intensities of valence vibrations of C-O-C groups comparing with initial PGU and by absence of prominent changes in area of amide II. The band at 805 cm-1 responding to intercycle C-O-C bonds disappears. It is accompanied by increasing of intensity of the band responding to free OH-group (3430 cm-1) and C-H bonds (2850 and 2920 cm-1).

Another evidence of sufficient urethane bonds destruction in PGU20 in aggressive medium is seen on microphoto of PGU films after aggressive destruction (Fig. 11). There can be distinguished fibrous polymer elements, usual for fibrous of initial polysaccharide xanthan, which formed after urethane destruction. This effect is absent on the micro photos of PGU sample after destruction in natural microbial medium .

1 2 3

1 – initial polysaccharide fibers; 2 - effect of aggressive microbial medium; 3 – effect of natural microbial medium.

Figure 11. Micro images of PGU films after contact with natural and aggressive microbial mediums

IR spectra of PGU20-cellulose present the difference between this polymer biodestruction and biodestruction of PGU20-xanthan (Fig. 12)

In particular, the intensity of the band in area 1600 – 1660 cm^{-1} falls both for natural medium and aggressive one. (Fig.13 stectra 2 and 3). Intensity of 809 cm-1 band decreases (valence vibrations of intercycle bonds C-O-C) after exposure of PGU20-cellulose in natural microbial medium (Fig.12 spectrum 2) differing from intensity of this band in PGU20-cellulose spectrum after contact with aggressive medium. Bands corresponding to symmetric and asymmetric vibrations of saccharide groups C-O-C are nearly unchanged.

Efficient difference in biodestruction character of PGU20-xanthan and PGU20-cellulose allows conclude that xanthan based PGU is destructed mainly in its saccharide side-chains (containing D-mannose, D- glucuronic acid and D-glucose).

IR data correlate with results of TGA analysis. Table 3 shows difference in character of TG, DTG and DTA curves for PGU20-cellulose after contact with aggressive and natural medium. On DTG curves of PGU20-cellulose after natural and aggressive medium it is seen that stage of thermal destruction in temperature interval 200 – 300 oC is moved into district of higher temperatures while in PGU40-xanthan this stage is splitted and differs for natural and aggressive medium.

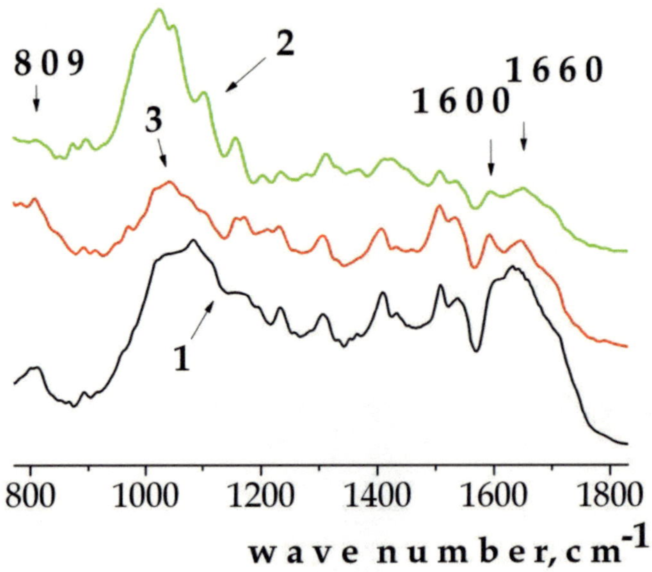

Figure 12. IR spectra PGU20-cellulose 1 – initial PGU20-cellulose; 2 – PGU after contact with natural microbial medium, 3 – PGU after contact with aggressive microbial medium.

System	Initial	Natural	Aggressive
PGU40	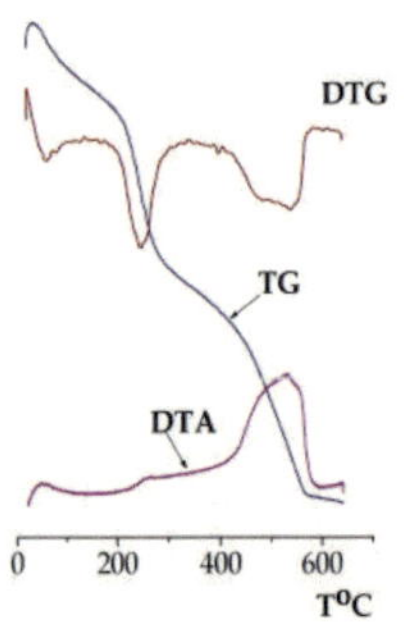	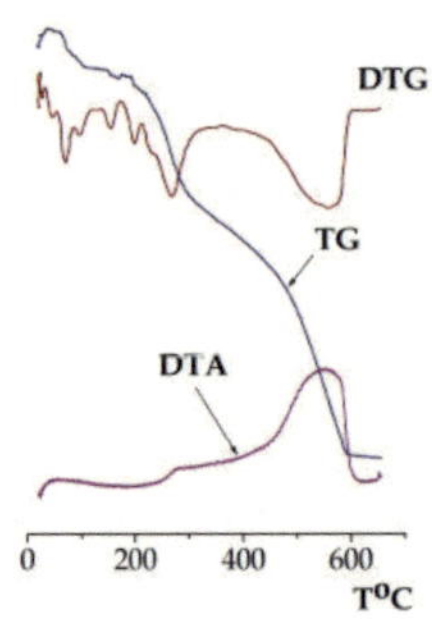	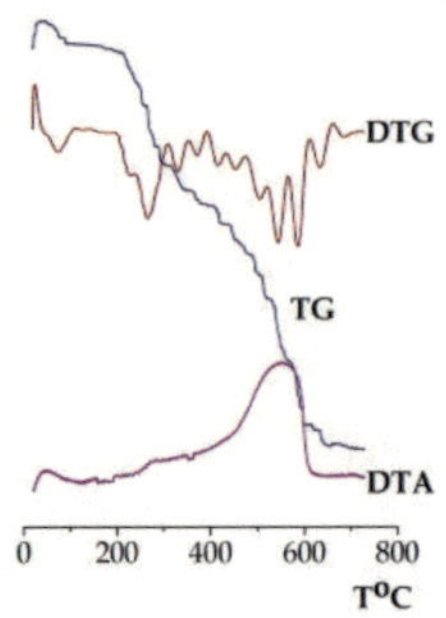
PGU-Cel-20	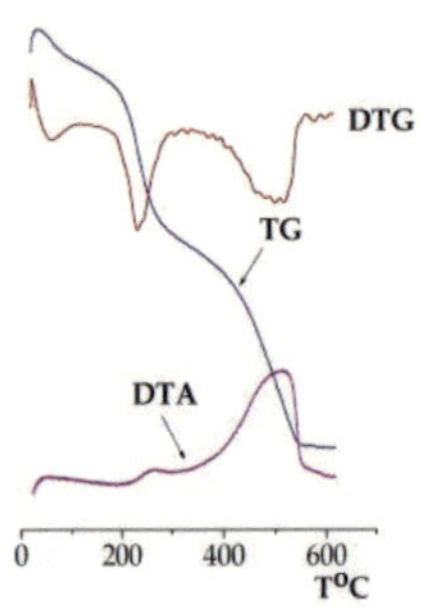	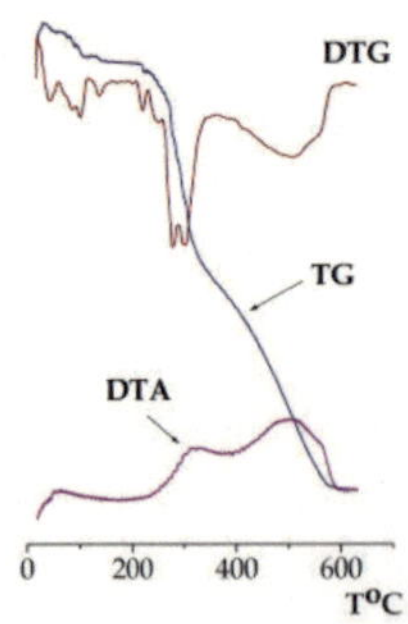	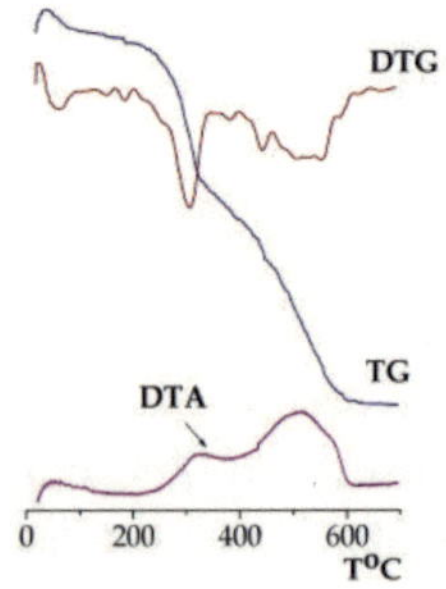

Table 3. IR spectra of PGU40 and PGU-cellulose-20 after biodestruction

6. Conclusions

New hydroxylcontaining polyurethane networks (polyglucanurethane) with various urethane group content were obtained based on microbial exopolysaccharide xanthan and latent blocked polyisocyanate using environment friendly technique. Obtained polyglukaneurethanes possess advanced thermal and chemical (both alkali and acid) resistance. Thermal resistance of PGU grows with increasing of urethane group amount. Were as with growth of urethane group content in the system the amount of absorbed water declines.

Study of properties of polyglucanuretanes reveals their ability to quantitative extraction of phenol and bivalent metal ions from water solution, while controlling sorption properties of the material with polysaccharide cross-linking degree. According to electron spectroscopy and EPR data the main mechanism of ion capture by polyglucanurethane consists in formation of "polymer-metal ion" chelate complexes.

Biodestruction research demonstrate that polyglucanurethane degradation in aggressive microbial media occurs via polysaccharide fragmentation due to urethane bonds cleavage. On the contrary microbial association that was isolated from pollution-free chernozem disintegrate the polysaccharide chains predominantly.

But both in aggressive microbial association and in natural microbial association that was isolated from pollution-free soil the direct relation was observed between destruction process intensity and percentage of polysaccharide OH-group substitution in PGU. The lower percentage of OH-groups was substituted, the more intensive destruction process was observed.

Author details

Nataly Kozak and Anastasyia Hubina
Institute of Macromolecular Chemistry National Academy of Sciences of Ukraine, Ukraine

Acknowledgement

The authors acknowledge Prof. Gvozdiak R.I, Dr. Dankevich L. and Dr. Vocelko S. (Institute of microbiology and virology of NAS of Ukraine) for help in the study of polyglukanurethane biodegradation in various microbial environments and for providing of various producents microbial exopolysaccharides.

7. References

Bergmann D., Furth G., Mayer Ch.Binding of bivalent cations by xanthan in aqueous solution. *International Journal of Biological Macromolecules*. 2008, Oct 1;43(3):245-51, 0141-8130

Crini G. Recent developments in polysaccharide-based materials used as absorbents in wastewater treatment. *Progress in Polymer Science*. 2005. V. 30, pp. 38 – 70, 0079-6700

Gvozdyak R.I., M.S. Matyshevskaya, Y.F. Grigoriev, O.A. Litvinchuk. (1989). *Microbial polysaccharide xanthan*, Naukova Dumka, 5120009670, Kyiv.

Hubina A.V., Kozak N.V., Nizelskii Yu.M. Hydroxyl containing polyurethane networks based on Xanthan and blocked polyisocianates and their interactin with phenol water solutions. *Polymer Journal*. 2009.-№1, pp. 58 – 61, 0203-327

Hubina A., Dankevich L., Kozak N., Yamborko N. 2009. Biodegradable Microbial Exopolysaccharide Based Polyurethane Networks for Phenols Sorption from Water Solutions. Odessa. 2009

Hubina A., Klepko V., Kozak N., Vasilkevich V., Slisenko V. Interaction of cross-linked polyglucanurethanes with transition metal salts water solutions. *Physics and Chemistry of Polymers (Tver)*. 2010. V. 16, pp. 214 - 219

Kozak N.V., Nizelskii Y.M. Polymer composition modification with blocked polyisocyanates. *Chemistry and chemical technology problems*, 2002.- №3, 0321-4095

State Standart (Ukraine) 12020-72

Chapter 20

Biobased Polyurethane from Palm Kernel Oil-Based Polyol

Khairiah Haji Badri

Additional information is available at the end of the chapter

http://dx.doi.org/10.5772/47966

1. Introduction

Polyurethanes are block copolymers containing segments of low molecular weight polyester or polyether bonded to a urethane group (-NHCO-O). Traditionally, these polymers are prepared by reacting three basic materials; polyisocyanates, hydroxyl-containing polymers (polyester or polyether polyol) and chain extender, normally low molecular weight diol or diamine (such as 1, 4-butanediol or 1, 4-dibutylamine).

Polyols are generally manufactured by one or two possible chemical routes, namely alkoxylation and esterification. Alkoxylation, by far is the most common route, involves the reaction between a hydroxyl or an amine-containing initiator (such as sucrose, glycerol) and either propylene- or ethylene oxide. A molecular weight of up to 6000 can be obtained by extending the polymer chain with the addition of alkylene oxide. This product is suitable for more flexible polyurethanes in cushioning and elastomeric applications. The alkylene oxide used in this process is derived from mineral oil via the petroleum industry. Propylene for instance, is derived from the petroleum cracking process and is then converted to propylene oxide before being further converted to polyol by reaction with an amine or hydroxyl-containing initiator such as glycerol.

At present, most polyols used in polyurethane industry are petroleum-based where crude oil and coal are used as starting raw materials. However, these materials have been escalating in price and rate of depletion is high as well as required high technology processing system. This necessitates a look at utilizing plants that can serve as alternative feed stocks of monomers for the polymer industry. Moreover, with increasing annual consumption of polyurethane, its industrial waste is a serious matter. In Europe and the United States of America for instance, government regulations encouraged recycling of materials to avoid excessive usage of landfill area. However, with thermosetting behavior of polyurethane the recycling activity is difficult and limited. The best alternative is biodegradation. Biodegradable polymers have widely been used in pharmaceutical industry such as suture usage, wound-dressings, surgical implants and medicine delivering system. But there are still some usage limitations either due to high production cost or its low performance. This performance can be achieved by chemical and physical modification of these materials through combination of biodegradable and non-biodegradable materials.

Polyurethane based on polyester has been known to be more biodegradable than from polyether. Utilization of renewable resources to replace petrochemicals in polyurethane industry has attracted attention of many technologists. Most of these renewable resources are forest products. Palmeri oil, vernonia oil, castor oil and cardanol oil (extracted from the cashew nut shell) have been used to synthesize polyurethane polyols with multiple functionality to replace the petrochemical-based polyols (Pourjavadi et al. 1998 and Bhunia et al. 1998). Castor oil has long been used in the polyurethane industry. Relatively, it is stable to hydrolysis due to its long fatty acid chain but sensitive to oxidation due to the present of unsaturated fatty acid. Commercially, it can only be used in the coating and adhesive industries.

Polyester polyols are generally consisted of adipic acid, phthalic anhydride, dimer acid (dimerized linoleic acid), monomeric glycol and triol. It has low acid number (normally 1-4 mg KOH/g) and low moisture content (less than 0.1%). These properties are not easily achieved unless a high-technology processing method is applied. Due to these industrials requirements, polyester polyols are usually supplied at higher price compared to polyether polyols. Polyether polyols on the other hand, are commercially produced from catalytic reaction of alkylene oxide i.e.propylene oxide or ethylene oxide to di- or polyfunctional alcohol. Its functionality is four and above and is useful in the production of rigid foam. It can also be produced with the presence of di- or polyfunctional amine i.e. diethanolamine when high reactivity is required (such as laminated continuous panel production). Important properties specified in polyurethane industry for polyols are as summarized in Table 1.

Classification	Flexible foam / Elastomer	Rigid / Structural foam
Molecular Weight	1,000 to 6,500	400 to 1,200
Hydroxyl value, mgKOH/g	28 to 160	250 to 1,000
Functionality	2.0 to 3.0	3.0 to 8.0

Table 1. Technical requirements for polyols used in polyurethane industry (Wood 1990).

The lower the equivalent weight of polyol is, the higher the rigidity of the polyurethane. These contributed to higher compressive strength, modulus, thermal stability and dimensional stability polyurethanes. If the equivalent weight is excessively low, the resulting polymer becomes more friable and required more isocyanate especially for the production of rigid polyurethane foam (Berlin and Zhitinkina 1982).

Natural occurring oils and fats are water-insoluble substances originated from vegetable, land or marine animal known as triglycerides. A triglyceride is the reaction product of one molecule of glycerol with three molecules of fatty acids to yield three molecules of water and one molecule of a triglyceride. The molecular weight of the glycerol portion (C_3H_5) of a triglyceride molecule is 41. The combined molecular weight of the fatty acid radicals (RCOO-) varies. Natural oils can undergo a number of chemical reactions such as hydrolysis, esterification, interesterification, saponification, hydrogenation, alkoxylation, halogenation, hydroxylation, Diels-Alder reaction and reaction with formaldehydes. Polyester is a high molecular weight chemical with ester group –O-C=O- as repeating unit. It is achieved by polycondensation and esterification of carboxylic acid with hydroxyl-containing compounds.

Lauric oil or better known as lauric acid is the main source of fatty acids. The only lauric oils available to the world market are coconut oil and palm kernel oil. The oil palm is a monocotyledon belonging to the Elaeis Guiness species. Palm kernel oil (PKO) is obtained from the kernel part of the oil palm fruit. The percentage of unsaturated fatty acids is much lower compared to palm oil as shown in Table 2. PKO consist of 80 percent saturated fatty acid and 10% of each polyunsaturated and unsaturated fatty acid. Palm oil on the other hand, consist of 53% saturated fatty acid, 10% polyunsaturated and 37% unsaturated fatty acids. The higher the unsaturated fatty acid contents the unstable it is when exposed to heat. The reactivity increases substantially if the double bond are conjugated (separated by one single bond) or methylene-interupted (separated by a –CH_2 unit). PKO contains only traces of carotene.

Vegetable Oil	Saturated Fatty Acid, %							Unsaturated Fatty Acid, %					
								Enoic				Dienoic	Trienoic
Carbon Chain	C8	C10	C12	C14	C16	C18	>C18	<C16	C16	C18	>C18	C18	C18
Palm Oil				1-6	32-47	1-6				40-52		2-11	
Palm Kernel Oil	2-4	3-7	45-52	14-19	6-9	1-3	1-2		0-1	10-18		1-2	

Table 2. Fatty acid contents in palm oil and palm kernel oil (Khairiah Haji Badri 2002).

Two major reactions occurred during polymerization of polyurethane. First, the reaction of isocyanate with water yields a disubstituted urea and generates carbon dioxide. This is called the blowing reaction because the carbon dioxide is acting as an auxiliary-blowing agent. The second reaction is between the polyfunctional alcohol (polyol) and the isocyanate (Fig. 1).

R-N=C=O + R'-O-H → R-NH-C(O)-O-R
Isocyanate Polyol — Polyurethane

R-N=C=O + H-O-H → R-NH-C(O)-O-R' + R-N-H + CO_2
Isocyanate Water — Polyurethane

R-N=C=O + R-N-H → R-N-C-N-R
Isocyanate

Figure 1. Addition polymerization of polyurethane

It generates a urethane linkage and this is referred to as the gelation reaction. The isocyanate reacts slowly with alcohols, water and the unstable amino products without the present of catalyst. However, for most commercial requirements the acceleration of these reactions is required.

One characteristic of amorphous polymeric systems is the glass transition temperature, T_g that defines the point where the polymer undergoes a change from glassy to rubbery behavior. Considerable attention has been devoted over the last several years to these studies: synthesis of polyurethane polyol from PKO and the production of oil palm empty fruit bunch fiber-filled PU composites (Badri et al. 1999, 2000[a], 2000[b], 2001; Khairiah Haji Badri 2002; Badri et al. 2004[a], 2004[b]; Badri et al. 2005; Badri & Mat Amin 2006; Badri et al. 2006[a], 2006[b]; Mat Amin et al. 2007, Norzali et al. 2011[a], 2011[b]; Liow et al.; Wong & Badri 2010, Badri & Redhwan 2010;). These include intensive evaluation on the chemical, mechanical, thermal and environmental stress on the synthesized polyol and PU foam by looking at various scopes:

- Synthesis of the palm kernel oil-based polyol from refined, bleached and deodorized (RBD) palm kernel oil via esterification and polycondensation.
- Preparation of the polyurethane foam from the RBD PKO-based polyol and evaluation of its chemical, mechanical and thermal decomposition and glass transition temperature of the foam.

2. Vegetable oil-based polyurethane polyol

Several reports have been published in producing polyurethane from vegetable oils and some of them have even been patented (Arnold 1983, Chittolini 1999 & Austin et al. 2000). Focus was given to utilization of mixture of vegetable oils in the polyurethane system and not as raw materials to produce the polyurethane. Vegetable oils that are frequently used are soybean oil, safflower oil, corn oil, sunflower seed oil, linseed oil, oititica, coconut oil, palm oil, cotton seed oil, peritta oil, olive oil, rape seed oil and nuts oil. Researches carried out using these oils were focusing on full usage of materials found abundance in certain area such as production of polyurethane foam from mixture of starch and triol polycaprolactone (Alfani et al. 1998) and mixture of starch, soybean oil and water *(Fantesk)* (Cunningham et al. 1997). Polyurethane products based on vegetable oils like nuts oil, soybean oil, corn oil, safflower oil, olive oil, canola oil and castor oil (Nayak et al. 1997, Bhunia et al. 1998, Mohapatra et al. 1998, Javni et al. 1999) exhibited high thermal stability. In Malaysia, the Malaysian Palm Oil Board (MPOB) has taken the initiative to produce polyol from the epoxidation and alcoholysis of palm oil (Ahmad et al. 1995, Siwayanan et al. 1999). An early finding has indicated that when natural oils or fats are epoxidized, they react with polyhydric alcohols to produce polyols. A study by Guthrie and Tait (2000) has successfully produced an ultraviolet (UV) curable coating from epoxidized and unprocessed palm oil, and epoxidised palm olein.

These researches however, are pointing to one direction that is synthesizing polyester. Polyester may be defined as heterochain macromolecules containing repeating ester groups (-

COO-) in the main chain of their skeletal structures. Most useful routes to polyester synthesis of carboxylic acids are step growth or direct polyesterification (condensation polymerization) and ring opening polymerization of lactones. The former is suitable for synthesis of aliphatic polyester where it utilizes primary and secondary glycols where the primary hydroxyl groups being esterified more readily. The removal of liberated water from the process is carried out by stirring and percolation of inert gas such as nitrogen, N_2. If a volatile monomer is used (i.e. glycol), an excess amount with respect to dicarboxylic acid (10 %w/w) should be added to compensate for losses caused by evaporation at high temperature. Side reactions may occur usually at 150°C and above which leads to changes in polymer structure and reduces molecular weight distribution of the polyester (Jedlinski 1992).

Esterification is one of many substitution reactions of carboxylic acids and their derivatives that involve tetrahedral addition intermediates. The extension of mechanism of carbonyl addition is as shown below. The best leaving group is the weakest base. In addition, reaction of ester with hydroxylamine (:NH_2OH) gives N-hydroxyamides (known as hydroxamic acids). This is the point where it is vital to add some reactivities to the existing polyester by addition of the amide group to form polyesteramide (Loudon 1988).

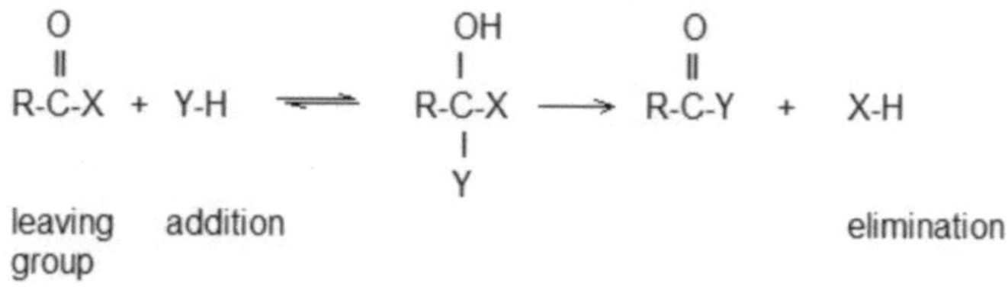

Polyurethanes are possible to decompose by prolonged contact with water, diluted acids or moist heat (causes swelling and slow hydrolysis, particularly in some ester-type polyurethanes), chlorine bleach solutions (may cause yellowing and decomposition) and prolonged exposure to light (discoloration of derivatives of aromatic isocyanates) (Roff et al. 1971). The dimensional stability of foams is a time-dependent property that receives considerable attention. Disregarding cold aging at -15±2°C, humid aging (70± 2°C at 95±5% relative humidity) is usually a prime property. Humid aging requirements (specifications) are determined by the end use of the foam. A foam that has expanded and the shrunk is considered, as a first approximation, to be caused by the effect of plasticization by heat and moisture that would allow the stresses built into the foam at the gel to relax, which will then allow the foam to return to a lower energy state. For urethane foams specifically, high thermal stability results in excellent dimensional stability over a large temperature range.

3. Green material and technology

The RBD palm kernel oil (viscosity of 65 cps, specific gravity of 0.99 g/ml, and moisture content of 0.02%) was obtained from Lee Oilmill Sdn Bhd, Kapar, Klang, Malaysia and was used as received without further purification. Polyhydric compounds consisted of dietanolamine, DEA (purity of 99.8%, hydroxyl value of 1057 mg KOH/g and functionality of 2, with viscosity of 236 cps and moisture content of 0.05%) and ethylene glycol, MEG (hydroxyl value of 1122 mg KOH/g) were supplied by Cosmopolyurethane (M) Sdn Bhd, Pelabuhan Klang, Malaysia with the inclusion of potassium acetate which was manufactured by Merck (M) Sdn Bhd, Shah Alam Malaysia. Chemicals used for the preparation of polyurethane foam were crude MDI (2,4-diphenylmethane diisocyanate), tetramethylhexanediamine (TMHDA) and pentamethyldiethyltriamine (PMDETA) (Cosmopolyurethane (M) Sdn Bhd, Port Klang, Malaysia) and silicon surfactant (Tegostab B8408, Th. Goldschmidth, Singapore). The blowing (foaming) agent used was tap water.

DEA, MEG and potassium acetate were mixed homogeneously with a ratio of 90:7:3 to form the polyhydric compound. A mixture of this polyhydric compound with RBD PKO at stochiometric ratio was continuously stirred in a 2-L glass reactor and was reacted separately at three different temperature ranges: 165-175°C, 175-185°C and 185-195°C, each for 30 minutes. The nitrogen gas was flushed into the system throughout the process. The reflux flask was connected to a condenser and a vacuum pump to withdraw the water and excess of reagent from the system. The progress of the reaction was monitored by sampling at intervals. The samples collected were then analyzed. At the end of the reaction, the polyol produced was kept in a sealed cap glass jar for further analysis. 140 g of crude MDI was poured into 100 g mixture of resin (Table 3 and Appendix A).

Composition	Part by weight, pbw
RBD PKO Polyol	100
Tegostab B8408	2
TMHDA	0.3
PMDETA	0.15
Water	4.5
Total pbw	106.95
Ratio of 100 parts to MDI	100:140

Table 3. Formulation of palm-based polyurethane foam system.

The mixture was agitated vigorously using a standard propeller at a speed of 200 rpm for 10 seconds at 20°C (Fig. 2).

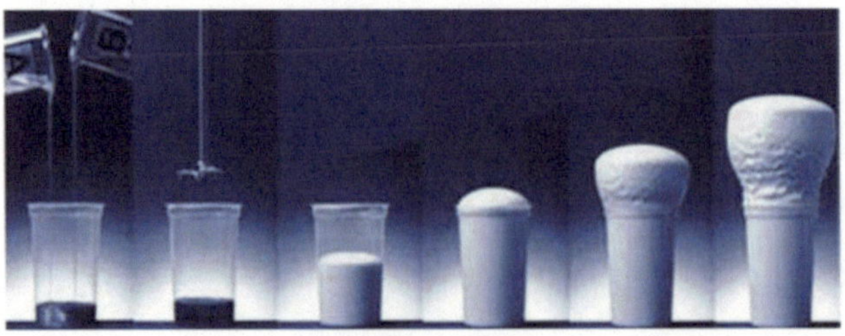

Figure 2. Polymerization of the palm-based polyurethane

The reaction time: cream time (CT), fiber/gel time (FT), tack-free time (TFT) and rise time (RT) was noted (Appendix B). The free-rise density (FRD) was calculated using equation (1).

$$\text{Free - rise density, FRD (kg / m}^3\text{)} = \frac{\text{(mass of foam and cup - mass of cup), kg}}{\text{capacity of cup, m}^3} \quad (1)$$

The mixture was poured into a waxed mold, covered and screwed tight. The foam was demolded after 10 minutes. The molded density was determined using equation (2). The molded foam was conditioned for 16 hours at 23 ± 2°C before further characterization of the polyurethane foam.

$$\text{Molded density, MD (kg/m}^3\text{)} = \frac{\text{mass of molded foam, kg}}{\text{volume of molded foam, m}^3} \quad (2)$$

Fourier Transform Infrared analysis of the RBD PKO polyol and palm-based PU was carried out on the Perkin Elmer Spectrum V-2000 spectrometer by Diamond Attenuation Total Reflectance (DATR) method. The samples collected during the intervals were scanned between 4000 and 600 cm^{-1} wavenumbers. For the former, two selected peaks (designated as peaks A and B) were used to monitor the progress of the reaction (derivatization).

Chromatography analyses were carried out on the former by thin layer chromatography followed by gas chromatography. A sample of 1 pph (part per hundred) by dilution in methanol was dropped on the silica plate with minimal diameter of about 0.5 mm and was applied 2 cm from the edge of the silica plate. The plate was removed once it traversed 2/3 of the length of the plate (normal length of a silica plate is 20 cm). The plate was placed in a chamber containing iodine crystals (iodine reacts with organic compounds to yield dark stain) after the methanol has all evaporated. The molecular weight was determined using gas chromatography coupled with mass spectrometer, GC-MS model Bruker 200 MHz with splitless inlet and HP5 (polar) column with flow rates of 1.0 μl/min. The oven was programmed to a temperature range of 100 to 280 °C at 6 °C/min.

Standard method ASTM D4274-88 (Standard Test Methods for Testing Polyurethane Raw Materials: Determination of Hydroxyl Numbers of Polyols) was used to determine the hydroxyl value of the polyol. The value calculated would be able to verify the FTIR peak ratio method for completion of derivatization process. The water content of the polyol was determined using the Karl Fischer Titrator model Metrohm KFT 701 series (ASTM D4672-00(2006) e1: Standard Test Methods for Polyurethane Raw Materials: Determination of Water Content of Polyols) while the viscosity of the polyol was determined using the Brookfield digital viscometer model DV-I (ASTM D4878-88: Standard Test Methods for Polyurethane Raw Materials- Determination of Viscosity of Polyols). The viscosity of the polyol is important in determining the flowability of the polyurethane resin in the foaming process where it is advantageous in the material consumption. The specific gravity was determined following ASTM D4669–07: Standard Test Method for Polyurethane Raw Materials: Determination of Specific Gravity of Polyols. Other physical characterizations were determination of cloud point, pH and solubility of polyol in methanol, benzene, acetone, ether and water.

The PU foams were characterized for their apparent molded and core densities, compression strength, dimensional stability and water absorption following standard method BS4370: Part 1:1988 (1996) Methods 1 to 5: Methods of test for rigid cellular materials. Foam samples were cut using into cubes of 100 mm × 100 mm × 100 mm in dimensions. A replicate of five specimens were used and carefully weighed using an analytical balance. The dimensions were measured following BS4370: Part 1:1988 (1996): Method 2. The apparent molded density was determined by using a simple mathematical equation, mass (kg)/volume (m^3). The core density is determined by the same method but using skinless foam. The compressive strength test was carried out on a Universal Testing Machine Model Testrometric Micro 350 following BS4370: Part 1:1988 (1996): Method 3 at 23 ±2°C. The specimens were cut into cubes of 50 mm × 50 mm × 50 mm in dimensions. The foam rise direction was marked and a crosshead speed of 50 mm/min was applied. The compression stress at 10% deflection, compression stress at 5% strain and compression modulus was noted. For the dimensional stability test, the specimens were cut into dimensions of 100 mm × 100 mm × 25 mm. The specimens were then put into a controlled temperature-humidity chamber each at –15 ± 2°C and 70 ± 2°C, 95 ± 5% relative humidity for 24 hours. Method 5A of BS4370: Part 1:1988 (1996) standard was followed. The specimens were remeasured and percentage of change in dimensions was calculated. These are then converted to percentage in volume change. The water sorption was carried out using method in Annex D BS6586: Part 1:1993. The specimens were cut into dimension of 50 mm × 50 mm × 50 mm.

The thermal decomposition of the polyurethane foam was measured using a thermogravimetric analyzer model Shimadzu TGA-50 with temperature ranging from room temperature to 600°C at heating rate of 10°C/min under nitrogen gas atmosphere. Samples were placed in alumina pan holders at a mass ranging from 5 to 15mg. The thermal property of the foam was determined using a Perkin Elmer Model DSC-7 differential scanning calorimeter interfaced to the Model 1020 Controller. The samples were analyzed from room temperature to 200°C at a heating rate of 10°C/min. Standard aluminum pans were used to analyze 10 mg samples under nitrogen gas atmosphere. The insulation value (k-factor or λ-value) of the polyurethane foam was determined using the Thermal Conductivity Analyzer model Anacon at testing temperature for cold plate at 25°C and hot plate at 35°C. The thickness of the specimens was 20-30 mm and method 7 of BS4370: Part 2: 1993 standard was followed.

The RBD PKO consists of triglycerides that when undergoes esterification form by products such as glycerol and other possible polyester network (Loudon 1988) as shown in Scheme 1 and Scheme 2. During the reaction, the acetate ion forms an intermediate, the carboxylic acids. These acids attack the lone pair in nitrogen atom in diethanolamine, DEA and formed the probable structure of the esteramide with hydroxyl terminal (Scheme 2)

R_1, R_2 and R_3 generally are represented by R and it is very common to have lauric-lauric-oleic composition of fatty acid in the carbon chains (Scheme 2).

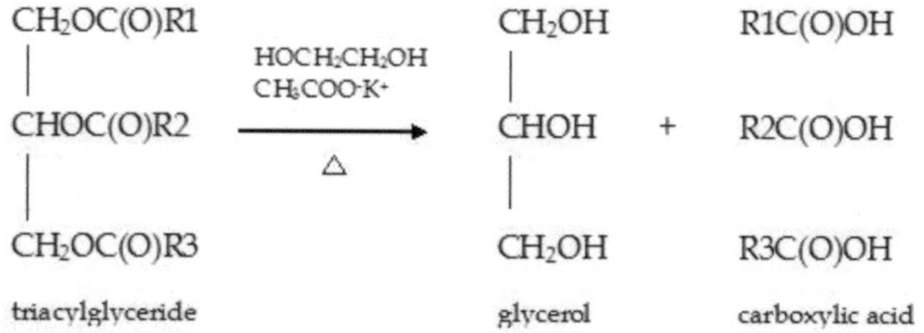

Scheme 1. Probable reaction mechanism between the RBD PKO and the hydroxyl-containing compound

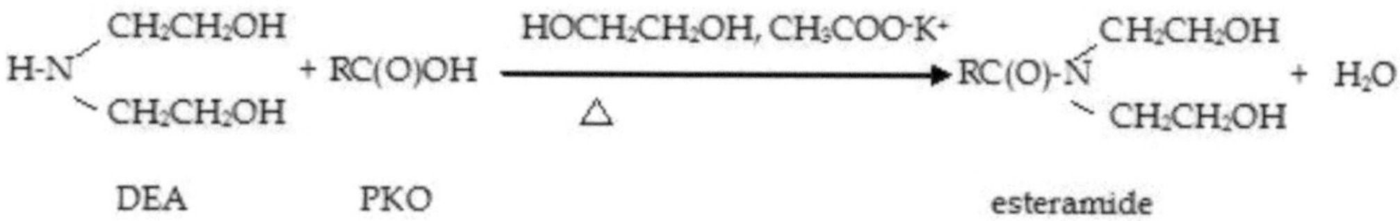

Scheme 2. Conversion of RBD PKO to the esteramide (RBD PKO-based polyol)

RBD PKO reacts with the polyhydroxyl compound in an alkaline medium (contributed by the potassium acetate). The alkalinity of the system ensured that the RBD PKO is fully reacted. The selection on polyhydroxyl compound being used is the critical part where it should offer highest hydroxyl value and functionality polyol possible to fully converting the RBD PKO into polyol (highest yield). Methods used in this study involved polycondensation and esterification where these are the only routes that offered low reaction temperature and short reaction time. It produced polyol (compound with functional group –OH) at high yield (almost 100%), low moisture content and no toxic vapor. The esteramide or PKO-based polyol is a monoester with OH terminal.

4. Properties of the PKO-based polyol

The derivatised RBD PKO-based polyol is a golden yellow liquid with a cloud point of 13°C. It has very low moisture content of 0.09% and low viscosity of 374 cps and specific gravity of 0.992 g/cm^3 at room temperature. Low water content and liquidy nature of the polyol are advantageous in formulating the polyurethane system especially when processing of end product is concerned. Less viscous polyol offers less viscous polyol resin which leads to system with good flowability. The viscosity increases as the degree of polycondensation and branching increases (Wood 1990). The physical properties of the PKO-based polyol are summarized in details in Table 4. It is important to note that raw RBD PKO solidified at room temperature with cloud point of about 23-24°C whilst the derivatized polyol solidified only at 13°C (cloud point). Polyol heating system is not required here as what is being used by other studies (Parthiban et al. 1999 and Ahmad et al. 1995).

Parameters	Result
State at 25°C	Liquid
Color	Golden yellow
Odor	Odorless
Density at 25°C, g/cm^3	0.992
Solubility	Alcohol, Ketone, Ether, Alkane, Water
Cloud Point, °C	13
Viscosity at 25°C, cps	374
pH	9-10
Moisture content at 25°C, %	0.09

Table 4. Physical properties of the derivatised RBD PKO-based polyol.

4.1. Chemical analysis

a. Fourier Transform Infrared Spectroscopy (FTIR)

The RBD PKO, a chain of fatty acid with carboxylic acid group displays intense C=O stretching bands of acids absorb at 1711 cm^{-1} as shown in Fig. 3 (a). The C-H stretches at 2932 and 2855 cm^{-1}. Two bands arising from C-O stretching and O-H bending appear in the spectra of RBD PKO near 1320-1210 and 1440-1395 cm^{-1} respectively. Both of these bands involve some interactions between C-O stretching and in-plane C-O-H bending. The C-O-H bending band near 1440-1395 cm^{-1} is of moderate intensity and occurs in the same region as the CH_2 scissoring vibration of the CH_2 group adjacent to the carbonyl (Silverstein et al. 1991).

The FTIR spectrum of the derivatized RBD PKO was obtained from samples taken at 175-180°C (Fig. 3(b)) during the esterification process. The spectrum was evaluated at peak 3351 cm^{-1} (designated as peak A) and 1622 cm^{-1} (designated as peak B). Peak A and B, which are the hydroxyl (-OH) and carbamate (O=C=N-) peaks respectively (assigned by IR Mentor Pro Classes, Sadtler Division Bio-Rad Laboratories 1990 and Silverstein et al. 1991). These peaks do not appear in the spectra of the raw RBD PKO (Fig. 3(a)). A vague trace of the hydroxyl peak was observed when PKO is mixed with the hydroxyl compound. Further increase in the reaction temperature and reaction time changed the percentage of transmittance for both peaks A and B significantly. It also indicated a formation of ester cleavage at 1710 cm^{-1}. The sharp absorption bands in the region of 1750-1700 cm^{-1} are characteristic of carbonyl group of ester (C=O) stretching vibrations (Silverstein et al. 1991).

Transmittance ratio of both peaks, the OH and the carbamate peaks (% transmittance of peak A divided by the % transmittance of peak B) was plotted as in Fig. 4. It was used to identify the progress of the derivatization process (Chian and Gan 1998). Fig. 4 also showed that the hydroxyl value (OHV) reached to a constant at 350-370 mg KOH/g sample at intervals of 175-180°C for 15-30 minutes of reaction time. The FTIR spectrum and hydroxyl value (OHV) curves both demonstrated that 175-180°C at 15-30 minutes as optimum temperature and reaction time respectively. Both methods are advantageous in the identification of optimum processing parameters assuming that Beer's Law is applicable here. However, OHV determination method is slow and time-consuming. Therefore, FTIR method is more preferable in determining the completion of reaction for the RBD PKO-based polyol (Chian and Gan 1998).

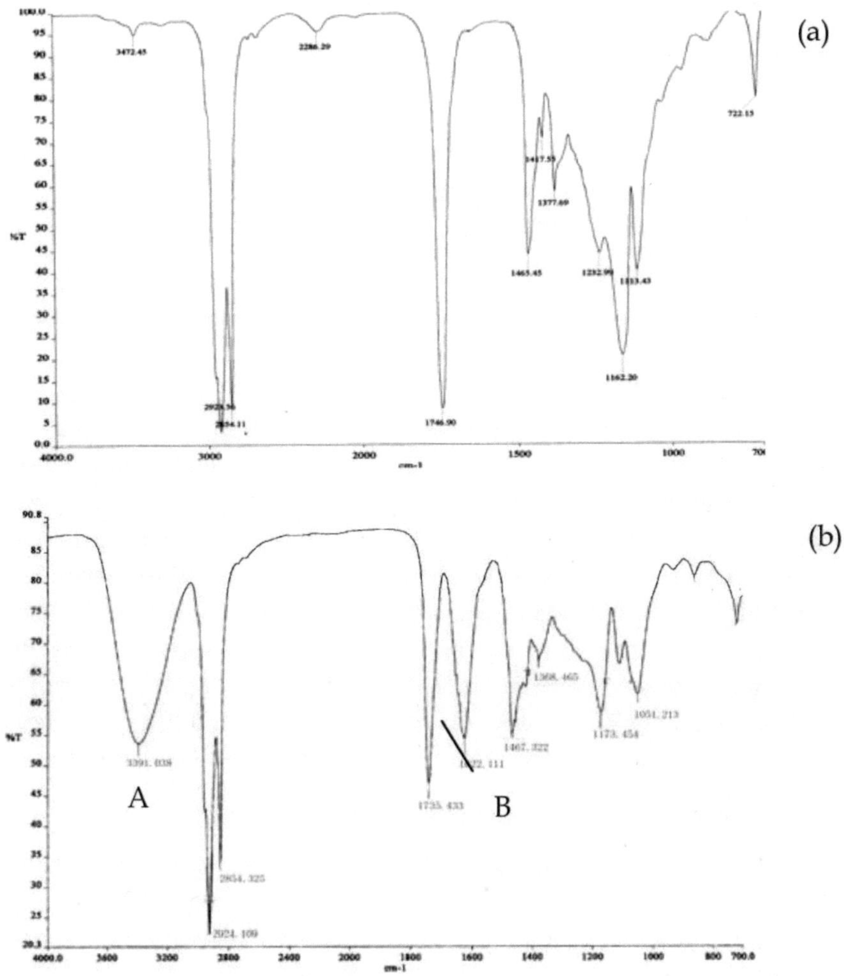

Figure 3. FTIR spectra of (a) the raw RBD PKO and (b) the palm-based esteramide

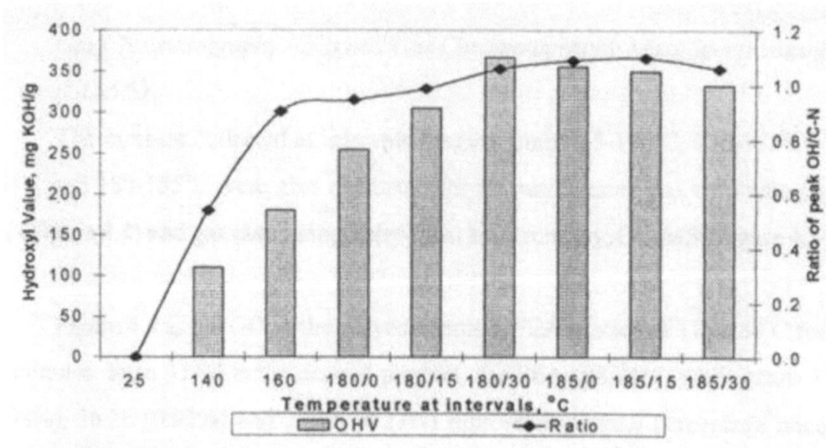

Note:
25 refers to derivatized RBD PKO at ambient temperature, 25°C
140 refers to derivatized RBD PKO at 140°C
160 refers to derivatized RBD PKO at 160°C
180/0 refers to derivatized RBD PKO at starting of 180°C
180/15 refers to derivatized RBD PKO at 180°C for 15 minutes
180/30 refers to derivatized RBD PKO at 180°C for 30 minutes
185/0 refers to derivatized RBD PKO at starting of 180°C
185/15 refers to derivatized RBD PKO at 185°C for 15minutes
185/30 refers to derivatized RBD PKO at 185°C for 30 minutes

Figure 4. Curve of ratio of OH peak to the C-N peak and the OHV curve of the blend at intervals

b. Thin Layer Chromatography

The thin layer chromatography (TLC) test on the desired products obtained at intervals of reaction time at 175-180°C (0, 15 and 30 minutes) indicated a clear qualitative separation. These separations were compared to TLC carried out on individual ingredients: The RBD PKO, diethanolamine (DEA), the catalyst-potassium acetate in monoethylene glycol and standard lauric acid (Athawale et al. 2000). There were three separation peaks, identify as the PKO, DEA and small trace of the catalyst up to 175-180°C at 0 minute. At 175-180°C for 15 minutes, only two separation peaks were observed and finally at 175-180°C for 30 minutes, only one separation peak was observed (Fig. 5). The result is parallel to the gas chromatography (GC) peaks of the final product, the RBD PKO-based polyol (Fig. 6)).

c. Gas Chromatography-Mass Spectrometry (GC-MS)

The samples collected at intervals ranging from 165-170°C, 170-175°C, 175-180°C and 180-185°C were also evaluated for its purity using gas chromatography, GC coupled with mass spectrometry, GC-MS. Fig. 6 is the GC of the RBD PKO-based polyol reacted at 175-180°C for 15-30 minutes. The signal at retention time of 31.92 min is the desired product, the RBD PKO-based polyol (98.24%) while signals at retention time of 13.37 (0.08%), 16.36 (0.92%) and 27.91 (0.27%) representing small percentage traces of MEG, glycerol (by-product of esterification) and DEA (C:\ DATABASE\WILEY275.L). Others (0.49%) are traces of oligomeric polyester components from C_{14} and C_{18} chains. The GC-MS scan of the RBD PKO-based polyol showed an estimated molecular weight of 477. Molecular weight obtained at 165-170 and 170-175°C of reaction temperature was 296 and 355 respectively. Thus, molecular weight obtained at 175-180°C is considered to be the most desirable molecular weight for this study. The functionality of the RBD PKO-based polyol derived from this molecular weight and the determined hydroxyl value (OHV of 350 to 370 mg KOH/g) is 2.98 to 3.15 calculated using the mathematical equation in equation 3.

$$\text{Functionality} = M_w \times \text{OHV} / 56100 \tag{3}$$

Note: M_w is the estimated molecular weight of the RBD PKO-based polyol obtained from GC-MS which is 477 OHV is the hydroxyl value of the RBD PKO-based polyol obtained using ASTM D4274-88 method, which is about 350-370 mg KOH/g sample

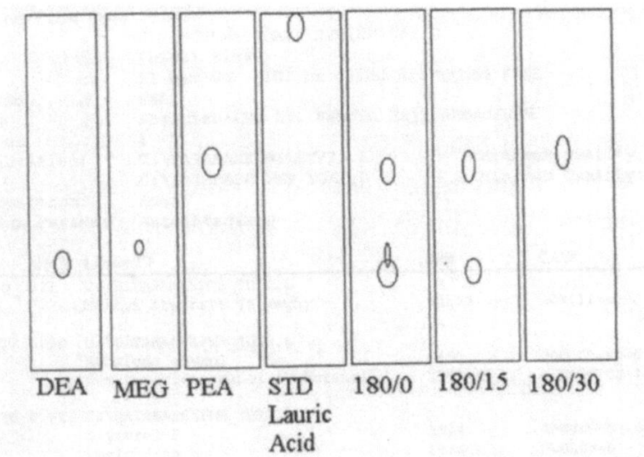

Note:

PEA	RBD PKO-based polyol
DEA	diethanolamine
MEG	monoethylene glycol
STD	standard lauric acid
180/0	derivatised RBD PKO at starting of 180°C
180/15	derivatised RBD PKO at 180°C for 15 minutes
180/30	derivatised RBD PKO at of 180°C for 30 minutes

Figure 5. The thin layer chromatography of the ingredients

This range of functionality is suitable for rigid foam application (Wood 1990).

Both FTIR (IR Mentor Pro 1990) and GC-MS approaches (Wiley MS-database) could be used to estimate the most probable molecular structure of the RBD PKO-based polyol at 175-180°C/30 minutes (optimum temperature and reaction time) as 2-hydroxy-undecanoamide as in Scheme 2 (library search on Wiley MS-database giving 98% quality match). There is no intention of purification of the synthesized RBD PKO-based polyol as all these hydroxyl-containing compounds would react with crude MDI.

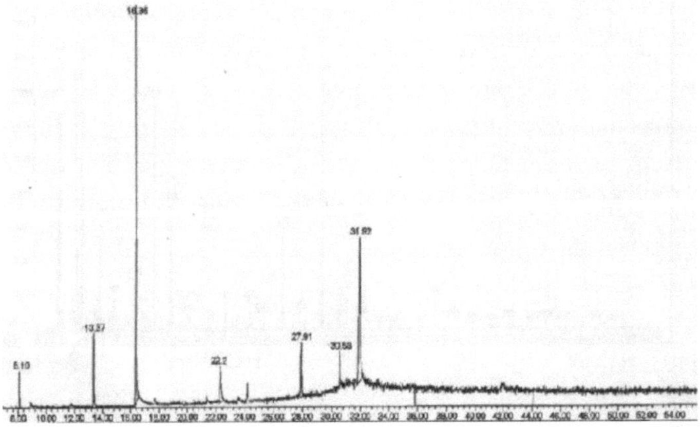

Figure 6. GC chromatogram of the RBD PKO-based polyol obtained at 175-180ºC for 30 minutes.

4.2. Thermal testing

The thermogram of the resulted RBD PKO-based polyol is as shown in Figure 7. Thermally, it is stable up to 167.6ºC and undergoes two stages decomposition at 167.6 to 406.3ºC with total weight loss of 99.41%. The initial 3.34% weight loss is contributed to the moisture content and other volatile impurities in the RBD PKO-based polyol (Oertel 1993). The initial decomposition is contributed by the degradation of RBD PKO-based polyol and traces of glycerol supported by the DTA curve which representing the softening temperature at 385ºC. Charred residue was obtained after testing.

Figure 7. TGA thermogram of the RBD PKO-based polyol obtained at 175-180ºC for 30 minutes

5. Properties of the PKO-based polyurethane foam

5.1. Physical properties

The PKO-based polyurethane foam (PUF) produced is a light yellow solid with skin thickness of about 1.5 mm. It is a stiff/rigid but brittle solid at 43-44kg/m^3 molded density and core density of 38-39 kg/m^3 with average void size of 0.10-0.15 mm (Fig. 8).

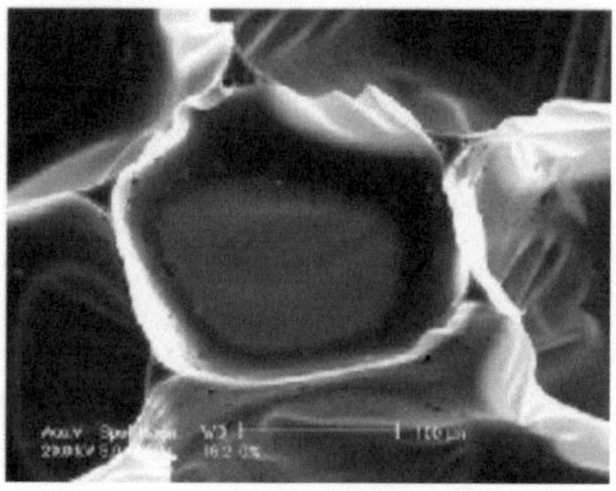

Figure 8. Scanning electron micrograph of the PUF at 250× magnification

5.2. FTIR analysis

The PUF is analysed by FTIR spectroscopy which showed the absence of the free OH groups and indicates a complete conversion of both –OH groups of the PEA to the urethane moiety (NH-C(O)-O). Typical FTIR spectrum of the PU is as shown in Fig. 9. The characteristic –NH stretching vibration of the –NH_2- (amide) is located at 3405 cm^{-1}, overlapping with the OH peak as a broad band. Bands at 2932 and 2894 cm^{-1} are the synchronous reflection of asymmetric and symmetric of CH_2 bridges, from the linkage of the urethane with the PEA. Bands at 1650 cm^{-1} is the overlapping of –N=C=O (urethane) and ester linkage of the PEA. Obviously, bands 1550, 1650 and 3350 cm^{-1} indicate complete conversion to urethane moiety (Silverstein et al. 1991).

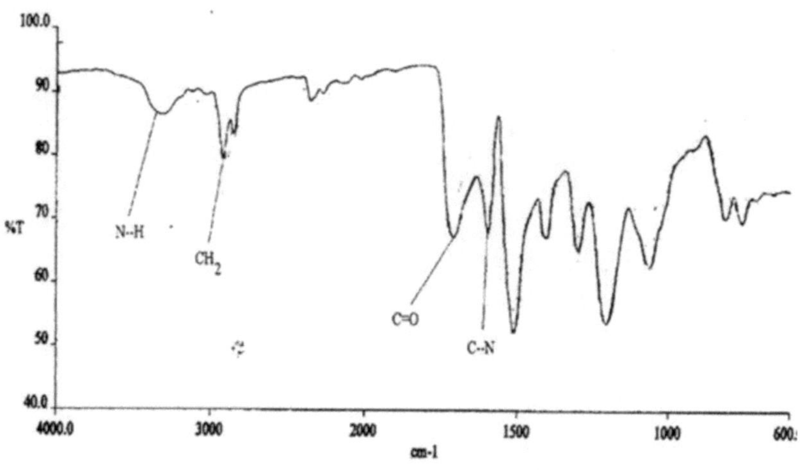

Figure 9. FTIR spectrum of the RBD PKO polyurethane foam

5.3. Thermal properties

The thermal instability of a PU may be defined as the ease by which heat produces changes in the chemical structure of the polymer network. These may involve simple bond-rupturing dissociation or reaction reversals and provide more volatile components, or they may result in extensive pyrolysis and fragmentation of the polymer. This characteristic provides a measure of fire hazard in that a more thermally stable polymer is less likely to ignite and contribute to a conflagration than a less stable one (Burgess, Jr. & Hilado 1973). Thermodynamic parameters such as decomposition temperatures, percentage of weight loss, melting temperature, T_m and glass transition temperature, T_g were determined by thermal analyses of the PU.

TGA thermogram of the PU is as shown in Fig. 10. Presence of three degradation stages implying the presence of three thermal degradation temperatures. It was thermally stable at 191.9ºC, a common stability temperature for PU (Hepburn 1991). The initial weight loss of about 41.24% commences at 191.9 to 396.9ºC. T_{max} from the DTA curve occurred at 275ºC attributed by carbon dioxide trapped in the sample. Degradation started at 396.3 to 498.4ºC, which was initially a fast process. The total weight loss up to 500ºC is 74%. This second stage of degradation rationalized the urethane linkage reported by Hepburn (1991).

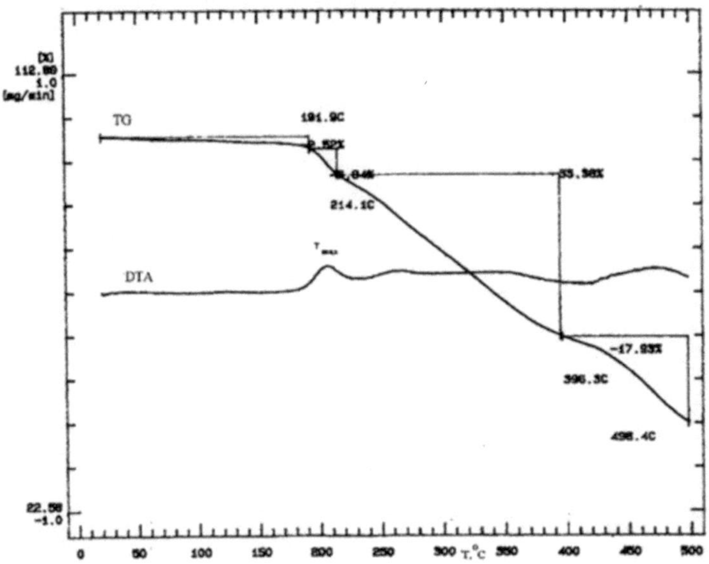

Figure 10. TGA thermogram of the RBD PKO PU foam

There is no indication of melting and crystallization temperatures curves in the DSC thermogram of the PU foam. Therefore, this polymer could be identified as an amorphous polymer (Badri et al. 2000). The glass transition temperature is 39.74ºC, a regular glass transition temperature for thermosetting polymers, with heat capacity of 33.0 J/g. Since the heat of evaporation of water is 2300J/g, moisture content of the PU was detected to be 1.43%.

However, the initial thermal conductivity of PU foam is found to be 0.0244 W/m-K. This is expected since it is a water-blown PU system where water has poor properties i.e. boiling point and k-factor compared to its industrial counterpart, chlorofluorocarbon, CFC (Crawford & Escarsega 2000). Low thermal conductivity is normally contributed by the low boiling point of the blowing agent such as CFC and finer cell structure of the foam (Hardings 1965, Frisch 1985, Hepburn 1991). However, another fact to be remembered is that water is a chemical blowing agent compared to CFC, a physical blowing agent. Water is capable of interfering in the polymerization of polyurethane by producing by-product such as urea and large amount of carbon dioxide when being used I larger quantity. Functionality of polyols also plays an important role in producing good insulated material (Wood 1990, Hass and Uhlig 1985).

5.4. Mechanical properties

The mechanical properties of the foam produced from the derivatized RBD PKO show comparable results (Table 5) to the British Standard requirement (practiced by industry such as building construction industry). It is expected for water-blown PU foam to have lower compressive stress at 5% strain and the compression due to irregular formation of cellular structure. This decreased the strength upon higher percentage of strain. Table 5 showed the summary of the mechanical properties of the PU foam.

The dimensional stability which is described in percentage of volume change indicated changes of -0.090% and 0.012% at -15±2ºC and 70±2ºC at 95±5% relative humidity for 24 hours respectively. A very minimum shrinkage and expansion problem was observed on the foam prepared from this palm oil-based polyurethane polyol in a water-borne system. Identical resin formulation was used using petroleum-based polyol to substitute the palm-based polyol. Major shrinkage and expansion problems were observed. Shrinkage and expansion problems are normally used as indicators of how good the foam is as an insulator. The mechanical properties could be enhanced by using low or high pressure dispersing machines (Oertel 1993). Better mechanical properties could also be achieved by introducing filler in the PU system (Rozman et al. 2001[a], 2001[b], 2000, 1998).

5.5. Rheological and kinetic properties

The PU system is polymerized kinetically using tetramethylhexadiamine, TMHDA as a gel/blow catalyst and pentamethyldiethylenetetramine, PMDETA as a blow catalyst. The addition of both catalysts is very minimum (0.05-0.10 pbw) in achieving an optimum kinetic reaction time (Tamano et al. 1996) especially when reactive RBD PKO-based polyol (Scheme 2) is used in the formulation. The cream time, gelling/fiber time, tack-free time and rise time (Appendix B) were 23, 71, 105 and 156 seconds respectively at 20ºC. The PUF is demolded after 10 minutes of mixing with skin thickness of about 1.5 mm. It has a flow index of 1.050 cm/g, a moderate flowability PU system (Colvin 1995). This is assumed to be helpful in reducing the consumption of raw materials, especially the RBD PKO-based polyol.

5.6. Resistance to environmental stress

The chemical resistance of the PU with normal closed-cell structures of rigid urethane foam prepared from the crude MDI and RBD PKO-based polyol is carried out to investigate the limitation of the interactions with surroundings to the surface layer in order to produce a chemically and physically stable material. Effects produced by chemical agents depend both on the chemicals and on the permeability of cell membranes. Solubility of the chemical in the foam affects both permeability and swelling. Results obtained are not representative of other temperatures, concentrations or exposure times.

Parameter	Method	Standard	Results
*Apparent molded density, kg/m^3	BS 4370:Part 1:1988 (Method 2)	Min 38	43.6 ±0.85
*Apparent density (core), kg/m^3	BS 4370:Part 1:1988 (Method 2)	Min 35	38.9 ±0.53
*Compressive strength to foam rise at 10% deflection, kPa	BS 4370:Part 1:1988 (Method 3)	Min 180	185.7 ±8.22
*Compressive stress at 5% strain, kPa	BS 4370:Part 2: 1993 (Method 6)	Min 140	105.4 ±2.41
Compressive modulus, N/m^2	BS 4370: Part 1: 1988 (Appendix A)	Not available	8.52 ±0.46
*Dimensional stability,%	BS 4370: Part 1: 1988 (Method 5B)		
	At -15 ±2°C for 24h	Maximum 1.0	Length: -0.151 ±0.03 Width: -0.433 ±0.03 Thickness: 1.373 ±0.06
	At 70 ±2°C, 95 ±5% r.h. for 24h	Maximum 3.0	Length: 0.359 ±0.25 Width: 0.017 ±0.04 Thickness: 1.654 ±0.09
*Apparent water absorption,%	BS 6586: Part 1: 1993 (Annex D)	Maximum 6.5	2.25 ±0.89
Shore A Hardness	ASTM D 2240	Not Available	29.0 ±1.4

Note: * Physical property requirements following BS6586: Part 1: 1993 industrial standard.

Table 5. The mechanical properties of the PU foam synthesized from the RBD PKO-based polyol.

Fig. 11 illustrates the compressive strength at 10% deflection and 5% strain as well as its compression modulus upon exposure to stress. All resistivity test medium being used result in a major increment in the strength at 10% deflection. Readings of above 0.20MPa (compared to the control foam) with maximum compressive strength are observed in benzene at about 0.34MPa, followed by PUF at ambient temperature (0.30MPa), freeze-thaw condition (0.26MPa), 10% NaOH (0.25MPa), saltwater (0.20MPa) and finally 10% HNO_3 (0.19MPa). The same trend is observed in compressive strength at 5% strain where the maximum value is encountered at freeze-thaw condition followed by at ambient temperature, 10% NaOH and finally benzene. The compression modulus reaches as high as 11.0MPa and others are in the range of 8.0 to 9.0MPa.

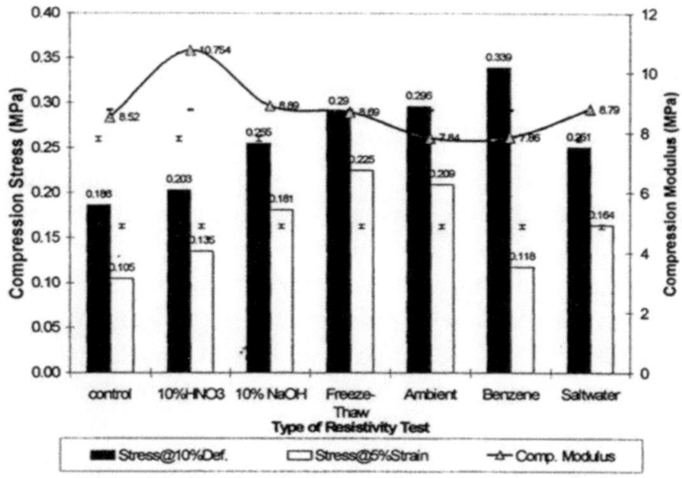

Figure 11. Effect of various environmental stresses on the compressive strength and compression modulus of the RBD PKO-based PU foam

Practically, the absorption of chemicals into the foam results in swelling of the cell faces, which apparently increases the compressive strength. Weathering conditions (ambient and freeze-thaw) however are very much dependence on the diffusion rate of carbon dioxide being replaced by the air which causes expansion of the foam and increases the compressive strength (Wood 1990). The foams are found to be unaffected by the test medium basically due to the mixture of organic components (RBD PKO-based polyol and MDI). Rigid PU foam is stable in the present of most solvents such as found in binders and sealers (Oertel 1993).

Physically, the foam becomes spongy with the formation of waxy material on the surface of the foam, as a result of prolonged exposure to benzene as an aromatic hydrocarbon. It is important to note that ester-based polyurethanes are easily attacked by hot aqueous alkali or moderately concentrated mineral acids, swollen by aromatic hydrocarbons and decomposed by prolonged contact with water, diluted acids and moist heat (causes swelling and slow hydrolysis) (Roff et al 1971).

The compression modulus of the PUF ranges from 7.8 to 10.8MPa. the compression modulus for the control PUF is at 8.5MPa which is lower compared to the modulus in 10%HNO_3,

10%NaOH, under freeze-thaw condition, and in saltwater but higher if compared to the modulus of the rest of the resistivity test.

Rigid PU prepared has high resistivity to the action of most organic solvents and are seriously degraded only by strong acid, oxidizing agent and corrosive chemicals. Only stronger polar solvents, which significantly swelled the polymer, led to shrinkage of the foam structure. Evaporation of the solvent normally returns the polymer to its original state (Oertel 1993).

6. Conclusion

Several advantages are foreseen from this study. Some important advantages are being identified through this method of polyol production. Firstly, it is attractive and economical. RBD PKO-based polyol is a naturally formed macromolecules found in Malaysia. It is extremely plentiful, easy to process and refine, capable of being cultivated with minimum capital investment and suitable for conversion to quality polyols using an inexpensive reaction process. Secondly is the simplicity of the process, which requires only a few reactors for producing the polyol as well as formulating the resin. Commercially, the process acquires only a few personnel to produce consistently good quality polyols. Thirdly, compare to the manufacturing of the petrochemical-based polyols, the process is relatively safe, where it involves the usage of hazardous chemicals. Generally, it is non-toxic and of low volatility.

Two major environmental advantages can be realized. Firstly, the source of oil is truly renewable, where it does not lead to permanent depletion of resources which has a limited global availability. Secondly, the amount of energy required to convert the natural oils to polyol is considerably less than using the conventional process. The foam made from this RBD PKO-based polyol is low in density, light in color, high in strength but low in water sorption. The produced RBD PKO-based polyurethane foam in this study also has other advantages as tabulated in Table 6.

Property	Rating	Consequence	Benefit
Thermal Insulation	Highest	Thinnest Section	Space
Rigidity	High	Added Strength	Structural
Adhesion	High	No glue-line	Manufacturing
Dimensional Stability	High	Non-sag, non-heave	Maintenance
Density	Low	Lightweight	Handling
Water Vapor Transmission	Low	Less Condensation	Construction

Table 6. The advantages of producing RBD PKO-based polyurethane.

The PUF meets the British Standard requirements in any medium of the tested environmental stress test. This ester-type polyurethanes are easily attacked by prolonged contact with water, diluted acids and moist heat (causes swelling and slow hydrolysis) and swollen by aromatic hydrocarbons. These rigid PUs either the PUF, are resistant to the action of most organic solvents and are seriously degraded only by strong acids, oxidizing agents and corrosive chemicals. Only polar solvents, which significantly swell the polymer, lead to shrinkage of the foam structure. Evaporation of the solvent normally returns the polymer to its original state.

In terms of application, these composites are most suitable in structures where stiffness and dimensional stability are of prime importance but is only a secondary choice to areas where structural strength is more vital than the component rigidity.

Author details

Khairiah Haji Badri
Polymer Research Center, Faculty of Science and Technology,
Universiti Kebangsaan Malaysia, Selangor, Malaysia

Acknowledgement

These works on the production of the RBD PKO-based polyol and other ranges of polyurethane polyols have been at present being produced at larger scale and ready to depart January 2012. This is being brought into realization with the support of Universiti Kebangsaan Malaysia under its entities School of Chemical Sciences and Food Technology, Polymer Research Center and Faculty of Science and Technology (UKM-OUP-FST-2012) for all the facilities provided. Thank you to Ministry of Higher Education, Ministry of Science, Technology and Innovation (previously known as Ministry of Science, Technology and Environment) and Yayasan Felda for the financial supports. Major contributions definitely came from graduates and colleagues of Universiti Kebangsaan Malaysia. For special individuals who initiated this project, Zulkefly Othman and in memory Haji Badri Haji Zakaria, my greatest thanks to both of you.

7. References

Ahmad, S., Siwayanan,P. & Wiese, D. 1995. Porim and INTERMED Sdn.Bhd. Malaysian Patent Application Number. PI9502302. Filling Date: 7 August, 1995.

Alfani, R., Iannace, S.&Nicolois, L. 1998. Synthesis and Characterization of Starch Based Polyurethane Foams. *J. Appl. Polym*. Sci. 68 (5) : 739-745

Apukhtina, N.P. 1973. Methods for Increasing the Thermal Stability of Polyurethanes: Soviet Urethane Technology, Ed. Schiller, A.M. pp. 198-210. Connecticut: Technomic Publishing Co., Inc.

Arnold, J.M. 1983. *Vegetable Oil Extended Polyurethane System.* US 4375521

Athawale, V.D., Rathi, S.C. & Bhabe, M.D. 2000. Novel Method For Separating Fatty Ester From Partial Glycerides in Biocatalytic Transesterification Of Oils, *Separation and Purification Technology*, 18:3:209-215.

Austin, P.E., Derderian, E.J.& Kayser, R.A. 2000. Hydrosilation in High Boiling Natural Vegetable Oils. US 6071977.

Norzali N.R.A., Badri, K.H. & Nawawi, M.Z. 2011[a]. Loading Effect of Aluminum Hydroxide onto the Mechanical, Thermal Conductivity, Acoustical and Burning Properties of the Palm-based Polyurethane Composites, Sains Malaysiana 40(7): 737-742

Norzali N.R.A., Badri, K.H. & Nawawi, M.Z. 2011[b]. The Effect Of Aluminium Hydroxide Loading On The Burning Property of The Palm-Based Polyurethane Hybrid Composite, Sains Malaysiana 40(4):385–390.

Liow, C.H., Badri, K.H. & Ahmad, S.H. 2011. Mechanical and Thermal Properties of Palm-Based Polyurethane Composites Filled With Fe3O4, PANI and PANI/F e3O4, Sains Malaysiana 40(4): 379–384.

Wong, C. S. & Badri, K.H. 2010. Sifat Terma Dan Kerintangan Api Poliuretana Berasaskan Minyak Isirung Sawit Dan Minyak Kacang Soya, Sains Malaysiana, 39 (5): 775-784.

Badri, K.H. & Redhwan, A. M. 2010. The effect of phosphite loading on the mechanical, thermal and fire properties of palm-based polyurethane, Sains Malaysiana, 39 (5): 769-774.

Mat Amin, K. A., Badri, K.H. & Othman, Z. 2007. Oil Palm-Based Hybrid Biocomposites with Kaolinite. Journal of Applied Polymer Science 105:2488-2496.

Badri, K.H., Ujar, A. H., Othman, Z. & Sahaldin, F. H. 2006[a]. Shear Strength of Wood-to-Wood Adhesive Based on Palm Kernel Oil, Journal of Applied Polymer Sciences, 100(3): 1750-1759

Badri, K.H., Mat Amin, K. A., Khalid, N. K., Othman, Z. & Abdul Manaf, K. 2006[b]. Effect Of Filler-To-Matrix Ratio On The Mechanical Strength Of Palm-Based Biocomposite Board, Polymer International, 55: 190-195

Badri, K.H. & Mat Amin, K. A. 2006[c]. Oil Palm-Based Biocomposites. Journal of Oil Palm Research, (Special Issue-April 2006):103-113

Badri, K.H., Othman, Z. & Mohd Razali, I. 2005. Mechanical properties of polyurethane composites from oil palm resources. Iranian Polymer Journal, 14 (5): 987-993

Badri, K.H., Othman, Z. & Ahmad, S.H. 2004[a]. Rigid Polyurethane Foams From Oil Palm Resources, Journal Of Materials Science. 39(16-17):5541-5542

Badri, K.H., Shahaldin, F. H. & Othman, Z. 2004[b]. Indigenous Coating Material From Palm Oil-Based Polyamide. J. Mater. Sci. Letters,39 (13):4331-4333.

Khairiah Haji Badri. 2002. Preparation and Charaterization of Polyurethane Foam from RBD Palm Kernel Oil-Based Polyurethane Polyol and Oil Palm Empty Fruit Bunch Fiber As Filler. *Proceeding of National Science Fellowship (NSF) Workshop*, pp 114-120.

Badri, K.H., Ahmad, S.H & Zakaria , S. 2000[a]. Development of Zero ODP Rigid Polyurethane Foam From RBD Palm Kernel Oil: *J. Mater. Sci. Letters*, 19: 1355-1356.

Badri, K.H, Ahmad, S.H. & Zakaria, S. 2000[b]. Thermal, Crystallinity and Morphological Studies on the Filled RBD Palm Kernel Oil Polyurethane Foam: *Nuclear Science Journal of Malaysia*, 18 (2): 57-62.

Badri, K.H., Ahmad, S.H. & Zakaria, S. 2001[a]. Production of a High-Functionality RBD Palm Kernel Oil-Based Polyester Polyol. *J. Appl. Polym. Sci.*, 81 (2): 384-389.

Benli,S., Yilmazer, U., Pekel, F. & Ozkar, S. 1998. Effect of Fillers on Thermal and Mechanical Properties of Polyurethane Elastomer, *J. Appl. Polym. Sci*. 68: 1057-1065.

Berlin, A.A. & Zhitinkina, A.K. 1982. Foam Based on Reactive Oligomers, Polyurethane Foams, pp. 51-111. London: Howard Publishing Inc.

Bhunia, H.P., Jana, R.N., Basak A., Lenka, S. & Nando, G.B. 1998. Synthesis of Polyurethane From Cashew Nut Shell Liquid (CNSL), A Renewable Resource. *J. Appl. Polym. Sci.* 36 (3): 391-400.

Burgess, Jr., P.E & Hilado, C.J. 1973. Thermal Decomposition and Flammability of Foams: *Plastic Foams* Part II. Ed. Frisch, K.C & Saunders, J.H. pp. 855-871. New York: Marcel Dekker, Inc.

C:\Database\wiley275.1.1999. Library Search of Acquisition Method. University of Malaya.

Chian, K.S. & Gan, L.H. 1998. Development of a Rigid Polyurethane Foam From Palm Oil. *J. Appl. Polym. Sci.*, 68 (3): 509-515

Chittolini, C. 1999. *Polyurethane Foam-Mixing Isocyanate Component and Polyol Component Including Pentane and Dialkanolamine Derived from Vegetable Oil or Fat to Make Polyurethane Foam*. US 5859078

Colvin, B.G. 1995. Low Cost Polyols From Natural Oils, *U'tech Asia '95*. 36: 1-10

Crawford, D.M. & Escarsega, J.A. 2000. Dynamic Mechanical Analysis of Novel Polyurethane Coating for Military Applications. *Thermochimica Acta*, 357-358: 161-168

Cunningham, R.L., Gordon, S.H., Felker, F.C. & Eskins, K. 1997. Jet-Cooked Starch Oil Composite in Polyurethane Foams. *J. Appl. Polym.Sci.* 64 (7): 1355-1361

Frisch, K.C. 1985. Fundamental Chemistry and Catalysis of Polyurethanes, *Polyurethane Technology*, Ed. Bruins, P.F. pp. 12-17. New York: Interscience Publishers.

Harding, R.H. 1965. Effect of Cell Geometry On PU Foam Performance, *J. Cell. Plastics*, 1: 224

Hass, P.F. & Uhlig, K. 1985. Additive and Auxiliary Materials, *Polyurethane Handbook*, 2nd ed. Ed. Oertel, G. pp. 98. New York: Hanser Publisher.

Hepburn, C. 1991. Polyurethane Elastomers. 2nd ed., pp. 441: Great Britain: Elsevier Science Publishers Ltd.

IR Mentor Pro Classes. 1990. Bio-Rad Laboratories, Sadtler Division, PEIM: 6

Javni,I., Petrovic, Z.S., Guo, A. & Fuller, R. 1999. Thermal Stability of Polyurethane-Based on Vegetable Oils. Annu. Tech. Conf.-Soc. Plast. Eng. 3: 3801-3805

Jedlinski, Z.J. 1992. Polyester: Handbook of Polymer Synthesis Part A, Ed. Kricheldorf, H.R. pp. 645-648. London: McGraw-Hill.

Loudon, G.M. 1988. Chemistry of Carboxylic Acids, *Organic Chemistry*. 2nd ed.pp. 816-817. London: Mc Graw Hill.

Mohapatra, D.K., Das, D., Nayak, P.L. & Lenka, S. 1998. Polymers From Renewable Resources. XX. Synthesis, Structure And Thermal Properties Of Semi- Intherpenetrating Polymer Networks Based On Cardanol-Formaldehyde Substituted Aromatic Compounds Copolymerized Resins And Castor Oil Polyurethanes. *J. Appl. Polym. Sci.* 70 (5): 837-842

Nayak, P., Mishra, D.K., Parida, D., Sahoo, K.C., Nanda, M. Lenka, S. & Nayak, P.L. 1997. Polymers From Renewable Resources. IX. Interpenetrating Polymer Networks Based On Castor Oil Polyurethane Poly(hydroxylethylmetacrylate): Synthesis, Chemical, Thermal and Mechanical Properties. *J. Appl. Polym. Sci.* 63 (5): 671-679

Oertel, G. 1993. *Polyurethane Handbook: Chemistry-Raw Material-Processing- Application-Properties. Cincinnati:* Hanser Gardner Publications, Inc.

Parthiban, S., Ooi, T.L., Kassim Shaari, N.Z., Ahmad, S., Wiese, D. & Chua, M.C. 1999. Polyurethane From Palm-Based Polyols, *Palm Oil Technical Bulletin*, September-October 1999: 4-6

Pourjavadi, A., Rezai, N. & Zohuriaan-M, M.J. 198. A Renewable Polyurethane: Synthesis and Characterization of the Interpenetrating Networks (IPNs) From Cardanol Oil. *J. Appl. Polym. Sci.*, 68: 173-183

Roff, W.J., Scott, J.R. & Pacitti, J. 1971. Fibres, *Films, Plastics and Rubbers: A Handbook of Common Polymers.* Pp. 446-457. New York: Butterworth & Co. (Publishers) Ltd.

Rozman, H.D. Tay, G.S., Kumar, R.N., Abusamah, A., Ismail. H. & Mohd Ishak, Z.A. 2001a. Polypropylene-Oil Palm Empty Fruit Bunch- Glass Fibre Hybrid Composites: A Preliminary Study on the Flexural and Tensile Properties. *European Polymer Journal.*37 (6): 1283-1291.

Rozman, H.D., Tay, G.S., Abubakar, A. & Kumar, R.N. 2001b. Tensile Properties of Oil Palm Empty Fruit Bunch- Polyurethane Composites. European Polym. Journ., 37: 1759-1765

Rozman, H.D., Lai, C.Y., Ismail, H. & Mohd Ishak, Z.A. 2000. The Effect Of Coupling Agents On the Mechanical And Physical Properties Of Oil Palm Empty Fruit Bunch-Polypropylene Composites. *Polym. Int*. 49 (11): 1273-1278

Rozman, H.D., Kon. B.K., Abusamah, A., Kumar, R.N. & Mohd Ishak, Z.A. 1998. Rubberwood-High Density Polyethylene Composites: Effect of Filler Size and Coupling Agents on Mechanical Properties. *J. Appl. Polym. Sci.* 69: 1993-2004

Silverstein, R.M., Bassler, G.C. & Morril, T.C. 1991. *Spectrometric Identification of Organic Compounds*, 5th ed., New York: John Wiley & Sons, Inc

Tamano, Y., Yoshimura, H., Ishida, M., Okuzono, S. & Lowe, D.W. 1996. The Characteristics and Role of tertiary Amine catalysts For Polyurethane Foams: Review of Tertiary Amine Catalysts "TEDA & TOYOCAT", *Conference Paper of UTECH '96*

Wood, G. 1990. The Chemistry and Materials of PU Manufacture, *The ICI Polyurethane Book, 2nd ed*. Ed. Genge, R. & Sparrow, D. pp. 41-42. New York: John Wiley & Sons.